MATHEMATICS OF PHYSICS AND ENGINEERING

MATHEMATICS OF PHYSICS AND ENGINEERING

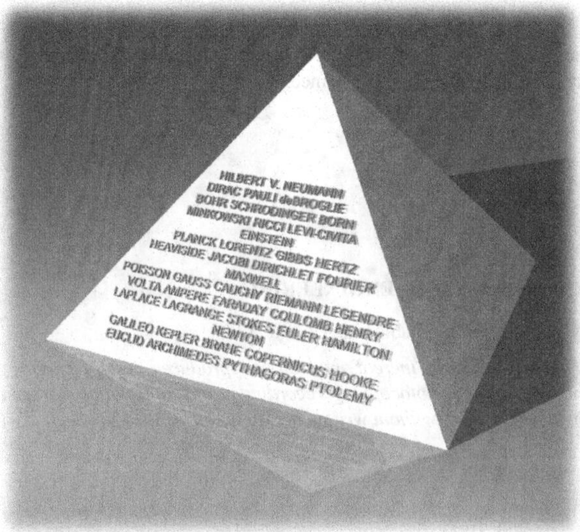

EDWARD K. BLUM
SERGEY V. LOTOTSKY
University of Southern California, USA

World Scientific

NEW JERSEY • LONDON • SINGAPORE • BEIJING • SHANGHAI • HONG KONG • TAIPEI • CHENNAI

Published by

World Scientific Publishing Co. Pte. Ltd.
5 Toh Tuck Link, Singapore 596224
USA office: 27 Warren Street, Suite 401-402, Hackensack, NJ 07601
UK office: 57 Shelton Street, Covent Garden, London WC2H 9HE

British Library Cataloguing-in-Publication Data
A catalogue record for this book is available from the British Library.

About the cover: The pyramid design by Martin Herkenhoff represents the pyramid of knowledge built up in levels over millenia by the scientists named, and others referred to in the text.

MATHEMATICS OF PHYSICS AND ENGINEERING

ISBN-13 978-981-256-621-8
ISBN-10 981-256-621-X

Printed in Singapore

To Lori, Debbie, Beth, and Amy.
To Kolya, Olya, and Lya.

Preface

What is mathematics of physics and engineering? An immediate answer would be "all mathematics that is used in physics and engineering", which is pretty much ... all the mathematics there is. While it is nearly impossible to present all mathematics in a single book, many books on the subject seem to try this.

On the other hand, a semester-long course in mathematics of physics and engineering is a more well-defined notion, and is present in most universities. Usually, this course is designed for advanced undergraduate students who are majoring in physics or engineering, and who are already familiar with multi-variable calculus and ordinary differential equations. The basic topics in such a course include introduction to Fourier analysis and partial differential equations, as well as a review of vector analysis and selected topics from complex analysis and ordinary differential equations. It is therefore useful to have a book that covers these topics — and nothing else. Besides the purely practical benefits, related to the reduction of the physical dimensions of the volume the students must carry around, the reduction of the number of topics covered has other advantages over the existing lengthy texts on engineering mathematics.

One major advantage is the opportunity to explore the connection between mathematical models and their physical applications. We explore this connection to the fullest and show how physics leads to mathematical models and conversely, how the mathematical models lead to the discovery of new physics. We believe that students will be stimulated by this interplay of physics and mathematics and will see mathematics come alive. For example, it is interesting to establish the connection between electromagnetism and Maxwell's equations on the one side and the integral theorems of vector calculus on the other side. Unfortunately, Maxwell's equations are often left out of an applied mathematics course, and the study of these equations in a physics course often leaves the mathematical part somewhat of a mystery. In our exposition, we maintain the full rigor of mathematics while always presenting the motivation from physics. We do this for the classical mechanics, electromagnetism, and mechanics of continuous medium, and introduce the main topics from the modern physics of relativity, both special and general, and quantum mechanics, topics usually omitted in conventional books on "Engineering Mathematics."

Another advantage is the possibility of further exploration through problems, as opposed to standard end-of-section exercises. This book offers a whole chapter, about 30 pages, worth of problems, and many of those problems can be a basis of a serious undergraduate research project.

Yet another advantage is the space to look at the historical developments of the subject. Who invented the cross product? (Gibbs in the 1880s, see page 3.) Who introduced the notation i for the imaginary unit $\sqrt{-1}$? (Euler in 1777, see page 79.) In the study of mathematics, the fact that there are actual people behind every formula is often forgotten, unless it is a course in the history of mathematics. We believe that historical background material makes the presentation more lively and should not be confined to specialized history books.

As far as the accuracy of our historical passages, a disclaimer is in order. According to one story, the Russian mathematician ANDREI NIKOLAEVICH KOLMOGOROV (1903–1987) was starting as a history major, but quickly switched to mathematics after being told that historians require at least five different proofs for each claim. While we tried to verify the historical claims in our presentation, we certainly do not have even two independent proofs for most of them. Our historical comments are only intended to satisfy, and to ignite, the curiosity of the reader.

An interesting advice for reading this, and any other textbook, comes from the Russian physicist and Nobel Laureate LEV DAVIDOVICH LANDAU (1908–1968). Rephrasing what he used to say, *if you do not understand a particular place in the book, read again; if you still do not understand after five attempts, change your major.* Even though we do not intend to force a change of major on our readers, we realize that some places in the book are more difficult than others, and understanding those places might require a significant mental effort on the part of the reader.

While writing the book, we sometimes followed the advice of the German mathematician CARL GUSTAV JACOB JACOBI (1804–1851), who used to say: "One should always generalize." Even though we tried to keep abstract constructions to a minimum, we could not avoid them altogether: some ideas, such as the separation of variables for the heat and wave equations, just ask to be generalized, and we hope the reader will appreciate the benefits of these generalizations. As a consolation to the reader who is not comfortable with abstract constructions, we mention that everything in this book, no matter how abstract it might look, is nowhere near the level of abstraction to which one can take it.

The inevitable consequence of unifying mathematics and physics, as we do here, is a possible confusion with notations. For example, it is customary in mathematics to denote a generic region in the plane or in space by G, from the German word *Gebiet*, meaning "territory." On the other hand, the same letter is used in physics for the universal gravitational constant; in our book, we use G to denote this constant (notice a slight difference between G and G). Since these two symbols never appear in the same formula, we hope the reader will not be confused.

We are not including the usual end-of-section exercises, and instead incorporate the exercises into the main presentation. These exercises act as speed-bumps, forcing the reader to have a pen and pencil nearby. They should also help the reader to follow the presentation better and, once solved, provide an added level of confidence. Each exercise is rated with a super-script A, B, or C; sometimes, different parts of the same exercise have different ratings. The rating is mostly the subjective view of the authors and can represent each of the following: (a) The level of difficulty, with C being the easiest; (b) The degree of importance for general understanding of the material, with C being the most important; (c) The aspiration of the student attempting the exercise. Our suggestion for the first reading is to understand the question and/or conclusion of every exercise and to attempt every C-rated exercise, especially those that ask to verify something. The *problems* are at the very end, in the chapter called "Further Developments," and are not rated. These problems provide a convenient means to give brief extensions of the subjects treated in the text (see, for example, the problems on special relativity).

A semester-long course using this book would most likely emphasize the chapters on complex numbers, Fourier analysis, and partial differential equations, with the chapters on vectors, mechanics, and electromagnetic theory covered only briefly while reviewing vectors and vector analysis. The chapters on complex numbers and Fourier analysis are short enough to be covered more or less completely, each in about ten 50-minute lectures. The chapter on partial differential equations is much longer, and, beyond the standard material on one-dimensional heat and wave equations, the selection of topics can vary to reflect the preferences of the instructor and/or the students. We should also mention that a motivated student can master the complete book in one 15-week semester by reading, on average, just 5 pages per day.

We extend our gratitude to our colleagues at USC: Todd Brun, Tobias Ekholm, Florence Lin, Paul Newton, Robert Penner, and Mohammed Ziane, who read portions of the manuscript and gave valuable suggestions. We are very grateful to Igor Cialenco, who carefully read the entire manuscript and found numerous typos and inaccuracies. The work of SVL was partially supported by the Sloan Research Fellowship, the NSF CAREER award DMS-0237724, and the ARO Grant DAAD19-02-1-0374.

E. K. Blum and S. V. Lototsky

Contents

Chapter 1

Euclidean Geometry and Vectors

1.1 Euclidean Geometry

1.1.1 *The Postulates of Euclid*

The two Greek roots in the word **geometry**, *geo* and *metron*, mean "earth" and "a measure," respectively, and until the early 19th century the development of this mathematical discipline relied exclusively on our visual, auditory, and tactile perception of the space in our immediate vicinity. In particular, we believe that our space is **homogeneous** (has the same properties at every point) and **isotropic** (has the same properties in every direction). The abstraction of our intuition about space is **Euclidean geometry**, named after the Greek mathematician and philosopher EUCLID, who developed this abstraction around 300 B.C.

The foundations of Euclidean geometry are five postulates concerning points and lines. A **point** is an abstraction of the notion of a position in space. A **line** is an abstraction of the path of a light beam connecting two nearby points. Thus, any two points determine a unique line passing through them. This is Euclid's **first postulate**. The **second postulate** states that a line segment can be extended without limit in either direction. This is rather less intuitive and requires an imaginative conception of space as being infinite in extent. The **third postulate** states that, given any straight line segment, a circle can be drawn having the segment as radius and one endpoint as center, thereby recognizing the special importance of the circle and the use of straight-edge and compass to construct planar figures. The **fourth postulate** states that all right angles are equal, thereby acknowledging our perception of perpendicularity and its uniformity. The **fifth and final postulate** states that if two lines are drawn in the plane to intersect a third line in such a way that the sum of the inner angles on one side is less than two right angles, then the two lines inevitably must intersect each other on that side if extended far enough. This postulate is equivalent to what is known as the `parallel postulate`, stating that, given a line and a point not on the line, there exists *one and only one straight line in the same plane* that passes through the point and never intersects the first line, no matter how far the lines are extended. For more information about the parallel postulate, see the book *Gödel, Escher, Bach: An Eternal Golden Braid* by D. R. Hofstadter, 1999. The parallel postulate is somewhat contrary to our physical perception of distance perspective, where in fact two lines constructed to run parallel seem to converge in the far distance.

While any geometric construction that does not exclusively rely on the five postulates of Euclid can be called non-Euclidean, the two basic `non-Euclidean geometries`, `hyperbolic` and `elliptic`, accept the first four postulates of Euclid, but use their own versions of the fifth. Incidentally, Euclidean geometry is sometimes called `parabolic`. For more information about the non-Euclidean geometries, see the book *Euclidean and Non-Euclidean Geometries: Development and History* by M. J. Greenberg, 1994.

The parallel postulate of Euclid has many implications, for example, that the sum of the angles of a triangle is 180°. Not surprisingly, this and other implications do not hold in non-Euclidean geometries. *Classical (Newtonian) mechanics assumes that the geometry of space is Euclidean.* In particular, our physical space is often referred to as the `three-dimensional Euclidean space` $\mathbb{R}^3$, with $\mathbb{R}$ denoting the set of the real numbers; the reason for this notation will become clear later, see page 7.

The development of Euclidean geometry essentially relies on our intuition that every line segment joining two points has a length associated with it. Length is measured as a multiple of some chosen unit (e.g. meter). A famous theorem that can be derived in Euclidean geometry is the theorem of Pythagoras: the square of the length of the hypotenuse of a right triangle is equal to the sum of the squares of the lengths of the other two sides. Exercise 1.1.4 outlines one possible proof. This theorem leads to the distance function, or `metric`, in Euclidean space when a cartesian coordinate system is chosen. The metric gives the distance between any two points by the familiar formula in terms of their coordinates (Exercise 1.1.5).

1.1.2 *Relative Position and Position Vectors*

Our intuitive conception and observation of position and motion suggest that the position of a point in space can only be specified relative to some other point, chosen as a reference. Likewise, the motion of a point can only be specified relative to some reference point.

The view that only relative motion exists and no meaning can be given to absolute position or absolute motion has been advocated by many prominent philosophers for many centuries. Among the famous proponents of this relativistic view were the Irish bishop and philosopher GEORGE BERKELEY (c.1685–1753), and the Austrian physicist and philosopher ERNST MACH (1836–1916). An opposing view of absolute motion also had prominent supporters, such as Sir ISAAC NEWTON (1642–1727). In 1905, the German physicist ALBERT EINSTEIN (1879–1955) and his theory of special relativity seemed to resolve the dispute in favor of the relativists (see Section 2.4 below).

Let us apply the idea of the relative position to points in the Euclidean space $\mathbb{R}^3$. We choose an arbitrary point O as a reference point and call it an `origin`. Relative to O, the position of every point P in $\mathbb{R}^3$ is specified by the directed line segment $\boldsymbol{r} = \overrightarrow{OP}$ from O to P. This line segment has length $\|\boldsymbol{r}\| = |OP|$, the distance from O to P, and is called the `position vector` of P relative to O (the Latin word *vector* means "carrier"). Conversely, any directed line segment starting at O determines a point P. This description does not require a coordinate system to locate P.

> In what follows, we denote vectors by bold letters, either lower or upper case: $\boldsymbol{u}$, $\boldsymbol{R}$. Sometimes, when the starting point O and the ending point P of the vector must be emphasized, we write $\overrightarrow{OP}$ to denote the corresponding vector.

The position vectors, or simply `vectors`, can be added and multiplied by real numbers. With these operations of addition and multiplication, the set of all vectors becomes a *vector space*. Because of the special geometric structure of $\mathbb{R}^3$, two more operations on vectors can be defined, the *dot product* and the *cross product*, and this was first done in the 1880s by the American scientist JOSIAH WILLARD GIBBS (1839–1903). We will refer to the study of the four operations on vectors (addition, multiplication by real numbers, dot product, cross product) as `vector algebra`. By contrast, `vector analysis` (also known as `vector calculus`) is the *calculus* on $\mathbb{R}^3$, that is, differentiation and integration of vector-valued functions of one or several variables. Vector algebra and vector analysis were developed in the 1880s, independently by Gibbs and by a self-taught British engineer OLIVER HEAVISIDE (1850–1925). In their developments, both Gibbs and Heaviside were motivated by applications to physics: many physical quantities, such as position, velocity, acceleration, and force, can be represented by vectors.

All constructions in vector algebra and analysis are not tied to any particular coordinate system in $\mathbb{R}^3$, and do not rely on the interpretation of vectors as position vectors. Nevertheless, it is convenient to depict a vector as a line segment with an arrowhead at one end to indicate direction, and think of the length of the segment as the magnitude of the vector.

Remark 1.1 *Most of the time, we will identify all the vectors having the same direction and length, no matter the starting point. Each vector becomes a representative of an* equivalence class *of vectors and can be moved around by parallel translation. While this identification is convenient to study abstract properties of vectors, it is not always possible in certain physical problems (Figure 1.1.1).*

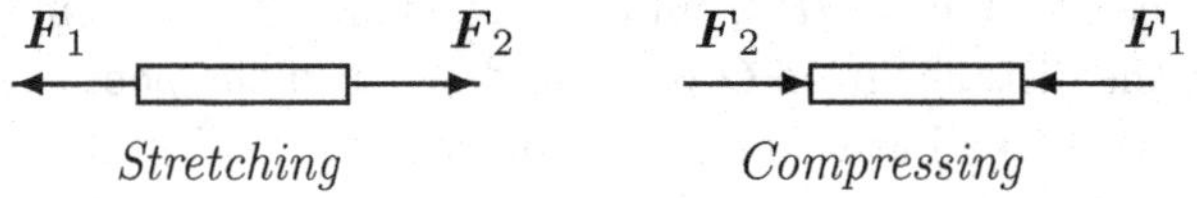

Fig. 1.1.1 Starting Point of a Vector Can Be Important!

1.1.3 *Euclidean Space as a Linear Space*

Consider the Euclidean space $\mathbb{R}^3$ and choose a point O to serve as the origin. In mechanics this is sometimes referred to as choosing a *frame of reference*, or *frame* for short. As was mentioned in Remark 1.1, we assume that all the vectors can be moved to the same starting point; this starting point defines the frame. Accordingly, in what follows, the word `frame` will have one of the three meanings:

- A fixed point;
- A fixed point with a fixed coordinate system (not necessarily Cartesian);
- A fixed point and a `vector bundle`, that is, the collection of *all* vectors that start at that point.

Let $\boldsymbol{r}$ be the position vector for a point P. Consider another frame with origin O'. Let $\boldsymbol{r}'$ be the position vector of P relative to O'. Now, let $\boldsymbol{v}$ be the position vector of O' relative to O. The three vectors form a triangle $OO'P$; see Figure 1.1.2. This suggests that we write $\boldsymbol{r} = \boldsymbol{v} + \boldsymbol{r}'$. To get from O to P we can first go from O to O' along $\boldsymbol{v}$ and then from O' to P along $\boldsymbol{r}'$. This can be depicted entirely with position vectors at O if we move $\boldsymbol{r}'$ parallel to itself and place its initial point at O. Then $\boldsymbol{r}$ is a diagonal of the parallelogram having sides $\boldsymbol{v}$ and $\boldsymbol{r}'$, all emanating from O. This is called the `parallelogram law` for vector addition. It is a geometric definition of $\boldsymbol{v} + \boldsymbol{r}'$. Note that the same result is obtained by forming the triangle $OO'P$.

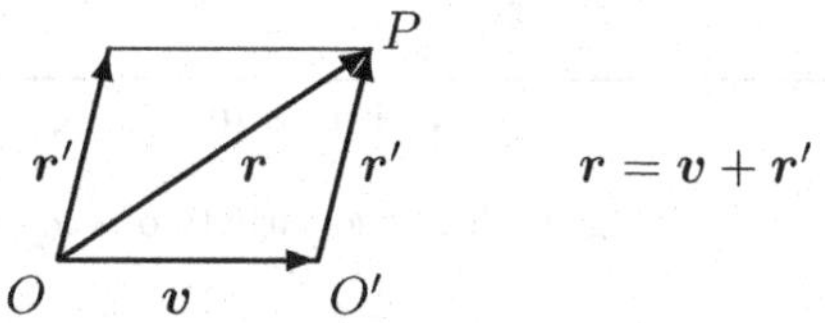

Fig. 1.1.2 Vector Addition

Now, consider three position vectors, $\boldsymbol{u}, \boldsymbol{v}, \boldsymbol{w}$. It is easy to see that the above definition of vector addition obeys the following algebraic laws:

$$
\begin{aligned}
\boldsymbol{u} + \boldsymbol{v} &= \boldsymbol{v} + \boldsymbol{u} && \text{(commutativity)} \\
(\boldsymbol{u} + \boldsymbol{v}) + \boldsymbol{w} &= \boldsymbol{u} + (\boldsymbol{v} + \boldsymbol{w}) && \text{(associativity; see Figure 1.1.3)} \quad (1.1.1) \\
\boldsymbol{u} + \mathbf{0} &= \mathbf{0} + \boldsymbol{u} = \boldsymbol{u}
\end{aligned}
$$

The zero vector $\mathbf{0}$ is the only vector with zero length and no specific direction.

Next consider two *real* numbers, λ and μ. In vector algebra, real numbers are called `scalars`. The vector $\lambda\boldsymbol{u}$ is the vector obtained from $\boldsymbol{u}$ by multiplying its length by $|\lambda|$. If $\lambda > 0$, then the vectors $\boldsymbol{u}$ and $\lambda\boldsymbol{u}$ have the same direction; if $\lambda < 0$, then the vectors have opposite directions. For example, $2\boldsymbol{u}$ points in the same direction as $\boldsymbol{u}$ but has twice its length, whereas $-\boldsymbol{u}$ has the same length as $\boldsymbol{u}$ and points in the opposite direction (Figure 1.1.4).

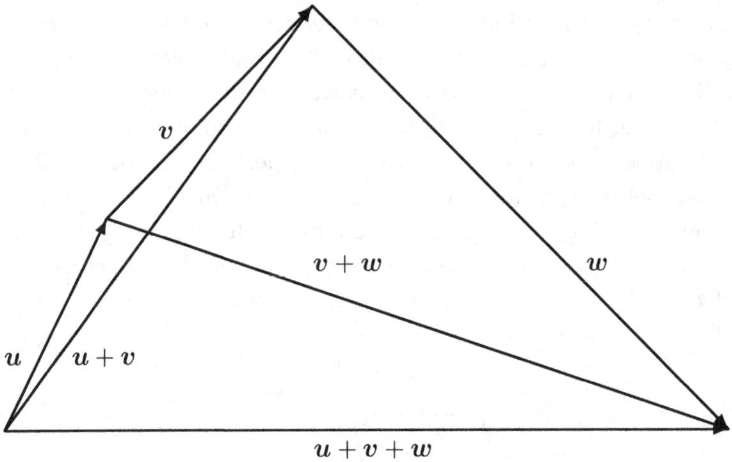

Fig. 1.1.3 Associativity of Vector Addition

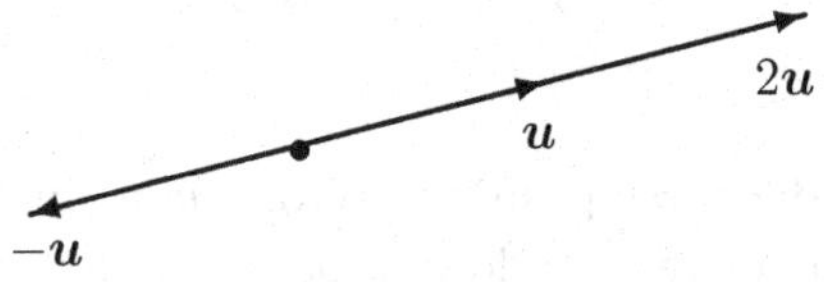

Fig. 1.1.4 Multiplication by a Scalar

Multiplication of a vector by a scalar is easily seen to obey the following algebraic rules:

$$\begin{aligned} \lambda(\boldsymbol{u}+\boldsymbol{v}) &= \lambda\boldsymbol{u}+\lambda\boldsymbol{v} && \text{(distributivity over vector addition)} \\ (\lambda+\mu)\boldsymbol{u} &= \lambda\boldsymbol{u}+\mu\boldsymbol{u} && \text{(distributivity over real addition)} \\ (\lambda\mu)\boldsymbol{u} &= \lambda(\mu\boldsymbol{u}) && \text{(a mixed associativity of multiplications)} \\ 1\cdot\boldsymbol{u} &= \boldsymbol{u}. \end{aligned} \tag{1.1.2}$$

In particular, two vectors are parallel if and only if one is a scalar multiple of the other.

Definition 1.1 A (real) `vector space` is any abstract set of objects, called vectors, with operations of vector addition and multiplication by (real) scalars obeying the seven algebraic rules (1.1.1) and (1.1.2).

Note that, while the set of position vectors is a vector space, the concepts of vector length and the angle between two vectors are not included in the general definition of a vector space. A vector space is said to be n-**dimensional** if the space has a set of n vectors, $\boldsymbol{u}_1, \ldots, \boldsymbol{u}_n$ such that any vector $\boldsymbol{v}$ can be represented as a linear combination of the $\boldsymbol{u}_i$, that is, in the form,

$$\boldsymbol{v} = x_1\,\boldsymbol{u}_1 + \cdots + x_n\,\boldsymbol{u}_n, \tag{1.1.3}$$

and the scalar **components** $x_1, \ldots, x_n$, are *uniquely* determined by $\boldsymbol{v}$. An n-dimensional real vector space is denoted by $\mathbb{R}^n$; with $\mathbb{R}$ denoting the set of real numbers, this notation is quite natural.

We say that the vectors $\boldsymbol{u}_i$, $i = 1, \ldots, n$, form a **basis** in $\mathbb{R}^n$. Notice that nothing is said about the length of the basis vectors or the angles between them: in an abstract vector space, these notions do not exist. The uniqueness of representation (1.1.3) implies that the basis vectors are **linearly independent**, that is, the equality $x_1\,\boldsymbol{u}_1 + \cdots + x_n\,\boldsymbol{u}_n = \boldsymbol{0}$ holds if and only if *all the numbers* $x_1, \ldots, x_n$ are equal to zero. It is not difficult to show that a vector space is n dimensional if and only if the space contains n linear independent vectors, and every collection of $n+1$ vectors is linear dependent; see Problem 1.7, page 411.

In the space $\mathbb{R}^3$ of position vectors, we do have the notions of length and angle. The standard basis in $\mathbb{R}^3$ is the **cartesian basis** $(\hat{\boldsymbol{\imath}}, \hat{\boldsymbol{\jmath}}, \hat{\boldsymbol{k}})$, consisting of the origin O and three mutually perpendicular vectors $\hat{\boldsymbol{\imath}}$, $\hat{\boldsymbol{\jmath}}$, $\hat{\boldsymbol{k}}$ of unit length with the common starting point O. In a cartesian basis, every position vector $\boldsymbol{r} = \overrightarrow{OP}$ of a point P is written in the form

$$\boldsymbol{r} = x\,\hat{\boldsymbol{\imath}} + y\,\hat{\boldsymbol{\jmath}} + z\,\hat{\boldsymbol{k}}; \tag{1.1.4}$$

the numbers (x, y, z) are called the **coordinates** of the point P with respect to the **cartesian coordinate system** formed by the lines along $\hat{\boldsymbol{\imath}}$, $\hat{\boldsymbol{\jmath}}$, and $\hat{\boldsymbol{k}}$. In the plane of $\hat{\boldsymbol{\imath}}$ and $\hat{\boldsymbol{\jmath}}$, the vectors $x\,\hat{\boldsymbol{\imath}} + y\,\hat{\boldsymbol{\jmath}}$ form a two-dimensional vector space $\mathbb{R}^2$. With some abuse of notation, we sometimes write $\boldsymbol{r} = (x, y, z)$ when (1.1.4) holds and the coordinate system is fixed.

The word "cartesian" describes everything connected with the French scientist RENÉ DESCARTES (1596–1650), who was also known by the Latin version of his last name, *Cartesius*. Beside the coordinate system, which he introduced in 1637, he is famous for the statement "I think, therefore I am."

Much of the power of the vector space approach lies in the freedom from any choice of basis or coordinates. Indeed, many geometrical concepts and results can be stated in vector terms without resorting to coordinate systems. Here are two examples:

(1) The line determined by two points in $\mathbb{R}^3$ can be represented by the position vector function

$$\boldsymbol{r}(s) = \boldsymbol{u} + s(\boldsymbol{v} - \boldsymbol{u}) = s\boldsymbol{v} + (1-s)\boldsymbol{u}, \quad -\infty < s < +\infty, \tag{1.1.5}$$

where $\boldsymbol{u}$ and $\boldsymbol{v}$ are the position vectors of the two points. More generally, a line passing through the point P_0 and having a `direction vector` $\boldsymbol{d}$ consists of the points with position vectors $\boldsymbol{r}(s) = \overrightarrow{OP_0} + s\,\boldsymbol{d}$.

(2) The plane determined by the three points having position vectors $\boldsymbol{u}, \boldsymbol{v}, \boldsymbol{w}$ is represented by the position vector function

$$\begin{aligned} \boldsymbol{r}(s,t) &= \boldsymbol{u} + s(\boldsymbol{v} - \boldsymbol{u}) + t(\boldsymbol{w} - \boldsymbol{u}) \\ &= s\boldsymbol{v} + t\boldsymbol{w} + (1-s-t)\boldsymbol{u}, \quad -\infty < s,\ t < +\infty. \end{aligned} \tag{1.1.6}$$

EXERCISE 1.1.1.^{B} *Verify that equations (1.1.5) and (1.1.6) indeed define a line and a plane, respectively, in* $\mathbb{R}^3$.

EXERCISE 1.1.2.^{B} *Let L_1 and L_2 be two parallel lines in $\mathbb{R}^3$. A line intersecting both L_1 and L_2 is called a* `transversal`.

(a) Let L be a transversal perpendicular to L_1. Prove that L is perpendicular to L_2. Hint: If not, then there is a right triangle with L as one side, the other side along L_1 and the hypotenuse lying along L_2. (b) Prove that the alternate angles made by a transversal are equal. Hint: Let A and B be the points of intersection of the transversal with L_1 and L_2 respectively. Draw the perpendiculars at A and B. They form two congruent right triangles.

EXERCISE 1.1.3.^{B} *Use the result of Exercise 1.1.2(b) to prove that the sum of the angles of a triangle equals a straight angle* (180°). *Hint: Let A, B, C be the vertices of the triangle. Through C draw a line parallel to side AB.*

EXERCISE 1.1.4.^{A} *Let a, b be the lengths of the sides of a right triangle with hypotenuse of length c. Prove that $a^2 + b^2 = c^2$ (Pythagorean Theorem). Hint: See Figure 1.1.5 and note that the acute angles A and B are complementary:* $A + B = 90°$.

EXERCISE 1.1.5. ^{C} *Use the result of Exercise 1.1.4 to derive the Euclidean distance formula:* $d(P_1, P_2) = [(x_1 - x_2)^2 + (y_1 - y_2)^2 + (z_1 - z_2)^2]^{1/2}$

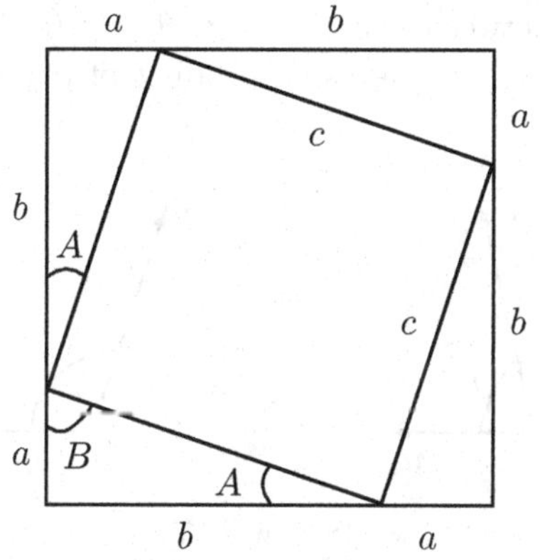

Fig. 1.1.5 Pythagorean Theorem

EXERCISE 1.1.6.[A] *Prove that the diagonals of a parallelogram intersect at their midpoints. Hint: let the vectors* $\boldsymbol{u}$ *and* $\boldsymbol{v}$ *form the parallelogram and let* $\boldsymbol{r}$ *be the position vector of the point of intersection of the diagonals. Argue that* $\boldsymbol{r} = \boldsymbol{u} + s(\boldsymbol{v} - \boldsymbol{u}) = t(\boldsymbol{u} + \boldsymbol{v})$ *and deduce that* $s = t = 1/2$.

1.2 Vector Operations

1.2.1 *Inner Product*

Euclidean geometry and trigonometry deal with lengths of line segments and angles formed by intersecting lines. In abstract vector analysis, lengths of vectors and angles between vectors are defined using the axiomatically introduced notions of norm and inner product.

In $\mathbb{R}^3$, where the notions of angle and length already exist, we use these notions to *define* the inner product $\boldsymbol{u} \cdot \boldsymbol{v}$ of two vectors. We denote the length of vector $\boldsymbol{u}$ by $\|\boldsymbol{u}\|$. A `unit vector` is a vector with length equal to one. If $\boldsymbol{u}$ is a non-zero vector, then $\boldsymbol{u}/\|\boldsymbol{u}\|$ is the unit vector with the same direction as $\boldsymbol{u}$; this unit vector is often denoted by $\widehat{\boldsymbol{u}}$. More generally, a hat $\widehat{\ }$ on top of a vector means that the vector has unit length. With the dot $\cdot$ denoting the inner product of two vectors, we will sometimes write $a.b$ to denote the product of two real numbers a, b.

Definition 1.2 Let $\boldsymbol{u}$ and $\boldsymbol{v}$ be vectors in $\mathbb{R}^3$. The `inner product` of $\boldsymbol{u}$ and $\boldsymbol{v}$, denoted $\boldsymbol{u} \cdot \boldsymbol{v}$, is defined by

$$\boldsymbol{u} \cdot \boldsymbol{v} = \|\boldsymbol{u}\|.\|\boldsymbol{v}\| \cos\theta, \tag{1.2.1}$$

where θ is the angle between $\boldsymbol{u}$ and $\boldsymbol{v}$, $0 \leq \theta \leq \pi$ (see Figure 1.2.1), and the notation $\|\boldsymbol{u}\|.\|\boldsymbol{v}\|$ means the usual product of two numbers. If $\boldsymbol{u} = \mathbf{0}$ or $\boldsymbol{v} = \mathbf{0}$, then $\boldsymbol{u} \cdot \boldsymbol{v} = 0$.

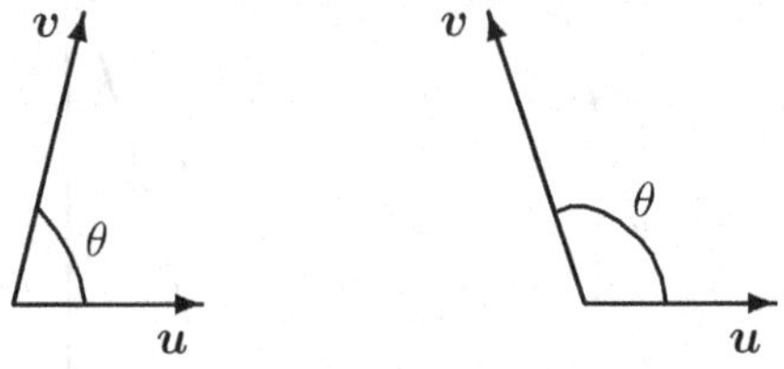

Fig. 1.2.1 Angle Between Two Vectors

Alternative names for the inner product are `dot product` and `scalar product`.

If $\boldsymbol{u}$ and $\boldsymbol{v}$ are non-zero vectors, then $\boldsymbol{u} \cdot \boldsymbol{v} = 0$ if and only if $\theta = \pi/2$. In this case, we say that the vectors $\boldsymbol{u}$ and $\boldsymbol{v}$ are `orthogonal` or `perpendicular`, and write $\boldsymbol{u} \perp \boldsymbol{v}$. Notice that

$$\boldsymbol{u} \cdot \boldsymbol{u} = \|\boldsymbol{u}\|^2 \geq 0. \tag{1.2.2}$$

In $\mathbb{R}^3$, a set of three unit vectors that are mutually orthogonal is called an `orthonormal set` or `orthonormal basis`. For example, the unit vectors $\hat{\imath}$, $\hat{\jmath}$, $\hat{\kappa}$ of a cartesian coordinate system make an orthonormal basis. Indeed, $\hat{\imath} \perp \hat{\jmath}$, $\hat{\imath} \perp \hat{\kappa}$ and $\hat{\jmath} \perp \hat{\kappa}$, $\hat{\imath} \cdot \hat{\jmath} = \hat{\imath} \cdot \hat{\kappa} = \hat{\jmath} \cdot \hat{\kappa} = 0$, and $\hat{\imath} \cdot \hat{\imath} = \hat{\jmath} \cdot \hat{\jmath} = \hat{\kappa} \cdot \hat{\kappa} = 1$.

The word "orthogonal" comes from the Greek *orthogonios*, or "right-angled"; the word "perpendicular" comes from the Latin *perpendiculum*, or "plumb line", which is a cord with a weight attached to one end, used to check a straight vertical position. The Latin word *norma* means "carpenter's square," another device to check for right angles.

The dot product simplifies the computations of the angles between two vectors. Indeed, if $\boldsymbol{u}$ and $\boldsymbol{v}$ are two unit vectors, then $\boldsymbol{u} \cdot \boldsymbol{v} = \cos\theta$. More generally, for two non-zero vectors $\boldsymbol{u}$ and $\boldsymbol{v}$ we have

$$\theta = \cos^{-1}\left(\frac{\boldsymbol{u} \cdot \boldsymbol{v}}{\|\boldsymbol{u}\|.\|\boldsymbol{v}\|}\right), \tag{1.2.3}$$

The notion of the dot product is closely connected with the ORTHOGONAL PROJECTION. If $\boldsymbol{u}$ and $\boldsymbol{v}$ are two non-zero vectors, then we can write $\boldsymbol{u} = \boldsymbol{u}_v + \boldsymbol{u}_p$, where $\boldsymbol{u}_v$ is parallel to $\boldsymbol{v}$ and $\boldsymbol{u}_p$ is perpendicular to $\boldsymbol{v}$ (see Figure 1.2.2).

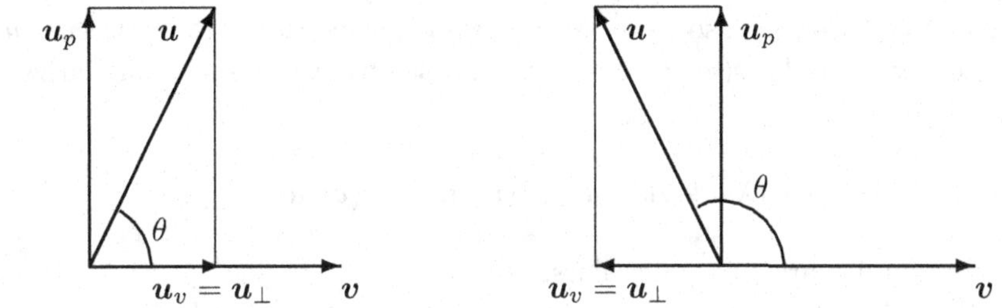

Fig. 1.2.2 Orthogonal Projection

It follows from the picture that $\|u_v\| = \|u\|.|\cos\theta|$ and u_v has the same direction as v if and only if $0 < \theta < \pi/2$. Comparing this with (1.2.1) we conclude that

$$u_v = \frac{u \cdot v}{\|v\|} \frac{v}{\|v\|}. \tag{1.2.4}$$

The vector u_v is called the **orthogonal projection** of u on v, and is denoted by $u_\perp$; the number $u \cdot v/\|v\|$ is called the **component** of u in the direction of v; note that $v/\|v\|$ is a unit vector. The verb "to project" comes from Latin "to through forward." *Let us emphasize that the orthogonal projection of a vector is also a vector.*

Let us now use the idea of the orthogonal projection to establish the PROPERTIES OF THE INNER PRODUCT.

Consider two non-zero vectors u and w and a unit vector v. Then $(u+w) \cdot v$ is the projection of $u+w$ on v. From Figure 1.2.3, we conclude that $(u+w) \cdot v = u \cdot v + w \cdot v$.

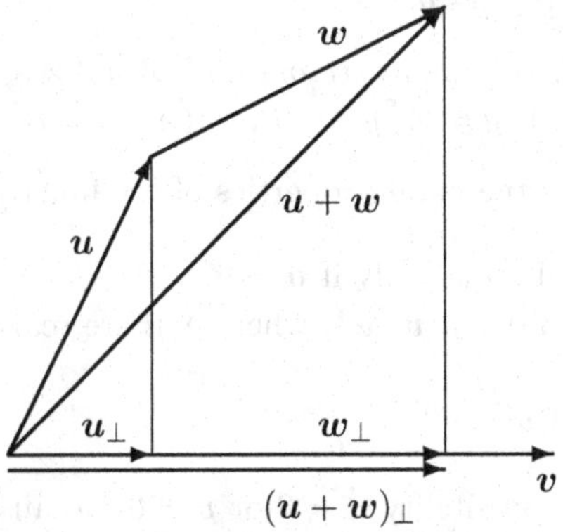

Fig. 1.2.3 Orthogonal Projection of Two Vectors

Furthermore, if λ is any real scalar, then $(\lambda \boldsymbol{u})\cdot \boldsymbol{v} = \lambda(\boldsymbol{u}\cdot\boldsymbol{v})$. For example $(2\boldsymbol{u})\cdot\boldsymbol{v} = 2(\boldsymbol{u}\cdot\boldsymbol{v})$. Also $(-\boldsymbol{u})\cdot\boldsymbol{v} = -(\boldsymbol{u}\cdot\boldsymbol{v})$, since the angle between $-\boldsymbol{u}$ and $\boldsymbol{v}$ is $\pi-\theta$ and $\cos(\pi-\theta) = -\cos\theta$. These observations are summarized by the formula

$$(\lambda\boldsymbol{u}+\mu\boldsymbol{v})\cdot\boldsymbol{w} = \lambda(\boldsymbol{u}\cdot\boldsymbol{w}) + \mu(\boldsymbol{v}\cdot\boldsymbol{w}), \tag{1.2.5}$$

where λ and μ are any real scalars.

Note that these properties of inner product are independent of any coordinate system.

Next, we will find an expression for the inner product in terms of the components of the vectors in cartesian coordinates. Let $\hat{\boldsymbol{\imath}}$, $\hat{\boldsymbol{\jmath}}$, $\hat{\boldsymbol{\kappa}}$ be an orthonormal set forming a cartesian coordinate system. Any position vector $\boldsymbol{x} = \overrightarrow{OP}$ can be expressed as $\boldsymbol{x} = x_1\hat{\boldsymbol{\imath}}+x_2\hat{\boldsymbol{\jmath}}+x_3\hat{\boldsymbol{\kappa}}$, where $x_1 = \boldsymbol{x}\cdot\hat{\boldsymbol{\imath}}$, $x_2 = \boldsymbol{x}\cdot\hat{\boldsymbol{\jmath}}$ and $x_3 = \boldsymbol{x}\cdot\hat{\boldsymbol{\kappa}}$ are the cartesian coordinates of the point P. If $\boldsymbol{y}$ is another vector, then $\boldsymbol{y} = y_1\hat{\boldsymbol{\imath}}+y_2\hat{\boldsymbol{\jmath}}+y_3\hat{\boldsymbol{\kappa}}$ and by (1.2.5) and the orthonormal property of $\hat{\boldsymbol{\imath}}$, $\hat{\boldsymbol{\jmath}}$, $\hat{\boldsymbol{\kappa}}$, we get

$$\boldsymbol{x}\cdot\boldsymbol{y} = x_1y_1 + x_2y_2 + x_3y_3. \tag{1.2.6}$$

This formula expresses $\boldsymbol{x}\cdot\boldsymbol{y}$ in terms of the coordinates of $\boldsymbol{x}$ and $\boldsymbol{y}$. Together with (1.2.3), we can use the result for computing the angle between two vector with known components in a given cartesian coordinate system.

In linear algebra and in some software packages, such as MATLAB, vectors are represented as `column vectors`, that is, as 3×1 matrices; for a summary of linear algebra, see page 451. If $\boldsymbol{x}$ and $\boldsymbol{y}$ are column vectors, then the transpose $\boldsymbol{x}^T$ is a `row vector` (1×3 matrix) and, by the rules of matrix multiplication, $\boldsymbol{x}\cdot\boldsymbol{y} = \boldsymbol{x}^T\boldsymbol{y}$.

EXERCISE 1.2.1.^{C} *Let* $\boldsymbol{x}$, $\boldsymbol{y}$ *be column vectors and,* A *a* 3×3 *matrix. Show that* $A\boldsymbol{x}\cdot\boldsymbol{y} = \boldsymbol{y}^TA\boldsymbol{x} = \boldsymbol{x}^TA^T\boldsymbol{y} = A^T\boldsymbol{y}\cdot\boldsymbol{x}$. *Hint:* $(AB)^T = B^TA^T$.

We can now summarize the main properties of the inner product:

(I1) $\boldsymbol{u}\cdot\boldsymbol{u}\geq 0$ and $\boldsymbol{u}\cdot\boldsymbol{u} = 0$ if and only if $\boldsymbol{u} = \mathbf{0}$.
(I2) $(\lambda\boldsymbol{u}+\mu\boldsymbol{v})\cdot\boldsymbol{w} = \lambda(\boldsymbol{u}\cdot\boldsymbol{w}) + \mu(\boldsymbol{v}\cdot\boldsymbol{w})$, where λ,μ are real numbers.
(I3) $\boldsymbol{u}\cdot\boldsymbol{v} = \boldsymbol{v}\cdot\boldsymbol{u}$.
(I4) $\boldsymbol{u}\cdot\boldsymbol{v} = 0$ if and only if $\boldsymbol{u}\perp\boldsymbol{v}$.

Property (I4) includes the possibility $\boldsymbol{u} = \mathbf{0}$ or $\boldsymbol{v} = \mathbf{0}$, because, by convention, the zero vector $\mathbf{0}$ does not have a specific direction and is therefore orthogonal to any vector. This is consistent with (I2): taking $\lambda = \mu = 1$ and $\boldsymbol{v} = \boldsymbol{0}$, we also find $\boldsymbol{w} \cdot \boldsymbol{0} = 0$ for every $\boldsymbol{w}$.

EXERCISE 1.2.2.^{C} *Prove the law of cosines: $a^2 = b^2 + c^2 - 2bc\cos\theta$, where a, b, c are the sides of a triangle and θ is the angle between b and c. Hint: Let $c = \|\boldsymbol{r}_1\|$, $b = \|\boldsymbol{r}_2\|$. Then $a^2 = \|\boldsymbol{r}_2 - \boldsymbol{r}_1\|^2 = (\boldsymbol{r}_1 - \boldsymbol{r}_2) \cdot (\boldsymbol{r}_1 - \boldsymbol{r}_2)$.*

We now discuss some APPLICATIONS OF THE INNER PRODUCT. We start with the EQUATION OF A LINE IN $\mathbb{R}^2$. Choose an origin O and drop the perpendicular from O to the line L; see Figure 1.2.4.

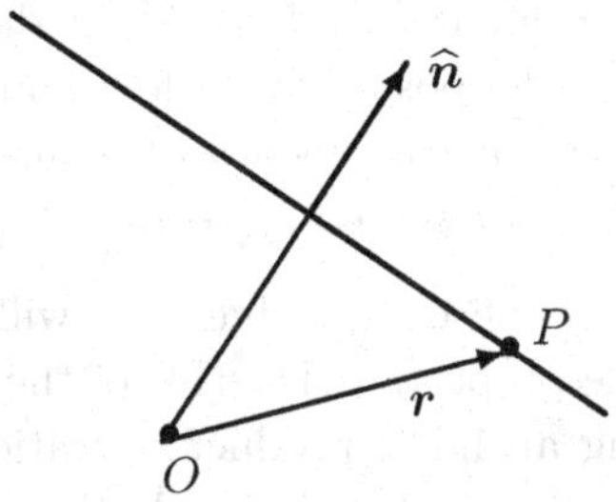

Fig. 1.2.4 Line in The Plane

Let $\boldsymbol{n}$ be a unit vector lying on this perpendicular. For any point P on L, the position vector $\boldsymbol{r}$ satisfies

$$\boldsymbol{r} \cdot \boldsymbol{n} = d, \tag{1.2.7}$$

where $|d|$ is the distance from O to L; indeed, $|\boldsymbol{r} \cdot \boldsymbol{n}|$ is the length of the projection of $\boldsymbol{r}$ on $\boldsymbol{n}$. In a cartesian coordinate system (x, y), $\boldsymbol{r} = x\,\hat{\imath} + y\,\hat{\jmath}$, and equation (1.2.7) becomes $ax + by = d$, where $\boldsymbol{n} = a\,\hat{\imath} + b\,\hat{\jmath}$. More generally, every equation of the form $a_1x + a_2y = a_3$, with real numbers a_1, a_2, a_3, defines a line in $\mathbb{R}^2$.

Similar arguments produce the EQUATION OF A PLANE IN $\mathbb{R}^3$. Let $\boldsymbol{n}$ be a unit vector perpendicular to the plane. For any point P in the plane, the equation (1.2.7) holds again; Figure 1.2.4 represents the view in the plane spanned by the vectors $\widehat{\boldsymbol{n}}$ and $\boldsymbol{r}$ and containing points O, P. In a cartesian coordinate system (x, y, z), $\boldsymbol{r} = x\,\hat{\imath} + y\,\hat{\jmath} + z\,\hat{\boldsymbol{k}}$, and equation (1.2.7) becomes $ax + by + cz = d$, where $\boldsymbol{n} = a\,\hat{\imath} + b\,\hat{\jmath} + c\,\hat{\boldsymbol{k}}$. More generally, every equation of the form $a_1x + a_2y + a_3z = a_4$ defines a plane in $\mathbb{R}^3$ with a (not necessarily unit) `normal vector` $a_1\,\hat{\imath} + a_2\,\hat{\jmath} + a_3\,\hat{\boldsymbol{k}}$. For alternative ways to represent a line and a plane see equations (1.1.5) and (1.1.6) on page 8.

EXERCISE 1.2.3.[C] *Using equation (1.2.7), write an equation of the plane that is 4 units from the origin and has the unit normal* $\widehat{\boldsymbol{n}} = (2, -1, 2)/3$. *How many such planes are there?*

EXERCISE 1.2.4.[C] *Let* $2x - y + 2z = 12$ *and* $x + y - z = 1$ *be the equations of two planes. Find the cosine of the angle between these planes.*

Yet another application of the dot product is to computing the WORK DONE BY A FORCE. Let $\boldsymbol{F}$ be a force vector acting on a mass m and moving it through a displacement given by vector $\boldsymbol{r}$. The work W done by $\boldsymbol{F}$ moving m through this displacement is $W = \boldsymbol{F} \cdot \boldsymbol{r}$, since $\|\boldsymbol{F}\| \cos\theta$ is the magnitude of the component of $\boldsymbol{F}$ along $\boldsymbol{r}$ and $\|\boldsymbol{r}\|$ is the distance moved.

We will see later that, beside the position and force, many other mechanical quantities (acceleration, angular momentum, angular velocity, momentum, torque, velocity) can be represented as vectors.

To conclude our discussion of the dot product, we will do some ABSTRACT VECTOR ANALYSIS. The properties (I1)–(I3) of the inner product can be taken as axioms defining an inner product operation in any vector space. In other words, an inner product is a rule that assigns to any pair $\boldsymbol{u}$, $\boldsymbol{v}$ of vectors a real number $\boldsymbol{u} \cdot \boldsymbol{v}$ so that properties (I1)–(I3) hold. With this approach, the definition and properties of the inner product are independent of coordinate systems.

Consider the vector space $\mathbb{R}^n$ with a basis $\mathfrak{U} = (\boldsymbol{u}_1, \ldots, \boldsymbol{u}_n)$; see page 7. We can represent every element $\boldsymbol{x}$ of $\mathbb{R}^n$ as an n-tuples $(x_1, \ldots, x_n)$ of the components of $\boldsymbol{x}$ in the fixed basis. Clearly, for $\boldsymbol{y} = (y_1, \ldots, y_n)$ and $\lambda \in \mathbb{R}$,

$$\boldsymbol{x} + \boldsymbol{y} = (x_1 + y_1, \ldots, x_n + y_n), \ \ \lambda\boldsymbol{x} = (\lambda x_1, \ldots, \lambda x_n).$$

We then *define*

$$\boldsymbol{x} \cdot \boldsymbol{y} = x_1 y_1 + \cdots + x_n y_n = \sum_{i=1}^{n} x_i y_i. \tag{1.2.8}$$

It is easy to verify that this definition satisfies (I1)–(I3). For $n = 3$ with a Cartesian basis, equation (1.2.6) is a special case of (1.2.8).

If an inner product is defined in a vector space, then in view of property (I1) we can define a `norm` or `length` of a vector by

$$\|\boldsymbol{u}\| = (\boldsymbol{u} \cdot \boldsymbol{u})^{1/2}. \tag{1.2.9}$$

While an inner product defines a norm, other norms in $\mathbb{R}^n$ exist that are not inner product-based; see Problem 1.8 on page 411.

An `orthonormal basis` in $\mathbb{R}^n$ is a basis consisting of pair-wise orthogonal vectors of unit length.

EXERCISE 1.2.5.[C] *Verify that, under definition (1.2.8), the corresponding basis $\boldsymbol{u}_1, \ldots, \boldsymbol{u}_n$ is necessarily orthonormal. Hint: argue that the basis vector $\boldsymbol{u}_k$ is represented by an n-tuple with zeros everywhere except the position k.*

EXERCISE 1.2.6.[B] *Prove the* `parallelogram law`*:*

$$\|\boldsymbol{u}+\boldsymbol{v}\|^2 + \|\boldsymbol{u}-\boldsymbol{v}\|^2 = 2\|\boldsymbol{u}\|^2 + 2\|\boldsymbol{v}\|^2. \tag{1.2.10}$$

Show that in $\mathbb{R}^3$ this equality can be stated as follows: in a parallelogram, the sum of the squares of the diagonals is equal to the sum of the squares of the sides (hence the name "parallelogram law").

Theorem 1.2.1 *The norm defined by (1.2.9) satisfies the* `triangle inequality`

$$\|\boldsymbol{u}+\boldsymbol{v}\| \le \|\boldsymbol{u}\| + \|\boldsymbol{v}\| \tag{1.2.11}$$

and the `Cauchy-Schwartz inequality`

$$|\boldsymbol{u}\cdot\boldsymbol{v}| \le \|\boldsymbol{u}\|.\|\boldsymbol{v}\|. \tag{1.2.12}$$

Proof. We first show that (1.2.11) follows from (1.2.12). Indeed,

$$\begin{aligned}
&\|\boldsymbol{u}+\boldsymbol{v}\|^2 = (\boldsymbol{u}+\boldsymbol{v})\cdot(\boldsymbol{u}+\boldsymbol{v}) = \|\boldsymbol{u}\|^2 + 2(\boldsymbol{u}\cdot\boldsymbol{v}) + \|\boldsymbol{v}\|^2 \\
&\le \|\boldsymbol{u}\|^2 + 2|\boldsymbol{u}\cdot\boldsymbol{v}| + \|\boldsymbol{v}\|^2 \le \|\boldsymbol{u}\|^2 + 2\|\boldsymbol{u}\|.\|\boldsymbol{v}\| + \|\boldsymbol{v}\|^2 = (\|\boldsymbol{u}\| + \|\boldsymbol{v}\|)^2.
\end{aligned}$$

To prove (1.2.12), first suppose $\boldsymbol{u}$ and $\boldsymbol{v}$ are unit vectors. By properties (I1)–(I3) of the inner product, for any scalar λ,

$$0 \le (\boldsymbol{u}+\lambda\boldsymbol{v})\cdot(\boldsymbol{u}+\lambda\boldsymbol{v}) = \boldsymbol{u}\cdot\boldsymbol{u} + 2\lambda\boldsymbol{u}\cdot\boldsymbol{v} + \lambda^2\boldsymbol{v}\cdot\boldsymbol{v} = 1 + 2\lambda\boldsymbol{u}\cdot\boldsymbol{v} + \lambda^2.$$

Now, take $\lambda = -(\boldsymbol{u}\cdot\boldsymbol{v})$. Then $0 \le 1 - 2(\boldsymbol{u}\cdot\boldsymbol{v})^2 + (\boldsymbol{u}\cdot\boldsymbol{v})^2 = 1 - (\boldsymbol{u}\cdot\boldsymbol{v})^2$. Hence, $|\boldsymbol{u}\cdot\boldsymbol{v}| \le 1$. On the other hand, for every non-zero vectors $\boldsymbol{u}$ and $\boldsymbol{v}$, $\boldsymbol{u} = \|\boldsymbol{u}\|\cdot\boldsymbol{u}/\|\boldsymbol{u}\|$ and $\boldsymbol{v} = \|\boldsymbol{v}\|\cdot\boldsymbol{v}/\|\boldsymbol{v}\|$. Since $\boldsymbol{u}/\|\boldsymbol{u}\|$ and $\boldsymbol{v}/\|\boldsymbol{v}\|$ are unit vectors, we have $|\boldsymbol{u}\cdot\boldsymbol{v}|/(\|\boldsymbol{u}\|\,\|\boldsymbol{v}\|) \le 1$, and so $|\boldsymbol{u}\cdot\boldsymbol{v}| \le \|\boldsymbol{u}\|\,\|\boldsymbol{v}\|$.

If either $\boldsymbol{u}$ or $\boldsymbol{v}$ is a zero vector, then (1.2.12) trivially holds. Theorem 1.2.1 is proved. □

Remark 1.2 *Analysis of the proof of Theorem 1.2.1 shows that equality in either (1.2.11) or (1.2.12) holds if and only if one of the vectors is a* *scalar multiple of the other:* $\boldsymbol{u} = \lambda\boldsymbol{v}$ *or* $\boldsymbol{v} = \lambda\boldsymbol{u}$ *for some real number* λ*; we have to write two conditions to allow either* $\boldsymbol{u}$ *or* $\boldsymbol{v}$*, or both, to be the zero vector.*

EXERCISE 1.2.7.^{C} *Choose a Cartesian coordinate system* (x, y, z) *with the corresponding unit basis vectors* $(\hat{\imath}, \hat{\jmath}, \hat{\kappa})$*. Let* P*,* Q*, be points with coordinates* $(1, -3, 2)$ *and* $(-2, 4, -1)$*, respectively. Define* $\boldsymbol{u} = \overrightarrow{OP}$*,* $\boldsymbol{v} = \overrightarrow{OQ}$*.*
(a) Compute $\overrightarrow{QP} = \boldsymbol{u} - \boldsymbol{v}$*,* $\|\boldsymbol{u}\|$*, and* $\|\boldsymbol{v}\|$*. Compute the angle between* $\boldsymbol{u}$ *and* $\boldsymbol{v}$*. Verify the Cauchy-Schwartz inequality and the triangle inequality.*
(b) Let $\boldsymbol{w} = 2\,\hat{\imath} + 4\,\hat{\jmath} - 5\,\hat{\kappa}$*. Check that the associative law holds for* $\boldsymbol{u}$*,* $\boldsymbol{v}$*,* $\boldsymbol{w}$*.*
(c) Suppose $\boldsymbol{u}$ *is a force vector. Compute the component of* $\boldsymbol{u}$ *in the* $\boldsymbol{v}$ *direction. Suppose* $\boldsymbol{v}$ *is the displacement of a unit mass acted on by the force* $\boldsymbol{u}$*. Compute the work done.*

Inequality (1.2.12) is also known as the Cauchy-Bunyakovky-Schwartz inequality, and all three possible combinations of any two of these three names can also refer to the same or similar inequality. This inequality is extremely useful in many areas of mathematics, and all three, Cauchy, Bunyakovky, and Schwartz, certainly deserve to be mentioned in connection with it. The Russian mathematician VIKTOR YAKOVLEVICH BUNYAKOVSKY (1804–1889) and the German mathematician HERMANN AMANDUS SCHWARZ (1843–1921) discovered a version of (1.2.12) for the integrals:

$$\int_a^b |f(x)g(x)|dx \le \left(\int_a^b f^2(x)dx\right)^{1/2} \left(\int_a^b g^2(x)dx\right)^{1/2}; \qquad (1.2.13)$$

Bunyakovsky published it in 1859, Schwartz, most probably unaware of Bunyakovsky's work, in 1884. The French mathematician AUGUSTIN LOUIS CAUCHY (1789–1857) has his name attached not just to (1.2.12) but to many other mathematical results. There are two main reasons for that: he was the first to introduce modern standards of rigor in the mathematical proofs, and he published a lot of papers (789 to be exact, some exceeding 300 pages), covering most ares of mathematics. We will be mentioning Cauchy a lot during our discussion of complex analysis. Throughout the rest of our discussions, we will refer to (1.2.12) and all its modifications as the `Cauchy-Schwartz inequality`.

EXERCISE 1.2.8.^{A} *(a) Use the same arguments as in the proof of (1.2.12) to establish (1.2.13). (b) Use the same arguments as in the proof of (1.2.12) to establish the following version of the Cauchy-Schwartz inequality:*

$$\sum_{k=1}^{\infty} |a_k b_k| \leq \left(\sum_{k=1}^{\infty} a_k^2\right)^{1/2} \left(\sum_{k=1}^{\infty} b_k^2\right)^{1/2}. \tag{1.2.14}$$

In both parts (a) and (b), assume all the necessary integrability and convergence.

We conclude this section with a brief discussion of *transformations* of a linear vector space. We will see later that a mathematical model of the motion of an object in space is a special transformation of $\mathbb{R}^3$.

Definition 1.3 A `transformation` A of the space $\mathbb{R}^n$, $n \geq 2$, is a rule that assigns to every element $\boldsymbol{x}$ of $\mathbb{R}^n$ a unique element $A(\boldsymbol{x})$ from $\mathbb{R}^n$. When there is no danger of confusion, we write $A\boldsymbol{x}$ instead of $A(\boldsymbol{x})$.
A transformation A is called an `isometry` if it preserves the distances between points: $\|A\boldsymbol{x} - A\boldsymbol{y}\| = \|\boldsymbol{x} - \boldsymbol{y}\|$ for all $\boldsymbol{x}$, $\boldsymbol{y}$ in $\mathbb{R}^n$.
A transformation A is called `linear` if $A(\lambda\,\boldsymbol{x} + \mu\,\boldsymbol{y}) = \lambda\,A(\boldsymbol{x}) + \mu\,A(\boldsymbol{y})$ for all $\boldsymbol{x}$, $\boldsymbol{y}$ from $\mathbb{R}^n$ and all real numbers λ, μ.
A transformation is called `orthogonal` if it is both a linear transformation and an isometry.

The two Latin roots in the word "transformation," *trans* and *forma*, mean "beyond" and "shape," respectively. The two Greek roots in the word "isometry", *isos* and *metron*, mean "equal" and "measure." We know from linear algebra that, in $\mathbb{R}^n$ with a fixed basis, every linear transformation is represented by a square matrix; see Exercise 8.1.4, page 453, in Appendix.

EXERCISE 1.2.9.^{A} *(a) Show that if A is a linear transformation, then $A(\mathbf{0}) = \mathbf{0}$. Hint: use that $\mathbf{0} = \lambda\,\mathbf{0}$ for all real λ.*
(b) Show that the transformation A is orthogonal if and only if it preserves the inner product: $(A\boldsymbol{x}) \cdot (A\boldsymbol{y}) = \boldsymbol{x} \cdot \boldsymbol{y}$ for all $\boldsymbol{x}$, $\boldsymbol{y}$ from $\mathbb{R}^n$. Hint: use the parallelogram law (1.2.10).

1.2.2 *Cross Product*

In the three-dimensional vector space $\mathbb{R}^3$, we use the Euclidean geometry and trigonometry to define the inner product of two vectors. This definition easily extends to every $\mathbb{R}^n$, $n \geq 2$. In $\mathbb{R}^3$, and only in $\mathbb{R}^3$, there exists another product of two vectors, called the `cross product`, or `vector product`.

Definition 1.4 Let $\boldsymbol{u}$ and $\boldsymbol{v}$ be two vectors in $\mathbb{R}^3$. Let θ be the angle between $\boldsymbol{u}$ and $\boldsymbol{v}$ ($0 \leq \theta \leq \pi$, see Figure 1.2.1). The cross product, $\boldsymbol{u} \times \boldsymbol{v}$, is the vector having magnitude $\|\boldsymbol{u} \times \boldsymbol{v}\| = \|\boldsymbol{u}\|.\|\boldsymbol{v}\| \sin\theta$ and lying on the line perpendicular to $\boldsymbol{u}$ and $\boldsymbol{v}$ and pointing in the direction in which a right-handed screw would move when $\boldsymbol{u}$ is rotated toward $\boldsymbol{v}$ through angle θ.

Sometimes, the symbol $\odot$ is used to represent a vector perpendicular to the plane and coming out of the plane toward the observer, while the symbol $\otimes$ represents a similar vector, but going away from the observer; see Figure 1.2.6.

The triple $(\boldsymbol{u}, \boldsymbol{v}, \boldsymbol{u} \times \boldsymbol{v})$ forms a *right-handed triad* (Figure 1.2.5). More generally, we say that an ordered triplet of vectors $(\boldsymbol{u}, \boldsymbol{v}, \boldsymbol{w})$ with a common origin in $\mathbb{R}^3$ is a `right-handed triad` (or right-handed triple) if the vectors are not in the same plane and the shortest turn from $\boldsymbol{u}$ to $\boldsymbol{v}$, as seen from the tip of $\boldsymbol{w}$, is *counterclockwise*.

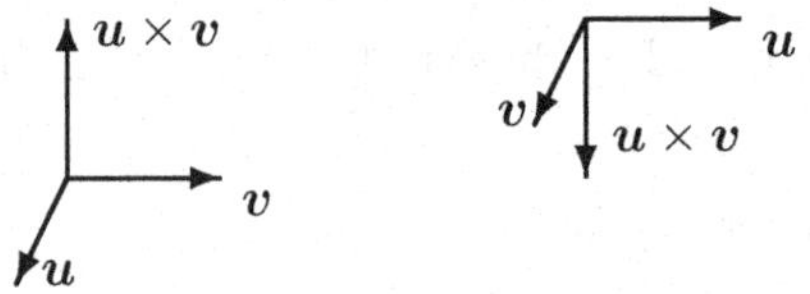

Fig. 1.2.5 The Cross Product I

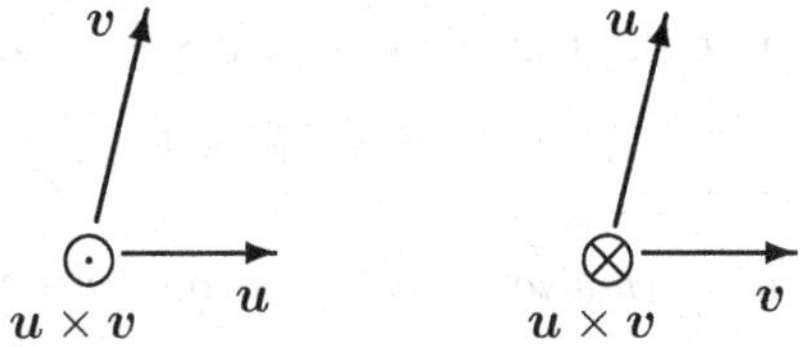

Fig. 1.2.6 The Cross Product II

An important application of cross-product in mechanics is the moment of a force about a point O. Suppose an object located at a point P is subjected to a force vector $\boldsymbol{F}$, applied at P. Let $\boldsymbol{r}$ be the position vector of P. The force $\boldsymbol{F}$ tends to rotate the object around O and exerts a `torque`, or `moment`, $\boldsymbol{T}$ around O. (The Latin verb *torquēre* means "to twist.") The magnitude of the torque $\boldsymbol{T}$ is $\|\boldsymbol{T}\| = \|\boldsymbol{r}\|.\|\boldsymbol{F}\| \sin\theta$, where θ is the angle between $\boldsymbol{r}$ and $\boldsymbol{F}$; recall that $a.b$ denotes the usual product of two numbers

a, b. The quantity $\|\boldsymbol{F}\| \sin\theta$ is the magnitude of the component of $\boldsymbol{F}$ perpendicular to $\boldsymbol{r}$. (The component of $\boldsymbol{F}$ along $\boldsymbol{r}$ has no rotational effect.) The magnitude $\|\boldsymbol{r}\|$ is called the `moment arm`. Our experience with levers convinces us that the torque magnitude is proportional to the moment arm and the magnitude of force applied perpendicular to the arm. Hence, we define the `torque` of $\boldsymbol{F}$ around O to be the vector $\boldsymbol{T} = \boldsymbol{r} \times \boldsymbol{F}$, where $\boldsymbol{r}$ is the position at which $\boldsymbol{F}$ is applied. The direction of $\boldsymbol{T}$ is perpendicular to $\boldsymbol{r}$ and $\boldsymbol{F}$ and $(\boldsymbol{r}, \boldsymbol{F}, \boldsymbol{T})$ is a right-handed triad.

PROPERTIES OF THE CROSS PRODUCT. From the definition it follows immediately that the vector $\boldsymbol{w} = \boldsymbol{u} \times \boldsymbol{v}$ has the following three properties:

(C1) $\|\boldsymbol{w}\| = \|\boldsymbol{u}\|.\|\boldsymbol{v}\| \sin\theta$.
(C2) $\boldsymbol{w} \cdot \boldsymbol{u} = \boldsymbol{w} \cdot \boldsymbol{v} = 0$.
(C3) $-\boldsymbol{w} = \boldsymbol{v} \times \boldsymbol{u}$.

A fourth property captures the geometry of the right-handed screw in algebraic terms. Choose any right-handed cartesian coordinate system given by three orthonormal vectors $\hat{\boldsymbol{\imath}}$, $\hat{\boldsymbol{\jmath}}$, $\hat{\boldsymbol{\kappa}}$. Suppose the components of the vectors $\boldsymbol{u}$, $\boldsymbol{v}$, $\boldsymbol{w} = \boldsymbol{u} \times \boldsymbol{v}$ in the basis $(\hat{\boldsymbol{\imath}}, \hat{\boldsymbol{\jmath}}, \hat{\boldsymbol{\kappa}})$ are, respectively, (u_1, u_2, u_3), (v_1, v_2, v_3), and (w_1, w_2, w_3). Then

(C4) $\det \begin{pmatrix} u_1 & u_2 & u_3 \\ v_1 & v_2 & v_3 \\ w_1 & w_2 & w_3 \end{pmatrix} > 0,$

where det is the determinant of the matrix; a brief review of linear algebra, including the determinants, is in Appendix. To prove (C4), choose $\hat{\boldsymbol{\kappa}}' = \boldsymbol{w}/\|\boldsymbol{w}\|$, $\hat{\boldsymbol{\jmath}}' = \boldsymbol{v}/\|\boldsymbol{v}\|$, and select a unit vector $\hat{\boldsymbol{\imath}}'$ orthogonal to both $\hat{\boldsymbol{\kappa}}'$ and $\hat{\boldsymbol{\jmath}}'$ to make $(\hat{\boldsymbol{\imath}}', \hat{\boldsymbol{\jmath}}', \hat{\boldsymbol{\kappa}}')$ a right-handed triad. In this new coordinate system, property (C4) becomes

$$\det \begin{pmatrix} u_1' & u_2' & 0 \\ 0 & \|\boldsymbol{v}\| & 0 \\ 0 & 0 & \|\boldsymbol{w}\| \end{pmatrix} = u_1' \|\boldsymbol{v}\|.\|\boldsymbol{w}\| > 0. \qquad (1.2.15)$$

Since $(\boldsymbol{u}, \boldsymbol{v}, \boldsymbol{w})$ is a right-handed triad, the choice of $\hat{\boldsymbol{\imath}}'$ implies that $u_1' > 0$, and (1.2.15) holds. For the system $\hat{\boldsymbol{\imath}}$, $\hat{\boldsymbol{\jmath}}$, $\hat{\boldsymbol{\kappa}}$ with the same origin as $(\hat{\boldsymbol{\imath}}', \hat{\boldsymbol{\jmath}}', \hat{\boldsymbol{\kappa}}')$, consider an orthogonal transformation that moves the basis vectors $\hat{\boldsymbol{\imath}}$, Vcj, $\hat{\boldsymbol{\kappa}}$ to the vectors $\hat{\boldsymbol{\imath}}'$, Vcj', $\hat{\boldsymbol{\kappa}}'$, respectively. If B is the matrix representing this transformation in the basis $(\hat{\boldsymbol{\imath}}, Vcj, \hat{\boldsymbol{\kappa}})$, then $\det B = 1$, and the two matrices, A in (C4) and A' in (1.2.15) are related by $A' = BAB^T$. Hence, $\det A = \det A' > 0$ and (C4) holds.

EXERCISE 1.2.10.[C] *Verify that $A' = BAB^T$. Hint: see Exercise 8.1.4 on page 453 in Appendix. Pay attention to the basis in which each matrix is written.*

The following theorem shows that the properties (C1), (C2), and (C4) define a unique vector $\boldsymbol{w} = \boldsymbol{u} \times \boldsymbol{v}$.

Theorem 1.2.2 *For every two non-zero, non-parallel vectors $\boldsymbol{u}, \boldsymbol{v}$ in $\mathbb{R}^3$, there is a unique vector $\boldsymbol{w} = \boldsymbol{u} \times \boldsymbol{v}$ satisfying (C1), (C2), (C4). If (u_1, u_2, u_3) and (v_1, v_2, v_3) are the components of $\boldsymbol{u}$ and $\boldsymbol{v}$ in a cartesian right-handed system $\hat{\imath}$, $\hat{\jmath}$, $\hat{\kappa}$, then the components w_1, w_2, w_3 of $\boldsymbol{u} \times \boldsymbol{v}$ are*

$$w_1 = u_2 v_3 - u_3 v_2, \quad w_2 = u_3 v_1 - u_1 v_3, \quad w_3 = u_1 v_2 - u_2 v_1. \tag{1.2.16}$$

Conversely, the vector with components defined by (1.2.16) has Properties (C1), (C2), and (C4).

Proof. Let $\boldsymbol{w}$ be a vector so that $\boldsymbol{w} \cdot \boldsymbol{u} = 0$ and $\boldsymbol{w} \cdot \boldsymbol{v} = 0$, that is, $\boldsymbol{w}$ is orthogonal to both $\boldsymbol{u}$ and $\boldsymbol{v}$. By the geometry of $\mathbb{R}^3$, there is such a vector. Choose a $\boldsymbol{w}$ with magnitude $\|\boldsymbol{w}\| = \|\boldsymbol{u}\| \, \|\boldsymbol{v}\| \sin\theta$, satisfying (C1). By (C2),

$$u_1 w_1 + u_2 w_2 + u_3 w_3 = 0, \tag{1.2.17}$$

$$v_1 w_1 + v_2 w_2 + v_3 w_3 = 0. \tag{1.2.18}$$

Multiply (1.2.17) by v_3 and (1.2.18) by u_3 and subtract to get

$$(\overbrace{u_1 v_3 - u_3 v_1}^{a}) w_1 = (\overbrace{u_3 v_2 - u_2 v_3}^{b}) w_2. \tag{1.2.19}$$

Similarly, multiply (1.2.17) by v_1 and (1.2.18) by u_1 and subtract to get

$$(\overbrace{v_2 u_1 - v_1 u_2}^{c}) w_2 = (\overbrace{u_3 v_1 - u_1 v_3}^{-a}) w_3. \tag{1.2.20}$$

Abbreviating, let $a = u_1 v_3 - u_3 v_1, b = u_3 v_2 - u_2 v_3$ and $c = u_1 v_2 - v_1 u_2$. Then (1.2.19) and (1.2.20) yield

$$w_1 = (b/a) w_2; \quad w_3 = (-c/a) w_2. \tag{1.2.21}$$

Hence, $\|\boldsymbol{w}\|^2 = (b/a)^2 w_2^2 + w_2^2 + (c/a)^2 w_2^2$ and

$$\|\boldsymbol{w}\|^2 = (1 + (b^2 + c^2)/a^2) w_2^2 = (a^2 + b^2 + c^2)(w_2^2/a^2). \tag{1.2.22}$$

Now, by simple algebra,

$$
\begin{aligned}
a^2 + b^2 + c^2 &= (u_1v_3 - u_3v_1)^2 + (u_3v_2 - u_2v_3)^2 + (u_1v_2 - v_1u_2)^2 \\
&= (u_1^2 + u_2^2 + u_3^2)(v_1^2 + v_2^2 + v_3^2) - (u_1v_1 + u_2v_2 + u_3v_3)^2 \\
&= \|\boldsymbol{u}\|^2\, \|\boldsymbol{v}\|^2 - (\boldsymbol{u} \cdot \boldsymbol{w})^2 = \|\boldsymbol{u}\|^2\, \|\boldsymbol{v}\|^2 (1 - \cos^2\theta) \\
&= \|\boldsymbol{u}\|^2\, \|\boldsymbol{v}\|^2 \sin^2\theta.
\end{aligned}
\tag{1.2.23}
$$

Applying Property (C1), we get $a^2+b^2+c^2 = \|\boldsymbol{w}\|^2$. Using (1.2.22), $w_2^2/a^2 = 1$. Hence, $w_2 = \pm\, a$ and by (1.2.21), $w_1 = \pm\, b$ and $w_3 = \mp\, c$. To determine the signs, consider the special case $\boldsymbol{u} = \hat{\boldsymbol{\imath}}$ and $\boldsymbol{v} = \hat{\boldsymbol{\jmath}}$. Then $u_1 = 1, u_2 = 0$ and $v_1 = 0, v_2 = 1$ and $c = 1 \cdot 1 - 0 \cdot 0 = 1$. On the other hand, the determinant in Property (C4) for this choice of $\boldsymbol{u}$ and $\boldsymbol{v}$ is

$$
\det \begin{vmatrix} 1 & 0 & 0 \\ 0 & 1 & 0 \\ \pm\, b & \pm\, a & \mp\, c \end{vmatrix} = \mp\, c,
$$

depending on whether $w_3 = -1$ or $w_3 = 1$. Since the determinant must be positive, we must take $w_2 = -a$, in order to make $w_3 = c = 1$ in this case. This implies $w_1 = -b$. Therefore, $\boldsymbol{w}$ is uniquely determined and has components $w_1 = -b, w_2 = -a, w_3 = c$. In other words, there exists a unique vector with the properties (C1), (C2), and (C4), and its components are given by (1.2.16).

Conversely, let $\boldsymbol{w}$ be a vector with components given by (1.2.16). Then direct computations show that $\boldsymbol{w}$ has the properties (C2) and (C4). After that, we repeat the calculations in (1.2.23) to establish Property (C1). The details of this argument are the subject of Problem 1.3 on page 410.

Theorem 1.2.2 is proved. □

Remark 1.3 *Formula (1.2.16) can be represented symbolically by*

$$
\boldsymbol{u} \times \boldsymbol{v} = \det \begin{vmatrix} \hat{\boldsymbol{\imath}} & \hat{\boldsymbol{\jmath}} & \hat{\boldsymbol{k}} \\ u_1 & u_2 & u_3 \\ v_1 & v_2 & v_3 \end{vmatrix}
\tag{1.2.24}
$$

and expanding the determinant by co-factors of the first row. Together with properties of the determinant, this representation implies Property (C3) of the cross product. Also, when combined with (C1), formula (1.2.24) can be used to compute the angle between two vectors with known components. Still, given the extra complexity of evaluating the determinant, the inner product formulas (1.2.3) and (1.2.6) are usually more convenient for angle computations.

Remark 1.4 *From (1.2.16) it follows that* $(\lambda\boldsymbol{u}) \times \boldsymbol{v} = \lambda(\boldsymbol{u} \times \boldsymbol{v}) = \boldsymbol{u} \times \lambda\boldsymbol{v}$ *for any scalar* λ. *Another consequence of (1.2.16) is the distributive property of the cross product:*

$$\boldsymbol{r} \times (\boldsymbol{u} + \boldsymbol{v}) = \boldsymbol{r} \times \boldsymbol{u} + \boldsymbol{r} \times \boldsymbol{v}. \tag{1.2.25}$$

Still, the cross product is not associative; *instead, the following identity holds:*

$$\boldsymbol{u} \times (\boldsymbol{v} \times \boldsymbol{w}) + \boldsymbol{v} \times (\boldsymbol{w} \times \boldsymbol{u}) + \boldsymbol{w} \times (\boldsymbol{u} \times \boldsymbol{v}) = \boldsymbol{0}. \tag{1.2.26}$$

EXERCISE 1.2.11.^{A} *Prove that*

$$\boldsymbol{u} \times (\boldsymbol{v} \times \boldsymbol{w}) = (\boldsymbol{u} \cdot \boldsymbol{w})\boldsymbol{v} - (\boldsymbol{u} \cdot \boldsymbol{v})\boldsymbol{w}. \tag{1.2.27}$$

Then use the result to verify (1.2.26). Hint: A possible proof of (1.2.27) is as follows (fill in the details). Choose an orthonormal basis $\hat{\boldsymbol{\imath}}$, $\hat{\boldsymbol{\jmath}}$, $\hat{\boldsymbol{k}}$ *so that* $\hat{\boldsymbol{\imath}}$ *is parallel to* $\boldsymbol{w}$ *and* $\hat{\boldsymbol{\jmath}}$ *is in the plane of* $\boldsymbol{w}$ *and* $\boldsymbol{v}$. *Then* $\boldsymbol{w} = w_1\hat{\boldsymbol{\imath}}$ *and* $\boldsymbol{v} = v_1\hat{\boldsymbol{\imath}} + v_2\hat{\boldsymbol{\jmath}}$ *and*

$$\boldsymbol{v} \times \boldsymbol{w} = \det\begin{vmatrix} \hat{\boldsymbol{\imath}} & \hat{\boldsymbol{\jmath}} & \hat{\boldsymbol{k}} \\ v_1 & v_2 & 0 \\ w_1 & 0 & 0 \end{vmatrix} = -v_2 w_1 \hat{\boldsymbol{k}};$$

$$\boldsymbol{u} \times (\boldsymbol{v} \times \boldsymbol{w}) = \det\begin{vmatrix} \hat{\boldsymbol{\imath}} & \hat{\boldsymbol{\jmath}} & \hat{\boldsymbol{k}} \\ u_1 & u_2 & u_3 \\ 0 & 0 & -v_2 w_1 \end{vmatrix} = -u_2 v_2 w_1 \hat{\boldsymbol{\imath}} + u_1 v_2 w_1 \hat{\boldsymbol{\jmath}};$$

$$(\boldsymbol{u} \cdot \boldsymbol{w})\boldsymbol{v} - (\boldsymbol{u} \cdot \boldsymbol{v})\boldsymbol{w} = u_1 w_1 (v_1\hat{\boldsymbol{\imath}} + v_2\hat{\boldsymbol{\jmath}}) - (u_1 v_1 + u_2 v_2) w_1 \hat{\boldsymbol{\imath}} = -u_2 v_2 w_1 \hat{\boldsymbol{\imath}} + u_1 v_2 w_1 \hat{\boldsymbol{\jmath}}.$$

While the properties (C1)–(C4) of the cross product are independent of the coordinate system, the definition does not generalize to $\mathbb{R}^n$ for $n \geq 4$ because in dimension $n \geq 4$ there are too many vectors orthogonal to two given vectors.

Property (C1) implies that $\|\boldsymbol{u} \times \boldsymbol{v}\|$ is the area of the parallelogram generated by the vectors $\boldsymbol{u}$ and $\boldsymbol{v}$. Accordingly, we have $\boldsymbol{u} \times \boldsymbol{v} = \boldsymbol{0}$ if and only if one of the vectors is a scalar multiple of the other. If P_1, P_2, P_3 are three points in $\mathbb{R}^3$, these points are `collinear` (lie on the same line) if and only if

$$\overrightarrow{P_1P_2} \times \overrightarrow{P_1P_3} = \boldsymbol{0}, \tag{1.2.28}$$

where $\overrightarrow{P_iP_j} = \overrightarrow{OP_j} - \overrightarrow{OP_i}$. If (x_i, y_i, z_i) are the cartesian coordinates of the point P_i, then the criterion for collinearity (1.2.28) becomes

$$\det \begin{vmatrix} \hat{\imath} & \hat{\jmath} & \hat{\kappa} \\ x_2 - x_1 & y_2 - y_1 & z_2 - z_1 \\ x_3 - x_1 & y_3 - y_1 & z_3 - z_1 \end{vmatrix} = \mathbf{0}. \qquad (1.2.29)$$

In the following three exercises, the reader will see how the mathematics of vector algebra can be used to solve problems in physics.

EXERCISE 1.2.12.[C] *Suppose two forces* $\boldsymbol{F}_1$, $\boldsymbol{F}_2$ *are applied at* P; $\boldsymbol{r} = \overrightarrow{OP}$. *Show that the total torque at* P *is* $\boldsymbol{T} = \boldsymbol{T}_1 + \boldsymbol{T}_2$, *where* $\boldsymbol{T}_1 = \boldsymbol{r} \times \boldsymbol{F}_1$ *and* $\boldsymbol{T}_2 = \boldsymbol{r} \times \boldsymbol{F}_2$.

EXERCISE 1.2.13.[A] *Consider a rigid rod with one end fixed at the origin* O *but free to rotate in any direction around* O *(say by means of a ball joint). Denote by* P *the other end of the rod;* $\boldsymbol{r} = \overrightarrow{OP}$. *Suppose a force* $\boldsymbol{F}$ *is applied at the point* P. *The rod will tend to rotate around* O.
(a) Let $\boldsymbol{r} = 2\,\hat{\imath} + 3\,\hat{\jmath} + \hat{\kappa}$ *and* $\boldsymbol{F} = \hat{\imath} + \hat{\jmath} + \hat{\kappa}$. *Compute the torque* $\boldsymbol{T}$. *(b) Let* $\boldsymbol{r} = 2\,\hat{\imath} + 4\,\hat{\jmath}$ *and* $\boldsymbol{F} = \hat{\imath} + \hat{\jmath}$, *so that the rotation is in the* $(\hat{\imath}, \hat{\jmath})$ *plane. Compute* $\boldsymbol{T}$. *In which direction will the rod start to rotate?*

EXERCISE 1.2.14.[A] *Suppose a rigid rod is placed in the* $(\hat{\imath}, \hat{\jmath})$ *plane so that the mid-point of the rod is at the origin* O, *and the two ends* P *and* P_1 *have position vectors* $\boldsymbol{r} = \hat{\imath} + 2\,\hat{\jmath}$ *and* $\boldsymbol{r}_1 = -\hat{\imath} - 2\,\hat{\jmath}$. *Suppose the rod is free to rotate around* O *in the* $(\hat{\imath}, \hat{\jmath})$ *plane. Let* $\boldsymbol{F} = \hat{\imath} + \hat{\jmath}$ *and* $\boldsymbol{F}_1 = -\hat{\imath} - \hat{\jmath}$ *be two forces applied at* P *and* P_1, *respectively. Compute the total torque around* O. *In which direction will the rod start to rotate?*

1.2.3 *Scalar Triple Product*

The `scalar triple product` $(\boldsymbol{u}, \boldsymbol{v}, \boldsymbol{w})$ of three vectors is defined by

$$(\boldsymbol{u}, \boldsymbol{v}, \boldsymbol{w}) = \boldsymbol{u} \cdot (\boldsymbol{v} \times \boldsymbol{w}).$$

Using (1.2.24) it is easy to see that, in cartesian coordinates,

$$(\boldsymbol{u}, \boldsymbol{v}, \boldsymbol{w}) = \det \begin{vmatrix} u_1 & u_2 & u_3 \\ v_1 & v_2 & v_3 \\ w_1 & w_2 & w_3 \end{vmatrix}.$$

From the properties of determinants it follows that

$$(\boldsymbol{u}, \boldsymbol{v}, \boldsymbol{w}) = -(\boldsymbol{v}, \boldsymbol{u}, \boldsymbol{w}) = (\boldsymbol{v}, \boldsymbol{w}, \boldsymbol{u}) = (\boldsymbol{w}, \boldsymbol{u}, \boldsymbol{v}).$$

Thus,

$$\boldsymbol{u}\cdot(\boldsymbol{v}\times\boldsymbol{w}) = \boldsymbol{w}\cdot(\boldsymbol{u}\times\boldsymbol{v}) = (\boldsymbol{u}\times\boldsymbol{v})\cdot\boldsymbol{w}. \tag{1.2.30}$$

In other words, the scalar triple product does not change under cyclic permutation of the vectors or when $\cdot$ and $\times$ symbols are switched.

EXERCISE 1.2.15.[C] *Verify that the ordered triplet of non-zero vectors $\boldsymbol{u}, \boldsymbol{v}, \boldsymbol{w}$ is a right-handed triad if and only if $(\boldsymbol{u}, \boldsymbol{v}, \boldsymbol{w}) > 0$.*

Recall that $\|\boldsymbol{v}\times\boldsymbol{w}\| = \|\boldsymbol{v}\|\cdot\|\boldsymbol{w}\|\sin\theta$ is the area of the parallelogram formed by $\boldsymbol{v}$ and $\boldsymbol{w}$. Therefore, $|\boldsymbol{u}\cdot(\boldsymbol{v}\times\boldsymbol{w})|$ is the volume of the parallelepiped formed by $\boldsymbol{u}, \boldsymbol{v}$, and $\boldsymbol{w}$. Accordingly, $(\boldsymbol{u}, \boldsymbol{v}, \boldsymbol{w}) = 0$ if and only if the three vectors are *linearly dependent*, that is, one of them can be expressed as a linear combination of the other two. Similarly, four points $P_i, i = 1, \ldots, 4$ are `co-planar` (lie in the same plane) if and only if

$$(\overrightarrow{P_1P_2}, \overrightarrow{P_1P_3}, \overrightarrow{P_1P_4}) = 0, \tag{1.2.31}$$

where $\overrightarrow{P_iP_j} = \overrightarrow{OP_j} - \overrightarrow{OP_i}$. If (x_i, y_i, z_i) are the cartesian coordinates of the point P_i, then (1.2.31) becomes

$$\det\begin{vmatrix} x_2-x_1 & y_2-y_1 & z_2-z_1 \\ x_3-x_1 & y_3-y_1 & z_3-z_1 \\ x_4-x_1 & y_4-y_1 & z_4-z_1 \end{vmatrix} = 0. \tag{1.2.32}$$

Notice a certain analogy with (1.2.28) and (1.2.29).

EXERCISE 1.2.16.[C] *Let $\boldsymbol{u} = (1, 2, 3)$, $\boldsymbol{v} = (-2, 1, 2)$, $\boldsymbol{w} = (-1, 2, 1)$. (a) Compute $\boldsymbol{u}\times\boldsymbol{v}$, $\boldsymbol{v}\times\boldsymbol{w}$, $(\boldsymbol{u}\times\boldsymbol{v})\times(\boldsymbol{v}\times\boldsymbol{w})$. (b) Compute the area of the parallelogram formed by $\boldsymbol{u}$ and $\boldsymbol{v}$. (c) Compute the volume of the parallelepiped formed by $\boldsymbol{u}$, $\boldsymbol{v}$, $\boldsymbol{w}$ using the triple product $(\boldsymbol{u}, \boldsymbol{v}, \boldsymbol{w})$.*

1.3 Curves in Space

1.3.1 *Vector-Valued Functions of a Scalar Variable*

To study the mathematical kinematics of moving bodies in $\mathbb{R}^3$, we need to define the velocity and acceleration vectors. The rigorous definition of these vectors relies on the concept of the derivative of a vector-valued function with respect to a scalar. We consider an idealized object, called a `point mass`, with all mass concentrated at a single point.

Choose an origin O and let $\boldsymbol{r}(t)$ be the position vector of the point mass at time t. The collection of points $P(t)$ so that $\overrightarrow{OP}(t) = \boldsymbol{r}(t)$ is the trajectory of the point mass. This trajectory is a **curve** in $\mathbb{R}^3$. More generally, a curve C is defined by specifying the position vector of a point P on C as a function of a scalar variable t.

Definition 1.5 A curve C in a frame O in $\mathbb{R}^3$ is the collection of points defined by a vector-valued function $\boldsymbol{r} = \boldsymbol{r}(t)$, for t in some interval I in $\mathbb{R}$, bounded or unbounded. A point P is on the curve C if an only if $\overrightarrow{OP} = \boldsymbol{r}(t_0)$ for some $t_0 \in I$. A curve is called **simple** if it does not intersect or touch itself. A curve is called **closed** if it is defined for t in a bounded closed interval $I = [a, b]$ and $\boldsymbol{r}(a) = \boldsymbol{r}(b)$. For a simple closed curve on $[a, b]$, we have $\boldsymbol{r}(t_1) = \boldsymbol{r}(t_2)$, $a \leq t_1 < t_2 \leq b$ if and only if $t_1 = a$ and $t_2 = b$.

By analogy with the elementary calculus, we say that the vector function $\boldsymbol{r}$ is **continuous** at t_0 if

$$\lim_{t \to t_0} \|\boldsymbol{r}(t) - \boldsymbol{r}(t_0)\| = 0. \tag{1.3.1}$$

Accordingly, we say that the curve C is **continuous** if the vector function that defines C is continuous.

Similarly, the **derivative** at t_0 of a vector-valued function $\boldsymbol{r}(t)$ is, by definition,

$$\frac{d\boldsymbol{r}}{dt}\Big|_{t=t_0} \equiv \boldsymbol{r}\,'(t_0) = \lim_{\Delta t \to 0} \frac{\boldsymbol{r}(t_0 + \Delta t) - \boldsymbol{r}(t_0)}{\Delta t}. \tag{1.3.2}$$

We say that $\boldsymbol{r}$ is **differentiable** at t_0 if the derivative $\boldsymbol{r}\,'(t)$ exists at t_0; we say that $\boldsymbol{r}$ is *differentiable on* (a, b) if $\boldsymbol{r}\,'(t)$ exists for all $t \in (a, b)$. We say that the curve is smooth if the corresponding vector function is differentiable and the derivative is not a zero vector.

Yet another notation for the derivative $\boldsymbol{r}\,'(t)$ is $\dot{\boldsymbol{r}}(t)$, especially when the parameter t is interpreted as time. For a scalar function of time $x = x(t)$, the same notations for the derivative are used:

$$\frac{dx}{dt} \equiv x\,'(t) \equiv \dot{x}(t).$$

Note that $\boldsymbol{r}(t+\Delta t) - \boldsymbol{r}(t) = \Delta\boldsymbol{r}(t)$ is a vector in the same frame O. The limits in (1.3.1) and (1.3.2) are defined by using the distance, or metric, for vectors. Thus, $\lim_{t \to t_0} \boldsymbol{r}(t) = \boldsymbol{r}(t_0)$ means that $\|\boldsymbol{r}(t) - \boldsymbol{r}(t_0)\| \to 0$ as $t \to t_0$. The derivative $\boldsymbol{r}\,'(t)$, being the limit of the difference quotient $\Delta\boldsymbol{r}(t)/\Delta t$ as $\Delta t \to 0$, is also a vector.

Given a fixed frame O, the formulas of differential calculus for vector functions in this frame are easily obtained by following the corresponding derivations for scalar functions in ordinary calculus. As in ordinary calculus, there are several rules for computing derivatives of vector-valued functions. All these rules follow directly from the definition (1.3.2).

The derivative of a sum:

$$\frac{d}{dt}(\boldsymbol{u}(t)+\boldsymbol{v}(t)) = \boldsymbol{u}'(t)+\boldsymbol{v}'(t). \tag{1.3.3}$$

Product rule for multiplication by a scalar: if $\lambda(t)$ is a scalar function, then

$$\frac{d}{dt}(\lambda(t)\boldsymbol{r}(t)) = \lambda'(t)\boldsymbol{r}(t)+\lambda(t)\boldsymbol{r}'(t). \tag{1.3.4}$$

Product rules for scalar and cross products:

$$\frac{d}{dt}(\boldsymbol{u}(t)\cdot\boldsymbol{v}(t)) = \frac{d\boldsymbol{u}}{dt}\cdot\boldsymbol{v}+\boldsymbol{u}\cdot\frac{d\boldsymbol{v}}{dt}, \tag{1.3.5}$$

and

$$\frac{d}{dt}(\boldsymbol{u}(t)\times\boldsymbol{v}(t)) = \frac{d\boldsymbol{u}}{dt}\times\boldsymbol{v}+\boldsymbol{u}\times\frac{d\boldsymbol{v}}{dt}. \tag{1.3.6}$$

The chain rule: If $t=\phi(s)$ and $\boldsymbol{r}_1(s)=\boldsymbol{r}(\phi(s))$, then

$$\frac{d\boldsymbol{r}_1}{ds} = \frac{d\boldsymbol{r}}{dt}\cdot\frac{d\phi}{ds}. \tag{1.3.7}$$

From the two rules (1.3.3) and (1.3.4), it follows that if $(\hat{\boldsymbol{\imath}},\ \hat{\boldsymbol{\jmath}},\ \hat{\boldsymbol{\kappa}})$ are constant vectors in the frame O so that $\boldsymbol{r}(t)=x(t)\,\hat{\boldsymbol{\imath}}+y(t)\,\hat{\boldsymbol{\jmath}}+z(t)\hat{\boldsymbol{\kappa}}$, then $\boldsymbol{r}'(t)=x'(t)\,\hat{\boldsymbol{\imath}}+y'(t)\,\hat{\boldsymbol{\jmath}}+z'(t)\,\hat{\boldsymbol{\kappa}}$.

Remark 1.5 *The underlying assumption in the above rules for differentiation of vector functions is that all the functions are defined in the* same *frame. We will see later that these rules for computing derivatives can fail if the vectors are defined in different frames and the frames are moving relative to each other.*

Lemma 1.1 *If $\boldsymbol{r}$ is differentiable on (a,b) and $\|\boldsymbol{r}(t)\|$ does not depend on t for $t\in(a,b)$, then $\boldsymbol{r}(t)\perp\boldsymbol{r}'(t)$ for all $t\in(a,b)$. In other words, the derivative of a constant-length vector is perpendicular to the vector itself.*

Proof. By assumption, $\boldsymbol{r}(t)\cdot\boldsymbol{r}(t)$ is constant for all t. By the product rule (1.3.5), $2\boldsymbol{r}'(t)\cdot\boldsymbol{r}(t)=0$ and the result follows. □

EXERCISE 1.3.1.[A] *(a) Show that if $\boldsymbol{r}$ is differentiable at t_0, then $\boldsymbol{r}$ is continuous at t_0, but the converse is not true. (b) Does continuity of $\boldsymbol{r}$ imply continuity of $\|\boldsymbol{r}\|$? Does continuity of $\|\boldsymbol{r}\|$ imply continuity of $\boldsymbol{r}$? (c) Does differentiability of $\boldsymbol{r}$ imply differentiability of $\|\boldsymbol{r}\|$? Does differentiability of $\|\boldsymbol{r}\|$ imply differentiability of $\boldsymbol{r}$?*

The complete description of every curve consists of two parts: (a) the set of its points in $\mathbb{R}^3$, (b) the ordering of those points relative to the ordering of the parameter set. For some curves, this complete description is possible in purely vector terms, that is, without choosing a particular coordinate system in the frame O. For other curves, a purely vector description provides only the set of points, while the ordering of that set is impossible without the selection of the particular coordinate system. We illustrate this observation on two simple curves: a straight line and a circle.

A straight line is described by $\boldsymbol{r}(t) = \boldsymbol{r}_2 - \phi(t)(\boldsymbol{r}_1 - \boldsymbol{r}_2)$, $\quad -\infty < t < \infty$, where $\boldsymbol{r}_1$ and $\boldsymbol{r}_2$ are the position vectors of two distinct points on the line and $\phi(t)$ is a scalar function whose range is all of $\mathbb{R}$. The function ϕ determines the ordering of the points on the line. For example, if $\phi(t) = t$, then the point $\boldsymbol{r}(t_2)$ follows $\boldsymbol{r}(t_1)$ in time if $t_2 > t_1$.

The circle as a set of points in $\mathbb{R}^3$ is defined by the two conditions, $\|\boldsymbol{r}(t)\| = R$ and $\boldsymbol{r}(t) \cdot \widehat{\boldsymbol{n}} = 0$, where $\widehat{\boldsymbol{n}}$ is the unit normal to the plane of the circle. Direct computations show that these conditions do not determine the function $\boldsymbol{r}(t)$ uniquely, and so do not give an ordering of points on the circle. To specify the ordering, we can, for example, fix one point $\boldsymbol{r}(t_0)$ on the circle at a reference time t_0 and define the angle between $\boldsymbol{r}(t)$ and $\boldsymbol{r}(t_0)$ as a function of t. But this is equivalent to choosing a polar coordinate system in the plane of the circle.

1.3.2 *The Tangent Vector and Arc Length*

Let $\boldsymbol{r} = \boldsymbol{r}(t)$ define a curve in $\mathbb{R}^3$. If $\overrightarrow{OP} = \boldsymbol{r}(t_0)$ and $\boldsymbol{r}'(t_0) \neq \boldsymbol{0}$, then, *by definition*, the `unit tangent vector` $\widehat{\boldsymbol{u}}$ at P is:

$$\widehat{\boldsymbol{u}}(t_0) = \frac{\boldsymbol{r}'(t_0)}{\|\boldsymbol{r}'(t_0)\|} \tag{1.3.8}$$

Note that the vector $\Delta \boldsymbol{r} = \boldsymbol{r}(t_0 + \Delta t) - \boldsymbol{r}(t_0)$ defines a line through two points on the curve; similar to ordinary calculus, definition (1.3.2) suggests that the vector $\boldsymbol{r}'(t_0)$ should be parallel to the tangent line at P.

The `equation of the tangent line` at point P is

$$\boldsymbol{R}(s) = \boldsymbol{r}(t_0) + s\widehat{\boldsymbol{u}}(t_0). \tag{1.3.9}$$

EXERCISE 1.3.2.[C] *Let C be a planar curve defined by the vector function $\boldsymbol{r}(t) = \cos t\,\hat{\imath} + \sin t\,\hat{\jmath}$, $-\pi < t < \pi$. Compute the tangent vector $\boldsymbol{r}\,'(t)$ and the unit tangent vector $\widehat{\boldsymbol{u}}(t)$ as functions of t. Compute $\boldsymbol{r}\,'(0)$ and $\widehat{\boldsymbol{u}}(0)$. Draw the curve C and the vectors $\boldsymbol{r}\,'(0)$, $\widehat{\boldsymbol{u}}\,'(0)$. Verify your results using a* `computer algebra system`, *such as MAPLE, MATLAB, or MATHEMATICA.*

EXERCISE 1.3.3.[C] *Let C be a spatial curve defined by the vector function $\boldsymbol{r}(t) = \cos t\,\hat{\imath} + \sin t\,\hat{\jmath} + t\,\hat{\kappa}$. Compute the tangent vector $\boldsymbol{r}\,'(t)$, the unit tangent vector $\widehat{\boldsymbol{u}}(t)$ and the vector $\widehat{\boldsymbol{u}}\,'(t)$. Compute $\boldsymbol{r}\,'(\pi/2)$. Draw the curve C for $0 \le t \le \pi/2$ and draw $\widehat{\boldsymbol{u}}\,'(\pi/2)$ at the point $\boldsymbol{r}(\pi/2)$. Verify your results using your favorite computer algebra system.*

Definition 1.6 A curve C, defined by a vector function $\boldsymbol{r}(t)$, $a < t < b$, is called `smooth` if the unit tangent vector $\widehat{\boldsymbol{u}} = \widehat{\boldsymbol{u}}(t)$ exists and is a continuous function for all $t \in (a, b)$. If the curve is closed, then, additionally, we must have $\boldsymbol{r}\,'(a) = \boldsymbol{r}\,'(b)$. The curve is called `piece-wise smooth` if it is continuous and consists of finitely many smooth pieces.

EXERCISE 1.3.4.[A] *Give an example of a non-smooth curve C defined by a vector function $\boldsymbol{r}(t)$, $-1 < t < 1$, so that the derivative vector $\boldsymbol{r}\,'(t)$ exists and is continuous for all $t \in (-1, 1)$.*

EXERCISE 1.3.5.[C] *Explain how the graph of a function $y = f(x)$ can be interpreted as a curve in $\mathbb{R}^3$. Show that this curve is smooth if and only if the function $f = f(x)$ has a continuous derivative, and show that, at the point $(x_0, f(x_0), 0)$, formula (1.3.9) defines the same line as $y = f(x_0) + f\,'(x_0)(x - x_0), z = 0$.*

Given a curve C and two points with position vectors $\boldsymbol{r}(c), \boldsymbol{r}(d)$, $a < c \le d < b$, on the curve, we define the distance between the two points *along the curve* using a limiting process. The construction is similar to the definition of the Riemann integral in ordinary calculus.

For each $n \ge 2$, choose points $c = t_0 < t_1 < \cdots < t_n = d$ and form the sums $L_n = \sum_{i=0}^{n-1} \|\triangle \boldsymbol{r}_i\|$, where $\triangle \boldsymbol{r}_i = \boldsymbol{r}(t_{i+1}) - \boldsymbol{r}(t_i)$. Assume that $\max_{0 \le i \le n-1}(t_{i+1} - t_i) \to 0$ as $n \to \infty$. If the limit $\lim_{n\to\infty} L_n$ exists for all $a < c < d < b$, and does not depend on the particular choice of the points t_k, then the curve C is called `rectifiable`. By definition, the distance

$L_{\mathcal{C}}(c,d)$ between the points $\boldsymbol{r}(c)$ and $\boldsymbol{r}(d)$ along a rectifiable curve $\mathcal{C}$ is

$$L_{\mathcal{C}}(c,d) = \lim_{n\to\infty} L_n,$$

Theorem 1.3.1 *Assume that $\boldsymbol{r}'(t)$ exists for all $t \in (a,b)$ and the vector function $\boldsymbol{r}'(t)$ is continuous. Then the curve $\mathcal{C}$ is rectifiable and*

$$L_{\mathcal{C}}(c,d) = \int_c^d \|\boldsymbol{r}'(t)\| dt. \tag{1.3.10}$$

Proof. It follows from the assumptions of the theorem and from relation (1.3.2) that $\Delta \boldsymbol{r}_i = \boldsymbol{r}'(t_i)\,\Delta t_i + \boldsymbol{v}_i$, where $\Delta t_i = t_{i+1} - t_i$ and the vectors $\boldsymbol{v}_i$ satisfy $\max_{0\le i\le n-1} \|\boldsymbol{v}_i\|/\Delta t_i \to 0$ as $\max_{0\le i\le n-1} \Delta t_i \to 0$. Therefore,

$$\|\Delta \boldsymbol{r}_i\| = \|\boldsymbol{r}'(t_i)\|\,\Delta t_i + \varepsilon_i \Delta t_i, \tag{1.3.11}$$

$$\sum_{i=0}^{n-1} \|\Delta \boldsymbol{r}_i\| = \sum_{i=0}^{n-1} \|\boldsymbol{r}'(t_i)\|\,\Delta t_i + \sum_{i=0}^{n-1} \varepsilon_i \Delta t_i,$$

where the numbers ε_i satisfy $\max_{0\le i\le n-1} \varepsilon_i \to 0$, $n \to \infty$. Then (1.3.10) follows after passing to the limit. □

EXERCISE 1.3.6.[B] *(a) Verify (1.3.11). Hint: use the triangle inequality to estimate $\|\boldsymbol{r}'(t_i)\,\Delta t_i + \boldsymbol{v}_i\| - \|\boldsymbol{r}'(t_i)\|\,\Delta t_i$. (b) Show that a piece-wise smooth curve is rectifiable. Hint: apply the above theorem to each smooth piece separately, and then add the results.*

EXERCISE 1.3.7. [C] *Interpreting the graph of the function $y = f(x)$ as a curve in $\mathbb{R}^3$, and assuming that $f'(x)$ exists and is continuous, show that the length of this curve from $(c, f(c), 0)$ to $(d, f(d), 0)$, as given by (1.3.10), is $\int_c^d \sqrt{1+|f'(x)|^2}dx$; the derivation of this result in ordinary calculus is similar to the derivation of (1.3.10).*

Given a point $\boldsymbol{r}(c)$ on a rectifiable curve $\mathcal{C}$, we define the `arc length` function $\mathfrak{s} = \mathfrak{s}(t)$, $t \ge c$, as

$$\mathfrak{s}(t) = L_{\mathcal{C}}(c,t)$$

It follows that $d\mathfrak{s}/dt = \|\boldsymbol{r}'(t)\| \ge 0$. We call $d\mathfrak{s} = \|\boldsymbol{r}'(t)\|dt$ the `line element` of the curve $\mathcal{C}$. If $\boldsymbol{r}(t) = x(t)\,\hat{\boldsymbol{\imath}} + y(t)\,\hat{\boldsymbol{\jmath}} + z(t)\,\hat{\boldsymbol{\kappa}}$, where $(\hat{\boldsymbol{\imath}}, \hat{\boldsymbol{\jmath}}, \hat{\boldsymbol{\kappa}})$ is a cartesian coordinate system at O, then

$$\left(\frac{d\mathfrak{s}}{dt}\right)^2 = \left\|\frac{dr}{dt}\right\|^2 = \left(\frac{dx}{dt}\right)^2 + \left(\frac{dy}{dt}\right)^2 + \left(\frac{dz}{dt}\right)^2. \tag{1.3.12}$$

If the curve is smooth, then $d\mathfrak{s}/dt > 0$ and $\mathfrak{s}$ is a monotone function of t so that t is a well-defined function of $\mathfrak{s}$. Hence, $\boldsymbol{r}(t(\mathfrak{s}))$ is a function of $\mathfrak{s}$, and is called the `canonical parametrization` of the smooth curve by the arc length. By the rules of differentiation,

$$\frac{d\boldsymbol{r}}{d\mathfrak{s}} = \frac{d\boldsymbol{r}}{dt}\frac{dt}{d\mathfrak{s}} = \frac{d\boldsymbol{r}}{dt}\frac{1}{d\mathfrak{s}/dt} = \frac{\boldsymbol{r}'(t)}{\|\boldsymbol{r}'(t)\|} = \widehat{\boldsymbol{u}}(t).$$

EXERCISE 1.3.8.[C] *Consider the* `right-handed circular helix`

$$\boldsymbol{r}(t) = a\cos t\,\hat{\boldsymbol{\imath}} + a\sin t\,\hat{\boldsymbol{\jmath}} + t\,\hat{\boldsymbol{k}},\ a > 0. \tag{1.3.13}$$

Re-write the equation of this curve using the arc length $\mathfrak{s}$ as the parameter.

1.3.3 *Frenet's Formulas*

In certain frames, called inertial, the `Second Law of Newton` *postulates* the following relation between the force $\boldsymbol{F} = \boldsymbol{F}(t)$ acting on the point mass m and the point's trajectory $\mathcal{C}$, defined by a curve $\boldsymbol{r} = \boldsymbol{r}(t)$:

$$m\,\frac{d^2\boldsymbol{r}(t)}{dt^2} = \boldsymbol{F}(t). \tag{1.3.14}$$

A detailed discussion of inertial frames and Newton's Laws is below on page 43. When $\boldsymbol{F}(t)$ is given, the solution of the differential equation (1.3.14) is the trajectory $\boldsymbol{r}(t)$. However, to get a unique solution of (2.1.1), we must start at some time t_0 and provide two initial conditions $\boldsymbol{r}'(t_0)$ and $\boldsymbol{r}(t_0)$ to determine a specific path. In other words, $\boldsymbol{r}(t_0)$ and $\boldsymbol{r}'(t_0)$ are reference vectors for the motion. At every time $t > t_0$, the vectors $\boldsymbol{r}(t)$ and $\boldsymbol{r}'(t)$ have a well-defined geometric orientation relative to the initial vectors $\boldsymbol{r}(t_0)$, $\boldsymbol{r}'(t_0)$. The three *Frenet formulas* provide a complete description of this orientation. In what follows, we assume that the curve $\mathcal{C}$ is smooth, that is, the unit tangent vector $\widehat{\boldsymbol{u}}$ exists at every point of the curve.

To write the formulas, we need several new notions: *curvature, principal unit normal vector, unit binormal vector*, and *torsion*. We will use the canonical parametrization of the curve by the arc length $\mathfrak{s}$ measured from some reference point P_0 on the curve.

Let $\widehat{\boldsymbol{u}} = \widehat{\boldsymbol{u}}(\mathfrak{s})$ be the unit tangent vector at P, where the parameter $\mathfrak{s}$ is the arc length from P_0 to P. By Lemma 1.1 on page 26, the derivative $\widehat{\boldsymbol{u}}'(\mathfrak{s})$ of $\widehat{\boldsymbol{u}}(\mathfrak{s})$ with respect to $\mathfrak{s}$ is orthogonal to $\widehat{\boldsymbol{u}}$. By definition, the `curvature` $\kappa(\mathfrak{s})$ at P is

$$\kappa(\mathfrak{s}) = \|\widehat{\boldsymbol{u}}'(\mathfrak{s})\|;$$

the **principal unit normal vector** at P is

$$\widehat{\boldsymbol{p}} = \frac{1}{\kappa}\,\widehat{\boldsymbol{u}}\,'(\mathfrak{s}); \tag{1.3.15}$$

the **unit binormal vector** at P is

$$\widehat{\boldsymbol{b}}(\mathfrak{s}) = \widehat{\boldsymbol{u}}(\mathfrak{s}) \times \widehat{\boldsymbol{p}}(\mathfrak{s}).$$

EXERCISE 1.3.9.^{C} *Parameterizing the circle by the arc length, verify that the curvature of the circle of radius R is $1/R$.*

To define the torsion, we derive the relation between $\widehat{\boldsymbol{b}}\,'(\mathfrak{s})$ and the vectors $\widehat{\boldsymbol{u}}, \widehat{\boldsymbol{p}}, \widehat{\boldsymbol{b}}$. Using Lemma 1.1 once again, we conclude that $\widehat{\boldsymbol{b}}\,'(\mathfrak{s})$ is orthogonal to $\widehat{\boldsymbol{b}}(\mathfrak{s})$. Next, we differentiate the relation $\widehat{\boldsymbol{b}}(\mathfrak{s}) \cdot \widehat{\boldsymbol{u}}(\mathfrak{s}) = 0$ with respect to $\mathfrak{s}$ and use the product rule (1.3.5) to find $\widehat{\boldsymbol{b}}\,'(\mathfrak{s})\cdot\widehat{\boldsymbol{u}}(\mathfrak{s})+\widehat{\boldsymbol{b}}(\mathfrak{s})\cdot\widehat{\boldsymbol{u}}\,'(\mathfrak{s}) = 0$. By construction, the unit vectors $\widehat{\boldsymbol{u}}, \widehat{\boldsymbol{p}}, \widehat{\boldsymbol{b}}$ are mutually orthogonal, and then the definition (1.3.15) of the vector $\widehat{\boldsymbol{p}}$ implies that $\widehat{\boldsymbol{b}}(\mathfrak{s}) \cdot \widehat{\boldsymbol{u}}\,'(\mathfrak{s}) = 0$. As a result, $\widehat{\boldsymbol{b}}\,'(\mathfrak{s}) \cdot \widehat{\boldsymbol{u}}(\mathfrak{s}) = 0$. Being orthogonal to both $\widehat{\boldsymbol{u}}(\mathfrak{s})$ and $\widehat{\boldsymbol{b}}(\mathfrak{s})$, the vector $\widehat{\boldsymbol{b}}\,'(\mathfrak{s})$ must then be parallel to $\widehat{\boldsymbol{p}}(\mathfrak{s})$. We therefore define the **torsion** of the curve C at point P as the number $\tau = \tau(\mathfrak{s})$ so that

$$\widehat{\boldsymbol{b}}\,'(\mathfrak{s}) = -\tau(\mathfrak{s})\,\widehat{\boldsymbol{p}}(\mathfrak{s}); \tag{1.3.16}$$

the choice of the negative sign ensures that the torsion is positive for the right-handed circular helix (1.3.13).

Note that the above definitions use the canonical parametrization of the curve by the arc length $\mathfrak{s}$; the corresponding formulas can be written for an arbitrary parametrization as well; see Problem 1.11 on page 412.

Relations (1.3.15) and (1.3.16) are two of the Frenet formulas. To derive the third formula, note that $\widehat{\boldsymbol{p}}(\mathfrak{s}) = \widehat{\boldsymbol{b}}(\mathfrak{s}) \times \widehat{\boldsymbol{u}}(\mathfrak{s})$. Differentiation with respect to $\mathfrak{s}$ yields $\widehat{\boldsymbol{p}}\,' = \widehat{\boldsymbol{b}} \times \widehat{\boldsymbol{u}}\,' + \widehat{\boldsymbol{b}}\,' \times \widehat{\boldsymbol{u}} = \widehat{\boldsymbol{b}} \times \kappa\,\widehat{\boldsymbol{p}} - \tau\,\widehat{\boldsymbol{p}} \times \widehat{\boldsymbol{u}}$, and

$$\widehat{\boldsymbol{p}}\,'(\mathfrak{s}) = -\kappa\,\widehat{\boldsymbol{u}}(\mathfrak{s}) + \tau\,\widehat{\boldsymbol{b}}(\mathfrak{s}). \tag{1.3.17}$$

Different sources refer to relations (1.3.15) – (1.3.17) as either the **Frenet** or the **Frenet-Serret** formulas. In 1847, the French mathematician JEAN FRÉDÉRIC FRENET (1816–1900) derived two of these formulas in his doctoral dissertation. Another French mathematician, JOSEPH ALFRED SERRET (1819–1885), gave an independent derivation of all three formulas, but we could not find the exact time of his work. Of course, neither Frenet nor Serret used the modern vector notations in their derivations.

At every point P of the curve, the vector triple $(\widehat{\boldsymbol{u}}, \widehat{\boldsymbol{p}}, \widehat{\boldsymbol{b}})$ is a right-handed coordinate system with origin at P. We will call this coordinate system **Frenet's trihedron** at P. The choice of initial conditions $\boldsymbol{r}(t_0), \boldsymbol{r}'(t_0)$ means setting up a coordinate system in the frame with origin at P_0, where $\overrightarrow{OP_0} = \boldsymbol{r}(t_0)$. The coordinate planes spanned by the vectors $(\widehat{\boldsymbol{u}}, \widehat{\boldsymbol{p}})$, $(\widehat{\boldsymbol{p}}, \widehat{\boldsymbol{b}})$, and $(\widehat{\boldsymbol{b}}, \widehat{\boldsymbol{u}})$ are called, respectively, the **osculating**, **normal**, and **rectifying** (**binormal**) planes. The word *osculating* comes from Latin *osculum*, literally, *a little mouth*, which was the colloquial way of saying "a kiss". Not surprisingly, of all the planes that pass through the point P, the osculating plane comes the closest to containing the curve $\mathcal{C}$.

EXERCISE 1.3.10.^{B} *A curve is called* **planar** *if all its points are in the same plane. Show that a planar curve other than a line has the same osculating plane at every point and lies entirely in this plane (for a line, the osculating plane is not well-defined).*

The curvature and torsion uniquely determine the curve, up to its position in space. More precisely, if $\kappa(\mathfrak{s})$ and $\tau(\mathfrak{s})$ are given continuous functions of $\mathfrak{s}$, we can solve the corresponding equations (1.3.15)–(1.3.17) and obtain the vectors $\widehat{\boldsymbol{u}}(\mathfrak{s}), \widehat{\boldsymbol{p}}(\mathfrak{s}), \widehat{\boldsymbol{b}}(\mathfrak{s})$ which determine the shape of a family of curves. To obtain a particular curve $\mathcal{C}$ in this family, we must specify initial values $(\widehat{\boldsymbol{u}}(\mathfrak{s}_0), \widehat{\boldsymbol{p}}(\mathfrak{s}_0), \widehat{\boldsymbol{b}}(\mathfrak{s}_0))$ of the trihedron vectors and an initial value $\boldsymbol{r}(\mathfrak{s}_0)$ of a position vector at a point P_0 on the curve. These four vectors are all in some frame with origin O. To obtain $\boldsymbol{r}(\mathfrak{s})$ at any point of $\mathcal{C}$ we solve the differential equation $d\boldsymbol{r}/d\mathfrak{s} = \widehat{\boldsymbol{u}}(\mathfrak{s})$, with initial condition $\boldsymbol{r}(\mathfrak{s}_0)$, together with (1.3.15)–(1.3.17). Note that the curvature is always non-negative, and the torsion can be either positive or negative.

EXERCISE 1.3.11.^{A} *For the right circular helix (1.3.13) compute the curvature, torsion, and the Frenet trihedron at every point. Show that the right circular helix is the only curve with constant curvature and constant positive torsion.*

As the point P moves along the curve, the trihedron executes three rotations. These rotations about the unit tangent, principal unit normal, and unit binormal vectors are called **rolling**, **yawing**, and **pitching**, respectively. Rolling and yawing change direction of the unit binormal vector $\widehat{\boldsymbol{b}}$, rolling and pitching change the direction of the principal unit normal vector $\widehat{\boldsymbol{p}}$, yawing and pitching change the direction of the unit tangent vector $\widehat{\boldsymbol{u}}$. To visualize these rotations, consider the motion of an airplane. Intuitively, it is clear that the tangent vector $\widehat{\boldsymbol{u}}$ points along the fuselage from the tail to the nose, and the normal vector $\widehat{\boldsymbol{p}}$ points up perpendicular to the wings (draw a picture!) In this construction, the vector $\widehat{\boldsymbol{b}}$ points along the wings to make $\widehat{\boldsymbol{u}}, \widehat{\boldsymbol{p}}, \widehat{\boldsymbol{b}}$ a right-handed triple. The center of mass of the plane is the natural common origin of the three vectors. The rolling of the plane, the rotation around $\widehat{\boldsymbol{u}}$, lifts one side of the plane relative to the other and is controlled by the *ailerons* on the back edges of the wings. Yawing, the rotation around $\widehat{\boldsymbol{p}}$, moves the nose left and right and is controlled by the *rudder* on the vertical part of the tail. Pitching of the plane, the rotation around $\widehat{\boldsymbol{b}}$, moves the nose up and down and is controlled by the *elevators* on the horizontal part of the tail.

Note that rolling and pitching are the main causes of motion sickness.

1.3.4 *Velocity and Acceleration*

Let the curve $\mathcal{C}$, defined by the vector function $\boldsymbol{r} = \boldsymbol{r}(t)$, be the trajectory of a point mass in some frame O. Between times t and $t + \Delta t$ the point moves through the arc length $\Delta\mathfrak{s} = \mathfrak{s}(t + \Delta t) - \mathfrak{s}(t)$, and therefore $d\mathfrak{s}(t)/dt$ is the `speed` of the point along $\mathcal{C}$. As we derived on page 30,

$$\frac{d\boldsymbol{r}}{dt} = \frac{d\mathfrak{s}}{dt}\frac{d\boldsymbol{r}}{d\mathfrak{s}} = \frac{d\mathfrak{s}}{dt}\,\widehat{\boldsymbol{u}}(t). \tag{1.3.18}$$

Therefore, we define the `velocity` $\boldsymbol{v}(t)$ as

$$\boldsymbol{v}(t) = d\boldsymbol{r}/dt.$$

In particular, $\|\boldsymbol{v}\| = |d\mathfrak{s}/dt| = d\mathfrak{s}/dt$, that is, *the speed is the magnitude of the velocity*; recall that the arc length $\mathfrak{s} = \mathfrak{s}(t)$ is a non-decreasing function of t. This mathematical definition of velocity agrees with our physical intuition of speed in the direction of the tangent line, while making the physical concept of velocity precise, as required in a quantitative science. The definition also works well in practical problems of motion. Indeed, precise physics is mathematical physics.

Similarly, the `acceleration` $\boldsymbol{a}(t)$ of the point mass is, by definition,

$$\boldsymbol{a}(t) = \boldsymbol{v}'(t) = \boldsymbol{r}''(t).$$

Since $d\boldsymbol{v}/dt = d\big((d\mathfrak{s}/dt)\,\widehat{\boldsymbol{u}}(t)\big)/dt$, the product rule (1.3.4) implies

$$\frac{d\boldsymbol{v}}{dt} = \frac{d^2\mathfrak{s}}{dt^2}\,\widehat{\boldsymbol{u}}(t) + \frac{d\mathfrak{s}}{dt}\frac{d\widehat{\boldsymbol{u}}}{d\mathfrak{s}}\frac{d\mathfrak{s}}{dt}$$

or

$$\boldsymbol{a}(t) = \frac{d^2\mathfrak{s}}{dt^2}\,\widehat{\boldsymbol{u}}(t) + \left(\frac{d\mathfrak{s}}{dt}\right)^2 \frac{d\widehat{\boldsymbol{u}}(\mathfrak{s})}{d\mathfrak{s}}. \tag{1.3.19}$$

Equation (1.3.19) shows that the acceleration $\boldsymbol{a}(t)$ has two components: the `tangential acceleration` $(d^2\mathfrak{s}/dt^2)\,\widehat{\boldsymbol{u}}(t)$ and the `normal acceleration` $(d\mathfrak{s}/dt)^2\,(d\widehat{\boldsymbol{u}}(\mathfrak{s})/d\mathfrak{s})$. By Lemma 1.1, page 26, the derivative of a unit vector is always orthogonal to the vector itself, and so the tangential and normal accelerations are mutually orthogonal. The derivation also shows that the decomposition (1.3.19) of the acceleration into the tangential and normal components does not depend on the coordinate system.

EXERCISE 1.3.12.[C] *In (1.3.20) below, $\boldsymbol{r} = \boldsymbol{r}(t)$ represents the position of point mass m at time t in the Cartesian coordinate system:*

$$\begin{aligned} \boldsymbol{r}(t) &= t^2\,\hat{\boldsymbol{\imath}} + 2t^2\,\hat{\boldsymbol{\jmath}} + t^2\,\hat{\boldsymbol{\kappa}}; & \boldsymbol{r}(t) &= 2\cos\pi t\,\hat{\boldsymbol{\imath}} + 2\sin\pi t\,\hat{\boldsymbol{\jmath}};\\ \boldsymbol{r}(t) &= 2\cos t^2\,\hat{\boldsymbol{\imath}} + 2\sin t^2\,\hat{\boldsymbol{\jmath}}; & \boldsymbol{r}(t) &= \cos t^2\,\hat{\boldsymbol{\imath}} + 2\sin t^2\,\hat{\boldsymbol{\jmath}}. \end{aligned} \tag{1.3.20}$$

For each function $\boldsymbol{r} = \boldsymbol{r}(t)$,

- *Sketch the corresponding trajectory;*
- *Compute the velocity and acceleration vectors as functions of t;*
- *Draw the trajectory for $0 \le t \le 1$ and draw the vectors $\boldsymbol{r}'(1)$, $\boldsymbol{r}''(1)$;*
- *Compute the normal and tangential components of the acceleration and draw the corresponding vectors when $t = 1$;*
- *Verify your results using a computer algebra system.*

We will now write the decomposition (1.3.19) for the CIRCULAR MOTION IN A PLANE. Let $\mathcal{C}$ be a circle with radius R and center at the point O. Assume a point mass moves along $\mathcal{C}$. Choose the cartesian coordinates $\hat{\boldsymbol{\imath}}, \hat{\boldsymbol{\jmath}}, \hat{\boldsymbol{\kappa}}$ with origin at O and $\hat{\boldsymbol{\imath}}, \hat{\boldsymbol{\jmath}}$ in the plane of the circle. Denote by $\theta(t)$ the angle between $\hat{\boldsymbol{\imath}}$ and the position vector $\boldsymbol{r}(t)$ of the point mass. Suppose that the function $\theta = \theta(t)$ has two continuous derivatives in t, $|\theta'(t)| > 0$, $t > 0$, and $\theta(0) = 0$. Then

$$\begin{aligned} \boldsymbol{r}(t) &= R\cos\theta(t)\,\hat{\boldsymbol{\imath}} + R\sin\theta(t)\,\hat{\boldsymbol{\jmath}},\\ \boldsymbol{v} = \boldsymbol{r}'(t) &= -\theta'(t)R\sin\theta(t)\,\hat{\boldsymbol{\imath}} + \theta'(t)R\cos\theta(t)\,\hat{\boldsymbol{\jmath}},\\ \|\boldsymbol{v}\| &= (\boldsymbol{r}'\cdot\boldsymbol{r}')^{\frac{1}{2}} = R|\theta'(t)|, \end{aligned}$$

and $\boldsymbol{v} \cdot \boldsymbol{r} = 0$. So $\boldsymbol{v}$ is tangent to the circle. The acceleration $\boldsymbol{a}$ is

$$\begin{aligned}\boldsymbol{a}(t) = \boldsymbol{v}'(t) = &- R\big(\theta''(t)\sin\theta(t) - (\theta'(t))^2\cos\theta(t)\big)\,\hat{\boldsymbol{\imath}} \\ &+ R\big(\theta''(t)\cos\theta(t) - (\theta'(t))^2\sin\theta(t)\big)\,\hat{\boldsymbol{\jmath}}\end{aligned}$$

or

$$\boldsymbol{a} = -(\theta')^2\,\boldsymbol{r} + \big(\theta''/\theta'\big)\,\boldsymbol{v}. \tag{1.3.21}$$

Thus, the tangential component of $\boldsymbol{a}$ is $\big(\theta''/\theta'\big)\,\boldsymbol{v}$, and the normal component, also known as the `centripetal acceleration`, is $-(\theta')^2\,\boldsymbol{r}$. Also, $\|\boldsymbol{a}\| = R\sqrt{(\theta')^4 + (\theta'')^2}$.

EXERCISE 1.3.13.^{B} *Verify that (1.3.21) coincides with (1.3.19). Hint: First verify that* $ds/dt = R\theta'(t)$ *and* $d\widehat{\boldsymbol{u}}(t)/dt = -(\theta'(t)/R)\,\boldsymbol{r}(t)$.

If the rotation is uniform with constant angular speed ω, then $\theta(t) = \omega t$ and we have the familiar expressions $\|\boldsymbol{a}\| = \omega^2 R = \|\boldsymbol{v}\|^2/R$.

Note that the centripetal acceleration is in the direction of $-\boldsymbol{r}$, that is, in the direction *toward* the center. It is not a coincidence that the Latin verb *petere* means "to look for."

Next, we write the decomposition (1.3.19) for the GENERAL PLANAR MOTION IN POLAR COORDINATES (r, θ). Consider a frame with origin O and fixed cartesian coordinate system $(\hat{\boldsymbol{\imath}}, \hat{\boldsymbol{\jmath}}, \hat{\boldsymbol{\kappa}})$ so that the motion is in the $(\hat{\boldsymbol{\imath}}, \hat{\boldsymbol{\jmath}})$ plane. Recall that, for a point P with position vector $\boldsymbol{r}$, $r = \|\boldsymbol{r}\|$, and θ is the angle from vector $\hat{\boldsymbol{\imath}}$ to $\boldsymbol{r}$. Let $\widehat{\boldsymbol{r}} = \boldsymbol{r}/r$ be the unit radius vector and let $\widehat{\boldsymbol{\theta}}$ be the unit vector orthogonal to $\boldsymbol{r}$ so that $\widehat{\boldsymbol{r}} \times \widehat{\boldsymbol{\theta}} = \hat{\boldsymbol{\imath}} \times \hat{\boldsymbol{\jmath}}$; draw a picture or see Figure 2.1.3 on page 48 below. Then

$$\begin{cases} \widehat{\boldsymbol{r}} = \cos\theta\,\hat{\boldsymbol{\imath}} + \sin\theta\,\hat{\boldsymbol{\jmath}}, \\ \widehat{\boldsymbol{\theta}} = -\sin\theta\,\hat{\boldsymbol{\imath}} + \cos\theta\,\hat{\boldsymbol{\jmath}}. \end{cases} \tag{1.3.22}$$

The vectors $\widehat{\boldsymbol{r}}, \widehat{\boldsymbol{\theta}}$ are functions of θ. From (1.3.22) we get

$$\begin{cases} d\widehat{\boldsymbol{r}}/d\theta = -\sin\theta\,\hat{\boldsymbol{\imath}} + \cos\theta\,\hat{\boldsymbol{\jmath}} = \widehat{\boldsymbol{\theta}}, \\ d\widehat{\boldsymbol{\theta}}/d\theta = -\cos\theta\,\hat{\boldsymbol{\imath}} - \sin\theta\,\hat{\boldsymbol{\jmath}} = -\widehat{\boldsymbol{r}}. \end{cases} \tag{1.3.23}$$

Let $\boldsymbol{r}(t)$ be the position of the point mass m at time t. In polar coordinates, $\boldsymbol{r}(t) = r(t)\widehat{\boldsymbol{r}}(\theta(t))$. The velocity of m in the frame O is $\boldsymbol{v} = d\boldsymbol{r}/dt = d(r(t)\widehat{\boldsymbol{r}}(\theta(t)))/dt$. Using the rule (1.3.4) and the chain rule, we get $\boldsymbol{v} = (dr/dt)\,\widehat{\boldsymbol{r}} + r\,(d\widehat{\boldsymbol{r}}/d\theta)(d\theta/dt)$, or

$$\boldsymbol{v} = \dot{r}\widehat{\boldsymbol{r}} + r\dot{\theta}\widehat{\boldsymbol{\theta}} = \dot{r}\widehat{\boldsymbol{r}} + r\omega\widehat{\boldsymbol{\theta}}. \tag{1.3.24}$$

The velocity $\boldsymbol{v}$ is a sum of the radial velocity component $\dot{r}\widehat{\boldsymbol{r}}$ and the angular velocity component $r\omega\widehat{\boldsymbol{\theta}}$. We call $\dot{r}$ and $r\dot{\theta}$ the `radial` and `angular` speeds, respectively.

The acceleration $\boldsymbol{a}$ in the frame O is obtained by differentiating (1.3.24) with respect to t according to the rules (1.3.3), (1.3.4):

$$\boldsymbol{a} = d\boldsymbol{v}/dt = \ddot{r}\,\widehat{\boldsymbol{r}} + \dot{r}\,(d\widehat{\boldsymbol{r}}/d\theta)\dot{\theta} + (\dot{r}\dot{\theta} + r\ddot{\theta})\,\widehat{\boldsymbol{\theta}} + r\dot{\theta}\,(d\widehat{\boldsymbol{\theta}}/d\theta)\dot{\theta},$$

or

$$\boldsymbol{a} = (\ddot{r} - r\dot{\theta}^2)\,\widehat{\boldsymbol{r}} + (r\ddot{\theta} + 2\dot{r}\dot{\theta})\,\widehat{\boldsymbol{\theta}}. \tag{1.3.25}$$

The acceleration $\boldsymbol{a}$ is a sum of the `radial` component $\boldsymbol{a}_r$ and the `angular` component $\boldsymbol{a}_\theta$, where

$$\boldsymbol{a}_r = (\ddot{r} - r\omega^2)\,\widehat{\boldsymbol{r}} \ \text{ and } \ \boldsymbol{a}_\theta = (r\ddot{\theta} + 2\dot{r}\omega)\,\widehat{\boldsymbol{\theta}}. \tag{1.3.26}$$

EXERCISE 1.3.14.[B] *Verify that decomposition (1.3.26) of the acceleration is a particular case of (1.3.19).*

Now assume that the trajectory of the point mass is a circle with center at O and radius R. Then $r(t) = R$ for all t and $\dot{r}(t) = \ddot{r}(t) = 0$. Let $\dot{\theta}(t) = \omega(t)$. By (1.3.26),

$$\begin{cases} \boldsymbol{a}_r = -R\omega^2\,\widehat{\boldsymbol{r}} & \text{(centripetal acceleration)} \\ \boldsymbol{a}_\theta = R\dot{\omega}\,\widehat{\boldsymbol{\theta}} & \text{(angular acceleration).} \end{cases} \tag{1.3.27}$$

Also, by (1.3.24),

$$\boldsymbol{v} = R\omega\,\widehat{\boldsymbol{\theta}}. \tag{1.3.28}$$

EXERCISE 1.3.15.[B] *Verify that formula (1.3.27) is a particular case of the decomposition (1.3.21) of the acceleration, as derived on page 34.*

If we further assume that the angular speed is constant, that is, $\omega(t) = \omega_0$ for all t, then $\dot{\omega} = 0$, and, by (1.3.27),

$$\boldsymbol{a}_r = -R\omega_0^2\,\widehat{\boldsymbol{r}}, \quad \boldsymbol{a}_\theta = \boldsymbol{0}. \tag{1.3.29}$$

EXERCISE 1.3.16.[B] *Verify that if the acceleration of a point mass in polar coordinates is given by (1.3.29), then the point moves around the circle of radius R with constant angular speed ω_0. Hint: Combine (1.3.29) and (1.3.26) to get differential equations for r and θ. Solve the equations with initial conditions $r(0) = R,\ \dot{r}(0) = 0,\ \theta(0) = 0,\ \dot{\theta}(0) = \omega_0$ to get $r(t) = R,\ \theta(t) = \omega_0 t$.*

EXERCISE 1.3.17.[A] *Let $(r(t), \theta(t))$ be the polar coordinates of a 2-D motion of a point mass m in a fixed frame O. Let $r(t) = 3t$ and $\theta(t) = 2t$. Sketch the trajectory of the point in the frame O for $0 < t < 5$ and verify the result using a computer algebra system. Compute the velocity and acceleration vectors in the frame O in terms of the unit vectors $\widehat{\mathbf{r}}, \widehat{\boldsymbol{\theta}}$.*

Chapter 2

Vector Analysis and Classical and Relativistic Mechanics

2.1 Kinematics and Dynamics of a Point Mass

`Kinematics` is the study of motion without reference to forces; the Greek word *kinēma* means "motion." `Dynamics` is the study of motion under the action of forces; the Greek word *dýnamis* means "force." Also, the Greek word *mēchanikós* means "machine."

A curve C, defined by a vector-valued function of time $\boldsymbol{r} = \boldsymbol{r}(t)$ provides the mathematical description of the trajectory in $\mathbb{R}^3$ of a particle (point mass) so that the location of the particle at time t is at the end point of the vector $\boldsymbol{r}(t)$. The initial point O of the vector is the origin of the corresponding frame in which the motion is studied. It is clear that the same motion can be studied in different frames and in different coordinates. The Frenet trihedron (page 32) is an example of a coordinate system in which the particle is at rest, but the coordinate system is moving. The objective of this section is to derive the rules for describing the motion of a point mass in different coordinate systems.

Unless explicitly mentioned otherwise, we assume that the curve C is `smooth`, that is, the unit tangent vector $\widehat{\boldsymbol{u}}$ exists at every point of the curve; see page 27.

2.1.1 *Newton's Laws of Motion and Gravitation*

The motion of a point mass m is related to the net force $\boldsymbol{F}$ acting on the mass. In an *inertial frame*, this relation is made precise by Newton's three laws of motion, and conversely, every frame in which these laws hold is called `inertial`. These laws were first formulated by Newton around 1666, less than a year after he received his bachelor's degree from Cambridge

University.

Newton's First Law: *Unless acted upon by a force, a point mass is either not moving or moves in a straight line with constant speed.* This law is also called the `Law of Inertia` or `Galileo's Principle`.
Newton's Second Law: *The acceleration of the point mass is directly proportional to the net force exerted and inversely proportional to the mass.*
Newton's Third Law: *For every action, there is an equal and opposite reaction.*

Mathematically, the Second Law is

$$\frac{d^2\boldsymbol{r}(t)}{dt^2} = \frac{\boldsymbol{F}}{m}, \tag{2.1.1}$$

where $\boldsymbol{r} = \boldsymbol{r}(t)$ is the position of the point mass at time t.

EXERCISE 2.1.1.[C] *Show that the Second Law implies the First Law. In other words, show that if (2.1.1) holds, then the point mass m acted on by zero external force will move with constant velocity $\boldsymbol{v}$. Hint: find the general solution of equation (2.1.1) when $\boldsymbol{F} = \boldsymbol{0}$.*

The notion of *momentum* provides an alternative formulation of Newton's Second Law. Consider a point mass m moving along the path $\boldsymbol{r} = \boldsymbol{r}(t)$ with velocity $\dot{\boldsymbol{r}}(t)$ relative to a reference frame with origin at O. The (linear) `momentum` $\boldsymbol{p}$ is the vector

$$\boldsymbol{p} = m\dot{\boldsymbol{r}}. \tag{2.1.2}$$

If the reference frame is inertial, then (2.1.1) becomes

$$\dot{\boldsymbol{p}} = \boldsymbol{F}, \tag{2.1.3}$$

and the force is now interpreted as the rate of change of the linear momentum. Incidentally, the Latin word *mōmentum* means "motion" or "cause of motion." One advantage of (2.1.3) over (2.1.1) is the possibility of *variable* mass.

Similarly, the study of circular motion suggest the definition of the `angular momentum` about the point O as the vector

$$\boldsymbol{L}_O = m\boldsymbol{r} \times \dot{\boldsymbol{r}}. \tag{2.1.4}$$

Note that both $\boldsymbol{p}$ and $\boldsymbol{L}_O$ depend on the reference point O, but do not depend on the coordinate system.

Recall that a force $\boldsymbol{F}$ acting on the point mass m has a `torque`, or `moment`, about O equal to

$$\boldsymbol{T}_O = \boldsymbol{r} \times \boldsymbol{F}. \tag{2.1.5}$$

Applying the rule (1.3.6), page 26, to formula (2.1.4) we find

$$\frac{d\boldsymbol{L}_O}{dt} = m\,(\dot{\boldsymbol{r}} \times \dot{\boldsymbol{r}} + \boldsymbol{r} \times \ddot{\boldsymbol{r}}) = m\,\boldsymbol{r} \times \ddot{\boldsymbol{r}}.$$

If the frame O is inertial, then $\ddot{\boldsymbol{r}} = \boldsymbol{F}/m$ and

$$\frac{d\boldsymbol{L}_O}{dt} = \boldsymbol{r} \times \boldsymbol{F} = \boldsymbol{T}_O. \tag{2.1.6}$$

Relation (2.1.6) describes the rotational motion just as (2.1.3) describes the translational motion.

As an example, consider the SIMPLE RIGID PENDULUM, which is a massless thin rigid rod of length ℓ connected to a point mass m at one end. The other end of the rod is connected to a frictionless pin-joint at a point O, which is a zero-diameter bearing that permits rotation in a fixed plane. We select a cartesian coordinate system $(\hat{\imath},\, \hat{\jmath},\, \hat{\kappa})$ with center at O and $\hat{\imath}$, $\hat{\jmath}$ fixed in the plane of the rotation (Figure 2.1.1), and assume that the corresponding frame is inertial.

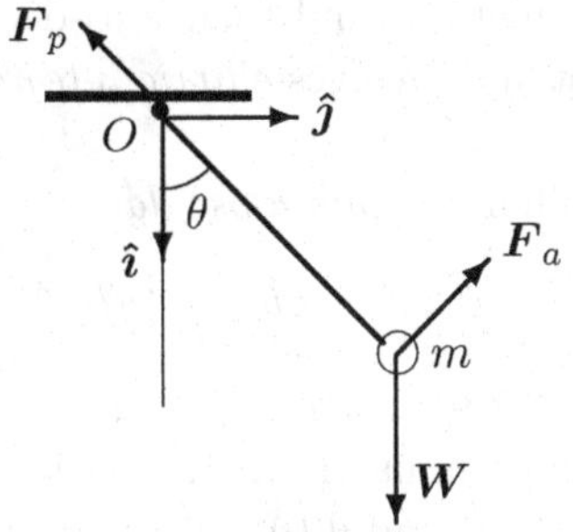

Fig. 2.1.1 Simple Rigid Pendulum

The motion of m is a circular rotation in the $(\hat{\imath},\, \hat{\jmath})$ plane, and is best described using polar coordinates (r, θ); see page 35. The rigidity assumption implies $r(t) = \ell$ for all t. Denote by $\theta = \theta(t)$ the angle from $\hat{\imath}$ to the rod at time t. Let $\boldsymbol{r}(t)$ be the position vector of the point mass in the frame O. Then $\boldsymbol{r}(t) = \ell\widehat{\boldsymbol{r}}(\theta(t))$, and by (1.3.24) on page 35, $\dot{\boldsymbol{r}} = \ell\,\dot{\theta}\,\widehat{\boldsymbol{\theta}}$. Hence, the angular momentum of the point mass about O is

$\boldsymbol{L} = m\,\boldsymbol{r} \times \dot{\boldsymbol{r}} = m(\ell\,\widehat{\boldsymbol{r}} \times \ell\dot{\theta}\,\widehat{\boldsymbol{\theta}}) = m\ell^2\dot{\theta}\,\hat{\boldsymbol{\kappa}}$, and

$$\frac{d\,\boldsymbol{L}}{dt} = m\ell^2\ddot{\theta}\,\hat{\boldsymbol{\kappa}}. \tag{2.1.7}$$

The forces acting on the pendulum are the weight $\boldsymbol{W}$ of m, the air resistance $\boldsymbol{F}_a$ and the force $\boldsymbol{F}_p$ exerted by the pin at O. Clearly, $\boldsymbol{W} = mg\,\hat{\boldsymbol{\imath}}$, where g is the acceleration of gravity. Physical considerations suggest that the force $\boldsymbol{F}_a$ on m may be assumed to act tangentially to the circular path and to be proportional to the tangential velocity: $\boldsymbol{F}_a = -c\ell\dot{\theta}\,\widehat{\boldsymbol{\theta}}$, where c is the damping constant. The total torque $\boldsymbol{T}$ about O exerted by these forces is

$$\boldsymbol{T} = \boldsymbol{r} \times \boldsymbol{W} + \boldsymbol{r} \times \boldsymbol{F}_a + \boldsymbol{0} \times \boldsymbol{F}_p, = \ell\,\widehat{\boldsymbol{r}} \times mg\,\hat{\boldsymbol{\imath}} - \ell\,\widehat{\boldsymbol{r}} \times c\,\ell\dot{\theta}\,\widehat{\boldsymbol{\theta}},$$

or

$$\boldsymbol{T} = -(mg\ell\sin\theta + \ell^2 c\dot{\theta})\,\hat{\boldsymbol{\kappa}}. \tag{2.1.8}$$

Since the frame O is inertial, equation (2.1.6) applies, and by (2.1.7) and (2.1.8) above, we obtain $m\ell^2\,\ddot{\theta} = -mg\ell\sin\theta - \ell^2 c\,\dot{\theta}$, or

$$m\ell\,\ddot{\theta} + c\ell\,\dot{\theta} = -mg\,\sin\theta. \tag{2.1.9}$$

Equation (2.1.9) is a nonlinear ordinary differential equation and cannot be *integrated in quadratures*, that is, its solution cannot be written using only elementary functions and their anti-derivatives. When $c = 0$, such a solution does exist and involves *elliptic integrals*; see Problem 2.3, page 417, if you are curious.

The more familiar *harmonic oscillator*

$$\ddot{\theta} = -(g/\ell)\theta \tag{2.1.10}$$

is obtained from (2.1.9) when $c = 0$ and θ is small so that $\sin\theta \approx \theta$; this equation should be familiar from the basic course in ordinary differential equations. If $\theta(0) = \theta_0$ and $\dot{\theta}(0) = 0$, then the solution of (2.1.10) is

$$\theta(t) = \theta_0\,\cos(\omega t),\ \text{where}\ \omega = (\ell/g)^{1/2}.$$

The period of the small undamped oscillations is $2\pi(\ell/g)^{1/2}$, and the value of ℓ can be adjusted to provide a desired ticking rate for a clock mechanism. The idea to use a pendulum for time-keeping was studied by the Italian scientist GALILEO GALILEI (1564–1642) during the last years of his life, but it was only in 1656 that the Dutch scientist CHRISTIAAN HUYGENS (1629–1695) patented the first pendulum clock.

EXERCISE 2.1.2.^{C} *Let point mass m move in a planar path C given by $\mathbf{r}(t)$, where $\mathbf{r}$ is the position vector with origin O.*
(a) Use formulas (1.3.24), page 35, and (2.1.4) to express the angular momentum of the point mass about O in the coordinate system $(\widehat{\boldsymbol{r}}, \widehat{\boldsymbol{\theta}})$.
(b) Suppose that the point mass moves in a circular path C with radius R and center O. Denote the angular speed by $\omega(t)$. (i) Compute the angular momentum $\boldsymbol{L}_O$ of the point mass about O and the corresponding torque. (ii) Find the force $\boldsymbol{F}$ that is required to produce this motion, assuming the frame O is inertial. (iii) Write $\boldsymbol{F}$ as a linear combination of $\dot{\boldsymbol{r}}$ and $\widehat{\boldsymbol{\theta}}$. (iv) How will the expressions simplify if $\omega(t)$ does not depend on time?

As we saw in Exercise 2.1.1, the Second Law of Newton implies the First Law. For further discussion of the logic of Newton's Laws see the book *Foundations of Physics* by H. Margenau and R. Lindsay, 1957. Regarding the First Law, they quote A. S. Eddington's remark from his book *Nature of the Physical World*, first published in the 1920s, that the law, in effect, says that "every particle continues in its state of rest or uniform motion in a straight line, except insofar as it doesn't." This is a somewhat facetious commentary on the logical circularity of Newton's original formulation, which depends on the notion of zero force acting, which can only be observed in terms of the motion being at constant velocity. The same logical difficulty arises in the definition of an `inertial frame` as a frame in which the three laws of Newton hold. We do not concentrate on these questions here and simply assume that the `primary inertial frame`, that is, a frame attached to far-away, and approximately fixed, stars is a good approximation of an inertial frame for all motions in the vicinity of the Earth. The idea of this frame goes back to the Irish bishop and philosopher G. Berkeley. The deep question "What is a force?" is also beyond the scope of our presentation; for the discussion of this question, see the above-mentioned book *Foundations of Physics* by H. Margenau and R. Lindsay, or else take as given that there are four basic kinds of forces: gravitational, electromagnetic, strong nuclear, and weak nuclear. In inertial frames, all other forces result from these four.

Newton discovered the Law of Universal Gravitation by combining his laws of motion with Kepler's Laws of Planetary Motion. The history behind this discovery is a lot more complex than the familiar legend about the apple falling from the tree and hitting Newton on the head. As many similar stories, this "apple incident" is questioned by modern historians. Below, we present some of the highlights of the actual development.

The basic ideas of modern astronomy go back to the Polish astronomer NICOLAUS COPERNICUS (1473–1543) and his heliocentric theory of the solar system. Copernicus was a canon (in modern terms, senior manager) of the cathedral at the town of Frauenburg (now Frombork) in northern Poland, and observed the stars and planets from his home. Around 1530, he came to the conclusion that planets in our solar system revolve around the Sun. He was hesitant to publish his ideas, both for fear of being charged with heresy and because of the numerous problems he could not resolve; his work, titled *De revolutionibus orbium coelestium* ("On the Revolutions of the Celestial Spheres") was finally published in 1543, apparently just a few weeks before he died.

It took some time to formalize the heliocentric ideas mathematically, and the key missing element was the empirical data. The main instrument for astronomical observations, the telescope, was yet to be invented: it was only in 1609 that Galileo Galilei made the first one. Without a telescope, collecting the data required a lot of time and patience, but the Danish scientist TYCHO BRAHE (1546–1601) had both. Brahe was the royal astronomer and mathematician to Rudolf II, the emperor of the Holy Roman Empire. At the observatory in Prague, the seat of the Holy Roman Empire at that time, Brahe compiled the world's first truly accurate and complete set of astronomical tables. His assistant, German scientist JOHANN KEPLER (1571–1630), had been a proponent of the heliocentric theory of Copernicus. After inheriting the position and all the astronomical data from Brahe in 1601, Kepler analyzed the data for the planet Mars and formulated his first two laws in 1609. Further investigations led him to the discovery of the third law in 1619.

Kepler's First Law: *The planets have elliptical orbits with the Sun at one focus.*

Kepler's Second Law: *The radius vector from the Sun to a planet sweeps over equal areas in equal time intervals.*

Kepler's Third Law: *For every planet p, the square of its period T_p of revolution around the Sun is proportional to the cube of the average distance R_p from the planet to the Sun. In other words, $T_p^2 = K_s R_p^3$, where the number K_s is the same for every planet.*

A planetary orbit has a very small eccentricity and so is close to a circle of some mean radius R. Kepler speculated that a planet is held in its orbit by a force of attraction between the Sun and the planet, and Newton quantified Kepler's qualitative idea. In modern terms, we can recover Newton's argument by combining equation (1.3.29), page 36, with his second law (2.1.1). Take an inertial frame with origin at the Sun and assume that a planet of mass m executes a circular motion around the origin with constant angular speed ω. Then (2.1.1) and (1.3.29) result in

$$\boldsymbol{F} = m\boldsymbol{a} = -m\omega^2\boldsymbol{r}(t) = -m\omega^2 R\widehat{\boldsymbol{r}}(t),$$

where $\widehat{\boldsymbol{r}} = \boldsymbol{r}/\|\boldsymbol{r}\|$ is the unit radius vector pointing from the Sun to the planet. On the other hand, $\omega = 2\pi/T$, and therefore the magnitude of $\boldsymbol{F}$ is $\|\boldsymbol{F}\| = m(4\pi^2/T^2)R$. Applying Kepler's Third Law, we obtain $\|\boldsymbol{F}\| = m(4\pi^2/K_sR^3)R = Cm/R^2$, where $C = 4\pi^2/K_s$. In other words, *the gravitational force exerted by the Sun on the planet of mass m at a distance R is proportional to m and R^{-2}.* By Newton's third law, there must be an equal and opposite force exerted by m on the Sun. By the same argument, we conclude that the magnitude of the force must also be proportional to M and R^{-2}, where M is the mass of the Sun. Therefore,

$$\|\boldsymbol{F}\| = \frac{\mathrm{G}Mm}{R^2}, \tag{2.1.11}$$

where G is a constant. Newton postulated that G is a `universal gravitational constant` , that is, has the same value for any two masses, and therefor (2.1.11) is a `Universal Law of Gravitation`. In 1798, the English scientist HENRY CAVENDISH (1731–1810), in his quest to determine the mass and density of Earth, verified the relation (2.1.11) experimentally and determined a numerical value of G: $\mathrm{G} \approx 6.67 \times 10^{-11}\ \mathrm{m}^3/(\mathrm{kg}\cdot \mathrm{sec}^2)$. Since then, the Universal Law of Gravitation has been tested and verified on many occasions. An extremely small discrepancy has been discovered in the orbit of Mercury that cannot be derived from (2.1.11), and is explained by Einstein's law of gravitation in the theory of general relativity; see Problem 2.2 on page 414.

In our derivation of (2.1.11), we implicitly used the *equivalence principle* that the two possible values of m, its *inertial* and *gravitational* masses, are equal. A priori, this is not at all obvious. Indeed, the mass m in equation (2.1.1) of Newton's Second Law, the `inertial mass`, expresses an object's resistance to external force: the larger the mass, the smaller the acceleration. The mass m in (2.1.11), the `gravitational mass`, expresses something completely different, namely, its gravitational attraction: the larger the mass, the stronger the gravitational attraction it produces. The equivalence principle is one of the foundations of Einstein's theory of *general relativity* and can be traced back to Galilei, who was among the first to study the motion of bodies under Earth's gravity. Even though many modern historians question whether indeed, around 1590, he was dropping different objects from the leaning tower of Pisa, in 1604 Galileo did conduct related experiments using an inclined plane; in 1608, he formulated mathematically the basic laws of accelerated motion under the gravitational force. The conjecture of Galilei that the acceleration due to gravity is essentially the same for all kinds of matter has been verified experimentally. Between 1905 and 1908, the Hungarian physicist VÁSÁROSNAMÉNYI BÁRÓ EÖTVÖS LORÁND (1848–1919), also known as ROLAND EÖTVÖS, measured a variation of about 5×10^{-9} in the Earth's pull on wood and platinum; somehow, the result was published only in 1922. In the 1950s, the American physicist ROBERT HENRY DICKE (1916–1997) measured a difference of $(1.3 \pm 1.0) \times 10^{-11}$ for the Sun's attraction of aluminum and gold objects.

Following the historical developments, we derived relation (2.1.11) from Newton's Second Law of Motion and Kepler's Third Law of Planetary Motion. Problem 2.1, page 413, presents a deeper insight into the problem. In particular, more detailed analysis shows that, in our derivation of (2.1.11), Kepler's Third Law can be replaced by his First Law, along with the assumption that the gravitational force is attracting and `central`, that is, acts along the line connecting the Sun and the planet. Moreover, the reader who completes Problem 2.1 will see that all three Kepler's laws follow from (2.1.1) and (2.1.11). This illustrates the power of mathematical models in reasoning about physical laws.

2.1.2 *Parallel Translation of Frames*

Recall that Newton's laws of motion hold only in inertial frames; see page 4 for the definition of frame. Ignoring the possible logical issues, we therefore say that an `inertial frame` is a frame in which a point mass maintains constant velocity in the absence of external forces. The same frame can be (approximately) inertial in some situations and not inertial in others. For example, a frame fixed to the surface of the Earth is inertial if the objective is to study the motion of a billiard ball on a pool table. The same frame is no longer inertial if the objective is to study the trajectory of an intercontinental ballistic missile: the inertial frame for this problem should not rotate with the Earth; the *primary inertial frame* fixed to the stars is a possible choice, see page 43. In this and the following two sections, we investigate how relation (2.1.1) changes if the frame is not inertial. The starting point is the analysis of the relative motion of frames.

The easiest motion is *parallel translation*, where the corresponding basis vectors in the frames stay parallel. Consider two such frames with origins O and O_1 respectively. Denote by $\boldsymbol{r}_{01}(t)$ the position vector of O_1 with respect to O. Let the position vectors of a point mass in frames O and O_1 be $\boldsymbol{r}_0(t)$ and $\boldsymbol{r}_1(t)$ respectively (Figure 2.1.2).

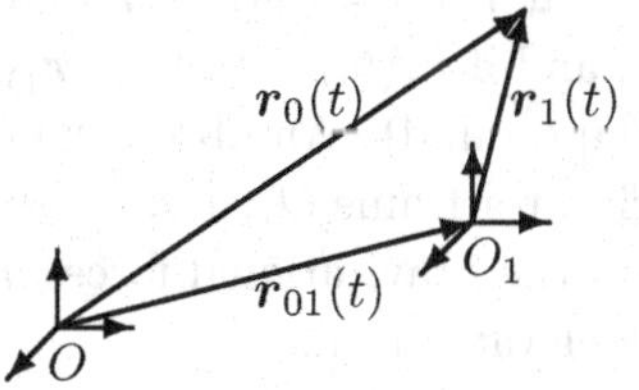

Fig. 2.1.2 Translation of Frames

Clearly, $\boldsymbol{r}_0(t) = \boldsymbol{r}_{01}(t) + \boldsymbol{r}_1(t)$, and the absence of relative rotation allows us to consider this equality in the frame O for each t We can identify the parallel vectors that have the same direction and length. Since position vectors in O and O_1 maintain their relative orientation when there is no relative rotation of the frames, the coordinate systems in the frames O and O_1 are the same. Then we can apply the rule for differentiating a sum (1.3.3) to obtain simple relations between the velocities and accelerations in the frames O and O_1:

$$\dot{\boldsymbol{r}}_0(t) = \dot{\boldsymbol{r}}_{01}(t) + \dot{\boldsymbol{r}}_1(t), \quad \ddot{\boldsymbol{r}}(t) = \ddot{\boldsymbol{r}}_{01}(t) + \ddot{\boldsymbol{r}}_1(t). \tag{2.1.12}$$

If the frame O is inertial, then, by Newton's Second Law, $m\ddot{\boldsymbol{r}}(t) = \boldsymbol{F}(t)$, where m is the mass of the point, and $\boldsymbol{F}$ is the sum of all forces acting on the point. The second equality in (2.1.12) then implies

$$m\ddot{\boldsymbol{r}}_1(t) = \boldsymbol{F}(t) - m\ddot{\boldsymbol{r}}_{01}(t) \tag{2.1.13}$$

In effect, there are two forces acting on m in the frame O_1. One is the force $\boldsymbol{F}$. The other, $-m\ddot{\boldsymbol{r}}_{01}$, is called a `translational acceleration force`. It is an example of an `apparent`, or `inertial`, force, that is, a force that appears because of the relative motion of frames and is not of any of the four types described on page 43. If $\dot{\boldsymbol{r}}_{01}(t)$ is constant, then $\ddot{\boldsymbol{r}}_{01}(t) = \boldsymbol{0}$ and the Second Law of Newton holds in O_1, that is, O_1 is also an inertial frame. Thus, all frames moving with constant velocity relative to an inertial frame are also inertial frames.

As an EXAMPLE, consider a golf ball in a moving elevator. We fix the frame O on the ground, and O_1, on the elevator, and select the usual cartesian coordinate systems in both frames so that the corresponding coordinate vectors are parallel. Assume that the elevator is falling down with the gravitational acceleration g, so that $\ddot{\boldsymbol{r}}_{01}(t) = -g\,\hat{\boldsymbol{\kappa}}$, and the ball is falling down inside the elevator, also with the gravitational acceleration g so that $\ddot{\boldsymbol{r}}_0(t) = -g\,\hat{\boldsymbol{\kappa}}$. Then (2.1.12) shows that $\ddot{\boldsymbol{r}}_1(t) = \ddot{\boldsymbol{r}}_0(t) - \ddot{\boldsymbol{r}}_{01}(t) = -g\,\hat{\boldsymbol{\kappa}} + g\,\hat{\boldsymbol{\kappa}} = \boldsymbol{0}$. Therefore, $\dot{\boldsymbol{r}}_1(t)$ is constant, and if $\dot{\boldsymbol{r}}_1(t_0) = \boldsymbol{0}$, then $\boldsymbol{r}_1'(t) = \boldsymbol{0}$ for all $t > t_0$ (or until the elevator hits the ground). An observer in the elevator would see the ball as fixed in the elevator frame O_1: the translational acceleration force $-m\,\ddot{\boldsymbol{r}}_{01}(t)$ compensates the gravitational force $m\,\ddot{\boldsymbol{r}}_0(t)$, and the ball behaves as weightless in the elevator frame.

2.1.3 *Uniform Rotation of Frames*

Note that it is the absence of rotation that allowed us to use the differentiation rule (1.3.3) in the derivation of relation (2.1.12). This and other rules of differentiation no longer apply if the frames are rotating relative to each other, and relation (2.1.12) must be modified.

We start with the analysis of `uniform rotation`, that is, rotation with constant angular speed around a fixed axis. As a motivational example, consider a car driving with constant angular speed ω_0 in a circle with radius R and center at O. Consider an object (a point mass m) moving inside the car with constant radial speed v_0 relative to O, and rotating together with the car with constant angular speed ω_0 around O. Introduce a new (non-inertial) rotating frame with origin O_1 inside the car and the coordinate basis vectors $\hat{\boldsymbol{\imath}}_1 = \hat{\boldsymbol{r}}$, $\hat{\boldsymbol{\jmath}}_1 = \hat{\boldsymbol{\theta}}$; see Figure 2.1.3. As before, let $\boldsymbol{r}_0$ and $\boldsymbol{r}_1$ be the position vectors of the point mass in the frames O and O_1, respectively, and denote $\overrightarrow{OO_1}$ by $\boldsymbol{r}_{01}$. Assume that $\boldsymbol{r}_1(0) = \boldsymbol{0}$.

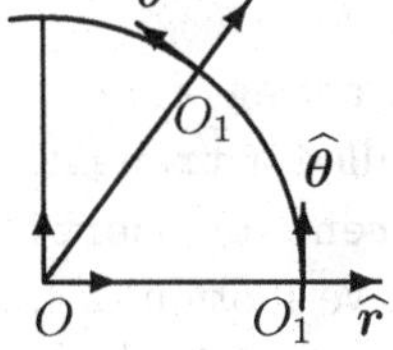

Fig. 2.1.3 Rotation of Frames

By construction, $\boldsymbol{r}_{01} = R\,\hat{\boldsymbol{r}}$. For a passenger riding in the car, the path $\boldsymbol{r}_1(t) = v_0 t\,\widehat{\boldsymbol{r}}$ of the point mass *relative to the car* is a straight line, since there is no angular displacement of m relative to the car. Thus, $\boldsymbol{r}_0(t) = (R + v_0 t)\,\widehat{\boldsymbol{r}}(t)$. Note that the polar coordinates of m in frame O are $(r(t), \theta(t))$, where $r(t) = R + v_0 t$. Hence $\dot{r}(t) = v_0$ and $\ddot{r}(t) = 0$. Also, $\dot{\theta}(t) = \omega_0$ and $\ddot{\theta}(t) = 0$. Formulas (1.3.26) on page 36 provide the acceleration $\boldsymbol{a} = \boldsymbol{a}_r + \boldsymbol{a}_\theta$ of the mass in the frame O, where, with $\ddot{r} = 0$ and $\ddot{\theta} = 0$,

$$\boldsymbol{a}_r = -(R + v_0 t)\omega_0^2\,\widehat{\boldsymbol{r}}, \quad \boldsymbol{a}_\theta = 2v_0\omega_0\,\widehat{\boldsymbol{\theta}} \tag{2.1.14}$$

By (1.3.27) on page 36, $\ddot{\boldsymbol{r}}_{01} = -R\omega_0^2\,\widehat{\boldsymbol{r}}$. In frame O_1, $\ddot{\boldsymbol{r}}_1(t) = \boldsymbol{0}$. Then (2.1.14) shows that the acceleration $\ddot{\boldsymbol{r}}_0(t)$ is

$$\ddot{\boldsymbol{r}}_0(t) = \ddot{\boldsymbol{r}}_{01}(t) + \ddot{\boldsymbol{r}}_1(t) - v_0 t\omega_0^2\,\widehat{\boldsymbol{r}} + 2v_0\omega_0\,\widehat{\boldsymbol{\theta}}. \tag{2.1.15}$$

We see that the simple relation (2.1.12) between the accelerations in translated frames does not correctly describe acceleration of the point mass in the frame O in terms of the acceleration in the frame O_1.

If O is an inertial frame and $\boldsymbol{F}$ is a force acting on the point mass in O to produce the motion, then, by the Newton's Second Law, $m\,\ddot{\boldsymbol{r}}_0 = \boldsymbol{F}$. According to (2.1.15),

$$m\,\ddot{\boldsymbol{r}}_1 = \boldsymbol{F} + (mR\omega_0^2 + mv_0 t\omega_0^2)\,\widehat{\boldsymbol{r}} - 2mv_0\omega_0\,\widehat{\boldsymbol{\theta}}. \tag{2.1.16}$$

Thus, frame O_1 is not inertial. Similar to (2.1.13), *inertial forces* appear as correctors to Newton's Second Law: the *centrifugal force* $\boldsymbol{F}_c = (mR\omega_0^2 + mv_0 t\omega_0^2)\,\widehat{\boldsymbol{r}}$ and the *Coriolis force* $\boldsymbol{F}_{cor} = -2mv_0\omega_0\,\widehat{\boldsymbol{\theta}}$. The centrifugal force prevents the mass from flying off at a tangent because of the rotational motion of the car and the mass. One component of this force, $mR\omega_0^2\,\widehat{\boldsymbol{r}} = -m\ddot{\boldsymbol{r}}_{01}(t)$, is related to the motion of the car causing the rotation of the origin of the frame O_1; the other, $mv_0 t\omega_0^2\,\widehat{\boldsymbol{r}}$, takes into account the outward radial motion of the point. Note that the direction of the centrifugal force is *in the direction of* $\widehat{\boldsymbol{r}}$, and is therefore *away* from the center and opposite to the direction of the centripetal acceleration (cf. page 35). Incidentally, the Latin verb *fugere* means "to run away."

The Coriolis force $-2mv_0\omega_0\,\widehat{\boldsymbol{\theta}}$ is somewhat less expected. This force is perpendicular to the linear path in O_1 and ensures that the trajectory of the point in the rotating frame is a straight radial line despite the rotation of the frame. This force was first described in 1835 by the French scientist GASPARD-GUSTAVE DE CORIOLIS (1792–1843). His motivation for the study came from the problems of the early 19th-century industry, such as the design of water-wheels. More familiar effects of the Coriolis force, such as rotation of the swing plane of the Foucault pendulum and the special directions of atmospheric winds, were discovered in the 1850s and will be discussed in the next section.

Recall that in our example $\ddot{\boldsymbol{r}}_1 = \mathbf{0}$. From (2.1.16) we conclude that $\boldsymbol{F} + \boldsymbol{F}_c + \boldsymbol{F}_{cor} = \mathbf{0}$. The *real* (as opposite to inertial) force $\boldsymbol{F}$ must balance the effects of the inertial forces to ensure the required motion of the object in the rotating frame. For a passenger sliding outward with constant velocity $v_0\,\widehat{\boldsymbol{r}}$ in a turning car, this real force is the reaction of the seat in the form of friction and forward pressure of the back of the seat. .

EXERCISE 2.1.3.[B] *Find the vector function describing the trajectory of m in the O frame. What is the shape of this trajectory? Verify your conclusion using a computer algebra system.*

EXERCISE 2.1.4.[A] *Suppose a point mass m is fixed at a point P in the O frame, that is, m remains at P in the O frame for all times. Find the vector function describing the trajectory of m in the O_1 frame. What is the shape of this trajectory? Verify your conclusion using a computer algebra system.* *Hint. This is the path of m relative to the car seen by a passenger riding in the car. Show that O is fixed in O_1.*

Coming back to Figure 2.1.3, note that the coordinate vectors in the frame O_1 spin around O_1 with constant angular speed ω_0, while the origin O_1 rotates around O with the same angular speed ω_0. This observation leads to further generalization by allowing different speeds of spinning and rotation.

EXERCISE 2.1.5.[A] *Suppose the origin O_1 rotates around the point O with angular speed ω_0, while the coordinate vectors $(\widehat{\boldsymbol{r}}, \widehat{\boldsymbol{\theta}})$ spin around O_1 with constant angular speed $2\omega_0$. Suppose a point mass is fixed at a point P in the O frame. Find the vector function describing the trajectory of m in the O_1 frame. What is the shape of this trajectory? Verify your conclusion using a computer algebra system.*

Our motivational example with the car illustrated some of the main effects that arise in rotating frames. The example was two-dimensional in nature, and now we move on to uniform rotations in space. There are many different ways to describe rotations in $\mathbb{R}^3$. We present an approach using vectors and linear algebra.

We start with a simple problem. Consider a point P moving around a circle of radius R with uniform angular velocity ω (Figure 2.1.4). Denote by $\boldsymbol{r}(t)$ the position vector of the point at time t and assume that the origin O of the frame is chosen so that $\|\boldsymbol{r}\|$ does not change in time. How to express $\dot{\boldsymbol{r}}(t)$ in terms of $\boldsymbol{r}(t)$ and ω?

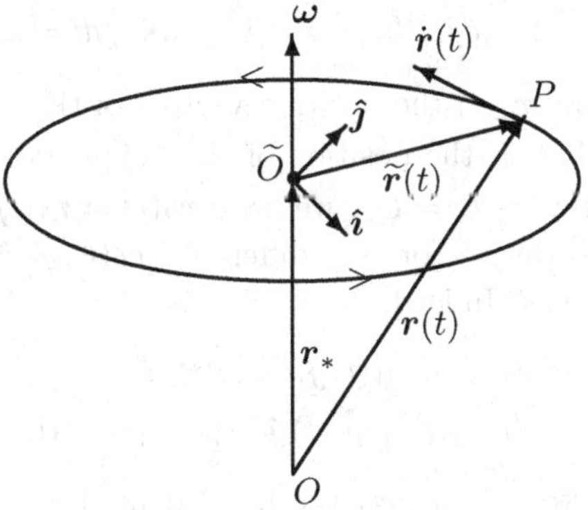

Fig. 2.1.4 Rotating Point

To solve this problem, consider the plane that contains the circle of rotation and define the `rotation vector` $\boldsymbol{\omega}$ as follows (see Figure 2.1.4). The vector $\boldsymbol{\omega}$ is perpendicular to the plane of the rotation; the direction of the vector $\boldsymbol{\omega}$ is such that the rotation is counterclockwise as seen *from the tip of the vector* (alternatively, the rotation is clockwise as seen *in the direction* of the vector); the length of the vector $\boldsymbol{\omega}$ is ω, the angular speed of the rotation. As seen from Figure 2.1.4, $\boldsymbol{r}(t) = \widetilde{\boldsymbol{r}}(t) + \boldsymbol{r}_*$ and the vector $\boldsymbol{r}_*$ does not change in time, so that $\dot{\boldsymbol{r}}(t) = \widetilde{\boldsymbol{r}}'(t)$. Consider a cartesian coordinate system $(\hat{\imath}, \hat{\jmath})$ with the origin at the center $\widetilde{O}$ of the circle in the plane of rotation so that $\widetilde{\boldsymbol{r}}(t) = R\cos\omega t\,\hat{\imath} + R\sin\omega t\,\hat{\jmath}$. Direct computations show that

$$\widetilde{\boldsymbol{r}}'(t) = (\hat{\imath} \times \hat{\jmath}) \times (\omega\,\widetilde{\boldsymbol{r}}(t)) = \boldsymbol{\omega} \times \widetilde{\boldsymbol{r}}(t) = \boldsymbol{\omega} \times \boldsymbol{r}(t), \tag{2.1.17}$$

$$\dot{\boldsymbol{r}}(t) = \boldsymbol{\omega} \times \boldsymbol{r}(t). \tag{2.1.18}$$

EXERCISE 2.1.6.^{C} *Verify all equalities in (2.1.17).*

We will now use (2.1.18) to derive the relation between the velocities and accelerations of a point mass m relative to two frames O and O_1, when the frame O_1 is rotating with respect to O. We assume that the two frames have the same origins: $O = O_1$. We also choose cartesian coordinate systems $(\hat{\imath}, \hat{\jmath}, \hat{\boldsymbol{k}})$ and $(\hat{\imath}_1, \hat{\jmath}_1, \hat{\boldsymbol{k}}_1)$ in the frames; see Figure 2.1.9

on page 62. Let the frame O_1 rotate relative to the frame O so that the corresponding rotation vector $\boldsymbol{\omega}$ is fixed in the frame O. Because of this rotation, the basis vectors in O_1 depend on time when considered in the frame O: $\hat{\boldsymbol{\imath}}_1 = \hat{\boldsymbol{\imath}}_1(t)$, $\hat{\boldsymbol{\jmath}}_1 = \hat{\boldsymbol{\jmath}}_1(t)$, $\hat{\boldsymbol{\kappa}}_1 = \hat{\boldsymbol{\kappa}}_1(t)$. By (2.1.18),

$$d\hat{\boldsymbol{\imath}}_1/dt = \boldsymbol{\omega} \times \hat{\boldsymbol{\imath}}_1, \quad d\hat{\boldsymbol{\jmath}}_1/dt = \boldsymbol{\omega} \times \hat{\boldsymbol{\jmath}}_1, \quad d\hat{\boldsymbol{\kappa}}_1/dt = \boldsymbol{\omega} \times \hat{\boldsymbol{\kappa}}_1. \tag{2.1.19}$$

Denote by $\boldsymbol{r}_0(t)$ and $\boldsymbol{r}_1(t)$ the position vectors of the point mass in O and O_1, respectively. If P is the position of the point mass, then $\boldsymbol{r}_0(t) = \overrightarrow{OP}$, $\boldsymbol{r}_1(t) = \overrightarrow{O_1P}$, and, with $O = O_1$, we have $\boldsymbol{r}_0(t) = \boldsymbol{r}_1(t)$ for all t. Still, the time derivatives of the vectors are different: $\dot{\boldsymbol{r}}_0(t) \neq \dot{\boldsymbol{r}}_1(t)$ because of the rotation of the frames. Indeed,

$$\begin{aligned} \boldsymbol{r}_1(t) &= x_1(t)\,\hat{\boldsymbol{\imath}}_1 + y_1(t)\,\hat{\boldsymbol{\jmath}}_1 + z_1(t)\,\hat{\boldsymbol{\kappa}}_1, \\ \boldsymbol{r}_0(t) &= x_1(t)\,\hat{\boldsymbol{\imath}}_1(t) + y_1(t)\,\hat{\boldsymbol{\jmath}}_1(t) + z_1(t)\,\hat{\boldsymbol{\kappa}}_1(t); \end{aligned} \tag{2.1.20}$$

recall that the vectors $\hat{\boldsymbol{\imath}}_1$, $\hat{\boldsymbol{\jmath}}_1$, $\hat{\boldsymbol{\kappa}}_1$ are fixed relative to O_1, but are moving relative to O. Let us differentiate both equalities in (2.1.20) with respect to time t. For the computations of $\dot{\boldsymbol{r}}_1(t)$, the basis vectors are constants. For the computations of $\dot{\boldsymbol{r}}_0(t)$, we use the product rule (1.3.4) and the relations (2.1.19). The result is

$$\dot{\boldsymbol{r}}_0(t) = \dot{\boldsymbol{r}}_1(t) + \boldsymbol{\omega} \times \boldsymbol{r}_1(t). \tag{2.1.21}$$

EXERCISE 2.1.7.^{C} *Verify (2.1.21).*

There is nothing in the derivation of (2.1.21) that requires us to treat $\boldsymbol{r}_0$ as a position vector of a point. Accordingly, an alternative form of (2.1.21) can be stated as follows. Introduce the notations D_0 and D_1 for the time derivatives in the frames O and O_1, respectively. Then, for every vector function $\boldsymbol{R} = \boldsymbol{R}(t)$, the derivation of (2.1.21) yields

$$D_0\boldsymbol{R}(t) = D_1\boldsymbol{R}(t) + \boldsymbol{\omega} \times \boldsymbol{R}(t). \tag{2.1.22}$$

Relation (2.1.21) is a particular case of (2.1.22), when $\boldsymbol{R}$ is the position vector of the point. We now use (2.1.22) with $\boldsymbol{R} = \dot{\boldsymbol{r}}_0$, the velocity of the point in the fixed frame, to get the relation between the accelerations. Then (i) $D_0\dot{\boldsymbol{r}}_0 = \ddot{\boldsymbol{r}}_0$; (ii) by (2.1.21), $D_1\dot{\boldsymbol{r}}_0 = \ddot{\boldsymbol{r}}_1 + \boldsymbol{\omega} \times \dot{\boldsymbol{r}}_1$; (iii) also by (2.1.21) $\boldsymbol{\omega} \times \dot{\boldsymbol{r}}_0 = \boldsymbol{\omega} \times (\dot{\boldsymbol{r}}_1 + \boldsymbol{\omega} \times \boldsymbol{r}_1)$. Collecting the terms in (2.1.22),

$$\ddot{\boldsymbol{r}}_0 = \ddot{\boldsymbol{r}}_1 + 2\,\boldsymbol{\omega} \times \dot{\boldsymbol{r}}_1 + \boldsymbol{\omega} \times (\boldsymbol{\omega} \times \boldsymbol{r}_1). \tag{2.1.23}$$

Therefore, the acceleration in the fixed frame has three components: the acceleration $\ddot{\boldsymbol{r}}_1$ in the moving frame, the `Coriolis acceleration` $\boldsymbol{a}_{cor} =$

$2\boldsymbol{\omega} \times \dot{\boldsymbol{r}}_1$, and the *centripetal acceleration* $\boldsymbol{a}_c = \boldsymbol{\omega} \times (\boldsymbol{\omega} \times \boldsymbol{r}_1)$. Note that $\boldsymbol{a}_c$ is orthogonal to both $\boldsymbol{\omega}$ and $\boldsymbol{\omega} \times \boldsymbol{r}_1$.

Assume that the fixed frame O is inertial, and let $\boldsymbol{F}$ be the force acting on the point mass m in O. By Newton's Second Law, we have $m\ddot{\boldsymbol{r}}_0 = \boldsymbol{F}$ in the inertial frame O and, by (2.1.23),

$$m\ddot{\boldsymbol{r}}_1 = \boldsymbol{F} - 2m\,\boldsymbol{\omega} \times \dot{\boldsymbol{r}}_1 - m\,\boldsymbol{\omega} \times (\boldsymbol{\omega} \times \boldsymbol{r}_1) \qquad (2.1.24)$$

in the rotating frame O_1. Similar to (2.1.16), *inertial forces* appear as corrections to Newton's Second Law in the non-inertial frame O_1. There are two such forces in (2.1.24): the *Coriolis force* $\boldsymbol{F}_{cor} = -2m\,\boldsymbol{\omega} \times \dot{\boldsymbol{r}}_1$ and the *centrifugal force* $\boldsymbol{F}_c = -m\,\boldsymbol{\omega} \times (\boldsymbol{\omega} \times \boldsymbol{r}_1)$.

As an example illustrating the relation (2.1.23), consider a point mass m moving on the surface of the Earth along a meridian (great circle through the poles) with constant angular speed γ; the axis of rotation goes through the North and South Poles. We place the origins O and O_1 of the fixed and rotating frames at the center of the Earth and assume that the frame O_1 is rotating with the Earth. (Figure 2.1.5).

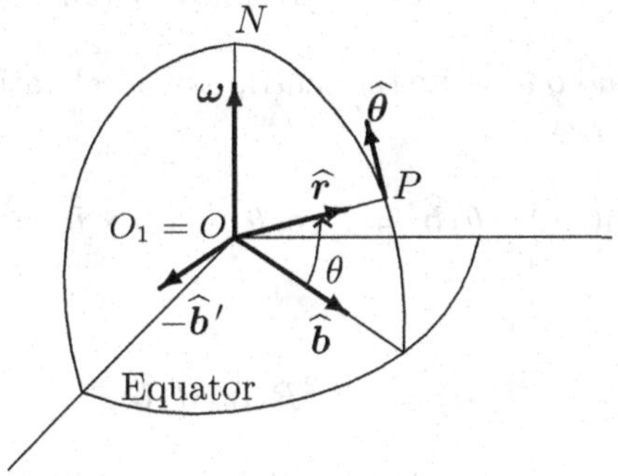

Fig. 2.1.5 Motion Along a Meridian

Denote by P the current position of the point, and consider the plane (NOP) Relative to the Earth, that is, in the frame O_1, the plane (NOP) is fixed, and the motion of m is a simple circular rotation in this plane with constant angular speed γ so that $\theta(t) = \gamma t$. Relative to the fixed frame O, the plane (NOP) is rotating, and the rotation vector is $\boldsymbol{\omega}$. We will determine the three components of the acceleration of the point in the frame O according to (2.1.23).

Introduce the polar coordinate vectors $\widehat{\boldsymbol{r}}, \widehat{\boldsymbol{\theta}}$ in the plane (NOP). By

(1.3.27), page 36, we find $\boldsymbol{a}_\theta = \boldsymbol{0}$ and so

$$\ddot{\boldsymbol{r}}_1 = \boldsymbol{a}_r = -\gamma^2 R\,\widehat{\boldsymbol{r}}, \tag{2.1.25}$$

where R is the radius of the Earth.

Similarly, we use (1.3.28), page 36, to find the velocity $\dot{\boldsymbol{r}}_1$, of the point in the frame O_1: $\dot{\boldsymbol{r}}_1 = \gamma R\widehat{\boldsymbol{\theta}}$, and then the Coriolis acceleration in O is $\boldsymbol{a}_{cor} = 2\,\boldsymbol{\omega} \times \dot{\boldsymbol{r}}_1 = 2\gamma R\,\boldsymbol{\omega} \times \widehat{\boldsymbol{\theta}}$.

To evaluate the product $\boldsymbol{\omega} \times \widehat{\boldsymbol{\theta}}$, we need some additional constructions. Note that both $\boldsymbol{\omega}$ and $\widehat{\boldsymbol{\theta}}$ are in the plane (NOP). Let $\widehat{\boldsymbol{b}}$ be the unit vector that lies both in the (NOP) and the equator planes as shown in Figure (2.1.5). This vector is rotating with the plane (NOP) and therefore changes in time; by (2.1.18), $\widehat{\boldsymbol{b}}'(t) = \boldsymbol{\omega} \times \widehat{\boldsymbol{b}}(t)$. With this construction, the vector $\boldsymbol{\omega} \times \widehat{\boldsymbol{\theta}}$ is in the plane of the equator and has the same direction as $-\widehat{\boldsymbol{b}}'(t)$. By definition, $\widehat{\boldsymbol{\theta}}$ is orthogonal to $\widehat{\boldsymbol{r}}$. By construction, the vectors $\boldsymbol{\omega}$ and $\widehat{\boldsymbol{b}}$ are also orthogonal, and so the angle θ between the vectors $\widehat{\boldsymbol{b}}$ and $\widehat{\boldsymbol{r}}$ is equal to the angle between the vectors $\boldsymbol{\omega}$ and $\widehat{\boldsymbol{\theta}}$. Since $\|\widehat{\boldsymbol{b}}'(t)\| = \omega = \|\boldsymbol{\omega}\|$ and $\|\widehat{\boldsymbol{\theta}}\| = 1$, we find $\boldsymbol{\omega} \times \widehat{\boldsymbol{\theta}} = -\sin\theta\,\widehat{\boldsymbol{b}}'$, and therefore

$$\boldsymbol{a}_{cor}(t) = -2\gamma R \sin(\gamma t)\,\widehat{\boldsymbol{b}}'(t). \tag{2.1.26}$$

Finally, we use θ and $\widehat{\boldsymbol{b}}$ to write the centripetal acceleration $\boldsymbol{a}_c = \boldsymbol{\omega} \times (\boldsymbol{\omega} \times \boldsymbol{r})$ of the point. We have

$$\boldsymbol{\omega} \times \boldsymbol{r} = R\sin(\pi/2 - \theta)\,\widehat{\boldsymbol{b}}' = R\cos\theta\,\widehat{\boldsymbol{b}}', \quad \boldsymbol{\omega} \times \widehat{\boldsymbol{b}}' = -\omega^2\,\widehat{\boldsymbol{b}}. \tag{2.1.27}$$

Hence,

$$\boldsymbol{a}_c = -(\omega^2 R \cos\gamma t)\,\widehat{\boldsymbol{b}}. \tag{2.1.28}$$

Combining (2.1.25), (2.1.26), and (2.1.28) in (2.1.23), we find the total acceleration $\ddot{\boldsymbol{r}}(t)$ of the point mass in the fixed frame O:

$$\ddot{\boldsymbol{r}}(t) = -\gamma^2 R\,\widehat{\boldsymbol{r}}(t) - 2\gamma R\sin(\gamma t)\,\widehat{\boldsymbol{b}}'(t) - \omega^2 R\cos(\gamma t)\,\widehat{\boldsymbol{b}}(t). \tag{2.1.29}$$

EXERCISE 2.1.8.^{C} *Verify all the equalities in (2.1.27).*

We now use (2.1.24) to study the dynamics of the point mass m moving on (or close to) the surface of the Earth. Let $\boldsymbol{r}_1 = \boldsymbol{r}_1(t)$ be the trajectory of the point in the frame O_1. Unlike the discussion leading to (2.1.29), we will no longer assume that the point moves along a meridian. Suppose that O is an inertial frame. If $\boldsymbol{F}$ is the force acting on m in the frame O, then

Newton's Second Law implies $m\,\ddot{\boldsymbol r} = \boldsymbol F$. By (2.1.24),

$$m\,\ddot{\boldsymbol r}_1 = \boldsymbol F - 2m\,\boldsymbol\omega \times \dot{\boldsymbol r}_1 - m\,\boldsymbol\omega \times (\boldsymbol\omega \times \boldsymbol r_1). \tag{2.1.30}$$

The force $\boldsymbol F$ is the sum of the Earth's gravitational force $\boldsymbol F_G = -(Km/\|\boldsymbol r_1\|^2)\,\widehat{\boldsymbol r}_1$ and the net "propulsive" force $\boldsymbol F_P(t)$ that ensures the motion of the point. If we prescribe the trajectory $\boldsymbol r_1(t)$ of m in the frame O_1, then the force $\boldsymbol F_P$ necessary to produce this trajectory is given by

$$\boldsymbol F_P(t) = m\,\ddot{\boldsymbol r}_1 + 2m\omega\,\widehat{\boldsymbol\omega} \times \dot{\boldsymbol r}_1 + m\,\boldsymbol\omega \times (\boldsymbol\omega \times \boldsymbol r_1) - \boldsymbol F_G. \tag{2.1.31}$$

Conversely, if the force $\boldsymbol F_P = \boldsymbol F_P(t)$ is specified, then the resulting trajectory is determined by solving (2.1.30) with the corresponding initial conditions $\boldsymbol r_1(0)$, $\dot{\boldsymbol r}_1(0)$ and with $\boldsymbol F = \boldsymbol F_G + \boldsymbol F_P$.

Note that (2.1.30) can be written as

$$m\,\ddot{\boldsymbol r}_1 = \boldsymbol F - m\,\boldsymbol a_{cor} - m\,\boldsymbol a_c = \boldsymbol F_G + \boldsymbol F_P + \boldsymbol F_{cor} + \boldsymbol F_c.$$

Both forces $\boldsymbol F_G$ and $\boldsymbol F_c$ act in the meridian plane (NOP). Indeed, by the Law of Universal Gravitation, the gravitational force $\boldsymbol F_G$ acts along the line OP, where P is the current location of the point. The centrifugal force $\boldsymbol F_c = m\,\boldsymbol\omega \times (\boldsymbol\omega \times \boldsymbol r_1)$ acts in the direction of the vector $\widehat{\boldsymbol b}$, as follows from the properties of the cross product; see Figure 2.1.5.

To analyze the effects of the Coriolis force $\boldsymbol F_{cor} = -m\,\boldsymbol a_{cor} = -2m\,\boldsymbol\omega \times \dot{\boldsymbol r}_1$ on the motion of the point mass, we again assume that the *point is in the Northern Hemisphere and moves north along a meridian* with constant angular speed. According to (2.1.26), the force $\boldsymbol F_{cor}$ is *perpendicular* to the meridian plane and is *acting in the eastward direction.* The magnitude of the force is proportional to $\sin\theta$, with the angle θ measured from the equator; see Figure 2.1.5. In particular, *the force is the strongest on the North pole, and the force is zero on the equator.* By (2.1.25) and (2.1.31), to maintain the motion along a meridian, the force $\boldsymbol F_P$ must have a westward component to balance $\boldsymbol F_{cor}$. Thus, to move due North, the mass must be subject to a propulsive force $\boldsymbol F_P$ having a westward component. The other component of $\boldsymbol F_P$ is in the meridian plane.

The following exercise analyzes the Coriolis force when the motion is *parallel* to the equator.

EXERCISE 2.1.9. *Let $\widehat{\boldsymbol\kappa}$ be the northward vector along the axis through the North and South poles. We assume that the Earth rotates around this axis, and denote by $\omega\,\widehat{\boldsymbol\kappa}$ the corresponding rotation vector. Let O be the fixed inertial frame and O_1, the frame rotating with the Earth; the origins* *of both frames are at the Earth center. Suppose that a point mass m is in the Northern hemisphere and moves East along a* `parallel` *(a circle cut on the surface of the Earth by a plane perpendicular to the line through the poles). Denote by $\dot{\boldsymbol{r}}_1$ the velocity of the point relative to the frame O_1 so that $\dot{\boldsymbol{r}}_1$ is perpendicular to the meridian plane; see Figure 2.1.6. By (2.1.24), the Coriolis force acting on the point mass is $\boldsymbol{F}_{cor} = -2m\omega\,\hat{\boldsymbol{\kappa}} \times \dot{\boldsymbol{r}}_1$.*

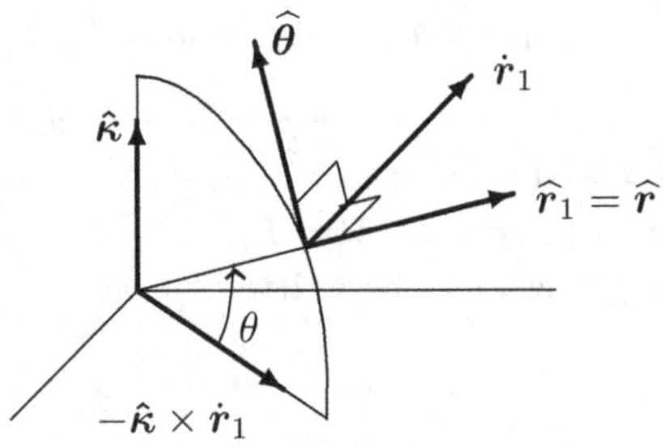

Fig. 2.1.6 Motion Along a Parallel

(a)[C] Assume that both $\|\dot{\boldsymbol{r}}_1\|$ and θ stay (approximately) constant during the travel and that there is zero propulsion force $\boldsymbol{F}_P$, as after a missile has been fired. Ignore air resistance.
(i) Show that the point mass is deflected to the South, and the magnitude of the deflection is $\omega\|\dot{\boldsymbol{r}}_1\|\sin\theta\, t^2$, where t is the time of travel. Hint: verify that $\boldsymbol{F}_{cor}\cdot\widehat{\boldsymbol{\theta}} = -2m\omega\|\dot{\boldsymbol{r}}_1\|\sin\theta$, and so $2m\omega\|\dot{\boldsymbol{r}}_1\|\sin\theta$ is the force pushing the point mass to the South. (ii) Suppose that $\theta = 41^\circ$, $\|\dot{\boldsymbol{r}}_1\| = 1000$ meters per second, and the point mass travels 1000 kilometers. Verify that the point mass will be deflected by about 50 kilometers to the South: if the target is due East, the missile will miss the target if aimed due East. Hint: $2\omega\sin\theta \approx 10^{-4}$.

(b)[A] Compute the deflection of the point mass taking into account the change of $\|\dot{\boldsymbol{r}}_1\|$ and θ caused by the Coriolis force.

We now summarize the effects of the Coriolis force on the motion of a point mass near the Earth.

- The force is equal to zero on the equator and is the strongest on the poles.
- For motion in the Northern Hemisphere:

Direction of Motion	Deflection of trajectory
North	East
South	West
East	South
West	North

- For motion in the Southern Hemisphere

Direction of Motion	Deflection of trajectory
North	West
South	East
East	North
West	South

EXERCISE 2.1.10.^{B} *Verify the above properties of the Coriolis force. Hint: The particular shape of the trajectory does not matter; all you need is the direction of* $\dot{\boldsymbol{r}}_1$, *similar to Exercise 2.1.9.*

EXERCISE 2.1.11. ^{A} *Because of the Coriolis force, an object dropped down from a high building does not fall along a straight vertical line and lands to the side. Disregarding the air resistance, compute the direction and magnitude of this deviation for an object dropped from the top of the Empire State Building (or from another tall structure of your choice).*

One of the most famous illustrations of the Coriolis force is the FOUCAULT PENDULUM, named after its creator, the French scientist JEAN BERNARD LEON FOUCAULT (1819–1869). Foucault was looking for an easy demonstration of the Earth's rotation around its axis, and around 1850 came up with the idea of a pendulum. He started with a small weight on a 6 feet long wire in his cellar, and gradually increased both the weight and the length of the wire. He also found a better location to conduct his experiment. The culmination was the year 1851, when he built a pendulum consisting of a 67 meter-long wire and a 28 kg weight swinging through a three-meter arc. The other end of the wire was attached to the dome of the Paris Pantheon and kept swinging via a special mechanism to compensate for the air resistance and to allow the swing in any vertical plane. Because of the rotation of the Earth around its axis, the Coriolis force was turning the plane of the swing by about 270 degrees every 24 hours, in the clockwise direction as seen from above.

EXERCISE 2.1.12.^{C} *Draw a picture and convince yourself that, if the pen-*

dulum starts to swing in the Northern Hemisphere in the meridian plane, then the Coriolis force will tend to turn the plane of the swing clockwise as seen from above.

Let us perform a simplified analysis of the motion of the Foucault pendulum in the Northern Hemisphere, away from the North Pole and the equator. Assume that the pendulum starts to swing in a meridian plane. Figure 2.1.7 presents a (grossly out-of-scale) illustration, with the Earth's surface represented by the semi-circle. In reality, the length of the support and the height of the supporting point O_1 are much smaller than the radius R of the Earth, so that $|OO_1| \approx |OP_0| \approx R$. Also, the amplitude of the swing is small compared to the length of the support, so that $|O_1P_0| \approx |O_1P_N| = |O_1P_S|$, and the linear distance $|P_NP_S|$ is approximately equal to the length of the corresponding circular swing arc.

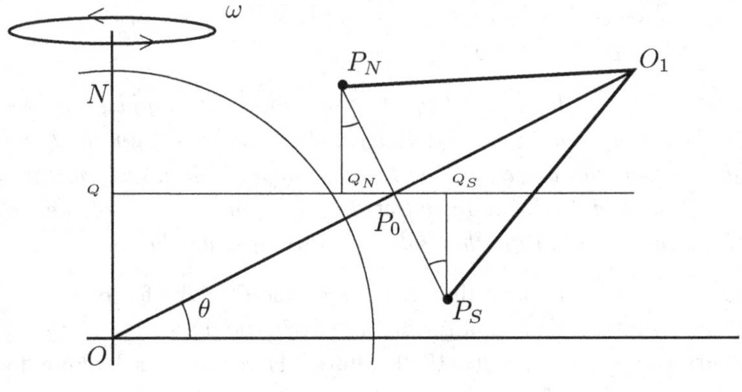

Fig. 2.1.7 Foucault Pendulum

The pendulum is suspended at the point O_1, and the points P_N, P_S are the two extreme positions of the weight. The angle θ is the *latitude* of the support point O_1. Denote the distance $|P_NP_S|$ by $2r$. As seen from the picture, the point P_N is closer to the axis ON of the rotation of the Earth than the point O_1, and the amount of this difference is $|Q_NP_0| = r\sin\theta$. Similarly, the point P_S is farther from the axis than O_1 by the same amount. Since the Earth is rotating around the axis ON with angular speed ω, the points P_N, O_1, and P_S will all move in the direction perpendicular to the meridian plane. The point P_N will move slower than O_1, and the point P_S, faster, causing the plane of the swing to turn. The speed of P_S relative to O_1 and of O_1 relative to P_N is $r\omega\sin\theta$. If we assume that these relative speeds stay the same throughout the revolution of the swing plane, then the weight will be rotating around the point P_0 with the speed $r\omega \sin\theta$, and we can find the time T of one complete turn. In time T, the weight will move the full circle of radius r, covering the distance $2\pi r$. The speed of this motion is $r\omega \sin\theta$, and so

$$T = \frac{2\pi r}{r\omega \sin\theta} = \frac{T_0}{\sin\theta}, \tag{2.1.32}$$

where $T_0 = 2\pi/\omega \approx 24$ hours is the period of Earth's revolution around its axis. The plane of the swing will rotate $2\pi \sin\theta$ radians every 24 hours. For the original Foucault pendulum in Paris, we have $\theta \approx 48.6°$, which results in $T = 32$ hours, or a 270° turn every 24 hours.

Note that the result (2.1.32) is true, at least formally, on the poles and on the equator. Still,

- On the poles, $T = 24$ hours as the Earth is turning under the pendulum, making a full turn every 24 hours.
- On the equator, where there is no Coriolis force, the points O_1, P_N, P_S are at the same distance from the axis ON (this is only approximately true if the swinging is not in the plane of the equator). As a result, the plane of the swing does not change: $T = +\infty$.

EXERCISE 2.1.13.[B] *Find the period T for the Foucault pendulum in your home town.*

The Coriolis force due to the Earth's rotation has greater effects on the motion than might be deduced from an intuitive approach based on the relative velocities of the moving object and the Earth. In particular, these effects must be taken into account when computing trajectories of long-range missiles. With all that, we must keep in mind that the effects of the Coriolis force due to the Earth rotation are noticeable only for large-scale motions. In particular, the Coriolis force contributes to the erosion of the river banks, but has nothing to do with the direction of water swirling in the toilet bowl.

The Coriolis force also influences the direction of the ATMOSPHERIC WINDS. This was first theorized in 1856 by the American meteorologist WILLIAM FERREL (1817–1891) and formalized in 1857 by the Dutch meteorologist CHRISTOPH HEINRICH DIEDRICH BUYS BALLOT (1817–1890).

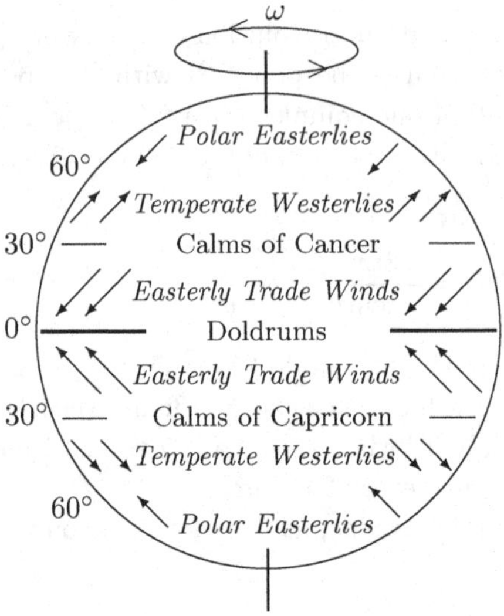

Fig. 2.1.8 Atmospheric Winds

The general wind pattern on the Earth is as follows (see Figure 2.1.8). Warm air rises vertically from the surface and is deflected by the Coriolis force, resulting in *easterly trade winds*, *temperate westerlies*, and *polar easterlies*. The deflection is to the right in the Northern Hemisphere and to the left in the Southern Hemisphere, so that the patterns in the two hemispheres are mirror images of each other. Three regions of relative calm form: the *doldrums* around the equator, *calms of Cancer* around the 30° parallel in the Northern Hemisphere, and *calms of Capricorn* around the 30° parallel in the Southern Hemisphere.

Let us discuss the formation of the easterly trade winds in the Northern Hemisphere. The Sun heats the surface of the Earth near the equator. The air near the equator also gets warm, becomes lighter, and moves up, creating the area of low pressure near the equator and causing the cooler air from the north to flow south. The flow of the cooler air from the North creates the area of low pressure at high altitudes, deflecting the rising warm air from the equator to the north. The Coriolis force deflects this flow to the East. At higher altitudes, the air cools down. Cooler, denser air descends around the 30° parallel and flows South back to the lower pressure area around the equator. The Coriolis force deflects this southward flow to the West. In the stationary regime, this circulation produces a steady wind from the North-East, the easterly trade winds.

EXERCISE 2.1.14.[A] *Explain the formation of the temperate westerlies and the polar easterlies.*

EXERCISE 2.1.15.[A] *The flight time from Los Angeles to Boston is usually different from the flight time from Boston to Los Angeles. Which flight takes longer? Which of the following factors contributes the most to this difference, and how: (a) The Earth's rotation under the airplane; (b) The Coriolis force acting on the airplane; (c) The atmospheric winds? Hint: If in doubt, check the schedules of direct flights between the two cities.*

The complete mathematical model of atmospheric physics is vastly more complicated and is outside the scope of this book; possible reference on the subject is the book *An Introduction to Dynamic Meteorology* by J. R. Holton, 2004, and some partial differential equations appearing in the modelling of flows of gases and liquids are discussed below in Section 6.3.5.

In 1963, while studying the differential equations of fluid convection, the American mathematician and meteorologist EDWARD NORTON LORENZ (b. 1917) discovered a chaotic behavior of the solution and a *strange attractor.* These Lorentz differential equations are a prime example of a chaotic flow. They also illustrate the intrinsic difficulty of accurate weather prediction. His book *The Nature And Theory of The General Circulation of The Atmosphere* 1967, is another standard reference in atmospheric physics.

2.1.4 *General Accelerating Frames*

The analysis in the previous section essentially relied on the equation (2.1.18) on page 51, which was derived for *uniformly* rotating frames. In what follows, we will use linear algebra to show that, with a proper definition of the vector $\boldsymbol{\omega}$, relation (2.1.18) continues to hold for arbitrary rotating frames.

Consider two cartesian coordinate systems: $(\hat{\boldsymbol{\imath}}, \hat{\boldsymbol{\jmath}}, \hat{\boldsymbol{k}})$ with origin O, and $(\hat{\boldsymbol{\imath}}_1, \hat{\boldsymbol{\jmath}}_1, \hat{\boldsymbol{k}}_1)$ with origin O_1. We assume that $O = O_1$; see Figure 2.1.9.

Consider a point P in $\mathbb{R}^3$. This point has coordinates (x, y, z) in $(\hat{\boldsymbol{\imath}}, \hat{\boldsymbol{\jmath}}, \hat{\boldsymbol{k}})$ and (x_1, y_1, z_1) in $(\hat{\boldsymbol{\imath}}_1, \hat{\boldsymbol{\jmath}}_1, \hat{\boldsymbol{k}}_1)$. Then

$$x\,\hat{\boldsymbol{\imath}} + y\,\hat{\boldsymbol{\jmath}} + z\,\hat{\boldsymbol{k}} = x_1\,\hat{\boldsymbol{\imath}}_1 + y_1\,\hat{\boldsymbol{\jmath}}_1 + z_1\,\hat{\boldsymbol{k}}_1. \tag{2.1.33}$$

We now take the dot product of both sides of (2.1.33) with $\hat{\boldsymbol{\imath}}$ to get

$$x = x_1(\hat{\boldsymbol{\imath}}_1 \cdot \hat{\boldsymbol{\imath}}) + y_1(\hat{\boldsymbol{\jmath}}_1 \cdot \hat{\boldsymbol{\imath}}) + z_1(\hat{\boldsymbol{k}}_1 \cdot \hat{\boldsymbol{\imath}}). \tag{2.1.34}$$

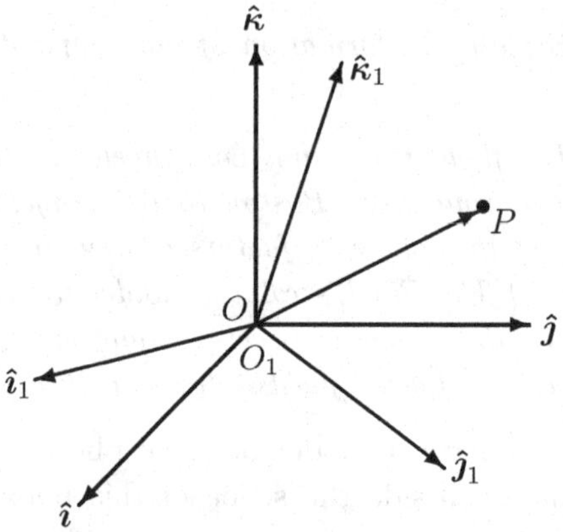

Fig. 2.1.9 3-D Rotation of Frames

Similarly, we take the dot product of both sides of (2.1.33) with $\hat{\jmath}$:

$$y = x_1(\hat{\imath}_1 \cdot \hat{\jmath}) + y_1(\hat{\jmath}_1 \cdot \hat{\jmath}) + z_1(\hat{\kappa}_1 \cdot \hat{\jmath}) \tag{2.1.35}$$

and with $\hat{\kappa}$:

$$z = x_1(\hat{\imath}_1 \cdot \hat{\kappa}) + y_1(\hat{\jmath}_1 \cdot \hat{\kappa}) + z_1(\hat{\kappa}_1 \cdot \hat{\kappa}). \tag{2.1.36}$$

The three equations (2.1.34)–(2.1.36) can be written as a single matrix - vector equation,

$$\begin{pmatrix} x \\ y \\ z \end{pmatrix} = \begin{pmatrix} \hat{\imath}_1 \cdot \hat{\imath} & \hat{\jmath}_1 \cdot \hat{\imath} & \hat{\kappa}_1 \cdot \hat{\imath} \\ \hat{\imath}_1 \cdot \hat{\jmath} & \hat{\jmath}_1 \cdot \hat{\jmath} & \hat{\kappa}_1 \cdot \hat{\jmath} \\ \hat{\imath}_1 \cdot \hat{\kappa} & \hat{\jmath}_1 \cdot \hat{\kappa} & \hat{\kappa}_1 \cdot \hat{\kappa} \end{pmatrix} \begin{pmatrix} x_1 \\ y_1 \\ z_1 \end{pmatrix}. \tag{2.1.37}$$

Consider the matrix

$$U = \begin{pmatrix} \hat{\imath}_1 \cdot \hat{\imath} & \hat{\jmath}_1 \cdot \hat{\imath} & \hat{\kappa}_1 \cdot \hat{\imath} \\ \hat{\imath}_1 \cdot \hat{\jmath} & \hat{\jmath}_1 \cdot \hat{\jmath} & \hat{\kappa}_1 \cdot \hat{\jmath} \\ \hat{\imath}_1 \cdot \hat{\kappa} & \hat{\jmath}_1 \cdot \hat{\kappa} & \hat{\kappa}_1 \cdot \hat{\kappa} \end{pmatrix}. \tag{2.1.38}$$

EXERCISE 2.1.16. *(a)*^{C} *Verify that the matrix U is orthogonal, that is $UU^T = U^TU = I$. Hint:* $1 = \hat{\imath}_1 \cdot \hat{\imath}_1 = (\hat{\imath} \cdot \hat{\imath}_1)^2 + (\hat{\jmath} \cdot \hat{\imath}_1)^2 + (\hat{\kappa} \cdot \hat{\imath}_1)^2$, $0 = \hat{\imath}_1 \cdot \hat{\jmath}_1 = (\hat{\imath}_1 \cdot \hat{\imath})(\hat{\jmath}_1 \cdot \hat{\imath}) + (\hat{\imath}_1 \cdot \hat{\jmath})(\hat{\jmath}_1 \cdot \hat{\jmath}) + (\hat{\imath}_1 \cdot \hat{\kappa})(\hat{\jmath}_1 \cdot \hat{\kappa})$. *(b)*^{A} *Verify that the determinant of the matrix U is equal to* 1. *(c)*^{A} *Verify that the matrix U is a representation,* in the basis $(\hat{\imath}_1, \hat{\jmath}_1, \hat{\kappa}_1)$, *of an orthogonal transformation (see Exercise 8.1.4, page 453, in Appendix). This transformation rotates the frame O_1 so that $(\hat{\imath}_1, \hat{\jmath}_1, \hat{\kappa}_1)$ moves into $(\hat{\imath}, \hat{\jmath}, \hat{\kappa})$. (d)*^{A} *Verify* *that the matrix* U^T *is a representation,* in the basis $(\hat{\imath}, \hat{\jmath}, \hat{\kappa})$, *of an orthogonal transformation that rotates the space so that* $(\hat{\imath}, \hat{\jmath}, \hat{\kappa})$ *moves into* $(\hat{\imath}_1, \hat{\jmath}_1, \hat{\kappa}_1)$.

Now assume that the picture on Figure 2.1.9 is changing in time as follows:

- Frame O and its coordinate system $(\hat{\imath}, \hat{\jmath}, \hat{\kappa})$ are fixed (not moving).
- $O_1(t) = O$ for all t.
- The coordinate system $(\hat{\imath}_1, \hat{\jmath}_1, \hat{\kappa}_1)$ in frame O_1 is moving (rotating) relative to $(\hat{\imath}, \hat{\jmath}, \hat{\kappa})$.
- The point P is fixed in frame O_1 relative to $(\hat{\imath}_1, \hat{\jmath}_1, \hat{\kappa}_1)$.

Thus, x_1, y_1, z_1 are constants and P is rotating in the $(\hat{\imath}, \hat{\jmath}, \hat{\kappa})$ frame. Then (2.1.33) becomes

$$x(t)\,\hat{\imath} + y(t)\,\hat{\jmath} + z(t)\,\hat{\kappa} = x_1\,\hat{\imath}_1(t) + y_1\,\hat{\jmath}_1(t) + z_1\,\hat{\kappa}_1(t),$$

Define the matrix $U = U(t)$ according to (2.1.38), and assume that the entries of the matrix U are differentiable functions of time; it is a reasonable assumption if the rotation is without jerking. Since $U(t)U^T(t) = I$ for all t, it follows that $d/dt(UU^T) = 0$, the zero matrix. The product rule applies to matrix differentiation and therefore

$$\dot{U}U^T + U\dot{U}^T = \dot{U}U^T + (\dot{U}U^T)^T = 0,$$

which means that $\Omega(t) = \dot{U}(t)U^T(t)$ is antisymmetric, that is, has the form

$$\Omega(t) = \begin{pmatrix} 0 & -\omega_3(t) & \omega_2(t) \\ \omega_3(t) & 0 & -\omega_1(t) \\ -\omega_2(t) & \omega_1(t) & 0 \end{pmatrix}.$$

We use the entries of the matrix $\Omega(t)$ to define mathematically the `instantaneous rotation vector` in the fixed frame O:

$$\boldsymbol{\omega}(t) = \omega_1(t)\,\hat{\imath} + \omega_2(t)\,\hat{\jmath} + \omega_3(t)\,\hat{\kappa}. \tag{2.1.39}$$

EXERCISE 2.1.17^C *(a) Verify that, for every vector* $\boldsymbol{R} = R_1\,\hat{\imath} + R_2\,\hat{\jmath} + R_3\,\hat{\kappa}$ *and each* t,

$$\Omega(t)\,\boldsymbol{R} = \boldsymbol{\omega}(t) \times \boldsymbol{R}. \tag{2.1.40}$$

Hint: direct computation. (b) *Consider the vector* $\overrightarrow{OP} = \boldsymbol{r}_0(t) = x(t)\,\hat{\boldsymbol{\imath}} + y(t)\,\hat{\boldsymbol{\jmath}} + z(t)\,\hat{\boldsymbol{\kappa}}$ *rotating in the frame O. Verify that, with the above definition of* $\boldsymbol{\omega}$, *we have*

$$\dot{\boldsymbol{r}}_0(t) = \boldsymbol{\omega}(t) \times \boldsymbol{r}_0(t). \tag{2.1.41}$$

Hint: write relation (2.1.37) as $\boldsymbol{r}_0(t) = U(t)\widetilde{\boldsymbol{r}}_1$, *where* $\widetilde{\boldsymbol{r}}_1 = x_1\,\hat{\boldsymbol{\imath}} + y_1\,\hat{\boldsymbol{\jmath}} + z_1\,\hat{\boldsymbol{\kappa}}$. *Then* $\widetilde{\boldsymbol{r}}_1 = U^T(t)\boldsymbol{r}_0(t)$ *and* $\dot{\boldsymbol{r}}_0(t) = \dot{U}(t)\widetilde{\boldsymbol{r}}_1$.

Note that (2.1.41) agrees with (2.1.18) on page 51 when $\boldsymbol{\omega}(t)$ is constant and justifies the above definition of $\boldsymbol{\omega}(t)$ as a rotation vector.

Denote by $\boldsymbol{r}_0(t)$ the position vector of a point P in the frame O, and by $\boldsymbol{r}_1(t)$, the position of the same point in the frame O_1. Since the frames have the same origin, we have $\boldsymbol{r}_0(t) = \boldsymbol{r}_1(t)$ for all t. On the other hand, because of the relative rotation of the frames, the values of $\dot{\boldsymbol{r}}_0(t)$ and $\dot{\boldsymbol{r}}_1(t)$ are different. As we did earlier on page 52, denote by D_0 and D_1 the derivatives with respect to time in the frames O and O_1, respectively. If the point P is fixed in the rotating frame O_1 and $\overrightarrow{O_1P} = \boldsymbol{r}_1 = x_1\,\hat{\boldsymbol{\imath}}_1 + y_1\,\hat{\boldsymbol{\jmath}}_1 + z_1\,\hat{\boldsymbol{\kappa}}_1$ is the position vector of P in O_1, then $D_1\boldsymbol{r}_0(t) = \dot{\boldsymbol{r}}_1(t) = \boldsymbol{0}$, and (2.1.41) implies $D_0\boldsymbol{r}_0(t) = \dot{\boldsymbol{r}}_0(t) = \boldsymbol{\omega} \times \boldsymbol{r}_0(t)$.

Similar to the derivation of (2.1.21), we can show that if the point P moves relative to the frame O_1, then

$$\dot{\boldsymbol{r}}_0(t) = \dot{\boldsymbol{r}}_1(t) + \boldsymbol{\omega} \times \boldsymbol{r}_0(t). \tag{2.1.42}$$

Therefore, for every vector $\boldsymbol{R} = \boldsymbol{R}(t)$, expressed as functions in frames O and O_1, both denoted by $\boldsymbol{R}$

$$D_0\boldsymbol{R}(t) = D_1\boldsymbol{R}(t) + \boldsymbol{\omega}(t) \times \boldsymbol{R}(t). \tag{2.1.43}$$

EXERCISE 2.1.18.^{C} *Verify (2.1.43). Hint: write* $\boldsymbol{R}(t) = x(t)\,\hat{\boldsymbol{\imath}}_1(t) + y(t)\,\hat{\boldsymbol{\jmath}}_1(t) + z(t)\,\hat{\boldsymbol{\kappa}}_1(t)$ *and differentiate this equality using the product rule. Since the vectors* $\hat{\boldsymbol{\imath}}_1, \hat{\boldsymbol{\jmath}}_1, \hat{\boldsymbol{\kappa}}_1$ *are fixed in the frame* O_1, *you can use equality (2.1.41) to compute the time derivatives of these vectors. Also, by definition,* $D_1\boldsymbol{R}(t) = \dot{x}(t)\,\hat{\boldsymbol{\imath}}_1(t) + \dot{y}(t)\,\hat{\boldsymbol{\jmath}}_1(t) + \dot{z}(t)\,\hat{\boldsymbol{\kappa}}_1(t)$.

Remark 2.1 *Let us stress that the vector* $\boldsymbol{r}_0(t) = x(t)\,\hat{\boldsymbol{\imath}} + y(t)\,\hat{\boldsymbol{\jmath}} + z(t)\,\hat{\boldsymbol{\kappa}}$ *is the same as the vector* $\boldsymbol{r}_1(t) = x_1(t)\,\hat{\boldsymbol{\imath}}_1(t) + y_1(t)\,\hat{\boldsymbol{\jmath}}_1(t) + z_1(t)\,\hat{\boldsymbol{\kappa}}_1(t)$: *both are equal to* $\overrightarrow{OP}$ *even if the point P moves relative to the frame* O_1. *As a result,* $\boldsymbol{r}_0(t) \neq U(t)\,\boldsymbol{r}_1(t)$. *There is no contradiction with (2.1.37), because the components of the vectors* $\boldsymbol{r}_0$, $\boldsymbol{r}_1$ *are defined in different frames and cannot be related by a matrix-vector product. What does follow from (2.1.37) is* *the equality* $\boldsymbol{r}_0(t) = U(t)\,\widetilde{\boldsymbol{r}}_1(t)$, *where* $\widetilde{\boldsymbol{r}}_1(t) = x_1(t)\,\hat{\boldsymbol{\imath}} + y_1(t)\,\hat{\boldsymbol{\jmath}} + z_1(t)\,\hat{\boldsymbol{k}}$. *Also keep in mind that, despite the equality of the vectors* $\boldsymbol{r}_0(t) = \boldsymbol{r}_1(t)$, *the curve* $\mathcal{C}$ *defined by* $\boldsymbol{r}_0$ *in the frame* O *is different from the curve* $\mathcal{C}_1$ *defined by* $\boldsymbol{r}_1$ *in the frame* O_1. *For example, if* P *is fixed in* O_1, *then* $\mathcal{C}_1$ *is jus a single point.*

EXERCISE 2.1.19.^{C} *Consider the special case of a uniform rotation of frame* O_1 *relative to frame* O *so that the origins of the frames coincide,* $\hat{\boldsymbol{k}} = \hat{\boldsymbol{k}}_1$, *and the rotation vector is* $\boldsymbol{\omega} = \omega_3\,\hat{\boldsymbol{k}}$. *Calculate* $U(t)$ *and* $\dot{U}(t)$. *Show that the matrix* $\Omega = \dot{U}(t)U^T(t)$ *has the form*

$$\Omega = \begin{pmatrix} 0 & -\omega_3 & 0 \\ \omega_3 & 0 & 0 \\ 0 & 0 & 0 \end{pmatrix}.$$

For the point P *fixed in the rotating frame and having the position vector in the fixed frame* $\boldsymbol{r}_0(t) = x(t)\,\hat{\boldsymbol{\imath}} + y(t)\,\hat{\boldsymbol{\jmath}} + z(t)\,\hat{\boldsymbol{k}}_1$ *show that*

$$\dot{\boldsymbol{r}}_0(t) = \Omega\,\boldsymbol{r}_0(t) = -\omega_3 y(t)\,\hat{\boldsymbol{\imath}} + \omega_3 x(t)\,\hat{\boldsymbol{\jmath}} = \boldsymbol{\omega} \times \boldsymbol{r}_0(t).$$

As a result, you recover relation (2.1.18) we derived geometrically on page 51.

To continue our analysis of rotation, assume that the vector function $\boldsymbol{\omega} = \boldsymbol{\omega}(t)$ is differentiable in t. Then we can set $\boldsymbol{R} = \dot{\boldsymbol{r}}_0 = D_0\boldsymbol{r}_0$ in (2.1.43) and use (2.1.42) to derive the relation between the accelerations of the point in the two frames:

$$\ddot{\boldsymbol{r}}_0(t) = \ddot{\boldsymbol{r}}_1(t) + 2\,\boldsymbol{\omega}(t)\times\dot{\boldsymbol{r}}_1(t) + \boldsymbol{\omega}(t)\times\big(\boldsymbol{\omega}(t)\times\boldsymbol{r}_0(t)\big) + \dot{\boldsymbol{\omega}}(t)\times\boldsymbol{r}_0(t); \quad (2.1.44)$$

as before, $\boldsymbol{r}_0$ and $\boldsymbol{r}_1$ are the position vectors of the point in the frames O and O_1, respectively.

EXERCISE 2.1.20.^{C} *(a) Verify (2.1.44). Hint:* $\ddot{\boldsymbol{r}}_0(t) = D_0\dot{\boldsymbol{r}}_0(t) = D_0(\dot{\boldsymbol{r}}_1 + \boldsymbol{\omega}\times\boldsymbol{r}_0) = D_1\dot{\boldsymbol{r}}_1 + \boldsymbol{\omega}\times\dot{\boldsymbol{r}}_1 + \dot{\boldsymbol{\omega}}\times\boldsymbol{r}_0 + \boldsymbol{\omega}\times(\dot{\boldsymbol{r}}_1 + \boldsymbol{\omega}\times\boldsymbol{r}_0)$. *Note that both* $\boldsymbol{\omega}$ *and* $\boldsymbol{r}_0$ *are defined in the same frame* O, *so the product rule (1.3.6), page 26, applies. (b) Verify that (2.1.44) can be written as*

$$\ddot{\boldsymbol{r}}_0(t) = \ddot{\boldsymbol{r}}_1(t) + 2\,\boldsymbol{\omega}(t)\times\dot{\boldsymbol{r}}_1(t) + \boldsymbol{\omega}(t)\times\big(\boldsymbol{\omega}(t)\times\boldsymbol{r}_1(t)\big) + \dot{\boldsymbol{\omega}}(t)\times\boldsymbol{r}_1(t). \quad (2.1.45)$$

Finally, assume that the point O_1 is moving relative to O so that the function $\boldsymbol{r}_{01}(t) = \overrightarrow{OO_1}$ is twice continuously differentiable. Then we have $\boldsymbol{r}(t) = \boldsymbol{r}_{01}(t) + \boldsymbol{r}_1(t)$. Consider the parallel translation of the frame O with the origin O' at O_1, and define the rotation vector $\boldsymbol{\omega}$ to describe the rotation of the frame O_1 relative to this translated frame O'. We can now combine relation (2.1.44) for rotation with relation (2.1.12) on page 47 for parallel translation to get

$$\begin{aligned}\ddot{\boldsymbol{r}}_0(t) &= \ddot{\boldsymbol{r}}_{01}(t) + \ddot{\boldsymbol{r}}_1(t) \\ &\quad + 2\,\boldsymbol{\omega}(t) \times \dot{\boldsymbol{r}}_1(t) + \boldsymbol{\omega}(t) \times \big(\boldsymbol{\omega}(t) \times \boldsymbol{r}_1(t)\big) + \dot{\boldsymbol{\omega}}(t) \times \boldsymbol{r}_1(t).\end{aligned} \tag{2.1.46}$$

EXERCISE 2.1.21^C *Verify (2.1.46). Hint: Apply (2.1.45) to* $\boldsymbol{r}_1$*, replacing frame* O *with* O'.

Suppose that the frame O is inertial, and a force $\boldsymbol{F}$ is acting on the point mass m. Then, by Newton's Second Law, $m\ddot{\boldsymbol{r}}_0(t) = \boldsymbol{F}$; to simplify the notations we will no longer write the time dependence explicitly. By (2.1.46),

$$m\,\ddot{\boldsymbol{r}}_1 = \boldsymbol{F} - m\,\ddot{\boldsymbol{r}}_{01} - 2m\,\boldsymbol{\omega} \times \dot{\boldsymbol{r}}_1 - m\,\boldsymbol{\omega} \times (\boldsymbol{\omega} \times \boldsymbol{r}_1) - m\,\dot{\boldsymbol{\omega}} \times \boldsymbol{r}_1. \tag{2.1.47}$$

As before in (2.1.13), page 47, and in (2.1.24), page 53, we have several corrections to Newton's Second Law in the non-inertial frame O_1. These corrections are the `translational acceleration force` $\boldsymbol{F}_{ta} = -m\,\ddot{\boldsymbol{r}}_{01}$, the `Coriolis force` $\boldsymbol{F}_{cor} = -2m\,\boldsymbol{\omega} \times \dot{\boldsymbol{r}}_1$, the `centrifugal force` $\boldsymbol{F}_c = -m\,\boldsymbol{\omega}\times(\boldsymbol{\omega}\times\boldsymbol{r}_1)$, and the `angular acceleration force` $\boldsymbol{F}_{aa} = -m\,\dot{\boldsymbol{\omega}}\times\boldsymbol{r}_1$.

2.2 Systems of Point Masses

The motion of a system of point masses can be decomposed into the motion of one point, the *center of mass*, and the rotational motion of the system around the center of mass. In what follows, we study this decomposition, first for a finite collection of point masses, and then for certain infinite collections, namely, rigid bodies.

2.2.1 *Non-Rigid Systems of Points*

Let S be a system of n point masses, $m_1, \dots, m_n$, and $M = \sum_{j=1}^{n} m_j$, the total mass of S. We assume that the system is *non-rigid*, that is, the distances between the points can change. Denote by $\boldsymbol{r}_j$ the position vector of m_j in some frame O. By definition, the `center of mass` (CM) of S is the point with position vector

$$\boldsymbol{r}_{CM} = \frac{1}{M}\sum_{j=1}^{n} m_j\,\boldsymbol{r}_j. \tag{2.2.1}$$

We will see that some information about the motion of S can be obtained by considering a single point mass M with position vector $\boldsymbol{r}_{CM}$.

EXERCISE 2.2.1. ^{C} *Verify that a change of the reference point O does not change either the location in space of the center of mass or formula (2.2.1) for determining the location: if O' is any other frame and $\widetilde{\boldsymbol{r}}_j$ is the position vector of m_j in O', then the position vector of CM in O' is $\widetilde{\boldsymbol{r}}_{CM} = (1/M)\sum_{j=1}^{n} m_j\,\widetilde{\boldsymbol{r}}_j$. Hint: $\widetilde{\boldsymbol{r}}_j = \boldsymbol{r}_j + \overrightarrow{O'O}$ and so $\widetilde{\boldsymbol{r}}_{CM} = \boldsymbol{r}_{CM} + \overrightarrow{O'O}$, which is the same point in space.*

EXERCISE 2.2.2. *(a)^B Show that the center of mass for three equal masses not on the same line is at the intersection of the medians of the corresponding triangle. (b)^A Four equal masses are at the vertices of a regular tetrahedron. Locate the center of mass.*

The velocity and acceleration of the center of mass are $\dot{\boldsymbol{r}}_{CM}$ and $\ddot{\boldsymbol{r}}_{CM}$, respectively. Differentiating (2.2.1) with respect to t, we obtain the relations

$$\dot{\boldsymbol{r}}_{CM} = \frac{1}{M}\sum_{j=1}^{n} m_j\,\dot{\boldsymbol{r}}_j, \tag{2.2.2}$$

$$\ddot{\boldsymbol{r}}_{CM} = \frac{1}{M}\sum_{j=1}^{n} m_j\,\ddot{\boldsymbol{r}}_j. \tag{2.2.3}$$

To study the motion of the center of mass, suppose that the reference frame O is inertial and denote by $\boldsymbol{F}_j$, $1 \le j \le n$, the force acting on the point mass m_j. Then $\boldsymbol{F}_j = m_j\,\ddot{\boldsymbol{r}}_j$, and, multiplying (2.2.3) by M, we get the relation

$$M\,\ddot{\boldsymbol{r}}_{CM} = \sum_{j=1}^{n} m_j\,\ddot{\boldsymbol{r}}_j = \sum_{j=1}^{m} \boldsymbol{F}_j = \boldsymbol{F}. \tag{2.2.4}$$

Equality (2.2.4) suggests that the total force $\boldsymbol{F}$ can be assumed to act on a point mass M at the position $\boldsymbol{r}_{CM}$. As a next step, we will study the structure of the force $\boldsymbol{F}$.

Typically, each $\boldsymbol{F}_j$ is a sum of an *external* force $\boldsymbol{F}_j^{(E)}$ from outside of S and an *internal* system force $\boldsymbol{F}_j^{(I)}$ exerted on m_j by the other $n-1$ point masses. Thus, $\boldsymbol{F}_j^{(I)} = \sum_{k \neq j} \boldsymbol{F}_{jk}^{(I)}$, where $\boldsymbol{F}_{jk}^{(I)}$ is the force exerted by m_k on m_j. Hence,

$$\boldsymbol{F} = \sum_{j=1}^{n} \boldsymbol{F}_j = \sum_{j=1}^{n} \boldsymbol{F}_j^{(E)} + \sum_{j=1}^{n} \boldsymbol{F}_j^{(I)} = \boldsymbol{F}^{(E)} + \boldsymbol{F}^{(I)}.$$

By Newton's Third Law, $\boldsymbol{F}_{jk}^{(I)} = -\boldsymbol{F}_{kj}^{(I)}$. It follows that $\boldsymbol{F}^{(I)} = \sum_{j=1}^{n} \boldsymbol{F}_j^{(I)} = \boldsymbol{0}$ and therefore, $\boldsymbol{F} = \sum_{j=1}^{n} \boldsymbol{F}_j^{(E)} = \boldsymbol{F}^{(E)}$. By (2.2.4), the motion of the center of mass is then determined by

$$M\,\ddot{\boldsymbol{r}}_{CM} = \boldsymbol{F}^{(E)}. \tag{2.2.5}$$

The (linear) `momentum` $\boldsymbol{p}_{CM}$ of the center of mass is, by definition,

$$\boldsymbol{p}_{CM} = M\,\dot{\boldsymbol{r}}_{CM}.$$

With this definition, equation (2.2.5) becomes $\dot{\boldsymbol{p}}_{CM} = \boldsymbol{F}^{(E)}$, and if the net external force $\boldsymbol{F}^{(E)}$ is zero, then $\boldsymbol{p}_{CM}$ is constant. By (2.2.2),

$$\boldsymbol{p}_{CM} = \sum_{j=1}^{n} m_j \dot{\boldsymbol{r}}_j = \sum_{j=1}^{n} \boldsymbol{p}_j, \tag{2.2.6}$$

where $\boldsymbol{p}_j = m_j\,\dot{\boldsymbol{r}}_j$ is the momentum of m_j. Thus, if $\boldsymbol{F}^{(E)} = \boldsymbol{0}$, then *the total linear momentum* $\boldsymbol{p}_S = \sum_{j=1}^{n} \boldsymbol{p}_j$ *of the system is conserved.*

Next, we consider the rotational motion of the system. By definition (see (2.1.4) on page 40), the angular momentum $\boldsymbol{L}_{O,j}$ of the point mass m_j about the reference point O is given by $\boldsymbol{L}_{O,j} = \boldsymbol{r}_j \times m_j\,\dot{\boldsymbol{r}}_j = \boldsymbol{r}_j \times \boldsymbol{p}_j$. Accordingly, we define the `angular momentum` $\boldsymbol{L}_O$ `of the system` S about O as the sum of the $\boldsymbol{L}_{O,j}$:

$$\boldsymbol{L}_O = \sum_{j=1}^{n} \boldsymbol{L}_{O,j} = \sum_{j=1}^{n} \boldsymbol{r}_j \times \boldsymbol{p}_j = \sum_{j=1}^{n} m_j\,\boldsymbol{r}_j \times \dot{\boldsymbol{r}}_j. \tag{2.2.7}$$

For the purpose of the definition, it is not necessary to assume that the frame O is inertial. Note that, unlike the relation (2.2.6) for the linear momentum, in general $\boldsymbol{L}_O \neq \boldsymbol{r}_{CM} \times M\,\dot{\boldsymbol{r}}_{CM}$.

From (2.2.7) it follows that

$$\frac{d\boldsymbol{L}_O}{dt} = \sum_{j=1}^{n} m_j \dot{\boldsymbol{r}}_j \times \dot{\boldsymbol{r}}_j + \sum_{j=1}^{n} m_j \boldsymbol{r}_j \times \ddot{\boldsymbol{r}}_j = \sum_{j=1}^{n} m_j \boldsymbol{r}_j \times \ddot{\boldsymbol{r}}_j,$$

since $\dot{\boldsymbol{r}}_j \times \dot{\boldsymbol{r}}_j = \boldsymbol{0}$. Now, we again assume that the frame O is inertial. Then $\boldsymbol{F}_j = m_j \ddot{\boldsymbol{r}}_j$ and $d\boldsymbol{L}_O/dt = \sum_{j=1}^{n} \boldsymbol{r}_j \times \boldsymbol{F}_j = \sum_{j=1}^{n} \boldsymbol{T}_j$, where $\boldsymbol{T}_j = \boldsymbol{r}_j \times \boldsymbol{F}_j$ is the torque about O of the force $\boldsymbol{F}_j$ acting on m_j. We define the *total torque* $\boldsymbol{T}_O = \sum_{j=1}^{n} \boldsymbol{T}_j$ and conclude that

$$\frac{d\boldsymbol{L}_O}{dt} = \boldsymbol{T}_O. \tag{2.2.8}$$

Equation (2.2.8) is an extension of (2.1.6), page 41, to finite systems of point masses. If $\boldsymbol{T}_O = \boldsymbol{0}$, then $\boldsymbol{L}_O$ is constant, that is, *angular momentum is conserved.*

In general, unlike the equation for the linear momentum (2.2.5), the torque $\boldsymbol{T}_O$ in (2.2.8) includes both the internal and external forces. If the internal forces are *central*, then only external forces appear in (2.2.8). Indeed, let us compute the total torque in the case of CENTRAL INTERNAL FORCES. The internal force $\boldsymbol{F}_j^{(I)}$ acting on particle j is $\boldsymbol{F}_j^{(I)} = \sum_{\substack{k=1\\k\neq j}}^{n} \boldsymbol{F}_{jk}^{(I)}$. By Newton's Third Law, we have $\boldsymbol{F}_{jk}^{(I)} = -\boldsymbol{F}_{kj}^{(I)}$. Then

$$\frac{d\boldsymbol{L}_0}{dt} = \sum_{j=1}^{n} \boldsymbol{r}_j \times \boldsymbol{F}_j = \sum_{j=1}^{n} \boldsymbol{r}_j \times \boldsymbol{F}_j^{(E)} + \sum_{j=1}^{n} \boldsymbol{r}_j \times \sum_{\substack{k=1\\k\neq j}}^{n} \boldsymbol{F}_{jk}^{(I)}.$$

The terms in the product $\sum_{j=1}^{n} \boldsymbol{r}_j \times \sum_{\substack{k=1\\k\neq j}}^{n} \boldsymbol{F}_{jk}^{(I)}$ can be arranged as a sum of pairs $\boldsymbol{r}_j \times \boldsymbol{F}_{jk}^{(I)} + \boldsymbol{r}_k \times \boldsymbol{F}_{kj}^{(I)}$ for each (j,k) with $j \neq k$. Also, $\boldsymbol{r}_j \times \boldsymbol{F}_{jk}^{(I)} + \boldsymbol{r}_k \times \boldsymbol{F}_{kj}^{(I)} = (\boldsymbol{r}_j - \boldsymbol{r}_k) \times \boldsymbol{F}_{jk}^{(I)}$. The vector $\boldsymbol{r}_k - \boldsymbol{r}_j$ is on the line joining m_j and m_k. If the forces $\boldsymbol{F}_{jk}^{(I)}$ are central, as in the cases of gravitational and electrostatic forces, then the vector $\boldsymbol{F}_{jk}^{(I)}$ is parallel to the vector $(\boldsymbol{r}_j - \boldsymbol{r}_k)$, and $(\boldsymbol{r}_j - \boldsymbol{r}_k) \times \boldsymbol{F}_{jk}^{(I)} = \boldsymbol{0}$. In other words, the internal forces do not contribute to the torque, and (2.2.8) becomes

$$\frac{d\boldsymbol{L}_0}{dt} = \sum_{j=1}^{n} \boldsymbol{r}_j \times \boldsymbol{F}_j^{(E)} = \sum_{j=1}^{n} \boldsymbol{T}_j^{(E)} = \boldsymbol{T}_0^{(E)}, \tag{2.2.9}$$

where $\boldsymbol{T}_j^{(E)} = \boldsymbol{r}_j \times \boldsymbol{F}_j^{(E)}$ is the external torque of $\boldsymbol{F}_j^{(E)}$ and $\boldsymbol{T}_0^{(E)}$ is the total torque on the system S by the external forces.

Next, we look at the ANGULAR MOMENTUM OF A SYSTEM RELATIVE TO THE CENTER OF MASS. Again, let O be an arbitrary frame of reference, let $\boldsymbol{r}_j(t)$ be the position of mass m_j, $1 \leq j \leq n$, and let $\boldsymbol{r}_{CM}(t)$ be the position of the center of mass. Define by $\mathfrak{r}_j$ the position of m_j relative to the center of mass:

$$\mathfrak{r}_j = \boldsymbol{r}_j - \boldsymbol{r}_{CM}. \tag{2.2.10}$$

Then $\dot{\boldsymbol{r}}_j = \dot{\boldsymbol{r}}_{CM} + \dot{\mathfrak{r}}_j$ and, according to (2.2.7),

$$\begin{aligned} \boldsymbol{L}_0 &= \sum_{j=1}^{n} m_j\, \boldsymbol{r}_j \times \dot{\boldsymbol{r}}_j = \sum_{j=1}^{n} m_j\, (\boldsymbol{r}_{CM} + \mathfrak{r}_j) \times (\dot{\boldsymbol{r}}_{CM} + \dot{\mathfrak{r}}_j) \\ &= \sum_{j=1}^{n} m_j\, \boldsymbol{r}_{CM} \times \dot{\boldsymbol{r}}_{CM} + \sum_{j=1}^{n} m_j\, \mathfrak{r}_j \times \dot{\boldsymbol{r}}_{CM} + \sum_{j=1}^{n} m_j\, \boldsymbol{r}_{CM} \times \dot{\mathfrak{r}}_j \quad (2.2.11) \\ &+ \sum_{j=1}^{n} m_j\, \mathfrak{r}_j \times \dot{\mathfrak{r}}_j. \end{aligned}$$

EXERCISE 2.2.3.[C] *(a) Verify that*

$$\sum_{j=1}^{n} m_j\, \mathfrak{r}_j = \mathbf{0}, \tag{2.2.12}$$

where $\mathfrak{r}_j$ is the position vector of the point mass m_j relative to the center of mass. Hint: $\sum_{j=1}^{n} m_j\, \mathfrak{r}_j = \sum_{j=1}^{n} m_j\, \boldsymbol{r}_j - \sum_{j=1}^{n} m_j\, \boldsymbol{r}_{CM} = \sum_{j=1}^{n} m_j\, \boldsymbol{r}_j - M\, \boldsymbol{r}_{CM} = \mathbf{0}.$
(b) Use (2.2.11) and (2.2.12) to conclude that

$$\boldsymbol{L}_0 = M\, \boldsymbol{r}_{CM} \times \dot{\boldsymbol{r}}_{CM} + \sum_{j=1}^{n} m_j\, \mathfrak{r}_j \times \dot{\mathfrak{r}}_j. \tag{2.2.13}$$

If we select the origin O of the frame at the center of mass of the system, then $\boldsymbol{r}_{CM} = \mathbf{0}$, and we get the expression for the `angular momentum around the center of mass`:

$$\boldsymbol{L}_{CM} = \sum_{j=1}^{n} m_j\, \mathfrak{r}_j \times \dot{\mathfrak{r}}_j, \tag{2.2.14}$$

where $\mathfrak{r}_j(t) = \boldsymbol{r}_j(t) - \boldsymbol{r}_{CM}(t)$. Hence, (2.2.13) becomes

$$\boldsymbol{L}_0 = M\,\boldsymbol{r}_{CM} \times \dot{\boldsymbol{r}}_{CM} + \boldsymbol{L}_{CM}, \tag{2.2.15}$$

that is, *the angular momentum of the system relative to a point O is equal to the angular momentum of the center of mass relative to that point plus the angular momentum of the system relative to the center of mass.* Once again, we see that the center of mass plays a very special role in the description of the motion of a system of points.

We emphasize that $\boldsymbol{L}_0 \neq M\,\boldsymbol{r}_{CM} \times \dot{\boldsymbol{r}}_{CM}$ as long as $\boldsymbol{L}_{CM} \neq \mathbf{0}$. Note also that the vector functions $\mathfrak{r}_j(t)$ and $\dot{\mathfrak{r}}_j(t)$ depend on the choice of the reference frame.

Let us now compute the time derivative of $\boldsymbol{L}_{CM}(t)$ using the differentiation rules of vector calculus (1.3.3), (1.3.4), (1.3.6) (see page 26). Since $\dot{\mathfrak{r}}_j \times \dot{\mathfrak{r}}_j = \mathbf{0}$, we have

$$\frac{d}{dt}\boldsymbol{L}_{CM} = \sum_{j=1}^{n} m_j\,\dot{\mathfrak{r}}_j \times \dot{\mathfrak{r}}_j + \sum_{j=1}^{n} m_j\,\mathfrak{r}_j \times \ddot{\mathfrak{r}}_j = \sum_{j=1}^{n} m_j\,\mathfrak{r}_j \times \ddot{\mathfrak{r}}_j. \tag{2.2.16}$$

To express $d\boldsymbol{L}_{CM}/dt$ in terms of the forces $\boldsymbol{F}_j$ acting on the m_j, we would like to use Newton's Second Law (2.1.1), and then we need the frame to be inertial. The frame at the center of mass is usually not inertial, because the center of mass can have a non-zero acceleration relative to an inertial frame. Accordingly, we choose a convenient inertial frame O and apply (2.2.4), page 67, in that frame:

$$\ddot{\mathfrak{r}}_j = \ddot{\boldsymbol{r}}_j - \ddot{\boldsymbol{r}}_{CM} = \boldsymbol{F}_j/m_j - \boldsymbol{F}/M,$$
$$m_j\,\mathfrak{r}_j \times \ddot{\mathfrak{r}}_j = \mathfrak{r}_j \times \boldsymbol{F}_j - (m_j/M)\,\mathfrak{r}_j \times \boldsymbol{F}.$$

By (2.2.16) above,

$$\frac{d\boldsymbol{L}_{CM}}{dt} = \sum_{j=1}^{n} \mathfrak{r}_j \times \boldsymbol{F}_j - \frac{1}{M}\left(\sum_{j=1}^{n} m_j\,\mathfrak{r}_j\right) \times \boldsymbol{F},$$

and then (2.2.12) implies

$$\frac{d\boldsymbol{L}_{CM}}{dt} = \sum_{j=1}^{n} \mathfrak{r}_j \times \boldsymbol{F}_j. \tag{2.2.17}$$

If the internal forces $\boldsymbol{F}_{jk}^{(I)}$ are central, then, by (2.2.9), these forces do not contribute to the total torque. Since $\mathfrak{r}_j - \mathfrak{r}_k = \boldsymbol{r}_j - \boldsymbol{r}_{CM} - (\boldsymbol{r}_k - \boldsymbol{r}_{CM}) =$

$\boldsymbol{r}_j - \boldsymbol{r}_k$, we therefore find

$$\frac{d\boldsymbol{L}_{CM}}{dt} = \sum_{j=1}^{n} \mathfrak{r}_j \times \boldsymbol{F}_j^{(E)} = \sum_{j=1}^{n} \boldsymbol{T}_{CMj}^{(E)} = \boldsymbol{T}_{CM}^{(E)}, \tag{2.2.18}$$

where $\boldsymbol{T}_{CMj}^{(E)}$ is the external torque of $\boldsymbol{F}_j^{(E)}$ about the center of mass and $\boldsymbol{T}_{CM}^{(E)}$ is the total torque by the external forces. Using (2.2.16) above, we find

$$\frac{d\boldsymbol{L}_{CM}}{dt} = \sum_{j=1}^{n} m_j \, \mathfrak{r}_j \times \ddot{\mathfrak{r}}_j = \boldsymbol{T}_{CM}^{(E)}. \tag{2.2.19}$$

Equations (2.2.5) and (2.2.19) provide a complete description of the motion of the system of point masses in an inertial frame.

As an EXAMPLE illustrating (2.2.5) and (2.2.19), let us consider BINARY STARS. A **binary**, or **double**, star is a system of two relatively close stars bound to each other by mutual gravitational attraction. Mathematically, a binary star is a system of $n = 2$ masses m_1 and m_2 that are close enough for the mutual gravitational attraction to be much stronger than the gravitational attraction from the other stars. In other words, we have $\boldsymbol{F}_{12}^{(I)} = -\boldsymbol{F}_{21}^{(I)}$ and $\boldsymbol{F}_1^{(E)} = \boldsymbol{F}_2^{(E)} = \boldsymbol{0}$. Using the equation for the linear momentum (2.2.5), page 68, we conclude that $\ddot{\boldsymbol{r}}_{CM} = \boldsymbol{0}$ and $\dot{\boldsymbol{r}}_{CM}$ is constant relative to every inertial frame O. We can therefore choose an inertial frame with origin at the center of mass of the two stars. Applying (2.2.19) in this frame, we find $d\boldsymbol{L}_{CM}/dt = \boldsymbol{0}$, and by (2.2.16),

$$m_1 \, \mathfrak{r}_1 \times \ddot{\mathfrak{r}}_1 + m_2 \, \mathfrak{r}_2 \times \ddot{\mathfrak{r}}_2 = \boldsymbol{0}. \tag{2.2.20}$$

EXERCISE 2.2.4.[B] *Assume that $m_1 = m_2$. Show that the two stars move in a circular orbit around their center of mass. Hint: you can complete the following argument. By (2.2.1), $\boldsymbol{r}_{CM} = (1/2)(\boldsymbol{r}_1 + \boldsymbol{r}_2)$. So CM is the midpoint between m_1 and m_2. Hence, $\mathfrak{r}_1 = -\mathfrak{r}_2$, $\dot{\mathfrak{r}}_1 = -\dot{\mathfrak{r}}_2$, $\ddot{\mathfrak{r}}_1 = -\ddot{\mathfrak{r}}_2$. By (2.2.20) above, $2\mathfrak{r}_1 \times \ddot{\mathfrak{r}}_1 = \boldsymbol{0}$. This implies that $\ddot{\mathfrak{r}}_1$ and $\mathfrak{r}_1$ are parallel (assuming $\ddot{\mathfrak{r}}_1 \neq 0$). Since $d\boldsymbol{L}_{CM}/dt = 0$, $\boldsymbol{L}_{CM}$ is constant. By (2.2.14) on page 70, $2m_1\mathfrak{r}_1 \times \dot{\mathfrak{r}}_1$ is constant as well. Together with $\mathfrak{r}_1 \times \ddot{\mathfrak{r}}_1 = 0$, this is consistent with equations (1.3.27), (1.3.28), and (1.3.29), page 36, for uniform circular motion.*

Binary stars provide one of the primary settings in which astronomers can directly measure the mass. It is estimated that about half of the fifty stars nearest to the Sun are actually binary stars. The term "binary star" was suggested in 1802 by the British astronomer Sir WILLIAM HERSCHEL

(1738–1822), who also discovered the planet Uranus (1781) and infra-red radiation (around 1800).

2.2.2 *Rigid Systems of Points*

A system S of point masses m_j, $1 \le j \le n$, is called **rigid** if the distance between every two points m_i, m_j never changes. Let O be a reference frame and let $\boldsymbol{r}_j(t)$ be the position in that frame of m_j at time t. The rigidity condition can be stated as

$$\|\boldsymbol{r}_j(t) - \boldsymbol{r}_i(t)\| = d_{ij} \quad \text{for all } t, \quad i,j = 1,\ldots,n, \tag{2.2.21}$$

where the d_{ij} are constants.

EXERCISE 2.2.5.^{C} *Verify that condition (2.2.21) is independent of the choice of the frame.*

EXERCISE 2.2.6.^{C} *Let S be a rigid system in motion. Prove that the norm $\|\boldsymbol{r}_j - \boldsymbol{r}_{CM}\|$ and the dot product $\boldsymbol{r}_j \cdot \boldsymbol{r}_{CM}$ remain constant over time for all $j = 1,\ldots,n$, that is, the center of mass of a rigid system is fixed relative to all m_j. Hence, the augmented system $m_1,\ldots,m_n$, M, with M located at the center of mass, is also a rigid system.*

We will now derive the equations of motion for a rigid system. If we consider a motion as a linear transformation of space, then condition (2.2.21) implies that the motion of a rigid system is an isometry. The physical reality also suggests that this motion is orientation-preserving, that is, if three vectors in a rigid system form a right-handed triad at the beginning of the motion, they will be a right-handed triad throughout the motion.

EXERCISE 2.2.7. *(a)^B Show that an orientation-preserving orthogonal transformation is necessarily a rotation. (b)^C Using the result of part (a) and the result of Problem 1.9 on page 412 conclude that the only possible motions of a rigid system are shifts (parallel translations) and rotations.*

Let O be a frame with a Cartesian coordinate system $(\hat{\boldsymbol{\imath}}, \hat{\boldsymbol{\jmath}}, \hat{\boldsymbol{k}})$. Let S be a rigid system moving relative to O. Denote by $\boldsymbol{r}_{CM}(t)$ the position of the center of mass of S in the frame O. We start by introducing two frames connected with the system. Let O_{CM} be the parallel translation of the frame O to the center of mass. Thus, O_{CM} moves with the center of mass of the system S but does not rotate relative to O. Let O_1 be the frame with $\overrightarrow{OO_1} = \boldsymbol{r}_{CM}(t)$ and with the cartesian basis $(\hat{\boldsymbol{\imath}}_1, \hat{\boldsymbol{\jmath}}_1, \hat{\boldsymbol{k}}_1)$ rotating with the system. Define the corresponding rotation vector $\boldsymbol{\omega}$ according to (2.1.39) on page 63. Let us apply equation (2.1.46), page 66, to the motions of the point mass m_j relative to frames O and O_1. Denote by $\boldsymbol{r}_j$ the position vector of m_j in the frame O (draw a picture!) Then $\mathfrak{r}_j = \boldsymbol{r}_j - \boldsymbol{r}_{CM}$ is the position vector of m_j in O_{CM}. Because of the rigidity condition, the position vector $\boldsymbol{r}_{1j}$ of m_j in the frame O_1 does not change in time, and so $\dot{\boldsymbol{r}}_{1j} = \mathbf{0}$ and $\ddot{\boldsymbol{r}}_{1j} = \mathbf{0}$. By (2.1.42) on page 64, $\dot{\boldsymbol{r}}_j = \dot{\boldsymbol{r}}_{CM} + \boldsymbol{\omega} \times \mathfrak{r}_j$, and then, by (2.1.46), $\ddot{\boldsymbol{r}}_j = \ddot{\boldsymbol{r}}_{CM} + \boldsymbol{\omega} \times (\boldsymbol{\omega} \times \mathfrak{r}_j) + \dot{\boldsymbol{\omega}} \times \mathfrak{r}_j$. Since the frame O_{CM} is not rotating relative to O, we have $\ddot{\mathfrak{r}} = \ddot{\boldsymbol{r}}_j - \ddot{\boldsymbol{r}}_{CM}$, and

$$\dot{\mathfrak{r}}_j = \boldsymbol{\omega} \times \mathfrak{r}_j, \tag{2.2.22}$$

$$\ddot{\mathfrak{r}}_j = \boldsymbol{\omega} \times (\boldsymbol{\omega} \times \mathfrak{r}_j) + \dot{\boldsymbol{\omega}} \times \mathfrak{r}_j. \tag{2.2.23}$$

Next, we use identity (1.2.27) on page 22 for the cross product:

$$\boldsymbol{\omega} \times (\boldsymbol{\omega} \times \mathfrak{r}_j) = (\boldsymbol{\omega} \cdot \mathfrak{r}_j)\,\boldsymbol{\omega} - (\boldsymbol{\omega} \cdot \boldsymbol{\omega})\,\mathfrak{r}_j = (\boldsymbol{\omega} \cdot \mathfrak{r}_j)\,\boldsymbol{\omega} - \omega^2\,\mathfrak{r}_j, \tag{2.2.24}$$

where $\omega = \|\boldsymbol{\omega}\|$. To compute the rate of change of the angular momentum, we will need the cross product $\mathfrak{r}_j \times \ddot{\mathfrak{r}}_j$. Applying (1.2.27) one more time,

$$\mathfrak{r}_j \times (\dot{\boldsymbol{\omega}} \times \mathfrak{r}_j) = \|\mathfrak{r}_j\|^2\,\dot{\boldsymbol{\omega}} - (\mathfrak{r}_j \cdot \dot{\boldsymbol{\omega}})\,\mathfrak{r}_j. \tag{2.2.25}$$

Putting everything together, we get

$$\mathfrak{r}_j \times \ddot{\mathfrak{r}}_j = (\boldsymbol{\omega} \cdot \mathfrak{r}_j)\,\mathfrak{r}_j \times \boldsymbol{\omega} + \|\mathfrak{r}_j\|^2\,\dot{\boldsymbol{\omega}} - (\mathfrak{r}_j \cdot \dot{\boldsymbol{\omega}})\,\mathfrak{r}_j. \tag{2.2.26}$$

After summation over all j, the right-hand side of the last equality does not look very promising, and to proceed we need some new ideas. Let us look at (2.2.26) in the most simple yet non-trivial situation, when the rotation axis is fixed in space, and all the point masses m_j are in the plane perpendicular to that axis. Then $\boldsymbol{\omega} \cdot \mathfrak{r}_j = 0$ for all j, and $\dot{\boldsymbol{\omega}} = \dot{\omega}\,\widehat{\boldsymbol{\omega}}$, where $\widehat{\boldsymbol{\omega}}$ is the unit vector in the direction of $\boldsymbol{\omega}$; note that since the rotation axis is fixed, the vectors $\boldsymbol{\omega}$ and $\dot{\boldsymbol{\omega}}$ are parallel. Accordingly, equality (2.2.24) becomes $\boldsymbol{\omega} \times (\boldsymbol{\omega} \times \mathfrak{r}_j) = -\omega^2 \mathfrak{r}_j$, and since $d(\omega\,\widehat{\boldsymbol{\omega}})/dt = \dot{\omega}\,\widehat{\boldsymbol{\omega}}$, equality (2.2.25) becomes $\mathfrak{r}_j \times (\dot{\boldsymbol{\omega}} \times \mathfrak{r}_j) = \|\mathfrak{r}_j\|^2 \dot{\omega}\,\widehat{\boldsymbol{\omega}}$. Summing over all j in (2.2.26) and taking into account these simplifications, we find from (2.2.16) on page 71

$$\frac{d\,\boldsymbol{L}_{CM}}{dt} = \sum_{j=1}^{n} m_j \mathfrak{r}_j \times \ddot{\mathfrak{r}}_j = \left(\sum_{j=1}^{n} m_j \|\mathfrak{r}_j\|^2 \right) \dot{\omega}\,\widehat{\boldsymbol{\omega}},$$

and it is therefore natural to introduce the quantity $I_{CM} = \sum_{j=1}^{n} m_j \|\mathfrak{r}_j\|^2$, which is called the *moment of inertia of S around the line that passes through the center of mass and is parallel to* $\widehat{\omega}$. Then

$$\frac{d\,\boldsymbol{L}_{CM}}{dt} = I_{CM}\dot{\omega}\,\widehat{\boldsymbol{\omega}}. \tag{2.2.27}$$

Our main goal is to extend equality (2.2.27) to a more general situation; the *moment of inertia* thus becomes the main object to investigate. To carry out this investigation, we backtrack a bit and look closely at the *angular momentum* of the rigid system around the center of mass $\boldsymbol{L}_{CM} = \sum_{j=1}^{n} m_j\, \mathfrak{r}_j \times \dot{\mathfrak{r}}_j$. By (2.2.22), we have $\mathfrak{r}_j \times \dot{\mathfrak{r}}_j = \mathfrak{r}_j \times (\boldsymbol{\omega} \times \mathfrak{r}_j)$. Similar to (2.2.25) we find $\mathfrak{r}_j \times (\boldsymbol{\omega} \times \mathfrak{r}_j) = \|\mathfrak{r}_j\|^2\, \boldsymbol{\omega} - (\mathfrak{r}_j \cdot \boldsymbol{\omega})\, \mathfrak{r}_j$ and

$$\boldsymbol{L}_{CM} = \left(\sum_{j=1}^{n} m_j \|\mathfrak{r}_j\|^2\right) \boldsymbol{\omega} - \sum_{j=1}^{n} m_j (\mathfrak{r}_j \cdot \boldsymbol{\omega})\, \mathfrak{r}_j. \tag{2.2.28}$$

As written, equality (2.2.28) does not depend on the basis in the frame O_{CM}. To *calculate* $\boldsymbol{L}_{CM}$, we now choose a cartesian coordinate system $(\hat{\imath}, \hat{\jmath}, \hat{\kappa})$ in the frame O_{CM}. Let $\mathfrak{r}_j(t) = x_j(t)\,\hat{\imath} + y_j(t)\,\hat{\jmath} + z_j(t)\,\hat{\kappa}$ and $\boldsymbol{\omega}(t) = \omega_x(t)\,\hat{\imath} + \omega_y(t)\,\hat{\jmath} + \omega_z(t)\,\hat{\kappa}$. From (2.2.28) above,

$$\begin{aligned}\boldsymbol{L}_{CM} = &\left(\sum_{j=1}^{n} m_j (x_j^2 + y_j^2 + z_j^2)\right) \boldsymbol{\omega} - \left(\sum_{j=1}^{n} m_j x_j\, \mathfrak{r}_j\right) \omega_x \\ &- \left(\sum_{j=1}^{n} m_j y_j\, \mathfrak{r}_j\right) \omega_y - \left(\sum_{j=1}^{n} m_j z_j\, \mathfrak{r}_j\right) \omega_z,\end{aligned} \tag{2.2.29}$$

where we omitted the time dependence notation to simplify the formula. Since $\boldsymbol{L}_{CM} = L_{CMx}\,\hat{\imath} + L_{CMy}\,\hat{\jmath} + L_{CMz}\,\hat{\kappa}$, to compute the x-component L_{CMx} of $\boldsymbol{L}_{CM}$, we replace the vector $\mathfrak{r}_j$ in (2.2.29) with x_j, and the vector $\boldsymbol{\omega}$, with ω_x:

$$L_{CMx} = \omega_x \sum_{j=1}^{n} m_j (y_j^2 + z_j^2) - \omega_y \sum_{j=1}^{n} m_j x_j y_j - \omega_z \sum_{j=1}^{n} m_j x_j z_j;$$

similar representations hold for L_{CMy} and L_{CMz}.

It is therefore natural to introduce the following notations:

$$I_{xx} = \sum_{j=1}^{n} m_j(y_j^2 + z_j^2), \quad I_{yy} = \sum_{j=1}^{n} m_j(x_j^2 + z_j^2), \quad I_{zz} = \sum_{j=1}^{n} m_j(x_j^2 + y_j^2),$$
$$I_{xy} = I_{yx} = \sum_{j=1}^{n} m_j x_j y_j, \quad I_{xz} = I_{zx} = \sum_{j=1}^{n} m_j x_j z_j,$$
$$I_{yz} = I_{zy} = \sum_{j=1}^{n} m_j y_j z_j. \tag{2.2.30}$$

With these notations, $L_{CMx} = \omega_x I_{xx} - \omega_y I_{xy} - \omega_z I_{xz}$.

EXERCISE 2.2.8.^{C} *Verify that*

$$L_{CMy} = -\omega_x I_{yx} + \omega_y I_{yy} - \omega_z I_{yz}, \quad L_{CMz} = -\omega_x I_{zx} - \omega_y I_{zy} + \omega_z I_{zz}.$$

EXERCISE 2.2.9.^{C} *Assume that all the point masses are in the* $(\hat{\imath}, \hat{\jmath})$ *plane. Show that* $I_{xz} = I_{yz} = 0$ *and* $I_{xx} + I_{yy} = I_{zz}$.

We can easily rewrite (2.2.28) in the matrix-vector form:

$$\mathcal{L}_{CM}(t) = I_{CM}(t)\,\boldsymbol{\Omega}(t), \tag{2.2.31}$$

where $\mathcal{L}_{CM}(t)$ is the column vector $(L_{CMx}(t), L_{CMy}(t), L_{CMz}(t))^T$, $\boldsymbol{\Omega}(t)$ is the column vector $(\omega_x(t), \omega_y(t), \omega_z(t))^T$, and

$$I_{CM}(t) = \begin{pmatrix} I_{xx} & -I_{xy} & -I_{xz} \\ -I_{yx} & I_{yy} & -I_{yz} \\ -I_{zx} & -I_{zy} & I_{zz} \end{pmatrix}. \tag{2.2.32}$$

The matrix I_{CM} is called the `moment of inertia matrix`, or `tensor of inertia`, of the system S around the center of mass in the basis $(\hat{\imath}, \hat{\jmath}, \hat{\kappa})$. The Latin word *tensor* means "the one that stretches," and, in mathematics, refers to abstract objects that change in a certain way from one coordinate system to another. All matrices are particular cases of tensors. For a summary of tensors, see page 457 in Appendix.

As much as we would like it, equality (2.2.31) is not the end of our investigation, and there two main reasons for that:

(1) It is not at all clear how to compute the entries of the matrix I_{CM}.
(2) Because the system S is rotating relative to the frame O, the entries of the matrix I_{CM} depend on time.

Remembering that our goal is an equation of the type (2.2.27), we have to continue the investigation of the moment of inertia.

Let us forget for a moment that we are dealing with a rotating system, and instead concentrate on the matrix (2.2.32). We know from linear algebra that every change of the coordinate system changes the look of the matrix; see Exercise 8.1.4, page 453 for a brief summary. For many purposes, including ours, the matrix looks the best when *diagonal*, that is, has zeros everywhere except on the main diagonal. While not all matrices can have this look, every *symmetric* matrix is diagonal in the basis of its normalized eigenvectors; see Exercise 8.1.5 on page 454.

By (2.2.30) and (2.2.32), the matrix I_{CM} is symmetric, and therefore there exists a cartesian coordinate system $(\hat{\boldsymbol{\imath}}^*, \hat{\boldsymbol{\jmath}}^*, \hat{\boldsymbol{k}}^*)$ in which I_{CM} has at most three non-zero entries $I^*_{11}, I^*_{22}, I^*_{33}$, and all other entries zero. In other words, there exists an orthogonal matrix U_* so that the matrix $I^*_{CM} = U_* I_{CM} U_*^T$ is diagonal. The matrix U_* is the representation in the basis $(\hat{\boldsymbol{\imath}}, \hat{\boldsymbol{\jmath}}, \hat{\boldsymbol{k}})$ of a linear transformation (rotation) that moves the vectors $\hat{\boldsymbol{\imath}}, \hat{\boldsymbol{\jmath}}, \hat{\boldsymbol{k}}$ to $\hat{\boldsymbol{\imath}}^*, \hat{\boldsymbol{\jmath}}^*, \hat{\boldsymbol{k}}^*$, respectively. The vectors $\hat{\boldsymbol{\imath}}^*, \hat{\boldsymbol{\jmath}}^*, \hat{\boldsymbol{k}}^*$ are called the `principal axes` of the system S. We will refer to the frame with the origin at the center of mass and the basis vectors $\hat{\boldsymbol{\imath}}^*, \hat{\boldsymbol{\jmath}}^*, \hat{\boldsymbol{k}}^*$ as the `principal axes frame`. Both the principal axes and the numbers $I^*_{11}, I^*_{22}, I^*_{33}$ depend only on the *configuration* of the rigid system S, that is, the positions of the point masses m_j relative to the center of mass. A matrix can have only one diagonal look, but in more than one basis: for example, the identity matrix looks the same in every basis. Accordingly, the principal axes might not be unique, but the numbers $I^*_{11}, I^*_{22}, I^*_{33}$ are uniquely determined by the configuration of the system, and, in particular, do not depend on time. We know from linear algebra that the numbers $I^*_{11}, I^*_{22}, I^*_{33}$ are the eigenvalues of the matrix I_{CM}, and the matrix U_* consists of the corresponding eigenvectors.

EXERCISE 2.2.10.A *Given an example of a symmetric* 3×3 *matrix whose entries depend on time, but whose eigenvalues do not. Can you think of a general method for constructing such a matrix?*

Formulas (2.2.30) for the entries of the matrix of inertia are true in every basis, and therefore can be used to compute the numbers $I^*_{11}, I^*_{22}, I^*_{33}$. Since these numbers do not depend on time, the coordinates x^*_j, y^*_j, z^*_j of m_j in the principal axes frame should not depend on time either. In other words, *the principal axes frame is not rotating relative to the system* S, but is fixed in S and rotates together with S.

EXERCISE 2.2.11.[C] *Consider four identical point masses m at the vertices of a square with side a. Denote by O the center of the square. Let $(\hat{\boldsymbol{\imath}}, \hat{\boldsymbol{\jmath}}, \hat{\boldsymbol{\kappa}})$ be a cartesian system at O, with the vectors $\hat{\boldsymbol{\imath}}$ and $\hat{\boldsymbol{\jmath}}$ along the diagonals of the square (draw a picture!). Verify the following statements: (a) The center of mass of the system is at O. (b) The vectors $\hat{\boldsymbol{\imath}}$, $\hat{\boldsymbol{\jmath}}$, $\hat{\boldsymbol{\kappa}}$ define the three principal axis of the system. (c) The diagonal elements of the matric I^*_{CM} in $(\hat{\boldsymbol{\imath}}, \hat{\boldsymbol{\jmath}}, \hat{\boldsymbol{\kappa}})$ are ma^2, ma^2, $2ma^2$. (d) If the system $(\hat{\boldsymbol{\imath}}, \hat{\boldsymbol{\jmath}}, \hat{\boldsymbol{\kappa}})$ is rotated by $\pi/4$ around $\hat{\boldsymbol{\kappa}}$ (does not matter clock- or counterclockwise), the result is again a principal axis frame for the system, and the matrix I^*_{CM} does not change.*

EXERCISE 2.2.12.[A] *When are some of the diagonal entries of I^*_{CM} equal to zero? Hint: not very often.*

We now summarize our excursion into linear algebra: *for every rigid system S of point masses, there exists a special frame, called the* `principal axes frame`, *in which the matrix of inertia I^*_{CM} of the system is diagonal and does not depend on time. This special frame is centered at the center of mass and is fixed (not rotating) relative to the system S.*

Let us go back to the analysis of the motion of the system S. With

$$\boldsymbol{L}_{CM} = L^*_{CMx}\,\hat{\boldsymbol{\imath}}^* + L^*_{CMx}\,\hat{\boldsymbol{\imath}}^* + L^*_{CMx}\,\hat{\boldsymbol{\imath}}^*, \quad \boldsymbol{\omega} = \omega^*_x\,\hat{\boldsymbol{\imath}}^* + \omega^*_y\,\hat{\boldsymbol{\jmath}}^* + \omega^*_z\,\hat{\boldsymbol{\kappa}}^*,$$

and after multiplying by U_* on the left and by U^T_* on the right, equation (2.2.31) becomes

$$\mathcal{L}^*_{CM} = I^*_{CM}\,\boldsymbol{\Omega}^*, \tag{2.2.33}$$

where $\mathcal{L}^*_{CM} = U_*\mathcal{L}_{CM}U^T_*$ is the column vector $(L^*_{CMx}, L^*_{CMx}, L^*_{CMz})^T$ and $\boldsymbol{\Omega}^* = U_*\boldsymbol{\Omega}U^T_*$ is the column-vector $(\omega^*_x, \omega^*_y, \omega^*_z)^T$.

By construction, the principal axes frame rotates relative to the frame O, and the corresponding rotation vector is $\boldsymbol{\omega}$. Denoting the time derivative in the frame O by D_0, and in the the principal axes frame, by D_*, and using the relation (2.1.43) on page 64, we find

$$D_0\boldsymbol{L}_{CM}(t) = D_*\,\boldsymbol{L}_{CM}(t) + \boldsymbol{\omega}(t) \times \boldsymbol{L}_{CM}(t). \tag{2.2.34}$$

To proceed, let us assume that the underlying frame O is inertial. Then relation (2.2.17) applies, and we find $D_0\boldsymbol{L}_{CM}(t) = \boldsymbol{T}_{CM}(t)$: the change of $\boldsymbol{L}_{CM}(t)$ in the inertial frame is equal to the torque of all forces about the center of mass. In the principal axes frame, we have $\boldsymbol{T}_{CM}(t) = T^*_{CMx}(t)\,\hat{\boldsymbol{\imath}}^* + T^*_{CMy}(t)\,\hat{\boldsymbol{\jmath}}^* + T^*_{CMz}(t)\,\hat{\boldsymbol{\kappa}}^*$. Also, since the basis vectors $\hat{\boldsymbol{\imath}}^*$, $\hat{\boldsymbol{\jmath}}^*$, $\hat{\boldsymbol{\kappa}}^*$ are fixed in the principal axes frame, we have $D_*\boldsymbol{L}_{CM}(t) = \dot{L}^*_{CMx}(t)\,\hat{\boldsymbol{\imath}}^* + \dot{L}^*_{CMy}(t)\,\hat{\boldsymbol{\jmath}}^* + \dot{L}^*_{CMz}(t)\,\hat{\boldsymbol{\kappa}}^*$. On the other hand, since the matrix I^*_{CM} does not depend on time, we use (2.2.33) to conclude that the column vector $\dot{\mathcal{L}}^*_{CM}$ of the components of $D_*\boldsymbol{L}_{CM}(t)$ satisfies $\dot{\mathcal{L}}^*_{CM} = I^*_{CM}\dot{\boldsymbol{\Omega}}^*$. Finally, we compute the cross product in (2.2.34) by writing the vectors in the principal axes frame and using the relation (2.2.33) for the components $\mathcal{L}^*_{CM}(t)$ of $\boldsymbol{L}_{CM}$ in that frame. The result is the three `Euler equations` describing the rotation of the rigid system about the center of mass:

$$\begin{cases} I^*_{xx}\dot{\omega}^*_x + \omega^*_y\omega^*_z(I^*_{zz} - I^*_{yy}) = T^*_{CMx}, \\ I^*_{yy}\dot{\omega}^*_y + \omega^*_x\omega^*_z(I^*_{xx} - I^*_{zz}) = T^*_{CMy}, \\ I^*_{zz}\dot{\omega}^*_z + \omega^*_x\omega^*_y(I^*_{yy} - I^*_{xx}) = T^*_{CMz}. \end{cases} \tag{2.2.35}$$

These equations were first published in 1765 by a Swiss mathematician LEONHARD EULER (1707–1783). Leonhard (or Léonard) Euler was the most prolific mathematician ever: extensive publication of his works continued for 50 years after his death and filled 80+ volumes; he also had 13 children. He introduced many modern mathematical notations, such as e for the base of natural logs (1727), $f(x)$ for a function (1734), Σ for summation (1755), and i for the square root of -1 (1777).

EXERCISE 2.2.13.^{B} *(a) Verify that (2.2.34) is indeed equivalent to (2.2.35). (b) Write (2.2.27) in an inertial frame and verify that the result is a particular case of (2.2.35).*

With a suitable definition of the numbers I^*_{xx}, I^*_{yy}, I^*_{zz} and the vector $\boldsymbol{T}^*_{CM}$, equations (2.2.35) also describe the motion of a rigid body. We study rigid bodies in the following section.

2.2.3 *Rigid Bodies*

Any collection of points, finite or infinite, can be a rigid system: if two points in the collection have trajectories $\boldsymbol{r}_1(t)$, $\boldsymbol{r}_2(t)$ in some frame, then the rigidity condition $\|\boldsymbol{r}_1(t) - \boldsymbol{r}_2(t)\| = \|\boldsymbol{r}_1(0) - \boldsymbol{r}_2(0)\|$ must hold for every two points in the collection.

Intuitively, a `rigid body` is a rigid system consisting of uncountably many points, each with infinitesimally small mass. Mathematically, a rigid body is described in a frame O by a `mass density function` $\rho = \rho(\boldsymbol{r})$, so that the volume $\triangle V$ of the body near the point with the position vector $\boldsymbol{r}_0$ has, approximately, the mass $\triangle m = \rho(\boldsymbol{r})\,\triangle V$. Even more precisely, if the body occupies the region $\mathcal{R}$ in $\mathbb{R}^3$, then the mass M of the body is given by the triple (or volume) integral

$$M = \iiint_{\mathcal{R}} \rho(\boldsymbol{r}) dV.$$

Without going into the details, let us note that rigid bodies can also be two-dimensional, for example, a (hard) spherical shell, or one-dimensional, for example, a piece of hard-to-bend wire. In these cases, we use surface and line integrals rather then volume integrals. In what follows, we focus on solid three-dimensional objects.

All the formulas for the motion of a rigid body can be derived from the corresponding formulas for a finite number of points by replacing m_j with the mass density function, and summation with integration. For example, the *center of mass* of a rigid body is the point with the position vector

$$\boldsymbol{r}_{CM} = \frac{1}{M} \iiint_{\mathcal{R}} \boldsymbol{r} \rho(\boldsymbol{r}) dV. \tag{2.2.36}$$

EXERCISE 2.2.14.^{C} *Show that both the mass and the location of the center of mass of a rigid body are independent of the frame O.*

For a rigid body $\mathcal{R}$ moving in space relative to a frame O, we denote by $\mathcal{R}(t)$ the part of the space occupied by the body at time t *relative to that frame O*. If the frame O is inertial, then an equation similar to (2.2.5) connects the trajectory $\boldsymbol{r}_{CM} = \boldsymbol{r}_{CM}(t)$ of the center of mass in the frame with the *external forces per unit mass* $\boldsymbol{F}^{(E)} = \boldsymbol{F}^{(E)}(\boldsymbol{r})$ acting on the points of the body:

$$M\, \ddot{\boldsymbol{r}}_{CM}(t) = \iiint_{\mathcal{R}(t)} \boldsymbol{F}^{(E)}(\boldsymbol{r}(t))\, \rho(\boldsymbol{r}(t))\, dV, \tag{2.2.37}$$

The angular momentum of $\mathcal{R}$ about O is, by definition,

$$\boldsymbol{L}_O(t) = \iiint_{\mathcal{R}(t)} \boldsymbol{r}(t) \times \dot{\boldsymbol{r}}(t)\, \rho(\boldsymbol{r}(t)) dV. \tag{2.2.38}$$

Similar to (2.2.15), page 71, we have

$$\boldsymbol{L}_O(t) = M\, \boldsymbol{r}_{CM}(t) \times \dot{\boldsymbol{r}}_{CM}(t) + \iiint_{\mathcal{R}(t)} \mathfrak{r}(t) \times \dot{\mathfrak{r}}(t)\, \rho(\boldsymbol{r}(t))\, dV, \tag{2.2.39}$$

where $\mathfrak{r}(t) = \boldsymbol{r}(t) - \boldsymbol{r}_{CM}(t)$, and is (2.2.14) replaced by

$$\boldsymbol{L}_{CM}(t) = \iiint\limits_{\mathcal{R}(t)} \mathfrak{r}(t) \times \dot{\mathfrak{r}}(t)\, \rho(\boldsymbol{r}(t))\, dV, \tag{2.2.40}$$

which is the angular momentum relative to the center of mass.

As with finite systems of points, consider the parallel translation of the frame O to the center of mass of the body. The rotation of the rigid body relative to this translated frame is described by the rotation vector $\boldsymbol{\omega}(t) = \omega_x(t)\,\hat{\boldsymbol{\imath}} + \omega_y(t)\,\hat{\boldsymbol{\jmath}} + \omega_z(t)\,\hat{\boldsymbol{\kappa}}$, where $(\hat{\boldsymbol{\imath}}, \hat{\boldsymbol{\jmath}}, \hat{\boldsymbol{\kappa}})$ is the cartesian basis in the translated frame. By analogy with (2.2.30), we write $\mathfrak{r}(t) = x(t)\,\hat{\boldsymbol{\imath}} + y(t)\,\hat{\boldsymbol{\jmath}} + z(t)\,\hat{\boldsymbol{\kappa}}$, $\boldsymbol{L}_{CM}(t) = L_{CMx}(t)\,\hat{\boldsymbol{\imath}} + L_{CMy}(t)\,\hat{\boldsymbol{\jmath}} + L_{CMz}(t)\,\hat{\boldsymbol{\kappa}}$, and define

$$\begin{aligned}
I_{xx} &= \iiint\limits_{\mathcal{R}(t)} (y^2(t) + z^2(t))\, \rho(\mathfrak{r}(t))\, dV, \ I_{yy} = \iiint\limits_{\mathcal{R}(t)} (x^2(t) + z^2(t))\, \rho(\mathfrak{r}(t))\, dV, \\
I_{zz} &= \iiint\limits_{\mathcal{R}(t)} (x^2(t) + y^2(t))\, \rho(\mathfrak{r}(t))\, dV, \\
I_{xy} &= I_{yx} = \iiint\limits_{\mathcal{R}(t)} x(t)y(t)\, \rho(\mathfrak{r}(t))\, dV, \ I_{xz} = I_{zx} = \iiint\limits_{\mathcal{R}(t)} x(t)z(t)\, \rho(\mathfrak{r}(t))\, dV, \\
I_{yz} &= I_{zy} = \iiint\limits_{\mathcal{R}(t)} y(t)z(t)\, \rho(\mathfrak{r}(t))\, dV.
\end{aligned} \tag{2.2.41}$$

Then we have relation (2.2.31), page 76, for rigid bodies:

$$\mathcal{L}_{CM}(t) = I_{CM}(t)\, \boldsymbol{\Omega}(t), \tag{2.2.42}$$

where $\mathcal{L}_{CM}(t)$ is the column vector $(L_{CMx}(t), L_{CMy}(t), L_{CMz}(t))^T$, $\boldsymbol{\Omega}$ is the column vector $(\omega_x(t), \omega_y(t), \omega_z(t))^T$, and

$$I_{CM}(t) = \begin{pmatrix} I_{xx} & -I_{xy} & -I_{xz} \\ -I_{yx} & I_{yy} & -I_{yz} \\ -I_{zx} & -I_{zy} & I_{zz} \end{pmatrix}. \tag{2.2.43}$$

The matrix I_{CM} is called the `moment of inertia matrix`, or `tensor of inertia`, of the rigid body $\mathcal{R}$ around the center of mass in the basis $(\hat{\boldsymbol{\imath}}, \hat{\boldsymbol{\jmath}}, \hat{\boldsymbol{\kappa}})$. As in the case of a finite rigid system of points, there exists a principal axes frame, in which the matrix I_{CM} is diagonal, and the diagonal elements $I^*_{xx}, I^*_{yy}, I^*_{zz}$ are uniquely determined by mass density function ρ. The principal axes frame, with center at the center of mass and basis vectors $\hat{\boldsymbol{\imath}}^*$, $\hat{\boldsymbol{\jmath}}^*$, $\hat{\boldsymbol{\kappa}}^*$, is attached to the body and rotates with it. If $\boldsymbol{\omega}(t) = \omega_x^* \hat{\boldsymbol{\imath}}^* + \omega_y^* \hat{\boldsymbol{\jmath}}^* + \omega_z^* \hat{\boldsymbol{\kappa}}^*$, then the Euler equations (2.2.35) describe the rotation of the body about the center of mass. Equations (2.2.37) and (2.2.35) provide a complete description of the motion of a rigid body.

As an example, consider the DISTRIBUTED RIGID PENDULUM. Recall that in the simple rigid pendulum (page 41), a point mass is attached to the end of a weightless rod. In the distributed rigid pendulum, a uniform rod of mass M and length ℓ is suspended by one end with a pin joint (Figure 2.2.1).

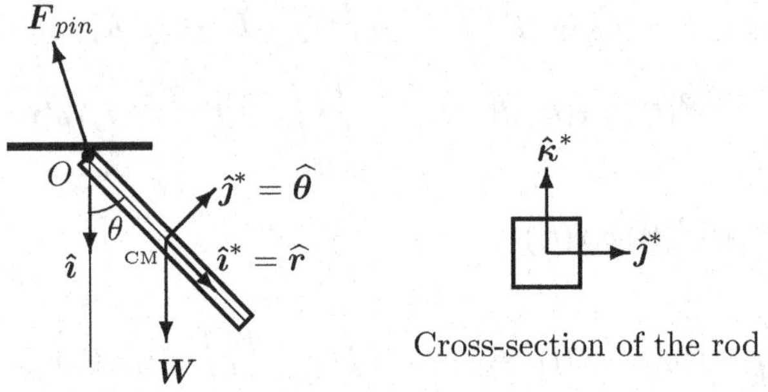

Fig. 2.2.1 Distributed Pendulum

We assume that the cross-section of the rod is a square with side a. Then the volume of the rod is $\ell\, a^2$ and the density is constant: $\rho(\boldsymbol{r}) = M/(\ell\, a^2)$.

EXERCISE 2.2.15.[C] *Verify that the center of mass CM of the rod is the mid-points of the axis of the rod.*

Consider the cartesian coordinates ($\hat{\boldsymbol{\imath}}^*$, $\hat{\boldsymbol{\jmath}}^*$, $\hat{\boldsymbol{\kappa}}^*$) with the origin at the center of mass (Figure 2.2.1). As usual, $\hat{\boldsymbol{\kappa}}^* = \hat{\boldsymbol{\imath}}^* \times \hat{\boldsymbol{\jmath}}^*$.

Let us compute I_{zz}^*, as this is the only entry of the matrix I_{CM}^* we will need:

$$\begin{aligned} I_{zz}^* &= \iiint\limits_{\mathcal{R}} (x^2 + y^2)\rho(\boldsymbol{r})\, dV \\ &= \frac{M}{\ell\, a^2}\left(\int_{-a/2}^{a/2} dy \int_{-a/2}^{a/2} dz \int_{-\ell/2}^{\ell/2} x^2 dx + \int_{-\ell/2}^{\ell/2} dx \int_{-a/2}^{a/2} dz \int_{-a/2}^{a/2} y^2 dy \right) \\ &= (\rho a^2 \ell^3/12) + (\rho \ell a^4/12) = (M/12)(\ell^2 + a^2). \end{aligned}$$

EXERCISE 2.2.16.^{C} *Verify that the vectors* $\hat{\imath}^*$, $\hat{\jmath}^*$, $\hat{\kappa}^*$ *define a principal axis frame. Hint:* $I^*_{xy} = I^*_{yz} = I^*_{xz} = 0$, *as seen from (2.2.41) and the symmetry of the rod.*

The motion of the rod is a 2-D rotation around the pin at O, and the vector $\hat{\kappa}^*$ is not rotating. *There are no internal forces in the rod to affect the motion.* As a result, the angular velocity vector is $\boldsymbol{\omega} = \omega^*_z\,\hat{\kappa}^* = \dot{\theta}\,\hat{\kappa}^*$, and the Euler equations (2.2.35) simplify to $I^*_{zz}\,\dfrac{d\omega^*_z}{dt} = T^{*(E)}_{CMz}$, or

$$\frac{M(\ell^2 + a^2)}{12}\,\ddot{\theta} = T^{*(E)}_{CMz}, \tag{2.2.44}$$

where $T^{*(E)}_{CMz}$ is the $\hat{\kappa}^*$-component of the external torque around the center of mass.

To simplify the analysis, we ignore air resistance. Then the external torque $\boldsymbol{T}^{*(E)}_{CM}$ around the center of mass is produced by two forces: the force of gravity $\boldsymbol{W}$ and the force $\boldsymbol{F}_{pin}$ exerted by the pin at O. Since the rod is uniform, the torque due to gravity is zero around the CM. To compute $\boldsymbol{F}_{pin}$ we now assume that the frame fixed at O is inertial and Newton's Second Law (2.2.5) applies. The external force on mass M at the CM is the sum of the total gravity force, $\boldsymbol{W} = Mg\,\hat{\imath}$, and the reaction of the pin $\boldsymbol{F}_{pin}$. Hence, $\boldsymbol{F}_{pin} = M\,\ddot{\boldsymbol{r}}_{CM} - \boldsymbol{W}$. Since CM moves around O in a circle of radius $\ell/2$ we use (1.3.27) on page 36 to obtain the components $\boldsymbol{a}_r$, $\boldsymbol{a}_\theta$ of $\ddot{\boldsymbol{r}}_{CM}$ in polar coordinates with origin at O: $\boldsymbol{a}_r = (-\ell/2)\omega^2\,\hat{\boldsymbol{r}}$, $\boldsymbol{a}_\theta = (\ell/2)\dot{\omega}\,\hat{\boldsymbol{\theta}}$. The point of application of $\boldsymbol{F}_{pin}$ *relative to* CM has the position vector $-\boldsymbol{r}_{CM} = -(\ell/2)\,\hat{\boldsymbol{r}}$. Then

$$\boldsymbol{T}^{*(E)}_{CM} = -\boldsymbol{r}_{CM} \times \boldsymbol{F}_{pin} + \boldsymbol{0} \times \boldsymbol{W} = -\frac{\ell}{2}\,\hat{\boldsymbol{r}} \times \boldsymbol{F}_{pin}.$$

Hence,

$$\boldsymbol{T}^{*(E)}_{CM} = \frac{-\ell}{2}\,\hat{\boldsymbol{r}} \times (M\ddot{\boldsymbol{r}}_{CM} - \boldsymbol{W}) = \frac{-\ell M}{2}\left(\frac{\ell}{2}\dot{\omega}\,\hat{\boldsymbol{r}} \times \hat{\boldsymbol{\theta}} - g\hat{\boldsymbol{r}} \times \hat{\imath}\right),$$
$$= \frac{-\ell M}{2}\left(\frac{\ell}{2}\dot{\omega} + g\sin\theta\right)\hat{\kappa}^*.$$

Substituting in (2.2.44), we get $(M(\ell^2 + a^2)/12)\ddot{\theta} = -(\ell M/2)((\ell/2)\dot{\omega} + g\sin\theta)$ or $((\ell^2 + a^2)/12)\ddot{\theta} + (\ell^2\dot{\omega}/4) + (g\ell\sin\theta/2) = 0$, or, with $\dot{\omega} = \ddot{\theta}$,

$$((4\ell^2 + a^2)/12)\ddot{\theta} = -(g\ell/2)\sin\theta. \tag{2.2.45}$$

If $a \approx 0$ and θ is so small that $\sin\theta \approx \theta$, then equation (2.2.45) becomes

$$(2\ell/3)\,\ddot{\theta} + g\,\theta = 0. \tag{2.2.46}$$

Comparing this with a similar approximation for the simple rigid pendulum (2.1.10), page 42, we conclude that a thin uniform stick of length ℓ suspended at one end oscillates at about the same frequency as a simple rigid pendulum of length $(2/3)\ell$.

The objective of the above example was to illustrate how the Euler equations work. Because of the simple nature of the problem, the system of three equations (2.2.35) degenerates to one equation (2.2.45). In fact, an alternative derivation of (2.2.45) is possible by avoiding (2.2.35) altogether; the details are in Problem 2.4, page 417.

Note that if the frame O is fixed on the Earth, then this frame is *not* inertial, and the Coriolis force will act on the pendulum, but a good pin joint can minimize the effects of this force.

For two more examples of rigid body motion, see Problems 2.7 and 2.8 starting on page 419.

2.3 The Lagrange–Hamilton Method

So far, we used the *Newton-Euler* method to analyze motion using forces and the three laws of Newton (it was L. Euler who, around 1737, gave a precise mathematical description of the method). An alternative method using *energy* and *work* was introduced in 1788 by the French mathematician Joseph-Louis Lagrange (1736–1813) and further developed in 1833 by the Irish mathematician Sir William Rowan Hamilton (1805–1865). This *Lagrange-Hamilton method* is sometimes more efficient than the Newton-Euler method, especially to study systems with *constraints*. An example of a constraint is the rigidity condition, ensuring that the distance between any two points is constant.

In what follows, we provide a brief description of the Lagrange-Hamilton method. The reader is assumed to be familiar with the basic tools of multivariable calculus, in particular, the chain rule and line integration.

2.3.1 *Lagrange's Equations*

We first illustrate Lagrange's method for a point mass moving in the plane. The modern methods of multi-variable calculus reduce the derivation of the main result (equation (2.3.17)) to a succession of simple applications of the chain rule.

Consider an inertial frame in $\mathbb{R}^2$ with origin O and basis vectors $\hat{\imath}$, $\hat{\jmath}$. Let $\boldsymbol{r} = x\,\hat{\imath} + y\,\hat{\jmath}$ be the position of point mass m moving under the action of force $\boldsymbol{F} = F_2\,\hat{\imath} + F_2\,\hat{\jmath}$. By Newton's Second Law (2.1.1),

$$F_1 = m\ddot{x}, \qquad F_2 = m\ddot{y}. \tag{2.3.1}$$

The *state* of m at any time t is given by the four-dimensional vector $(x, y, \dot{x}, \dot{y})$, that is, by the position and velocity. Knowledge of the state at a given time allows us to determine the state at all future times by solving equations (2.3.1).

Now consider a different pair (q_1, q_2) of coordinates in the plane, for example, for example, polar coordinates, so that

$$x = x(q_1, q_2), \quad y = y(q_1, q_2); \qquad q_1 = q_1(x, y), \; q_2 = q_2(x, y), \tag{2.3.2}$$

and all the functions are sufficiently smooth. Differentiating (2.3.2),

$$\dot{x} = \frac{\partial x}{\partial q_1}\dot{q}_1 + \frac{\partial x}{\partial q_2}\dot{q}_2, \quad \dot{y} = \frac{\partial y}{\partial q_1}\dot{q}_1 + \frac{\partial y}{\partial q_2}\dot{q}_2. \tag{2.3.3}$$

We call $q_1, q_2, \dot{q}_1, \dot{q}_2$ the **generalized coordinates**, since their values determine the state $(x, y, \dot{x}, \dot{y})$ by (2.3.2) and (2.3.3). Note that the partial derivatives $\partial x/\partial q_i$, $\partial y/\partial q_i$, $i = 1, 2$, depend only on q_1 and q_2. We then differentiate (2.3.3) to find

$$\frac{\partial \dot{x}}{\partial \dot{q}_1} = \frac{\partial x}{\partial q_1}, \; \frac{\partial \dot{x}}{\partial \dot{q}_2} = \frac{\partial x}{\partial q_2}, \; \frac{\partial \dot{y}}{\partial \dot{q}_1} = \frac{\partial y}{\partial q_1}, \; \frac{\partial \dot{y}}{\partial \dot{q}_2} = \frac{\partial y}{\partial q_2}. \tag{2.3.4}$$

Next, we apply the chain rule to the function $\partial x(q_1(t), q_2(t))/\partial q_1$ to find

$$\frac{d}{dt}\left(\frac{\partial x}{\partial q_1}\right) = \frac{\partial^2 x}{\partial q_1^2}\frac{dq_1}{dt} + \frac{\partial^2 x}{\partial q_2 \partial q_1}\frac{dq_2}{dt} = \frac{\partial^2 x}{\partial q_1^2}\dot{q}_1 + \frac{\partial^2 x}{\partial q_2 \partial q_1}\dot{q}_2. \tag{2.3.5}$$

From (2.3.3), differentiating with respect to the variable q_1,

$$\frac{\partial \dot{x}}{\partial q_1} = \frac{\partial^2 x}{\partial q_1^2}\dot{q}_1 + \frac{\partial^2 x}{\partial q_1 \partial q_2}\dot{q}_2 = \frac{d}{dt}\left(\frac{\partial x}{\partial q_1}\right). \tag{2.3.6}$$

Similarly,

$$\frac{d}{dt}\left(\frac{\partial x}{\partial q_1}\right)=\frac{\partial \dot{x}}{\partial q_1},\quad \frac{d}{dt}\left(\frac{\partial y}{\partial q_1}\right)=\frac{\partial \dot{y}}{\partial q_1},$$
$$\frac{d}{dt}\left(\frac{\partial x}{\partial q_2}\right)=\frac{\partial \dot{x}}{\partial q_2},\quad \frac{d}{dt}\left(\frac{\partial y}{\partial q_2}\right)=\frac{\partial \dot{y}}{\partial q_2}. \tag{2.3.7}$$

EXERCISE 2.3.1.^{C} *Verify (2.3.7).*

By definition, the kinetic energy $\mathcal{E}_K$ of m is

$$\mathcal{E}_K=\frac{m}{2}(\dot{x}^2+\dot{y}^2). \tag{2.3.8}$$

Replacing $\dot{x}$ and $\dot{y}$ by their functions of q_1, q_2 as given by (2.3.3), we obtain the function

$$\mathcal{E}_K=\mathcal{E}_K(q_1,q_2,\dot{q}_1,\dot{q}_2). \tag{2.3.9}$$

From (2.3.8), again by differentiation and using (2.3.4),

$$\frac{\partial \mathcal{E}_K}{\partial \dot{q}_1}=\frac{\partial \mathcal{E}_K}{\partial \dot{x}}\frac{\partial \dot{x}}{\partial \dot{q}_1}+\frac{\partial \mathcal{E}_K}{\partial \dot{y}}\frac{\partial \dot{y}}{\partial \dot{q}_1}=m\dot{x}\frac{\partial x}{\partial q_1}+m\dot{y}\frac{\partial y}{\partial q_1}.$$

Differentiating with respect to t and applying (2.3.7), we get

$$\frac{d}{dt}\left(\frac{\partial \mathcal{E}_K}{\partial \dot{q}_1}\right)=m\ddot{x}\frac{\partial x}{\partial q_1}+m\ddot{y}\frac{\partial y}{\partial q_1}+m\dot{x}\frac{\partial \dot{x}}{\partial q_1}+m\dot{y}\frac{\partial \dot{y}}{\partial q_1}. \tag{2.3.10}$$

From (2.3.8) we also get

$$\frac{\partial \mathcal{E}_K}{\partial q_1}=m\dot{x}\frac{\partial \dot{x}}{\partial q_1}+m\dot{y}\frac{\partial \dot{y}}{\partial q_1}. \tag{2.3.11}$$

Subtracting (2.3.11) from (2.3.10), we find that

$$\frac{d}{dt}\left(\frac{\partial \mathcal{E}_K}{\partial \dot{q}_1}\right)-\frac{\partial \mathcal{E}_K}{\partial q_1}=m\ddot{x}\frac{\partial x}{\partial q_1}+m\ddot{y}\frac{\partial y}{\partial q_1}. \tag{2.3.12}$$

EXERCISE 2.3.2.^{C} *Verify that*

$$\frac{d}{dt}\left(\frac{\partial \mathcal{E}_K}{\partial \dot{q}_2}\right)-\frac{\partial \mathcal{E}_K}{\partial q_2}=m\ddot{x}\frac{\partial x}{\partial q_2}+m\ddot{y}\frac{\partial y}{\partial q_2}. \tag{2.3.13}$$

To continue our derivation, we use equations (2.3.1) to rewrite (2.3.12) and (2.3.13) as

$$\frac{d}{dt}\left(\frac{\partial \mathcal{E}_K}{\partial \dot{q}_j}\right)-\frac{\partial \mathcal{E}_K}{\partial q_j}=F_1\frac{\partial x}{\partial q_j}+F_2\frac{\partial y}{\partial q_j},\quad j=1,2. \tag{2.3.14}$$

Recall that the work W done by the force $\boldsymbol{F}$ is the line integral $\int_C \boldsymbol{F} \cdot d\boldsymbol{r}$. Hence, the differential is $dW = \boldsymbol{F} \cdot d\boldsymbol{r}$. Since $d\boldsymbol{r} = dx\,\hat{\imath} + dy\,\hat{\jmath}$, we have

$$dW = F_1 dx + F_2 dy. \tag{2.3.15}$$

By the chain rule,

$$dx = (\partial x/\partial q_1)dq_1 + (\partial x/\partial q_2)dq_2, \quad dy = (\partial y/\partial q_1)dq_1 + (\partial y/\partial q_2)dq_2.$$

Substituting in (2.3.15), we get $dW = Q_1 dq_1 + Q_2 dq_2$, where

$$\begin{aligned} Q_1 &= F_1 \frac{\partial x}{\partial q_1} + F_2 \frac{\partial y}{\partial q_1} = \boldsymbol{F} \cdot \frac{\partial \boldsymbol{r}}{\partial q_1}, \\ Q_2 &= F_1 \frac{\partial x}{\partial q_2} + F_2 \frac{\partial y}{\partial q_2} = \boldsymbol{F} \cdot \frac{\partial \boldsymbol{r}}{\partial q_2}. \end{aligned} \tag{2.3.16}$$

The functions Q_1 and Q_2 are called `generalized forces` corresponding to the generalized coordinates q_1, q_2. Substituting in (2.3.14), we obtain the `Lagrange equations of motion`,

$$\frac{d}{dt}\left(\frac{\partial \mathcal{E}_K}{\partial \dot{q}_j}\right) - \frac{\partial \mathcal{E}_K}{\partial q_j} = Q_j, \quad j = 1, 2, \tag{2.3.17}$$

where Q_j is given by (2.3.16).

For the point mass moving in cartesian coordinates under the force $\boldsymbol{F} = F_1\,\hat{\imath} + F_2\,\hat{\jmath}$, we have $q_1 = x$, $q_2 = y$, $\mathcal{E}_K = m(\dot{x}^2 + \dot{y}^2)/2$, and equations (2.3.17) coincide with (2.3.1), so it might seem that we did not accomplish much. It is clear, though, that (2.3.17) is more general, and covers motions not only in cartesian, but also polar and any other coordinates that might exist in $\mathbb{R}^2$. In other words, equation (2.3.17) is at a higher level of abstraction than (2.3.1), and therefore has a higher mathematical value.

EXERCISE 2.3.3. *(a)^C Verify that for $q_1 = x$, $q_2 = y$, equations (2.3.17) indeed coincide with (2.3.1). (b)^A Write (2.3.17) in polar coordinates, with $q_1 = t$, $q_2 = \theta$.*

Next, we consider a system of n point masses in $\mathbb{R}^3$, and assume that there are k degrees of freedom, $k \leq 3n$. In other words, only k out of $3n$ coordinates can change independently of one another; the remaining $3n - k$ coordinates are uniquely determined by those k. The state vector is now

$$(x_1, y_1, z_1,\; x_2, y_2, z_2, \ldots, x_n, y_n, z_n, \dot{x}_1, \dot{y}_1, \dot{z}_1, \ldots, \dot{x}_n, \dot{y}_n, \dot{z}_n)$$

and the generalized coordinates are

$$(q_1, q_2, q_3, \ldots, q_k, \dot{q}_1, \ldots, \dot{q}_k),$$

two for each degree of freedom. Equations of the type (2.3.1)–(2.3.8) can be written for $x_i, q_i, \dot{x}_i, \dot{q}_i$. The kinetic energy is then given by

$$\mathcal{E}_K = \sum_{i=1}^{n}(m_i/2)(\dot{x}_i^2 + \dot{y}_i^2 + \dot{z}_i^2) \text{ or } \mathcal{E}_K = \mathcal{E}_K(q_1, \ldots, q_k, \dot{q}_1, \ldots, \dot{q}_k). \quad (2.3.18)$$

With forces $\boldsymbol{F}_i = (F_{1,i}, F_{2,i}, F_{3,i}),\ i = 1, \ldots, n$, equations (2.3.1) become

$$F_{1,i} = m_i\ddot{x}_i, \quad F_{2,i} = m_i\ddot{y}_i, \quad F_{3,i} = m_i\ddot{z}_i, \quad i = 1, \ldots, n,$$

equation (2.3.12) becomes

$$\frac{d}{dt}\left(\frac{\partial \mathcal{E}_K}{\partial \dot{q}_j}\right) - \frac{\partial \mathcal{E}_K}{\partial q_j} = \sum_{i=1}^{n} m_j\ddot{x}_i\frac{\partial x_i}{\partial q_j} + \sum_{i=1}^{n} m_i\ddot{y}_i\frac{\partial y_i}{\partial q_j} + \sum_{i=1}^{n} m_i\ddot{z}_i\frac{\partial z_i}{\partial q_j}, \quad (2.3.19)$$

equation (2.3.14) becomes

$$\frac{d}{dt}\left(\frac{\partial \mathcal{E}_K}{\partial \dot{q}_j}\right) - \frac{\partial \mathcal{E}_K}{\partial q_j} = \sum_{i=1}^{n}\left(F_{1,i}\frac{\partial x_i}{\partial q_j} + F_{2,i}\frac{\partial y_i}{\partial q_j} + F_{3,i}\frac{\partial z_i}{\partial q_j}\right), \quad (2.3.20)$$

$j = 1, \ldots, k$, and equation (2.3.15),

$$dW = \sum_{i=1}^{n} \boldsymbol{F}_i \cdot dr_i = \sum_{j=1}^{k} Q_j dq_j. \quad (2.3.21)$$

The generalized forces are

$$Q_j = \sum_{i=1}^{n}\left(F_{1,i}\frac{\partial x_i}{\partial q_j} + F_{2,i}\frac{\partial y_i}{\partial q_j} + F_{3,i}\frac{\partial z_i}{\partial q_j}\right) = \sum_{i} \boldsymbol{F}_i \cdot \frac{\partial \boldsymbol{r}_i}{\partial q_j}, \quad (2.3.22)$$

$j = 1, \ldots, k$. Thus, there are exactly k Lagrange's equations, one for each degree of freedom:

$$\frac{d}{dt}\left(\frac{\partial \mathcal{E}_K}{\partial \dot{q}_j}\right) - \frac{\partial \mathcal{E}_K}{\partial q_j} = Q_j, \quad j = 1, \ldots, k. \quad (2.3.23)$$

Let us now look at the particular case of CONSERVATIVE FORCE FIELDS. By definition, we say that the force $\boldsymbol{F}$ is `conservative` or defines a `conservative force field` if the vector $\boldsymbol{F}$ is a gradient of some scalar function: $\boldsymbol{F} = -\boldsymbol{\nabla} V$; the negative sign is used by convention. The function V is called the `potential` of the force field $\boldsymbol{F}$. The work done by a conservative force along a path $\boldsymbol{r}(t)$, $a \le t \le b$, is

$$\begin{aligned} W &= \int_a^b \boldsymbol{F} \cdot \dot{\boldsymbol{r}} dt = \int_a^b -\boldsymbol{\nabla} V \cdot \dot{\boldsymbol{r}}\, dt \\ &= -\int_a^b \left(\frac{\partial V}{\partial x}\frac{dx}{dt} + \frac{\partial V}{\partial y}\frac{dy}{dt} + \frac{\partial V}{\partial z}\frac{dz}{dt} \right) dt = -\int_a^b \frac{dV(\boldsymbol{r}(t))}{dt}\, dt \\ &= -\big(V(\boldsymbol{r}(b)) - V(\boldsymbol{r}(a))\big), \end{aligned} \tag{2.3.24}$$

That is, $dW = -dV$, and the work done by a conservative force equals the change in potential.

Let us assume that all the forces $\boldsymbol{F}_1, \ldots, \boldsymbol{F}_n$ are conservative, and denote by $V_1, \ldots, V_n$ the corresponding potentials. By construction, we have $\boldsymbol{F}_i = \boldsymbol{F}_i(x_i, y_i, z_i)$, $i = 1, \ldots, n$ and so $V_i = V_i(x_i, y_i, z_i)$. Define $V = V_1 + \cdots + V_n$. According to (2.3.21), $dW = \sum_{i=1}^n \boldsymbol{F}_i \cdot d\boldsymbol{r}_i = -\sum_{i=1}^n dV_i = -dV$. We now write V as a function of the *generalized coordinates*: $V = V(q_1, \ldots, q_k)$, and get $dW = -(\partial V/\partial q_1)dq_1 - \cdots - (\partial V/\partial q_k)dq_k$. Comparison of the last equality with (2.3.21) results in the conclusion $Q_j = -\partial V/\partial q_j$, $j = 1, \ldots, k$. Thus, *for conservative forces*, equations (2.3.23) become

$$\frac{d}{dt}\left(\frac{\partial \mathcal{E}_K}{\partial \dot{q}_j}\right) - \frac{\partial \mathcal{E}_K}{\partial q_j} = -\frac{\partial V}{\partial q_j}, \quad j = 1, \ldots, n. \tag{2.3.25}$$

We will transform (2.3.25) even further by noticing that V does not depend on the velocities, and therefore is independent of $\dot{q}_j$: $\partial V/\partial \dot{q}_j = 0$. We then define the `Lagrangian`

$$L = \mathcal{E}_K - V, \tag{2.3.26}$$

and re-write equation (2.3.25) as

$$\frac{d}{dt}\left(\frac{\partial L}{\partial \dot{q}_j}\right) - \frac{\partial L}{\partial q_j} = 0. \tag{2.3.27}$$

Equation (2.3.27) has at least three advantages over (2.3.23): (a) a more compact form; (b) a higher level of abstraction; (c) a possibility to include constraints.

To conclude this section on Lagrange's equations, let us look briefly at some NON-CONSERVATIVE FORCES. As before, $\boldsymbol{q} = (q_1, \ldots, q_k)$ is the vector of generalized coordinates and $\dot{\boldsymbol{q}} = (\dot{q}_1, \ldots, \dot{q}_k)$, the vector of the corresponding time derivatives, so that the vector $(\boldsymbol{q}, \dot{\boldsymbol{q}})$ uniquely determines the state of the system. We derived (2.3.27) under the assumption $\boldsymbol{Q} = (Q_1, \ldots, Q_k) = -\boldsymbol{\nabla} V$, where $V = V(\boldsymbol{q})$. Now *assume* that the generalized force $\boldsymbol{Q}$ is no longer conservative, and has a non-conservative component that depends only on $\dot{\boldsymbol{q}}$; many non-conservative forces, such as friction and some electromagnetic forces, indeed depend only on the velocity. In other words, we assume that

$$\boldsymbol{Q} = \boldsymbol{Q}^{(1)} + \boldsymbol{Q}^{(2)}, \; \boldsymbol{Q}^{(1)} = -\boldsymbol{\nabla} V_1, \; \boldsymbol{Q}^{(2)} = \frac{d}{dt}\boldsymbol{\nabla} V_2 \tag{2.3.28}$$

for some scalar functions $V_1 = V_1(\boldsymbol{q})$ and $V_2 = V_2(\dot{\boldsymbol{q}})$. Then

$$Q_j = -\frac{\partial V_1}{\partial q_j} + \frac{d}{dt}\frac{\partial V_2}{\partial \dot{q}_j}. \tag{2.3.29}$$

Define $V = V_1 + V_2$ so that $\partial V_1/\partial q_j = \partial V/\partial q_j$, $\partial V_2/\partial \dot{q}_j = \partial V/\partial \dot{q}_j$, and

$$Q_j = -\frac{\partial V}{\partial q_j} + \frac{d}{dt}\frac{\partial V}{\partial \dot{q}_j}. \tag{2.3.30}$$

If $L = \mathcal{E}_K - V$, then Lagrange's equations (2.3.27) follow after substituting (2.3.30) into (2.3.17). Alternatively, if we define *the conservative Lagrangian* $L_1 = \mathcal{E}_K - V_1$, then (2.3.17) and (2.3.30) imply

$$\frac{d}{dt}\frac{\partial L_1}{\partial q_1} - \frac{\partial L_1}{\partial q_j} = Q_j^{(2)}, \tag{2.3.31}$$

a modification of (2.3.17).

2.3.2 *An Example of Lagrange's Method*

Our goal in this section is to *verify* the Lagrange equations (2.3.17) for a points mass that moves in space, but has only one degree of freedom. The example below is adapted from the book *Introduction to Analytical Mechanics* by N. M. J. Woodhouse, 1987.

Consider a smooth wire in the shape of an elliptical helix. Choose an inertial frame with origin O and cartesian basis vectors $(\hat{\boldsymbol{\imath}}, \hat{\boldsymbol{\jmath}}, \hat{\boldsymbol{\kappa}})$, so that $\hat{\boldsymbol{\kappa}}$ is along the axis of the helix. In this frame, the helix is the set of points with coordinates (x, y, z) so that

$$\frac{x^2}{a^2} + \frac{y^2}{b^2} = 1, \quad z = c\cos^{-1}\left(\frac{x}{a}\right); \tag{2.3.32}$$

see also Exercise 2.3.4 below. A bead of mass m placed on the wire will slide down under the force of gravity. *Note that the two equations in (2.3.32), describing the helix, define two constraints and leave the bead with only one degree of freedom.* We assume that the friction force is negligible, so that the restraining force $\boldsymbol{N} = (N_1, N_2, N_3)$, exerted by the wire on the bead, is orthogonal to the velocity vector at every point of the motion. Our objective is to derive the equations of the motion of the bead along the wire.

First, let us look at what the Newton-Euler method will produce. By Newton's Second Law, we get three equations

$$m\ddot{x} = N_x, \quad m\ddot{y} = N_y, \quad m\ddot{z} = N_z - mg, \tag{2.3.33}$$

with the six unknowns x, y, z, N_1, N_2, N_3. Equations (2.3.32) are the two constraint equations on x, y, z. The sixth equation comes from the orthogonality condition for frictionless motion:

$$\dot{\boldsymbol{r}} \cdot \boldsymbol{N} = \dot{x}N_1 + \dot{y}N_2 + \dot{z}N_3 = 0. \tag{2.3.34}$$

To solve this system of six equations, it is natural to parameterize the helix by setting

$$x = a\cos q, \quad y = b\sin q, \quad z = cq, \tag{2.3.35}$$

where q is the generalized coordinate for this problem.

EXERCISE 2.3.4.[A] *Verify that equations (2.3.32) and (2.3.35) define the same set of points. How should one interpret the inverse cosine in (2.3.32)?*

The position vector of the bead is $\boldsymbol{r}(q) = x(q)\,\hat{\boldsymbol{\imath}} + y(q)\,\hat{\boldsymbol{\jmath}} + z(q)\,\hat{\boldsymbol{\kappa}}$, and the motion of the bead is determined by the function $q = q(t)$. Note that the tangent vector to the helix is $d\boldsymbol{r}/dq$. Using the six equation, the unknown reaction force $\boldsymbol{N}$ can be eliminated and a single second-order ordinary differential equation for $q = q(t)$ can be obtained; the details of this approach are the subject of Problem 2.9, page 421.

We now look more closely at the alternative approach using the Lagrange method. Let $\boldsymbol{p} = m\,\dot{\boldsymbol{r}}$ be the momentum of the sliding bead. Then equation (2.3.33) in vector form becomes

$$\dot{\boldsymbol{p}} = \boldsymbol{N} - mg\,\hat{\boldsymbol{\kappa}}. \tag{2.3.36}$$

Denote by $p_{\tan} = \boldsymbol{p} \cdot (d\boldsymbol{r}/dq)$ and $Q = -mg\,\hat{\boldsymbol{\kappa}} \cdot (d\boldsymbol{r}/dq)$ the tangential components of the momentum $\boldsymbol{p}$ and the gravitational force $-mg\,\hat{\boldsymbol{\kappa}}$. This definition of Q agrees with (2.3.22) on page 88. Differentiating p_{tan} with respect to t, we get

$$\dot{p}_{\text{tan}} = \dot{\boldsymbol{p}} \cdot \frac{d\boldsymbol{r}}{dq} + \boldsymbol{p} \cdot \frac{d^2\boldsymbol{r}}{dq^2}\,\dot{q}.$$

Using (2.3.36) and $\boldsymbol{p} = m\,\dot{\boldsymbol{r}} = m\,(d\boldsymbol{r}/dq)\,\dot{q}$, we find

$$\dot{p}_{\text{tan}} = (\boldsymbol{N} - mg\,\hat{\boldsymbol{\kappa}}) \cdot \frac{d\boldsymbol{r}}{dq} + m\,\frac{d\boldsymbol{r}}{dq} \cdot \frac{d^2\boldsymbol{r}}{dq^2}\,\dot{q}^2,$$

or, since $\boldsymbol{N} \cdot (d\boldsymbol{r}/dq) = 0$,

$$\dot{p}_{\text{tan}} = Q + m\frac{dr}{dq} \cdot \frac{d^2r}{dq^2}\,\dot{q}^2, \tag{2.3.37}$$

The kinetic energy is then

$$\mathcal{E}_K = \frac{1}{2}m\,\dot{\boldsymbol{r}} \cdot \dot{\boldsymbol{r}} = \frac{m}{2}\frac{d\boldsymbol{r}}{dq} \cdot \frac{d\boldsymbol{r}}{dq}\,\dot{q}^2, \tag{2.3.38}$$

and by (2.3.35),

$$\mathcal{E}_K = \frac{m}{2}(a^2 \sin^2 q + b^2 \cos^2 q + c^2)\,\dot{q}^2. \tag{2.3.39}$$

Hence, from (2.3.38),

$$\frac{\partial \mathcal{E}_K}{\partial \dot{q}} = m\,\frac{d\boldsymbol{r}}{dq} \cdot \frac{d\boldsymbol{r}}{dq}\dot{q} = m\,\frac{d\boldsymbol{r}}{dq} \cdot \frac{d\boldsymbol{r}}{dt} = p_{\text{tan}}.$$

From (2.3.39), using standard rules of partial differentiation,

$$\frac{\partial \mathcal{E}_K}{\partial q} = m(a^2 \sin q \cos q - b^2 \cos q \sin q)\,\dot{q}^2.$$

Now, $\boldsymbol{r}(q) = a\cos q\,\hat{\boldsymbol{\imath}} + b \sin q\,\hat{\boldsymbol{\jmath}} + cq\,\hat{\boldsymbol{\kappa}}$, so that

$$\frac{d\boldsymbol{r}}{dq} = -a \sin q\,\hat{\boldsymbol{\imath}} + b\cos q\,\hat{\boldsymbol{\jmath}} + c\,\hat{\boldsymbol{\kappa}}, \quad \frac{d^2\boldsymbol{r}}{dq^2} = -a\cos q\,\hat{\boldsymbol{\imath}} - b \sin q\,\hat{\boldsymbol{\jmath}}.$$

Therefore,

$$\frac{\partial \mathcal{E}_K}{\partial q} = m\frac{d\boldsymbol{r}}{dq} \cdot \frac{d^2\boldsymbol{r}}{dq^2}\,\dot{q}^2,$$

and equation (2.3.37) is equivalent to Lagrange's equation of motion

$$\frac{d}{dt}\left(\frac{\partial \mathcal{E}_K}{\partial \dot{q}}\right) - \frac{\partial \mathcal{E}_K}{\partial q} = Q, \tag{2.3.40}$$

where Q is the generalized force from (2.3.16), that is,

$$Q = \boldsymbol{F} \cdot d\boldsymbol{r}/dq = -mg\,\hat{\boldsymbol{\kappa}} \cdot d\boldsymbol{r}/dq. \tag{2.3.41}$$

We therefore accomplished our objective and verified the Lagrange method for our example by explicitly deriving equation (2.3.40). Once (2.3.39) and (2.3.41) are taken into account, equation (2.3.40) becomes a second-order ODE for the function $q = q(t)$. A reader interested in solving this equation should try Problem 2.9, page 121.

Notice that the unknown, constraint-induced, reaction force $\boldsymbol{N}$ appears in Newton's equation (2.3.33), but not in the Lagrange equation (2.3.40). Of course, the Newton-Euler method, requiring elimination of the unknown force $\boldsymbol{N}$, results in the same equation of motion; see Problem 2.9 for details.

2.3.3 *Hamilton's Equations*

Lagrange's equations (2.3.27) are a system of k second-order ODEs. A change of variables reduces this system to $2k$ first-order ODEs. While there are many changes of variables to achieve this reduction, one special change transforms (2.3.27) into a particularly elegant form, known as Hamilton's equations.

Given the Lagrange function (2.3.26), introduce new variables $\boldsymbol{p} = (p_1, \ldots, p_k)$ by

$$p_j = \frac{\partial L}{\partial \dot{q}_j}, \tag{2.3.42}$$

and assume that equations (2.3.42) are solvable in the form

$$\dot{\boldsymbol{q}} = \boldsymbol{f}(\boldsymbol{p}, \boldsymbol{q}) \tag{2.3.43}$$

for some vector function $\boldsymbol{f} = (f_1, \ldots, f_k)$. For example, if $L = \frac{1}{2}\sum_{j=1}^{k}(m_j\dot{q}_j^2 - h_j q_j^2)$, then (2.3.42) becomes $p_j = m_j\dot{q}_j$ or $\dot{q}_j = p_j/m_j$. In applications to mechanics, q_i usually has the dimension of distance and m_j is the mass, so that p_j has the dimension of momentum. This is why the variable p_j defined by (2.3.42) is called the `generalized momentum`.

Together with (2.3.42), equations (2.3.27) become a system of k first-order ODEs

$$\frac{\partial p_j}{\partial t} = \frac{\partial L}{\partial q_j},\ j = 1, \ldots, k. \tag{2.3.44}$$

To obtain another k equation, define the **Hamiltonian**

$$H(\boldsymbol{p}, \boldsymbol{q}) = \sum_{j=1}^{k} p_j \dot{q}_j - L(\boldsymbol{q}, \dot{\boldsymbol{q}}), \qquad (2.3.45)$$

where, according to (2.3.43), each $\dot{q}_j$ is a function of $\boldsymbol{p}$ and $\boldsymbol{q}$. Then direct calculations show that

$$\frac{\partial q_j}{\partial t} = \frac{\partial H}{\partial p_j}, \quad \frac{\partial p_j}{\partial t} = -\frac{\partial H}{\partial q_j}, \quad j = 1, \ldots, k. \qquad (2.3.46)$$

Equations (2.3.46) are a system of 2k first-order ODEs, known as **Hamilton's equations.**

EXERCISE 2.3.5.[B] *Derive equations (2.3.46) Hint: the first follows directly by computing $\partial H/\partial p_j$ and using (2.3.42), (2.3.43), and $\dot{q}_j = dq_j/dt$; for the second, note that $\partial L/\partial q_j = -\partial H/\partial q_j$ and use (2.3.44).*

EXERCISE 2.3.6.[C] *Show that if (2.3.31) is used instead of (2.3.27), then the second equation in (2.3.46) becomes*

$$\frac{\partial p_j}{\partial t} = -\frac{\partial H}{\partial q_j} + Q_j^{(2)}. \qquad (2.3.47)$$

We conclude the section by showing that, in a typical mechanical problem, when the vector $\boldsymbol{q}$ is the position and V is the potential energy, the Hamiltonian is the *total energy* of the system.

Theorem 2.3.1 *Consider a system of n point masses in $\mathbb{R}^3$. Assume that the position vectors $\boldsymbol{r}_j$ and the potential V of the system do not depend on $\dot{\boldsymbol{q}}$ and depend only on $\boldsymbol{q}$. Then the Hamiltonian of the system is equal to the total energy:*

$$H(\boldsymbol{p}, \boldsymbol{q}) = \mathcal{E}_K + V. \qquad (2.3.48)$$

Proof. By assumption,

$$\boldsymbol{r}_i = \boldsymbol{r}_i(\boldsymbol{q}), \; i = 1, \ldots, n. \qquad (2.3.49)$$

Then

$$\dot{\boldsymbol{r}}_i = \sum_{j=1}^{3k} \frac{\partial \boldsymbol{g}_i}{\partial q_j} \dot{q}_j. \qquad (2.3.50)$$

Substituting (2.3.50) into

$$\mathcal{E}_K = \frac{1}{2}\sum_{i=1}^{n} m_i \dot{\boldsymbol{r}}_i \cdot \dot{\boldsymbol{r}}_i$$

results in

$$\mathcal{E}_K = \sum_{j,\ell=1}^{3k} a_{j\ell}\dot{q}_i\dot{q}_\ell = \dot{\boldsymbol{q}}^T A\dot{\boldsymbol{q}}, \tag{2.3.51}$$

where $A = (a_{j,\ell},\ j,\ell = 1,\ldots,3k)$ is a symmetric matrix and

$$a_{ij} = \frac{1}{2}\sum_{i=1}^{n} m_i \frac{\partial \boldsymbol{r}_i}{\partial q_j}\cdot\frac{\partial \boldsymbol{r}_i}{\partial q_\ell}.$$

Clearly, if $F(\boldsymbol{x}) = \boldsymbol{x}^T B\boldsymbol{x} = \sum_{i,j} b_{ij}x_ix_j$ is a quadratic form with the coefficients b_{ij} independent of $\boldsymbol{x}$, then $\boldsymbol{\nabla} F = 2B\boldsymbol{x}$ and $\boldsymbol{\nabla} F \cdot \boldsymbol{x} = 2F$. Therefore, by (2.3.51),

$$\sum_{j=1}^{3k} \frac{\partial \mathcal{E}_K}{\partial \dot{q}_j}\dot{q}_j = 2\mathcal{E}_K. \tag{2.3.52}$$

Since V does not depend on $\dot{\boldsymbol{q}}$, it follows that $\partial\mathcal{E}_K/\partial\dot{q}_j = \partial L/\partial\dot{q}_j = p_j$, and (2.3.52) becomes

$$\sum_{i=1}^{3k} p_i\dot{q}_i = 2\mathcal{E}_K. \tag{2.3.53}$$

Since $H = \sum_{i=1}^{3k} p_i\dot{q}_i - L$ and $L = \mathcal{E}_K - V$, equality (2.3.48) follows. □

As with most equation related to the Lagrange-Hamilton method, the main significance of (2.3.46) is its mathematical elegance and high level of abstraction; it is the level of abstraction that makes (2.3.46) an extremely interesting object to study in various branches of mathematics. With all that, equations (2.3.46) simplify the analysis of many concrete problems in mechanics. Unfortunately, these topics fall outside the scope of this book.

2.4 Elements of the Theory of Relativity

Newtonian mechanics, as discussed in the previous sections, describes the motion of objects at speeds much smaller than the speed of light; at speeds close to the speed of light the description is provided by the theory of relativity.

Special relativity mechanics arose from Newtonian mechanics and the hypothesis that the mathematical equations describing laws of mechanics should have the same form in every two inertial frames that move with the constant velocity relative to each other. This hypothesis of *mathematical invariance* does not just express a mathematical aesthetic. From a physical standpoint, the hypothesis requires that two observers, one in a frame O and the other in a frame O_1 moving with constant velocity $\boldsymbol{v}$ relative to O, should be able to describe the motion of a particle using the same law. One such law could be Newton's Second Law (2.1.1) on page 40. Although Newton understood the position is relative to the observer in different frames, he assumed time to be an absolute physical quantity that is the same in different frames. Mathematically, these assumptions are expressed by the `Galilean transformation` $\boldsymbol{r}_1 = \boldsymbol{r} + \boldsymbol{v}t$, $t_1 = t$. Note that the Galilean transformation does leave (2.1.1) unchanged: $\ddot{\boldsymbol{r}}_1 = \ddot{\boldsymbol{r}}$, and so if $m\,\ddot{\boldsymbol{r}} = \boldsymbol{F}$, then $m\,\ddot{\boldsymbol{r}}_1 = \boldsymbol{F}$.

The assumption of absolute time contradicts the experimental result, not known at Newton's time, that light has the same speed in O and O_1. We will see that the constant speed of light implies that time is as relative as position, and the Galilean transformation must be replaced by that of Lorentz. Not surprisingly, the Lorentz transformation no longer leaves (2.1.1) unchanged, and we will see how the law of motion must be modified to satisfy the invariance hypothesis; see (2.4.15) on page 102 below. This new law of motion is a more general mathematical model of mechanics than (2.4.1) and leads to a number of remarkable conclusions, such as the famous relation $\mathcal{E} = mc^2$ between the mass and energy.

The theory of relativity is universally considered as one of the pinnacles of human intellectual achievements. Many also believe that understanding of this theory is beyond the ability of all but very few bright minds. Our main objective in the following few sections is to demystify the subject and to demonstrate that one only needs a basic knowledge of linear algebra and multi-variable calculus to understand the main ideas of the theory, to follow the computations that lead to the most famous implications of this theory, and to see how the mathematical models of relativity interact with the underlying physical phenomena.

We start with a short historical background, mostly related to the development of special relativity. Next, we present an elementary derivation of the Lorentz transformation, and finally discuss the main results of general relativity. An interested reader can follow up on some of the topics in Problem 2.2, page 414, and in Section 3.

2.4.1 *Historical Background*

As one legend puts it, back when he was a little boy, A. Einstein was curious about two questions: (1) What happens if you run after a ray of light and try to catch it? (2) What happens if you are in a moving elevator and suddenly somebody turns off the gravity? The answers to these questions later became the theory of special and general relativity.

We mentioned earlier that both the "relative" and "absolute" points of view had proponents from the early years of modern science. In 1707, the Irish bishop George Berkeley disputed the concept of absolute space as described in Newton's 1687 book *Principia.* Newton contended that an absolute rotation can be determined experimentally by measuring the effect of a centrifugal force as in his famous experiment with a rotating bucket of water in which the surface of the rotating water assumed a concave shape. Hence, Newton argued, there is no need for a second body relative to which the rotation is measured. Berkeley disputed this on philosophical grounds, saying that "motion cannot be understood without a determination of its direction, which in turn cannot be understood except in relation to our or some other body." Absolute space, he claimed, is unobservable. If a body is the only one in the universe, it is meaningless to speak of its rotation, and we need the existence of other bodies, relative to which the rotation can be measured. We now know mathematical formulas supporting this argument: for example, the centrifugal force $-m\boldsymbol{\omega}\times(\boldsymbol{\omega}\times\boldsymbol{r}_1)$ in (2.1.47) on page 66 involves the rotation vector $\boldsymbol{\omega}$ that must be defined in an inertial frame. For more about the dispute between Newton and Berkeley, see the book *The Physical Foundations of General Relativity* by D. Sciama, 1969.

The propagation of light was more difficult to understand. Up until the late 19th century, it was assumed that light propagated through the *aether* (or ether), a hypothetical substance believed to be filling all absolute space and serving as the medium for the propagation of electromagnetic waves. In 1887 in Ohio, two American physicists, ALBERT ABRAHAM MICHELSON (1852–1931) and EDWARD WILLIAMS MORLEY (1838–1923) conducted an experiment aimed at measuring the velocity of the Earth relative to the aether.

The relative propagation speed of other wave disturbances in a medium, like sound and water waves, depends on the velocity of the source. If the aether existed, then the Earth's motion around the Sun, at the speed of about 30km/sec, would affect the speed of light depending on the direction of the light beam. Michelson and Morley emitted a beam of light, first in the direction of the Earth's motion and then perpendicular to it. They used mirrors and an interferometer to measure the speed of light in each case, and failed to detect any significant difference; see Problem 3.3 on page 424 for details. As a result, contrary to its initial intent, the Michelson-Morley experiment provided the first evidence *against* the existence of the aether, and in that respect became the most famous failed experiment in history. The experiment certainly had a profound effect on the further development of natural sciences, and in 1907 Michelson became the first American to receive a Nobel prize in physics.

In 1905, Albert Einstein, who was 26 at the time, made the result of the Michelson-Morley experiment the basis of a new theory of relative motion, *special relativity.* He proposed two *postulates*:

(i) An observer measures the speed of light to be the same in any direction regardless of the observer's motion relative to other bodies; this is a consequence of the Michelson-Morley experiment.

(ii) Any two observers moving with constant relative velocity can agree on all observed data and mathematical relations (laws) between the data; this is the postulate of *mathematical invariance.*

According to Postulate (i), all observers measure the same value for the speed of light in a vacuum: $c = 3 \times 10^8$ meters/second. It follows that absolute motion (in some special frame) is unobservable, hence physically meaningless. In other words, Einstein sided with Berkeley against Newton on the question of relative motion. The letter c in the notation for the speed of light is for celerity (swiftness).

According to Postulate (ii), for two observers in frames O and O_1, mathematical relations which hold in the frame O are to be transformed into corresponding relations in the other frame O_1 by a transformation of the variables (x, y, z, t) to the variables (x_1, y_1, z_1, t_1); the experimental study of light propagation suggests that this transformation should be *linear.* The result, called the *Lorentz transformation,* is a necessary consequence of Einstein's two postulates.

2.4.2 The Lorentz Transformation and Special Relativity

Consider two frames, the fixed frame O and the moving frame O_1. We choose cartesian coordinate systems in both frames and assume that, at each time moment, the frame O_1 is a parallel translation of the frame O so that the vectors $\hat{\imath}$, $\hat{\imath}_1$ are always on the same line; see Figure 2.4.1. As usual, we represent the points in the frames O and O_1 with vectors of the form $x\,\hat{\imath} + y\,\hat{\jmath} + z\,\hat{\kappa}$ and $x_1\,\hat{\imath}_1 + y_1\,\hat{\jmath}_1 + z_1\,\hat{\kappa}_1$, respectively.

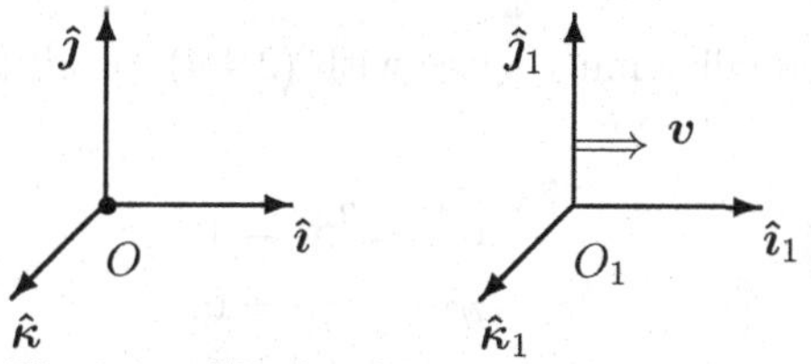

Fig. 2.4.1 Fixed and Moving Frames

Thus, O_1 is moving to the right relative to O with constant velocity $\boldsymbol{v}$. A linear transformation of the space-time coordinates is

$$x = \alpha x_1 + \beta t_1, \quad y = y_1, \quad z = z_1 \tag{2.4.1}$$

and

$$t = \gamma x_1 + \sigma t_1 \tag{2.4.2}$$

for some real numbers $\alpha, \beta, \gamma, \sigma$. Equation (2.4.2) means that time is no more absolute than space: each frame has its own time coordinate as well as space coordinates. Equations (2.4.1) and (2.4.2) also demonstrate that space and time are bound together in one `space-time` continuum. We will see that Postulate (i) cannot hold without relation (2.4.2) with appropriate γ and σ.

Consider point O_1, the origin of the moving frame. It has the $\hat{\imath}_1$ coordinate $x_1 = 0$ in the O_1 frame at all times t_1. In the fixed O frame, by (2.4.1), the $\hat{\imath}$ coordinate of O_1 is $x = \beta t_1$; by (2.4.2), we also have $t = \sigma t_1$. Note that $\beta > 0$. If $v = \|\boldsymbol{v}\|$ is the speed of O_1 relative to O, then $v = x/t = \beta/\sigma$. A light impulse emitted at the instant when $O = O_1$ will reach, at time t, a point P, where the coordinates (x, y, z) of P in O satisfy

$$x^2 + y^2 + z^2 = c^2 t^2. \tag{2.4.3}$$

By Postulate (i), if (x_1, y_1, z_1) are the coordinates of the same point P in the frame O_1, then

$$x_1^2 + y_1^2 + z_1^2 = c^2 t_1^2 \tag{2.4.4}$$

By Postulate (ii), and because of relations (2.4.1), (2.4.2) above, equation (2.4.3) becomes

$$(\alpha x_1 + \beta t_1)^2 + y_1^2 + z_1^2 = c^2(\gamma x_1 + \sigma t_1)^2, \text{ or}$$
$$(\alpha^2 - c^2\gamma^2)x_1^2 + y_1^2 + z_1^2 + 2(\alpha\beta - c^2\gamma\sigma)x_1 t_1 = c^2(\sigma^2 - \beta^2/c^2)t_1^2.$$

Since the last equality must agree with (2.4.4) for all (x_1, y_1, z_1, t_1) values, it follows that

$$\alpha^2 - c^2\gamma^2 = 1, \tag{2.4.5}$$
$$\alpha\beta - c^2\gamma\sigma = 0, \tag{2.4.6}$$
$$\sigma^2 - \beta^2/c^2 = 1. \tag{2.4.7}$$

These three equation together with $\beta/\sigma = v$ can be solved easily for $\alpha, \beta, \gamma, \sigma$. From (2.4.6), $\beta/\sigma = c^2\gamma/\alpha = v$. So $\gamma = v\alpha/c^2$. Substituting in (2.4.5), we get $\alpha = (1 - v^2/c^2)^{-\frac{1}{2}}$ and so $\gamma = (v/c^2)(1 - v^2/c^2)^{-\frac{1}{2}}$. In (2.4.7), substitute $\beta = v\sigma$ and solve for σ to get $\sigma = \alpha$. Therefore, (2.4.1) and (2.4.2) become $y = y_1$, $z = z_1$ and

$$x = \frac{x_1 + vt_1}{\sqrt{1 - \frac{v^2}{c^2}}}, \quad t = \frac{t_1 + \frac{v}{c^2}x_1}{\sqrt{1 - \frac{v^2}{c^2}}}. \tag{2.4.8}$$

In what follows, we will often use the notations

$$\alpha = \left(1 - \frac{v^2}{c^2}\right)^{-1/2}, \quad \gamma = \frac{v}{c^2}\alpha, \tag{2.4.9}$$

and write (2.4.8) as

$$x = \alpha x_1 + v\alpha t_1, \qquad t = \alpha t_1 + \gamma x_1. \tag{2.4.10}$$

Then the inverse transformation becomes

$$x_1 = \alpha x - v\alpha t, \qquad t_1 = \alpha t - \gamma x. \tag{2.4.11}$$

One can verify this directly or use the physical argument that the inverse transformation must be obtained from (2.4.10) by switching the roles of (x, t), (x_1, t_1) and replacing v with $-v$.

Equations (2.4.9), (2.4.10), and (2.4.11) are the `Lorentz transformation` equations that govern *special relativity* kinematics. If v is very small compared to c, then $1 - (v^2/c^2)$ is close to 1 and the Lorentz equations become (approximately) $x = x_1 + vt,\ t = t_1$. This is the traditional `Galilean transformation` relating two frames moving with relative velocity $\boldsymbol{v}$ in Newtonian kinematics. We always use the Galilean transformation, especially the $t = t_1$ part, in our investigations of the classical Newtonian mechanics. Let us stress again that, in Newtonian mechanics, time is the same in both frames.

Equations (2.4.11) were first proposed in 1904 by the Dutch scientist HENDRIK ANTOON LORENTZ (1853–1928) as a means of explaining the result of the Michelson-Morley experiment. The transformation became the main mathematical means of verifying whether a proposed equation can describe effects at speeds comparable with the speed of light: all one needs to check is whether the equation is invariant under the Lorentz transformation, or, in other words, whether the equation treats the space and time variables in a manner consistent with special relativity.

The Galilean transformation goes back to Galilei's 1638 book *Discorsi e Dimostrazioni Matematiche, intorno á due nuoue scienze*, known in the English translation as *Discourse on Two New Sciences.*

The Lorentz transformation implies a number of important, and sometimes counter-intuitive, results of special relativity. Let us summarize some of these results.

(i) RELATIVITY OF SPACE AND TIME INTERVALS: The length ℓ of a rod placed along the x-axis and moving with speed v, as measured by an observer in the fixed frame O, is smaller than the length ℓ_0 of the same rod measured in the frame O_1 that moves with the rod; see Figure 2.4.1. Similarly, the time interval Δt between two events in a frame moving with speed v, as measured in the fixed frame, is longer than the time interval Δt_0 between the same events, measured in the moving frame:

$$\ell = \ell_0\sqrt{1-(v^2/c^2)},\quad \Delta t = \frac{\Delta t_0}{\sqrt{1-(v^2/c^2)}}. \qquad (2.4.12)$$

(ii) RELATIVITY OF MASS: the `rest mass` m_0 of an object is either its inertial or gravitational mass measured in an inertial frame when the object is not moving relative to that frame. The `relativistic mass` m and the

total energy $\mathcal{E}$ of the same object moving with velocity v are

$$m = \frac{m_0}{\sqrt{1-(v^2/c^2)}}, \quad \mathcal{E} = mc^2. \tag{2.4.13}$$

EXERCISE 2.4.1.[B] *Verify (2.4.12). Hint: Use the Lorentz transformation. For* ℓ, *use (2.4.11), since* $\ell_0 = \Delta x_1 = (\Delta x - v\Delta t)\big(1-(v/c)^2)\big)^{-1/2}$ *with* $\Delta t = 0$, *because the difference of the coordinates is recorded at the same time in the fixed frame. Similarly, for* Δt, *use (2.4.8) with* $\Delta x_1 = 0$, *as the events happen at the same point of the moving frame.*

The relativity of mass (2.4.13) is a consequence of the relativity of time and is a necessary consequence of the required invariance of the laws of mechanics under the Lorentz transformation. This is one of the foremost examples of how a mathematical model, namely, the Lorentz transformation, leads to discoveries of new physical phenomena. Relations (2.4.13) have been verified experimentally on various particles. In 2005, the journal *Nature* reported an experimental verification of the relation $\mathcal{E} = mc^2$ to within 0.00004%.

We will now derive the main equation of relativistic kinematics. Recall that Newton's Second Law for the point mass m can be written as $\boldsymbol{F} = d\boldsymbol{p}/dt$, where $\boldsymbol{p} = m\dot{\boldsymbol{r}}$ is the momentum of m; see page 40. The extra requirement of invariance of the law under the Lorentz transformation leads to the definition of the **relativistic momentum**

$$\boldsymbol{p} = m\dot{\boldsymbol{r}} = \frac{m_0\dot{\boldsymbol{r}}}{\sqrt{1-(\|\dot{\boldsymbol{r}}\|/c)^2}}, \tag{2.4.14}$$

where m is the relativistic mass from (2.4.13). The relativistic form of Newton's Second Law is therefore

$$\boldsymbol{F} = \frac{d\boldsymbol{p}}{dt}, \text{ or } \boldsymbol{F} = m_0\frac{d}{dt}\left(\frac{\dot{\boldsymbol{r}}}{\sqrt{1-(\|\dot{\boldsymbol{r}}\|/c)^2}}\right). \tag{2.4.15}$$

Equation (2.4.15) is the main equation of relativistic kinematics. It is invariant under the Lorentz transformation, just as the non-relativistic Second Law of Newton is invariant under the Galilean transformation; see Problem 3.2 on page 423. Note that, in the limits $c \to \infty$ or $\|d\boldsymbol{r}\|/c \to 0$, meaning speed of motion much smaller than the speed of light, equation (2.4.14) becomes the familiar Second Law of Newton (2.1.1): $\boldsymbol{F} = m_0\,\ddot{\boldsymbol{r}}$.

EXERCISE 2.4.2. *(a)*^{C} *Verify that*

$$\|\boldsymbol{p}\|^2 + m_0^2c^2 = m^2c^2. \tag{2.4.16}$$

Hint: direct computation using (2.4.14).
(b)^B *Show that the quantity* $c^2(m - m_0)$ *is natural to interpret as the* relativistic kinetic energy, *that is,* $\mathcal{E} = mc^2 = \mathcal{E}_K + m_0c^2$. *Hint: if* $\mathcal{C}$ *is the trajectory,* $v = \|\dot{\boldsymbol{r}}\|$ *is the speed, and the initial speed is zero, then the work done by a force* $\boldsymbol{F}$ *is*

$$\mathcal{E}_K = \int_{\mathcal{C}} \boldsymbol{F} \cdot d\boldsymbol{r} = \int_0^v \dot{\boldsymbol{r}} \cdot d(m\,\dot{\boldsymbol{r}}) = \int_0^v y\, d\left(\frac{m_0 y}{\left(1 - (y^2/c^2)\right)^{1/2}}\right) = mc^2 - m_0c^2,$$

where the last equality follows after integration by parts.

One consequence of the second equality in (2.4.12) is the `Twin Paradox`, which is often described as follows. Consider two identical twin brothers, Peter and Paul, who live in an inertial frame O. Peter stays in that frame O, and Paul makes a round trip in a spaceship, travelling out and back in a fixed direction with fixed speed $v = (24/25)c$. Then the total travel time $\Delta t_0 = 2t_1$, as measured by Paul in the moving spaceship's frame, will be $\Delta t = 25\Delta t_0/7$ in the frame O as measured by Peter; the key relation here, beside (2.4.12), is $25^2 = 24^2 + 7^2$. In particular, if $\Delta t_0 = 14$ years, then $\Delta t = 50$ years, that is, Peter will be waiting for 50 years for Paul to return from what is 14-year long trip for Paul. If Paul's biological clock also obeys (2.4.12), then, by the end of the trip, Paul will be 26 year younger than Peter.

For further discussions of (2.4.12) and (2.4.13), including the Twin Paradox, see, for example, the books *Concepts of Modern Physics* by A. Beiser, 2002, and *Spacetime Physics* by E. F. Taylor and J. A. Wheeler, 1992.

We conclude this section with a brief discussion of the geometry of relativistic space-time. The reader will see that this geometry is not exactly Euclidean, thus getting a gentle introduction to *general relativity.*

`Relativistic space-time` is a combination of time with the three spacial dimensions of our physical space, and is modelled mathematically as a four-dimensional vector space. Fixing a coordinate system in this space represents every point by four coordinates (x, y, z, t). A curve $\boldsymbol{r}(s) = (x(s), y(s), z(s), t(s))$, $s_0 \le s \le s_1$, in this space is called a `world line`. If this were a truly Euclidean space with cartesian coordinates, then, by analogy with formula (1.3.10) for the arc length on page 29, the length of this world line would be $L_{s_1,s_2} = \int_{s_1}^{s_2} \sqrt{|x'(s)|^2 + |y'(s)|^2 + |z'(s)|^2 + |t'(s)|^2}\, ds$. Even if we disregard that t has the dimension of time and not distance, this is still not the correct way to measure the length: as the reader will soon discover, this formula is not invariant under the Lorentz transformation. Accordingly, let us try to find an invariant distance formula in the form $\int_{s_1}^{s_2} \sqrt{a_1|x'(s)|^2 + a_2|y'(s)|^2 + a_3|z'(s)|^2 + b|t'(s)|^2}\, ds$ for some real numbers a_1, a_2, a_3, b; the requirement of invariance takes priority over the requirement of positivity. Following formula (1.3.12) for the arc length on page 29, it is convenient to write the distance formula in the differential form:

$$(d\mathfrak{s})^2 = a_1(dx)^2 + a_2(dy)^2 + a_3(dz)^2 + b\,(dt)^2. \tag{2.4.17}$$

EXERCISE 2.4.3[B] *Verify that (2.4.17) is invariant under the Lorentz transformation if and only if $a_1 = a_2 = a_3 = -b/c^2$, where c is the speed of light. Hint: direct computations. Start with the frame moving along the x-axis so that $x = \alpha x_1 + \alpha v t_1$, $y = y_1$, $z = z_1$, $t = \alpha t_1 + \gamma x_1$, with α, γ from (2.4.9) on page 100. Compute $dx = \alpha dx_1 + \alpha v dt_1$, etc., and plug in (2.4.17). Then the invariance requirement $(d\mathfrak{s})^2 = a_1(dx_1)^2 + a_2(dy_1)^2 + a_3(dz_1)^2 + b(dt_1)^2$ implies $c^2 a_1 = -b$. Letting the frame move along y axis and then along z axis, conclude that $c^2 a_2 = -b$, $c^2 a_3 = -b$.*

As a result, the general form of $d\mathfrak{s}$, invariant under the Lorentz transformation, is $(d\mathfrak{s})^2 = a\big((dx)^2 + (dy)^2 + (dz)^2 - c^2(dt)^2\big)$, and the two natural choices of a are 1 and -1; other choices would introduce uniform stretching or compressing. Both $a = 1$ and $a = -1$ are used in different areas of physics. In what follows, our selection will be $a = 1$, corresponding to

$$(d\mathfrak{s})^2 = (dx)^2 + (dy)^2 + (dz)^2 - c^2(dt)^2. \tag{2.4.18}$$

To avoid negative distances, we have to modify the arc length formula (1.3.10) and compute the length of the world line as follows:

$$L(s_1, s_2) = \int_{s_1}^{s_2} \sqrt{\varepsilon\left(|x'(s)|^2 + |y'(s)|^2 + |z'(s)|^2 - c^2|t'(s)|^2\right)}\, ds, \tag{2.4.19}$$

where

$$\varepsilon = \begin{cases} 1, & \text{if } |x'(s)|^2 + |y'(s)|^2 + |z'(s)|^2 - c^2|t'(s)|^2 \geq 0 \text{ (space-like world line)}, \\ -1, & \text{if } |x'(s)|^2 + |y'(s)|^2 + |z'(s)|^2 - c^2|t'(s)|^2 < 0 \text{ (time-like world line)}. \end{cases}$$

This special way to measure distance is the main difference between relativistic space-time and ordinary Euclidean space. As a result, the vector space $\mathbb{R}^4$ in which the distance is measured by (2.4.18) is often called the `Minkowski space`, after the German mathematician HERMANN MINKOWSKI (1864–1909), who introduced relativistic space-time in 1907 as a part of a new theory of electrodynamics.

Mathematics takes (2.4.17) to a whole new level of abstraction by writing

$$(d\mathfrak{s})^2 = (dx, dy, dz, dt)\, \mathfrak{g}\, (dx, dy, dz, dt)^T, \tag{2.4.20}$$

where (dx, dy, dz, dt) is the row vector, $\mathfrak{g}$ is the `metric tensor`, that is, a 4×4 *symmetric* matrix with non-zero determinant, and $(dx, dy, dz, dt)^T$ is the column vector, Formula (2.4.20) defines a `metric` in $\mathbb{R}^4$, and $d\mathfrak{s}$ is called the `line element` corresponding to the metric $\mathfrak{g}$. As a result, the same vector space $\mathbb{R}^4$ can lead to different *metric spaces*, corresponding to different choices of the matrix $\mathfrak{g}$. While this construction is possible in any number of dimensions, and is the subject of `pseudo-Riemannian geometry` (the standard Riemannian geometry deals with positive-definite matrices $\mathfrak{g}$), we will only consider the four-dimensional case. The matrix $\mathfrak{g}$ can, in principle, be different at different points (x, y, z, t). A metric corresponding to a constant matrix $\mathfrak{g}$ is called `flat`; in particular, the metric (2.4.18) used in *special relativity* is often called the `flat Minkowski metric`. In `general relativity`, the metric is not flat, and the matrix $\mathfrak{g}$ is different at different points, that is, each element $\mathfrak{g}_{ij}$ of $\mathfrak{g}$ is a function of (x, y, z, t). We will discuss general relativity in Section 2.4.3.

If the matrix $\mathfrak{g}$ is the same at all points, then, after a suitable change of basis, we can assume that $\mathfrak{g}$ is diagonal, see Exercise 8.1.5 on page 454. For a non-constant matrix, a simultaneous reduction to a diagonal form at all points at once might not be possible. A non-constant matrix $\mathfrak{g}$ that is diagonal everywhere still gives rise to a non-flat metric.

EXERCISE 2.4.4. ^{C} *(a) What matrix corresponds to the usual Euclidean metric? Hint: the identity matrix. (b) Write the components of the matrix $\mathfrak{g}$ corresponding to the metric (2.4.18). Hint: the matrix is diagonal.*

2.4.3 *Einstein's Field Equations and General Relativity*

As we discussed on page 95, Einstein adopted the hypothesis that physical laws should be stated in a form that is invariant under coordinate transformations between frames. In his theory of special relativity, the frames move with constant relative velocity, and there is only one possible coordinate transformation: the Lorentz transformation. In his theory of general relativity, the frames can move with variable relative velocities, for example, rotate, and there are many admissible coordinate transformations. To achieve this general invariance, he used tensor analysis and non-Euclidean geometry. Thus,the mathematical foundation of general relativity is tensor analysis and pseudo-Riemannian geometry, and the main result is a system of ten nonlinear partial differential equations. This is more than enough to scare most people away from even trying to understand the subject. The good news is that, while deriving the equations indeed requires all this advanced knowledge, solving these equations often requires little beyond the basic theory of ordinary differential equations. Accordingly, we start by introducing the equations. We will then describe the meaning of all components of the equations, and finally derive and analyze one particular solution, known as the Schwarzschild solution. For the sake of completeness, we summarize the main facts about tensors in Section 8.3 in Appendix.

At the end of Section 2.4.2, we discussed the metric geometry of relativistic space-time. This geometry is characterized by the matrix, or the `metric tensor`, $\mathfrak{g}$ in equation (2.4.20), page 105. In special relativity, the metric is *flat*, that is, the matrix $\mathfrak{g}$ is the same at every point, see (2.4.18) on page 104.

In general relativity, the metric tensor can be different at different points and is determined by the *gravitational field.* The corresponding metric in space-time is no longer flat, but *curved.* This equivalence between gravitation and curvature of space-time is the mathematical expression of the main idea of general relativity. Let us emphasize that general relativity deals only with gravitational forces.

Einstein's field equations, also known as `Einstein's gravitation equations`, or the field equations of general relativity, describe the relation between the metric tensor $\mathfrak{g}$ and the gravitational field at every point of relativistic space-time. To state these equations, we switch from the (x, y, z, t) variables to (x^1, x^2, x^3, x^4) variables:

$$x = x^1,\ y = x^2,\ z = x^3,\ t = x^4,\ \ \boldsymbol{x} = (x^1, x^2, x^3, x^4). \tag{2.4.21}$$

Note the use of super-scripts rather than sub-scripts in this definition. The position of indices is important in tensor calculus; for example, in tensor calculus, $\mathfrak{g}^{ij}$ is an object very different from $\mathfrak{g}_{ij}$. *The reader should pay special attention to position of indices in all formulas in this section.*

The equations, introduced by A. Einstein in 1915, are written as follows:

$$R_{ij}[\mathfrak{g}] - \frac{1}{2}R[\mathfrak{g}]\,\mathfrak{g}_{ij} = \frac{8\pi \mathrm{G}}{c^4}\mathrm{T}_{ij}[\mathfrak{g}],\quad i,j = 1,\dots 4, \tag{2.4.22}$$

where R_{ij}, R, and T_{ij} are non-linear partial differential operators acting on the metric tensor $\mathfrak{g}$.

Our most immediate goal now is to understand the meaning of every component of (2.4.22).

• The metric tensor $\mathfrak{g} = \left(\mathfrak{g}_{ij}(\boldsymbol{x}),\ i,j = 1,\dots,4\right)$ is a symmetric 4×4 matrix defined at the points $\boldsymbol{x} = (x^1, x^2, x^3, x^4)$ of relativistic space-time. This matrix has a nonzero determinant and continuous second-order partial derivatives of all components $\mathfrak{g}_{ij}(\boldsymbol{x})$ for most $\boldsymbol{x}$ (say, all but finitely many). This matrix defines a metric on space-time by defining the `line element` ar every point $\boldsymbol{x}$ according to the formula

$$(d\mathfrak{s})^2(\boldsymbol{x}) = \sum_{i,j=1}^{4} \mathfrak{g}_{ij}(\boldsymbol{x})dx^i\,dx^j; \tag{2.4.23}$$

compare this with (2.4.20) on page 105. Summations of this kind will appear frequently in our discussion. To minimize the amount of writing, we will use `Einstein's summation convention`, assuming summation in an expression over an index from 1 to 4 if the index appears in the expression twice. With this convention, (2.4.23) becomes

$$(d\mathfrak{s})^2(\boldsymbol{x}) = \mathfrak{g}_{ij}(\boldsymbol{x})\,dx^i\,dx^j. \tag{2.4.24}$$

EXERCISE 2.4.5.^{C} *The* `Kronecker symbol` δ *is defined by*

$$\delta^i_j = \begin{cases} 1, & i = j, \\ 0, & i \neq j \end{cases} \tag{2.4.25}$$

Verify that, according to Einstein's summation convention, $\delta^i_i = 4$.

• The operators R_{ij} in (2.4.22) are defined as follows:

$$R_{ij}[\mathfrak{g}](\boldsymbol{x}) = \frac{\partial \Gamma^k_{ik}(\boldsymbol{x})}{\partial x^j} - \frac{\partial \Gamma^k_{ij}(\boldsymbol{x})}{\partial x^k} + \Gamma^m_{ik}(\boldsymbol{x})\Gamma^k_{jm}(\boldsymbol{x}) - \Gamma^m_{mk}(\boldsymbol{x})\Gamma^k_{ij}(\boldsymbol{x}), \tag{2.4.26}$$

where

$$\Gamma^m_{kj}(\boldsymbol{x}) = \frac{1}{2}\mathfrak{g}^{mn}(\boldsymbol{x})\left(\frac{\partial \mathfrak{g}_{nj}(\boldsymbol{x})}{\partial x^k} + \frac{\partial \mathfrak{g}_{nk}(\boldsymbol{x})}{\partial x^j} - \frac{\partial \mathfrak{g}_{jk}(\boldsymbol{x})}{\partial x^n}\right) \tag{2.4.27}$$

and $\mathfrak{g}^{mn}(\boldsymbol{x})$, $m,n = 1,\dots,4$, are the components of the *inverse* matrix $\mathfrak{g}^{-1}(\boldsymbol{x})$, that is, $\boxed{\mathfrak{g}^{mn}(\boldsymbol{x})\mathfrak{g}_{nj}(\boldsymbol{x}) = \delta^m_j}$ for every $\boldsymbol{x}$, with δ^m_j defined in

(2.4.25).

• The operator R in (2.4.22) is defined as follows:

$$R[\mathfrak{g}](\boldsymbol{x}) = \mathfrak{g}^{ij}(\boldsymbol{x}) R_{ij}[\mathfrak{g}](\boldsymbol{x}). \tag{2.4.28}$$

EXERCISE 2.4.6. *(a)*^{C} *With the summation convention now in force, write all the missing sums in (2.4.26), (2.4.27), and (2.4.28).* *(b)*^{B} *Verify that, for all* $\boldsymbol{x}$ *and all* i, j, k, m, $\Gamma^m_{kj} = \Gamma^m_{jk}$, *and so there are at most 40 distinct values of* Γ^m_{kj}. *Hint: all you need is that* $\mathfrak{g}_{ij}(\boldsymbol{x}) = \mathfrak{g}_{ji}(\boldsymbol{x})$. *(c)*^{A} *Verify that* $R_{ij}[\mathfrak{g}](\boldsymbol{x}) = R_{ji}[\mathfrak{g}](\boldsymbol{x})$. *Hint: find a function* f *so that* $\Gamma^m_{mj} = \partial f/\partial x^j$.
(d)^B *Verify that*

$$\mathfrak{g}^{ij}(\boldsymbol{x}) R_{ij}[\mathfrak{g}](\boldsymbol{x}) - \frac{1}{2}\mathfrak{g}^{ij}(\boldsymbol{x}) R[\mathfrak{g}](\boldsymbol{x})\, \mathfrak{g}_{ij}(\boldsymbol{x}) = -R[\mathfrak{g}](\boldsymbol{x}). \tag{2.4.29}$$

• The collection of operators T_{ij}, $i, j = 1, \ldots, 4$, in (2.4.22) is called the `stress-energy tensor` and is determined by a given distribution of the gravitational matter in a region of space. This tensor is symmetric: $\mathrm{T}_{ij} = \mathrm{T}_{ji}$ (it has to be, since the left-hand side of (2.4.22) is symmetric). At this level of generality, no specific representation for T_{ij} can be provided: the particular form of T_{ij} depends on the particular problem. In what follows, we will concentrate on the `vacuum solutions` of the field equations, that is, the solutions of (2.4.22) corresponding to $\mathrm{T}_{ij} = 0$; accordingly, we will not discuss the stress-energy tensor any further.
• The remaining letters in (2.4.22) have the following meaning: $\mathrm{G} \approx 6.67 \cdot 10^{-11}$ m^3/(kg$\cdot$ sec^2) is the universal gravitational constant, $c \approx 3 \cdot 10^8$ m/sec is the speed of light, $\pi \approx 3.14$.

EXERCISE 2.4.7.^{C} *Verify that (2.4.22) is a system of ten equations for the ten components* $\mathfrak{g}_{ij}$, $1 \le i \le j \le 4$ *of the metric tensor.*

The system of equations (2.4.22) does not depend on the particular choice of the coordinate system (x^1, x^2, x^3, x^4) in space-time; in fact, the functions $\mathfrak{g}_{ij}$, Γ^m_{nk}, $R_{ij}[\mathfrak{g}]$, and $R[\mathfrak{g}]$ can be defined in a coordinate-free way, and the formulas (2.4.26)–(2.4.28) turn out the same in every coordinate system (of course, the particular form of $\mathfrak{g}_{ij}$ will depend on the coordinate system). The resulting independence of the field equations (2.4.22) of the coordinate system is the very foundation of general relativity.

Equations (2.4.22) are too complicated to justify as a postulate or as a generalization of experimental facts. Instead, *the main postulate of gen-*

eral relativity is that the physical laws are independent of a particular frame, inertial or not, in space-time, that is, the equations are invariant under general coordinate transformations from one frame to another; what makes special relativity *special* is the restriction of this invariance to inertial frames. When combined with differential geometry and tensor calculus, whose constructions do not depend on a particular coordinate system and are valid in every space, Euclidean or not, this invariance assumption motivates the formulation of equations (2.4.22). Before formulating the equations in 1915, Einstein studied Riemannian geometry and also learned tensor calculus directly from one of the creators of the theory, the Italian mathematician TULLIO LEVI-CIVITA (1873–1941). We give a brief and elementary summary of tensors on page 457 in Appendix.

In classical Newtonian theory, the gravitational field is described by the partial differential equation $\varphi_{xx} + \varphi_{yy} + \varphi_{zz} = 4\pi \mathrm{G}\rho$, where φ is the potential of the field and ρ is the density of the gravitating mass; see Exercise 3.3.8 on page 168 below. Einstein started with a relation, $A_{ij}[\mathfrak{g}] = 4\pi \mathrm{G}\mathrm{T}_{ij}[\mathfrak{g}]$, with $\mathfrak{g}$ instead of φ and the stress-energy tensor instead of the usual mass density, and assumed that the tensor A_{ij} depends linearly on the second-order partial derivatives of $\mathfrak{g}$. In tensor analysis, it is proved that such a tensor $A_{ij}[\mathfrak{g}]$ must have the form

$$A_{ij}[\mathfrak{g}] = \alpha\, R_{ij}[\mathfrak{g}] + \beta\, R[\mathfrak{g}]\mathfrak{g}_{ij} + \gamma\, \mathfrak{g}_{ij}$$

for some real numbers α, β, γ, with R_{ij} and R defined in (2.4.26) and (2.4.28), respectively. Physical considerations, such as dimension analysis and conservation of energy and momentum, result in the values $\alpha = -2\beta = c^4/2$, $\gamma = 0$. For more details, see the book *Theory of Relativity* by W. Pauli, 1981.

To complete the general discussion of (2.4.22), let us mention some other related terms and names:

• The collection of $R_{ij}[\mathfrak{g}]$, $i, j = 1, \ldots, 4$, is called the `Ricci curvature tensor` (also *Ricci curvature* or *Ricci tensor*), after the Italian mathematician GREGORIO RICCI-CURBASTRO (1853–1925), whose most famous publication *The Absolute Differential Calculus*, written together with his former student T. Levi-Civita, was published with his name truncated to Ricci.

• The number $R[\mathfrak{g}]$ is called the `scalar curvature` or `Ricci scalar`.

• Each Γ^i_{jk} is called a `Christoffel symbol`, after the German mathematician ELWIN BRUNO CHRISTOFFEL (1829–1900). Because of the way the numbers Γ^i_{jk} change when the coordinate system is changed, the collection of Christoffel symbols is not a tensor.

• The tensor $R_{ij}[\mathfrak{g}] - (1/2)R[\mathfrak{g}]\,\mathfrak{g}_{ij}$, which is the left-hand side of (2.4.22), is called `Einstein's tensor`.

Even though equations (2.4.22) are complicated, let us assume that we were able to solve them and found the metric tensor $\mathfrak{g}$. In classical Newtonian mechanics, the ultimate goal is to compute the trajectory of a moving object under the action of forces. The equation of the trajectory is provided by the Second Law of Newton. In general relativity, the Second Law of Newton is replaced by a postulate that a point mass in a curved space moves along a `geodesic`. A geodesic in a curved space is what a straight line is in a flat space. Once you think about it, a straight line in a flat space has two special properties: (a) it defines the path of shortest distance between two points; (b) it is straight in the sense that the tangent vectors at each point are parallel, all being parallel to the direction vector of the line. Accordingly, each of these properties could be used to define a geodesic in a curved space, and it turns out that the second property results in a more convenient definition. By this definition, it is shown in differential geometry that $x^i = x^i(s)$, $s \geq 0$, $i = 1, \ldots, 4$, is the vector representation of a geodesic if and only if

$$\frac{d^2\,x^i(s)}{ds^2} + \Gamma^i_{jk}(\boldsymbol{x}(s))\,\frac{dx^j(s)}{ds}\,\frac{dx^k(s)}{ds} = 0, \tag{2.4.30}$$

where the Γ^i_{jk} are from (2.4.27) on page 107. Below, we outline the proof that a curve defined by (2.4.30) is a path of shortest distance between two points.

Therefore, a trajectory of a moving object in a gravitational field in general relativity is computed as follows:
(a) Find the metric tensor $\mathfrak{g}$ by solving the field equations (2.4.22) for the specified stress-energy tensor T_{ij};
(b) Compute the functions Γ^i_{jk} according to (2.4.27);
(c) Solve the system of equations (2.4.30).

This summarizes the general relativity mechanics.

Let us now look more closely at equations (2.4.30). Without any gravitation, the space is flat so that each $\mathfrak{g}_{ij}$ is independent of $\boldsymbol{x}$, which makes each $\Gamma^i_{jk} = 0$. Then the geodesic equations (2.4.30) become $d^2\,x^i(s)/ds^2 = 0$ so that $x^i(s) = x^i(0) + v^i s$, which is a straight line, corresponding to the motion with constant velocity.

It turns out that a geodesic is always parameterized by its arc length $\mathfrak{s}$, although not necessarily canonically. In other words, equations (2.4.30) imply that $d\mathfrak{s} = \alpha ds$ for some constant $\alpha > 0$; canonical parametrization corresponds to $\alpha = 1$, see page 30.

EXERCISE 2.4.8.^{A} *(a) Verify that, if a vector function $\boldsymbol{x} = \boldsymbol{x}(s)$ satisfies (2.4.30), then*

$$\frac{d}{ds}\left(\mathfrak{g}_{ij}(\boldsymbol{x}(s))\frac{dx^i(s)}{ds}\frac{dx^j(s)}{ds}\right) = 0. \qquad (2.4.31)$$

Hint: differentiate and use (2.4.27). (b) Conclude that $\mathfrak{s} = \alpha\, s + \beta$, where $\mathfrak{s}$ is the arc length of the curve and α, β are real numbers. Hint: the left-hand side of (2.4.31) is $d^2\mathfrak{s}(s)/ds^2$.

By following the steps below, the reader can prove that the geodesic, as defined by (2.4.30), can be interpreted as the shortest path between two points in a curved space.
Step 1. It is easier to do the basic computations in an abstract setting in $\mathbb{R}^n$. Let A and B be two fixed points in $\mathbb{R}^n$ and $F = F(\boldsymbol{q}, \boldsymbol{p})$, a smooth $\mathbb{R}$-valued function of $2n$ variables $q^1, \ldots, q^n, p^1, \ldots, p^n$; we keep the convention of this section and write the indices as superscripts. Let $\boldsymbol{x} = \boldsymbol{x}(s)$, $a \leq s \leq b$, be a smooth function with values in $\mathbb{R}^n$. Consider the `functional`

$$L(\boldsymbol{x}) = \int_a^b F(\boldsymbol{x}(s), \boldsymbol{x}'(s))ds,$$

where $\boldsymbol{x}'(s) = d\boldsymbol{x}(s)/ds$; being a rule assigning a number to a function, L is indeed a functional. Later, we will take

$$F(\boldsymbol{q}, \boldsymbol{p}) = \sqrt{\mathfrak{g}_{ij}(\boldsymbol{q})p^ip^j} \text{ (remember the summation convention)}, \qquad (2.4.32)$$

so that $L(\boldsymbol{x})$ is the distance from A to B along the curve $\boldsymbol{x}(s)$.
(a) Let $\boldsymbol{y} = \boldsymbol{y}(s)$, $a \leq s \leq b$, be a smooth function with values in $\mathbb{R}^n$ so that $\boldsymbol{y}(a) = \boldsymbol{y}(b) = \boldsymbol{0}$ and let $\widetilde{\boldsymbol{x}}(s)$ be a function for which the value $L(\boldsymbol{x})$ is the smallest (assume that such a function exists.) For a real number λ, define $f(\lambda) = L(\widetilde{\boldsymbol{x}} + \lambda\, \boldsymbol{y})$. Argue that $f'(0) = 0$. (b) Show that

$$f'(0) = \int_a^b \left(\frac{\partial F(\widetilde{\boldsymbol{x}}(s), \widetilde{\boldsymbol{x}}'(s))}{\partial q^i} y^i(s) + \frac{\partial F(\widetilde{\boldsymbol{x}}(s), \widetilde{\boldsymbol{x}}'(s))}{\partial p^i}\frac{dy^i(s)}{ds}\right) ds.$$

(c) Integrate by parts and use $y^i(a) = y^i(b) = 0$ to conclude that

$$f'(0) = \int_a^b \left(\frac{\partial F(\widetilde{\boldsymbol{x}}(s), \widetilde{\boldsymbol{x}}'(s))}{\partial q^i} - \frac{d}{ds} \frac{\partial F(\widetilde{\boldsymbol{x}}(s), \widetilde{\boldsymbol{x}}'(s))}{\partial p^i} \right) y^i(s)\, ds.$$

(d) Argue that, since $f'(0) = 0$ and $\boldsymbol{y} = \boldsymbol{y}(s)$ is an arbitrary smooth function, the expression in the big parentheses in the last equality must vanish for all s and so, for all $i = 1, \ldots, n$,

$$\frac{\partial^2 F(\widetilde{\boldsymbol{x}}(s), \widetilde{\boldsymbol{x}}'(s))}{\partial q^i \partial q^j} \frac{d\widetilde{x}^j}{ds} + \frac{\partial^2 F(\widetilde{\boldsymbol{x}}(s), \widetilde{\boldsymbol{x}}'(s))}{\partial q^i \partial p^j} \frac{d^2 \widetilde{x}^j}{ds^2} - \frac{\partial F(\widetilde{\boldsymbol{x}}(s), \widetilde{\boldsymbol{x}}'(s))}{\partial q^i} = 0. \tag{2.4.33}$$

System (2.4.33) is an example of the **Euler-Lagrange equations** of variational calculus.

Step 2. Verify that, with F as in (2.4.32) and $s = \mathfrak{s}$, equation (2.4.33) becomes (2.4.30).

While in most situations of practical importance the geodesic does indeed define the shortest path between two points, and the path is unique up to a re-parametrization, there are two main technical problems with *defining* the geodesic as the distance-minimizing path: (i) The a priori possibility that several different paths have this property, and (ii) The possibility that the sign of (2.4.32) is different in different parts of the space (even the flat Minkowski metric (2.4.18) can change sign.) Definition (2.4.30) avoids these problems by generalizing a different property of the straight line, namely, that the tangent vectors at every two points of the curve are parallel. This property formalizes the intuitive notion of going straight, and the functions Γ^i_{jk} define the parallel transport of a vector along a curve by means of the **Levi-Civita connection**. For more details, an interested reader should consult a specialized reference on Riemannian geometry, such as Chapter 10 of the book *Differentiable Manifolds: A First Course* by L. Conlon, 1993.

The rest of the section will be an investigation of the important special case of (2.4.22), corresponding to $\mathrm{T}_{ij} = 0$, namely, the equation $R_{ij}[\mathfrak{g}] - (1/2)R\,\mathfrak{g}_{ij} = 0$. The solutions of this equation are called the **vacuum solutions** of (2.4.22), because $\mathrm{T}_{ij} = 0$ in empty space.

EXERCISE 2.4.9.^{C} *Verify that* $R_{ij}[\mathfrak{g}] - (1/2)R\,\mathfrak{g}_{ij} = 0$ *if and only if*

$$R_{ij}[\mathfrak{g}] = 0. \tag{2.4.34}$$

Hint: use (2.4.28) and (2.4.29); the matrices $\mathfrak{g}$ *and* $\mathfrak{g}^{-1}$ *are non-singular.*

From now on, we will consider (2.4.34), which is still a system of 10 equations, and call this system the **vacuum field equations**. Every flat metric, that is, a constant non-singular matrix $\mathfrak{g}$, is a *trivial solution* of (2.4.34). Indeed, for constant $\mathfrak{g}$, the numbers Γ^i_{kj} are all equal to zero, and then so are R_{ij}. A *relativistic* vacuum solution must be invariant under the Lorentz transformation. By Exercise 2.4.3 on page 104, such a solution $\mathfrak{g}$ is a constant multiple of the **flat Minkowski metric** (2.4.18) on page 104.

EXERCISE 2.4.10.^{C} *Verify the above assertion that every trivial solution of (2.4.34), invariant under the Lorentz transformation, is a constant multiple of the flat Minkowski metric (2.4.18).*

In 1916, the German physicist KARL SCHWARZSCHILD (1873–1916) found the first non-trivial solution of the vacuum field equations. Vacuum field equations do not necessarily describe a space without any gravitation whatsoever; gravitating matter can be present, but in other parts of the space, where we are not trying to solve the field equations. Accordingly, in the derivation of the Schwarzschild solution, it is assumed that there is only a *stationary spherical object of mass M somewhere in space.*

Introduce spherical coordinates (r, θ, φ) in the space part of space-time (x, y, z, t): $x = r\cos\theta\sin\varphi$, $y = r\sin\theta\sin\varphi$, $z = r\cos\varphi$. Then we set $x^1 = r$, $x^2 = \theta$, $x^3 = \varphi$, $x^4 = t$. **Schwarzschild's solution**, or **Schwarzschild's metric**, $\mathfrak{g}$ is a diagonal matrix so that

$$d(\mathfrak{s})^2 = \frac{1}{1-(R_\circ/r)}(dr)^2 + r^2\big(\sin^2\varphi\,(d\theta)^2 + (d\varphi)^2\big) - c^2\big(1-(R_\circ/r)\big)(dt)^2, \tag{2.4.35}$$

where

$$R_\circ = \frac{2GM}{c^2} \tag{2.4.36}$$

is called the **Schwarzschild radius** of the mass M. Schwarzschild found this metric by looking for a spherically symmetric solution of (2.4.34); see below for an outline of the corresponding computations.

EXERCISE 2.4.11. ^{C} *(a) Verify that $R_\circ$ indeed has the dimension of the length. (b) Compute the Schwarzschild radius of the following objects: (i) The Sun (take $M = 2\cdot 10^{30}$ kg), (ii) The Earth, (iii) yourself. (c) Find the mass and the Schwarzschild radius of the ball whose density is $\rho = 1000$ kg/m^3 and whose radius is $R_\circ$. Hint: you have $R_\circ = 2G(4\pi/3)R_\circ^3\rho/c^2$; find $R_\circ$ and then $M = (4\pi/3)R_\circ^3\rho$. (d) Repeat part (c) when $\rho = 10^{18}$ kg/m^3, the density of a typical atomic nucleus. (e) Verify that, as $r\to\infty$, the metric* *in (2.4.35) approaches the flat Minkowski metric in spherical coordinates* $(dr)^2 + r^2\big(\sin^2\varphi\,(d\theta)^2 + (d\varphi)^2\big) - c^2(dt)^2$.

The reader can derive (2.4.35) by following the steps below. The computations are not hard, but Steps 2 and 3 require a lot of care and patience.
Step 1. Argue that spherical symmetry and stationarity imply $ds^2 = F(r)(dr)^2 + r^2\big(\sin^2\varphi(d\theta)^2 + (d\varphi)^2\big) + H(r)(dt)^2$ for some functions F, H that are not equal to zero. With $x^1 = r$, $x^2 = \theta$, $x^3 = \varphi$, $x^4 = t$, we have the components of the metric tensor $\mathfrak{g}_{11} = F(r)$, $\mathfrak{g}_{22} = r^2\sin^2\varphi$, $\mathfrak{g}_{33} = r^2$, $\mathfrak{g}_{44} = H(r)$ and all other $\mathfrak{g}_{ij}$ are zero. Similarly, for the inverse tensor, $\mathfrak{g}^{11} = 1/F(r)$, $\mathfrak{g}^{22} = 1/(r^2\sin^2\varphi)$, $\mathfrak{g}^{33} = 1/r^2$, $\mathfrak{g}^{44} = 1/H(r)$ and all other $\mathfrak{g}^{ij}$ are zero.
Step 2. Using formulas (2.4.27), verify that, of the 40 values of Γ^i_{jk}, only nine are non-zero:

$$\Gamma^1_{11} = \frac{F'(r)}{2F(r)},\ \Gamma^1_{22} = -\frac{r\sin^2\varphi}{F(r)},\ \Gamma^1_{33} = -\frac{r}{F(r)},\ \Gamma^1_{44} = -\frac{H'(r)}{2F(r)};$$
$$\Gamma^2_{12} = \frac{1}{r},\ \Gamma^2_{23} = \frac{\cos\varphi}{\sin\varphi};\ \Gamma^3_{13} = \frac{1}{r}, \Gamma^3_{22} = -\sin\varphi\,\cos\varphi;\ \Gamma^4_{14} = \frac{H'(r)}{2H(r)}. \tag{2.4.37}$$

Step 3. Using formulas (2.4.26), verify that, out of 10 values of R_{ij}, only four are non-zero:

$$\begin{aligned}
R_{11} &= \frac{H''(r)}{2H(r)} - \frac{(H'(r))^2}{4H^2(r)} - \frac{F'(r)}{rF(r)} - \frac{F'(r)H'(r)}{4F(r)H(r)};\\
R_{22} &= \left(-1 + \frac{1}{F(r)} - \frac{rF'(r)}{2F^2(r)} + \frac{rH'(r)}{F(r)H(r)}\right)\sin^2\varphi;\\
R_{33} &= -1 + \frac{1}{F(r)} - \frac{rF'(r)}{2F^2(r)} + \frac{rH'(r)}{F(r)H(r)} = \frac{R_{22}}{\sin^2\varphi};\\
R_{44} &= \frac{H''(r)}{2F(r)} - \frac{F'(r)H'(r)}{4F^2(r)} + \frac{H'(r)}{rF(r)} - \frac{(H'(r))^2}{4F(r)H(r)}.
\end{aligned} \tag{2.4.38}$$

Step 4. (a) Multiply by $4rF(r)H^2(r)$ the equation $R_{11} = 0$, multiply by $4rF^2(r)H(r)$ the equation $R_{44} = 0$, and take the difference to conclude that $4H(r)\big(H(r)F(r)\big)' = 0$ or $F(r)H(r) = A$ for some real number A. (b) Multiply by $2F^2(r)H(r)$ the equation $R_{33} = 0$ and use $F(r)H(r) = A$ to conclude that $rF'(r) = F(r)(1 - F(r))$ or $F(r) = (1 + 1/(Br))^{-1}$ for some real number B.
Step 5. We have $\lim_{r\to\infty} F(r) = \lim_{r\to\infty} H(r) = 1$ and, by (2.4.18) on page 104, far away from the gravitating mass the metric should be flat, that is

$(d\mathfrak{s})^2 = (dr)^2 + r^2 \sin^2\varphi\,(d\theta)^2 + r^2(d\varphi)^2 - c^2(dt)^2$. Conclude that $A = -c^2$, where c is the speed of light. Keep in mind that $(dx)^2 + (dy)^2 + (dz)^2 = (dr)^2 + r^2 \sin^2\varphi\,(d\theta)^2 + r^2(d\varphi)^2$.
Step 6. Using (2.4.30) and (2.4.37), verify that the geodesic equations for the trajectory $(r(s), \theta(s), \varphi(s), t(s))$ are as follows:

$$\begin{aligned}
&r''(s) + \frac{F'(r(s))}{2F(r(s))}(r'(s))^2 - \frac{r(s)}{F(r(s))}(\theta'(s))^2 - \frac{H'(s)}{2F(s)} = 0;\\
&\theta''(s) + \frac{2}{r}r'(s)\theta'(s) = 0;\\
&\varphi''(s) + \frac{1}{r(s)}r'(s)\varphi'(s) - \sin\varphi(s)\,\cos\varphi(s)(\theta'(s))^2 = 0;\\
&t''(s) + \frac{H'(s)}{H(s)}r'(s)t'(s) = 0.
\end{aligned} \tag{2.4.39}$$

Keep in mind that, since $\Gamma_{ij}^k = \Gamma_{ji}^k$, the terms in (2.4.30) often come in pairs.
Step 7. Consider the circular trajectory of radius R in the equatorial plane ($\varphi = \pi/2$) in the Newtonian approximation, so that $r(s) = R$, $\varphi(s) = \pi/2$, $t'(s) = 1$; the absolute time $t'(s) = 1$ is the key feature of Newtonian mechanics. Use the function $H(r) = -c^2(1 + 1/(Br))$ we found in Steps 4 and 5, to conclude from the first equation in (2.4.39) $(\theta'(s))^2 = -H'(R)/(2R) = -c^2/(2BR^3)$. On the other hand, Newton's laws imply that, for the circular trajectory in the gravitational field of mass M, $(\theta'(s))^2 = M\mathrm{G}/R^3$. Conclude that $B = -c^2/(2MG) = -1/R_\circ$. This completes the derivation of (2.4.35).

It is always a major accomplishment to find an explicit non-trivial solution of a nonlinear partial differential equation. The importance of the Schwarzschild solution is further enhanced by the following facts:
• This solution leads to a direct verification of general relativity in three important problems, namely, perihelion precession of planets, gravitational deflection of light, and gravitational red shift; we discuss these problems below);
• The solution is a lot more general than it looks: in 1927, the American mathematician GEORGE DAVID BIRKHOFF (1884–1944) proved that, after a suitable change of coordinates, *every spherically symmetric solution of the vacuum field equations becomes a Schwarzschild solution.*
• The solution formula (2.4.35) leads to a natural, and mathematically precise, definition of a `black hole` as an object whose size is smaller than or

equal to the corresponding $R_\circ$. As the preceding exercise shows, a black hole can be produced either by compression or by vast accumulation of uncompressed matter. Once you do Exercise 2.4.11, you realize that neither you nor the Earth nor the Sun is likely to become a black hole.

The first experimental verification of general relativity was the calculation of the precession of the perihelion (the point closest to the Sun) for the planet Mercury. Astronomical observations of the precession had been carried out since the late 18th century, but the Newtonian theory of gravitation was unable to calculate the observed precession correctly. In 1916, Einstein used the Schwarzschild solution, with $R_\circ$ equal to the Schwarzschild radius of the Sun, to compute the correct value of the precession; see Problem 2.2 on page 414.

The reader can derive equation (7.2.7) on page 415, describing the trajectory of Mercury (or any other planet, for that matter) by following the steps below.

Step 1. Write three equations for the geodesic involving $\theta(s)$, $\varphi(s)$, and $t(s)$ using (2.4.30) and (2.4.37); use $\dot{g}$ to denote $dg(s)/ds$, and Q', to denote $dQ(r)/dr$, for the appropriate functions g, Q:

$$\ddot{\theta}(s) + \frac{2}{r(s)}\,\dot{\theta}(s)\dot{r}(s) = 0; \tag{2.4.40}$$

$$\ddot{\varphi}(s) + \frac{1}{r(s)}\,\dot{\varphi}(s)\dot{r}(s) - \sin\varphi(s)\cos\varphi(s)\,\dot{\theta}^2(s) = 0; \tag{2.4.41}$$

$$\ddot{t}(s) + \frac{H'(r(s))}{H(r(s))}\,\dot{t}(s)\,\dot{r}(s) = 0. \tag{2.4.42}$$

Step 2. From (2.4.40) conclude that $r^2(s)\dot{\theta}(s) = \alpha$, with the real number α independent of s; from (2.4.41), that $\varphi(s) = \pi/2$ is a possible solution; from (2.4.42), that $\dot{t}(s) = \beta F(r(s))$, with β independent of s.

Step 3. Keeping in mind that the geodesic is parameterized by the arc length, take $d\mathfrak{s} = c\,ds$ to have s in time units, and then, with $\varphi(s) = \pi/2$, conclude that (2.4.35) implies

$$(ds)^2 = c^2H(r(s))(dt(s))^2 - F(r(s))(dr(s))^2 - r^2(s)(d\theta(s))^2,$$
$$1 = c^2H(r(s))\,(\dot{t}(s))^2 - F(r(s))(\dot{r}(s))^2 - r^2(s)(\dot{\theta}(s))^2,$$
$$(\dot{r}(s))^2 + \frac{\alpha^2}{r^2(s)}\left(1 - \frac{R_\circ}{r(s)}\right) - \frac{c^2R_\circ}{r(s)} = c^2(\beta^2 - 1). \tag{2.4.43}$$

Similar to (2.4.19) on page 104, we have to switch the sign in (2.4.35) to ensure that the expression for $(ds)^2$ is positive.

Step 4. Put $u(\theta(s)) = 1/r(s)$, using $\dot{r}(s) = -(u'(\theta)/u^2(\theta))\dot{\theta}(s) = -\alpha u'(\theta)$, to get

$$(u'(\theta))^2 + u^2(\theta)\big(1 - R_\circ u(\theta)\big) - c^2 R_\circ u(\theta)/\alpha^2 = c^2(\beta^2 - 1)/\alpha^2.$$

Step 5. Differentiate the last equality with respect to θ to get

$$u''(\theta) + u(\theta) = \frac{c^2 R_\circ}{2\alpha^2} + \frac{3R_\circ}{2} u^2(\theta),$$

which is exactly equation (7.2.7) on page 415, with $A_0 = c^2 R_\circ/(2\alpha^2)$.

Another experimental verification of general relativity is the predicted GRAVITATIONAL DEFLECTION OF LIGHT. For the flat Minkowski metric (2.4.18), page 104, the trajectories of photons are lines satisfying $x^2 + y^2 + z^2 = c^2 t^2$ or $d\mathfrak{s} = 0$. In a curved space, a photon geodesic is not necessarily a line and can be curved (deflected) by the gravitational mass. Because of the properties of $\mathfrak{g}$, this geodesic must still satisfy $d\mathfrak{s} = 0$.

EXERCISE 2.4.12.^{B} *Verify that the trajectory of a photon satisfies $d\mathfrak{s} = 0$ in every curved space. Hint: change the metric on the flat space; the zero on the right will stay zero.*

Using the Schwarzschild metric (2.4.35), Einstein calculated the deflection of light rays passing near the Sun. The experimental validation of the calculation came in 1919, when, during a total solar eclipse, the British astronomer Sir ARTHUR STANLEY EDDINGTON (1882–1944) showed that a star whose light passed close to the Sun appeared displaced by the amount that corresponded to the deflection value calculated by Einstein.

EXERCISE 2.4.13.^{C} *Consider a photon moving radially, that is, $d\theta = d\varphi = 0$. Use (2.4.35) to conclude that $d\mathfrak{s}^2 = 0$ implies that, for $r > R_\circ$, $t = \pm(r + r_0 + R_\circ \ln\big((r/R_\circ) - 1\big)$. Therefore, the equation of trajectory (world line) of the photon is different from the equation of the straight line in four-dimensional space-time.*

Yet another experimental verification of general relativity is GRAVITATIONAL RED SHIFT, that is, the increase of wavelength of photons that are mowing away from the source of the gravitational field Red light has the longest wavelength in the optical spectrum, whence the term *red* shift. On a deep level, the gravitational red shift illustrates two major results of general relativity, the slow-down of time and shortening of length in a gravitational field; both these results can be deduced from the Schwarzschild solution. On an intuitive level, this shift can be explained without using the Schwarzschild solution or any general relativity. Indeed, let a photon of wavelength λ be emitted from a far-away star of mass M and radius R. At the moment of the emission, the photon has radiation energy $h\nu$, mass $m = h\nu/c^2$, and potential energy $-mMG/R$. Far away from the star, with no other gravitating objects around, the potential energy of the photon is approximately zero. This increase in potential energy of the photon must be compensated by a decrease of its radiation energy. Thus, $mMG/R = -h\Delta\nu$ or $\Delta\nu/\nu = -MG/(Rc^2) = -R_\circ/(2R)$. Assuming that the relative changes are small, and using $\nu = c/\lambda$, we can write the result as

$$\frac{\Delta\lambda}{\lambda} = \frac{R_\circ}{2R}. \tag{2.4.44}$$

EXERCISE 2.4.14. *(a)^C Estimate the gravitational red shift for the Sun. (take $M = 2 \cdot 10^{30}$ kg, $R = 7 \cdot 10^9$ m). (b)^C Consider the radiation emitted at the surface of the Earth and received H meters above the surface, with H much smaller than the radius of the Earth. Denote by g the gravitational acceleration on the Earth surface. Show that the observed red shift should be*

$$\frac{\Delta\lambda}{\lambda} = \frac{gH}{c^2}. \tag{2.4.45}$$

(c)^A Explain how the slow-down of the clock and shrinking of length in the gravitational field can be deduced from the Schwarzschild solution, and connect these effects with the gravitational red shift. Hint: for example, with $d\theta = d\varphi = dt = 0$ in (2.4.35), we have $d\mathfrak{s} = (1 - (R_\circ/r))^{-1/2}dr$.

Einstein predicted the gravitational red shift eight years before formulating the theory of general relativity, but, because of the need for very high accuracy, both in the generation of the radiation and in the measurements, it was only in the 1960s that the gravitational red shift was observed in an experiment (for $H = 90$m, the right-hand side of (2.4.45) is about 10^{-14}). The famous `Pound-Rebka-Snider experiment` was first conducted in 1960 at Harvard University by R. V. Pound and G. A. Rebka Jr.; in 1964, R. V. Pound and J. L. Snider carried out a more accurate experiment; as of 2005, Robert V. Pound is Emeritus Professor of Physics at Harvard.

For an observer on the Earth, the clock on an orbiting satellite will appear to be running slow because of the special relativity effects, see formula (2.4.12) on page 101, and because of the gravitational red shift: the gravitational field of the Earth is much weaker in space. With communication signals travelling at the speed of light, even a small discrepancy in the clocks can cause problems. For example, a 10-meter error in space can be caused by a $3.3 \cdot 10^{-8}$-second error in time (the time it take a radio-signal to travel 10 meters). The relativity effects can be 10^{-6} seconds per day or more. The appropriate adjustment of the clock is necessary, and is indeed implemented, on many navigation satellites.

EXERCISE 2.4.15.$^{A+}$ *Estimate the necessary time correction per day for a satellite on a geostationary orbit (that is, a satellite moving in the equatorial plane of the Earth in the direction of the rotation of the Earth and with period of revolution around the Earth equal to the Earth period of rotation about its axis). Take into account both special and general relativity effects.*

The gravitational red shift is different from the **cosmological red shift**, which is attributed to the expansion of the Universe, and which was first observed in 1929 by the American astronomer EDWIN POWELL HUBBLE (1889–1953). The cosmological red shift is consistent with general relativity, and is connected with *non-vacuum* solutions of the field equations; it is a popular topic of research in modern cosmology. The idea is to assume a particular form of the stress-energy tensor T_{ij} on the right-hand side of the Einstein field equations (2.4.34) so that the solution has a prescribed form. A popular form of the solution produces the following line element:

$$(d\mathfrak{s})^2 = c^2(dt)^2 - R(t)\left((dr)^2 + r^2\left(\sin^2\varphi\,(d\theta)^2 + (d\varphi)^2\right)\right) - c^2(dt)^2,$$

where $R(t)$ represents the radius of the Universe.

EXERCISE 2.4.16.^{C} *(a) Verify that the radial motion of the photon, corresponding to above metric, with* $d\mathfrak{s} = 0$ *and* $\theta = \varphi = 0$*, satisfies*

$$r(t_1) - r(t_0) = c\int_{t_0}^{t_1} \frac{dt}{R(t)}. \tag{2.4.46}$$

(b) Assume that a pulse of electromagnetic radiation, lasting Δt_0 *seconds, is emitted from a point in space at time moments* t_0*; the value of* $1/\Delta t_0$ *can be interpreted as the frequency of the emitted radiation. The pulse is received at a different point during the time interval* $[t_1, t_1 + \Delta t_1]$*, where* $t_1 > t_0 + \Delta t_0$*; the value* $1/\Delta t_1$ *can be interpreted as the frequency of the received radiation. Assuming that both* Δt_0 *and* Δt_1 *are small, use (2.4.46)*

to conclude that

$$\frac{\Delta t_0}{R(t_0)} \approx \frac{\Delta t_1}{R(t_1)}. \tag{2.4.47}$$

Hint: $r(t_1) - r(t_0) = r(t_1 + \Delta t_1) - r(t_0 + \Delta t_0)$.

By observing the light from distant galaxies, Hubble determined that $\Delta t_0 < \Delta t_1$ or $1/\Delta t_0 > 1/\Delta t_1$, that is, the frequency of the received radiation shifts toward the lower-frequency, or red, part of the spectrum. By (2.4.47), this observation implies $R(t_0) < R(t_1)$, that is, the Universe is *expanding.*

Chapter 3

Vector Analysis and Classical Electromagnetic Theory

3.1 Functions of Several Variables

3.1.1 *Functions, Sets, and the Gradient*

In the previous chapter, we saw how the tools of vector analysis work in modelling the mechanics of point masses and rigid bodies. In this chapter we will see how the tools of vector analysis work in continuum mechanics, that is in the study of continuous media such as fluids, heat flow, and electromagnetic fields.

Mathematically, a continuous medium is described using certain functions, called scalar and vector fields. Recall that a `function` is a correspondence between two sets A and B so that to every element $A \in \mathbb{A}$ there corresponds *at most one* element $f(A) \in \mathbb{B}$. Despite its simplicity, it was only in 1837 that the German mathematician JOHANN PETER GUSTAV LEJEUNE DIRICHLET (1805–1859) introduced this definition of a function. We write $f : \mathbb{A} \to \mathbb{B}$.

For most functions in this chapter, the set $\mathbb{A}$ will be either $\mathbb{R}^2$ or $\mathbb{R}^3$, and the set $\mathbb{B}$, $\mathbb{R}$, $\mathbb{R}^2$, or $\mathbb{R}^3$. When $\mathbb{B} = \mathbb{R}$, the collection of pairs $\{(P, f(P)),\ P \in \mathbb{A}\}$ is called a `scalar field`; the function f is also called a scalar field. When $\mathbb{B} = \mathbb{A}$, that is, when $f(P)$ is a vector, the collection of pairs $\{(P, f(P)),\ P \in \mathbb{A}\}$ is called a `vector field`; the function f is also called a vector field. Following our convention from the previous chapters, we will denote vector fields with bold-face letters.

Let us review the main definitions related to the sets in $\mathbb{R}^2$ and $\mathbb{R}^3$. Recall that the distance between two points A, B in $\mathbb{R}^n$, $n = 2, 3$, is denoted by $|AB|$. A `neighborhood` of a point A is the set $\{B : |AB| < b\}$ for some $b > 0$, that is, an `open disk` in $\mathbb{R}^2$ and an `open ball` in $\mathbb{R}^3$, with center at A and radius b. A point in a set is called `interior` if there exists a neighborhood of the point that lies entirely in the set.

A point P is called a **boundary point of the set** G if every neighborhood of the point contains at least one point that is not in G, and at least one point that belongs to G and is different from P. The collection of all boundary points of G is called the **boundary of** G and is denoted by ∂G. FOR EXAMPLE, the boundary points of the set $G = \{B : |AB| < 1\}$ are exactly the points P satisfying $|AP| = 1$, so that $\partial G = \{P : |AP| = 1\}$ is the circle with center at A and radius 1.

A point P in a set is called **isolated** if there exists a neighborhood of P in which P is the only point belonging to the set.

A set is called

- **Bounded**, if it lies entirely inside an open ball of sufficiently large radius.
- **Closed**, if it contains all its boundary points.
- **Connected**, if every two points in the set can be connected by a continuous curve lying completely in the set.
- **Open**, if for every point in the set there exists a neighborhood of this point that is contained in the set. In other words, all points of an open set are interior points.
- **Domain**, if it is open and connected.
- **Simply connected**, if it has no holes. More precisely, consider a curve that lies entirely in the set and assume that this curve is simple, closed, and continuous (see page 25). The set is simply connected if there is a surface that lies entirely in the set and has this curve as the boundary. For a planar set, this condition means that every simple closed continuous curve in the set encloses a domain that lies entirely in the set.

The **closure of a set** is the set together with all its boundary points. The **complement** of a set is the set of all points that are not in the set.

FOR EXAMPLE, the closure of the open set $\{P : |OP| < 1\}$ is $\{P : |OP| \le 1\}$, and the complement of that open set is the closed set $\{P : |OP| \ge 1\}$. Every disk or ball is both connected and simply connected, while the set $\{P : 0 < |OP| < 1\}$ is connected, and is simply connected in $\mathbb{R}^3$, but not simply connected in $\mathbb{R}^2$ (make sure you understand why).

EXERCISE 3.1.1.[A] *Give an example of a set in $\mathbb{R}^3$ that is not bounded, is neither open nor closed and is neither connected nor simply connected.*

Intuitively, it is obvious that a simple, closed, continuous curve (also known as a **Jordan curve**) divides the plane into two domains, and the domain enclosed by the curve is simply connected, but the rigorous proof of this statement is surprisingly hard. The French mathematician MARIE ENNEMOND CAMILLE JORDAN (1838–1922) was the first to realize the importance and difficulty of the issue. Even though Jordan's proof later turned out to be wrong, the result is known as the *Jordan curve theorem.* It was the American mathematician OSWALD VEBLEN (1880-1960), who, in 1905, produced the first correct proof. We will take the Jordan curve theorem for granted. The analog of this result in space, that a closed continuous **simple** (that is, without self-intersections) surface in $\mathbb{R}^3$ separates $\mathbb{R}^3$ into interior and exterior domains, is even harder to prove, and we take it for granted as well.

We will now discuss continuity of functions on $\mathbb{R}^n$, $n = 2, 3$. A scalar field f is called **continuous at the point** A if f is defined in some neighborhood of A and

$$\lim_{|AB|\to 0} |f(A) - f(B)| = 0. \tag{3.1.1}$$

Since the point A in the above definition is fixed, the condition $|AB| \to 0$ means that B is approaching A: $B \to A$. Then (3.1.1) can be written in the form that is completely analogous to the definition of continuity for functions of a real variable:

$$\lim_{B\to A} f(B) = f(A).$$

Similarly, a vector field $\boldsymbol{F}$ is called continuous at a point A if $\boldsymbol{F}$ is defined in some neighborhood of A and

$$\lim_{|AB|\to 0} \|\boldsymbol{F}(A) - \boldsymbol{F}(B)\| = 0. \tag{3.1.2}$$

Next, we discuss the question of differentiability for scalar fields. Let $f : \mathbb{A} \mapsto \mathbb{R}$ be a scalar field, where $\mathbb{A}$ is either $\mathbb{R}^2$ or $\mathbb{R}^3$. Let $\widehat{\boldsymbol{u}}$ be a unit vector and, for a real t, define the point B_t so that $\overrightarrow{AB_t} = t\widehat{\boldsymbol{u}}$. A function f is called **differentiable at the point** A **in the direction of** $\widehat{\boldsymbol{u}}$ if the function $g(t) = f(B_t)$ is differentiable for $t = 0$. We also call $g'(0)$ the **(directional) derivative of** f **at** A **in the direction of** $\widehat{\boldsymbol{u}}$ and denote this directional derivative by $D_{\widehat{\boldsymbol{u}}}f(A)$.

The function f is called **differentiable at the point** A if f is defined in some neighborhood of A and there exists a vector $\boldsymbol{a}$ so that

$$\lim_{|AB|\to 0} \frac{|f(B) - f(A) - \boldsymbol{a}\cdot\overrightarrow{AB}|}{|AB|} = 0. \tag{3.1.3}$$

Alternatively, for every unit vector $\widehat{\boldsymbol{u}}$,

$$\lim_{t\to 0} \frac{|f(B_t) - f(A) - t\,\boldsymbol{a}\cdot\widehat{\boldsymbol{u}}|}{t} = 0, \tag{3.1.4}$$

where $\overrightarrow{AB_t} = t\,\widehat{\boldsymbol{u}}$. If f is differentiable at the point A, then the corresponding vector $\boldsymbol{a}$ in (3.1.3) is denoted by $\boldsymbol{\nabla} f(A)$ or $\operatorname{grad} f(A)$ and is called the **gradient** of f at A. We say that the function f is **continuously differentiable in a domain** G if the vector field $\boldsymbol{\nabla} f(A)$ is defined at all points A in G and the vector field $\boldsymbol{\nabla} f$ is continuous in G. Similarly, a vector field $\boldsymbol{F}$ is called **continuously differentiable** in G if, for every unit vector $\widehat{\boldsymbol{n}}$, the scalar field $\boldsymbol{F}\cdot\widehat{\boldsymbol{n}}$ is continuously differentiable in G. A scalar or vector field is called **smooth** if it is continuously differentiable infinitely many times.

EXERCISE 3.1.2. *(a)*^{C} *Verify that if f is differentiable at the point A, then f is differentiable at A in every direction $\widehat{\boldsymbol{u}}$, and*

$$D_{\widehat{\boldsymbol{u}}} f(A) = \boldsymbol{\nabla} f(A)\cdot\widehat{\boldsymbol{u}}. \tag{3.1.5}$$

(b)^A *What does it mean for a vector field $\boldsymbol{F}$ to be three times continuously differentiable in a domain G?*

By (3.1.5) and the definition of the dot product (1.2.1), page 9, we have

$$-\|\boldsymbol{\nabla} f(A)\| \le D_{\widehat{\boldsymbol{u}}} f(A) \le \|\boldsymbol{\nabla} f(A)\|,$$

which shows that the most rapid rate of increase of f at the point A is in the direction of $\boldsymbol{\nabla} f(A)$, and the most rapid rate of decrease of f at the point A is in the direction of $-\boldsymbol{\nabla} f(A)$.

Given a point A, a direction at A can be defined by another point B with the unit vector $\widehat{\boldsymbol{u}}_B = \frac{\overrightarrow{AB}}{|AB|}$ in the direction of $\overrightarrow{AB}$. Then (3.1.3) becomes

$$f(B) = f(A) + \big(\boldsymbol{\nabla} f(A)\cdot\widehat{\boldsymbol{u}}_B\big)\,|AB| + \varepsilon(B), \tag{3.1.6}$$

where $\lim_{|AB|\to 0} |\varepsilon(B)|/|AB| = 0$. The number $\boldsymbol{\nabla} f(A)\cdot\widehat{\boldsymbol{u}}_B$ is the **(directional) derivative of** f **at point** A **in the direction of point** B.

EXERCISE 3.1.3.[B] *Show that if $\nabla f = \mathbf{0}$ in a domain G, then f is constant in G. Hint: For point $P \in G$ and a unit vector $\hat{\mathbf{u}}$, define the point B_t, $t \in \mathbb{R}$, so that $\overrightarrow{PB_t} = t\mathbf{u}$. Using the definition of open set show that B_t is in G for sufficiently small $|t|$. Apply the mean-value theorem to $g(t) = f(B_t)$ to show that $f(B_t) = f(P)$. Given another point P_1 in G, connect P and P_1 with a continuous curve in G, and apply the above argument several times as you move along the curve from P to P_1.*

Let us now introduce cartesian coordinates in the space $\mathbb{A}$. Then every point P will have coordinates (x, y) or (x, y, z) and we write $f(P) = f(x, y)$ or $f(P) = f(x, y, z)$. Recall that the `partial derivative` f_x is computed by differentiating f with respect to x and treating all other variables as constants:

$$f_x(x_0, y_0) = \lim_{\Delta x \to 0} \frac{f(x_0 + \Delta x, y_0) - f(x_0, y_0)}{x} \tag{3.1.7}$$

EXERCISE 3.1.4.[C] *Verify that, in cartesian coordinates, $f_x = D_{\hat{\imath}} f$.*

An alternative notation for f_x is $\partial f/\partial x$. The partial derivatives f_y, f_z are defined similar to (3.1.7) and coincide with the directional derivatives in the $\hat{\jmath}$ and $\hat{\kappa}$ directions, respectively. The higher-order partial derivatives are defined as partial derivatives of the corresponding lower-order partial derivatives:

$$f_{xx} = \frac{\partial f_x}{\partial x}, \; f_{xy} = \frac{\partial f_x}{\partial y}, \; f_{yx} = \frac{\partial f_y}{\partial x}, \text{ etc.}$$

EXERCISE 3.1.5.[C] *For two scalar fields f, g, verify the following properties of the gradient, generalizing the product and quotient rules from one-variable calculus:*

$$\nabla(fg) = f\nabla g + g\nabla f, \;\; \nabla(f/g) = (g\nabla f - f\nabla g)/g^2.$$

The next two results should be familiar from a course in multi-variable calculus:

- *If the function $f = f(x, y, z)$ has partial derivatives f_x, f_y, f_z in some neighborhood of the point A and the derivatives are continuous at A, then the function is differentiable at A and*

$$\nabla f(A) = f_x(A)\,\hat{\imath} + f_y(A)\,\hat{\jmath} + f_z(A)\,\hat{\kappa}. \tag{3.1.8}$$

• *If the partial derivatives* f_{xy} *and* f_{yx} *exist and are continuous in some neighborhood of the point* A*, then* $f_{xy}(A) = f_{yx}(A)$; the result is known as `Clairaut's theorem`, after the French mathematician ALEXIS CLAUDE CLAIRAUT (1713–1765).

EXERCISE 3.1.6.^{C} *Let* $\boldsymbol{r} = \boldsymbol{r}(t)$ *be a vector-valued function, and* f*, a scalar field. Define the function* $g(t) = f(\boldsymbol{r}(t))$*, where* $f(\boldsymbol{r}(t)) = f(P)$ *if* $\boldsymbol{r}(t) = \overrightarrow{OP}$ *for some fixed reference point* O*. Assume that* $\boldsymbol{r}$ *is differentiable at the point* $t = t_0$ *and the function* f *is differentiable at the point* A*, where* $\overrightarrow{OA} = \boldsymbol{r}(t_0)$*. Show that the function* g *is differentiable at the point* $t = t_0$ *and*

$$g'(t_0) = \boldsymbol{\nabla} f(A) \cdot \boldsymbol{r}'(t_0). \tag{3.1.9}$$

Try to argue directly by the definitions without using partial derivatives. Hint: by (3.1.6), $g(t) - g(t_0) \approx \boldsymbol{\nabla} f(A) \cdot (\boldsymbol{r}(t) - \boldsymbol{r}(t_0))$*.*

If $\boldsymbol{r} = \boldsymbol{r}(t)$ is a vector-valued function of one variable and $\boldsymbol{G}$ is a vector field, then $\boldsymbol{R}(t) = \boldsymbol{G}(\boldsymbol{r}(t))$ is a vector-valued function of one variable. How to find the derivative $\boldsymbol{R}'(t)$ in terms of $\boldsymbol{r}$ and $\boldsymbol{G}$? In cartesian coordinates, $\boldsymbol{G} = G_1\,\hat{\boldsymbol{\imath}} + G_2\,\hat{\boldsymbol{\jmath}} + G_3\,\hat{\boldsymbol{k}}$, and G_1, G_2, G_3 are scalar fields. By (3.1.9), we conclude that $\boldsymbol{R}'(t) = \boldsymbol{r}'(t) \cdot \boldsymbol{\nabla} G_1(\boldsymbol{r}(t))\,\hat{\boldsymbol{\imath}} + \boldsymbol{r}'(t) \cdot \boldsymbol{\nabla} G_2(\boldsymbol{r}(t))\,\hat{\boldsymbol{\jmath}} + \boldsymbol{r}'(t) \cdot \boldsymbol{\nabla} G_3(\boldsymbol{r}(t))\,\hat{\boldsymbol{k}}$. To make the construction coordinate-free, we define the vector $(\boldsymbol{r}'(t) \cdot \boldsymbol{\nabla})\boldsymbol{G}(\boldsymbol{r}(t))$ so that, for every fixed unit vector $\widehat{\boldsymbol{n}}$,

$$\big((\boldsymbol{r}'(t) \cdot \boldsymbol{\nabla})\boldsymbol{G}(\boldsymbol{r}(t))\big) \cdot \widehat{\boldsymbol{n}} = \boldsymbol{r}'(t) \cdot \big(\boldsymbol{\nabla}(\boldsymbol{G} \cdot \widehat{\boldsymbol{n}})(\boldsymbol{r}(t))\big).$$

With this definition,

$$\frac{d}{dt}\boldsymbol{G}(\boldsymbol{r}(t)) = (\boldsymbol{r}'(t) \cdot \boldsymbol{\nabla})\boldsymbol{G}(\boldsymbol{r}(t)). \tag{3.1.10}$$

More generally, for two vector fields $\boldsymbol{F}$, $\boldsymbol{G}$, we define the vector field $(\boldsymbol{F} \cdot \boldsymbol{\nabla})\boldsymbol{G}$ so that, for every fixed unit vector $\widehat{\boldsymbol{n}}$,

$$\widehat{\boldsymbol{n}} \cdot \big((\boldsymbol{F} \cdot \boldsymbol{\nabla})\boldsymbol{G}\big) = \boldsymbol{F} \cdot \big(\boldsymbol{\nabla}(\boldsymbol{G} \cdot \widehat{\boldsymbol{n}})\big). \tag{3.1.11}$$

The expressions of the type $(\boldsymbol{F} \cdot \boldsymbol{\nabla})\boldsymbol{F}$ often appears in the equations describing the motion of fluids.

Given a scalar field f and a real number c, the set of points $\{P : f(P) = c\}$ is called a `level set` of f. In two dimensions, that is, for $\mathbb{A} = \mathbb{R}^2$, the level sets of f are *curves*. In three dimensions ($\mathbb{A} = \mathbb{R}^3$), the level sets of

f are *surfaces.* This observation leads to several possible ways to define curves and surfaces.

Consider the space $\mathbb{R}^2$ with cartesian coordinates (x, y). Then each of the following defines a curve in $\mathbb{R}^2$:

- A graph $y = g(x)$ of a function of one variable.
- A level set $f(x, y) = c$ of a function of two variables.
- A vector-valued function $\boldsymbol{r}(t) = x(t)\,\hat{\boldsymbol{\imath}} + y(t)\,\hat{\boldsymbol{\jmath}}$, $t_0 < t < t_1$, of one variable.

EXERCISE 3.1.7. *(a)*^{B} *Give an example of a curve that cannot be represented as a graph* $y = g(x)$. *(b)*$^{A+}$ *Is there a continuous curve that cannot be represented as a level set of a function of two variables?*

Consider the space $\mathbb{R}^3$ with cartesian coordinates (x, y, z). Then each of the following defines a surface in $\mathbb{R}^3$:

- A graph $z = f(x, y)$ of a function of two variables;
- A level set $F(x, y, z) = c$ of a function of three variables;
- A a vector-valued function $\boldsymbol{r}(u, v) = x(u, v)\,\hat{\boldsymbol{\imath}} + y(u, v)\,\hat{\boldsymbol{\jmath}} + z(u, v)\,\hat{\boldsymbol{\kappa}}$. of two variables; this is known as a `parametric representation` of the surface.

EXERCISE 3.1.8. *(a)*^{B} *Give an example of a surface that cannot be represented as a graph* $z = f(x, y)$. *(b)*^{B} *Think of at least two different ways to represent a curve in* $\mathbb{R}^3$. *(c)*$^{A+}$ *Is there a surface that can be represented parametrically using a continuous function* $\boldsymbol{r} = \boldsymbol{r}(u, v)$ *but cannot be represented as a level set of a function of three variables?*

EXERCISE 3.1.9.^{C} *Below, we summarize some geometric properties of the gradient that are usually studied in a vector calculus class.*

(a) Let $f(x, y) = c$ *be a curve. Let* (x_0, y_0) *be a point on the curve and assume that the function* f *is differentiable at* (x_0, y_0). *(i) Show that the curve is smooth at the point* (x_0, y_0) *if and only if* $\boldsymbol{\nabla} f(x_0, y_0) \neq 0$. *(ii) Use (3.1.9) to show that* $\boldsymbol{\nabla} f(x_0, y_0)$ *is perpendicular to the unit tangent vector of the curve at the point* (x_0, y_0). *Hint: in this case,* $g(t) = c$ *and so* $g'(t_0) = 0$. *(iii) Write the equation of the* `normal line` *to the curve at the point* (x_0, y_0). *Hint: the line has* $\boldsymbol{\nabla} f(x_0, y_0)$ *as the* `direction vector`.

(b) Let $F(x, y, z) = c$ *be a surface. Let* $P_0 = (x_0, y_0, z_0)$ *be a point on the surface, and* C*, a smooth curve on the surface passing through the point* P_0. *Assume that the function* F *is differentiable at* P_0. *(i) Show that* $\boldsymbol{\nabla} F(P_0)$ *is perpendicular to the unit tangent vector to* C *at* P_0. *Hint: same arguments as in part (a). (ii) Write the equation of the* `tangent plane` *to the surface at the point* P_0. *Hint: the plane has* $\boldsymbol{\nabla} F(P_0)$ *as the* `normal vector`.

(c) Let $\boldsymbol{r} = \boldsymbol{r}(u, v)$ *be a parametric representation of the surface. Assume that the partial derivatives* $\boldsymbol{r}_u$ *and* $\boldsymbol{r}_v$ *exist, are continuous, and are not equal to zero at a point* (u_0, v_0)*. Let* P_0 *be the point on the surface corresponding to the position vector* $\boldsymbol{r}(u_0, v_0)$*, and* C*, a smooth curve on the surface passing through the point* P_0*. (i) Verify that the vector* $\boldsymbol{r}_u(u_0, v_0) \times \boldsymbol{r}_v(u_0, v_0)$ *is perpendicular to the unit tangent vector to* C *at* P_0*. Hint:* C *is represented by a vector function* $\boldsymbol{r}(u(t), v(t))$ *for suitable functions* u, v*; at the point* P_0 *this curve has a tangent vector* $\boldsymbol{r}_u u' + \boldsymbol{r}_v v'$*. (ii) Write the equation of the* `tangent plane` *to the surface at the point* P_0*. Hint: the plane has* $\boldsymbol{r}_u(u_0, v_0) \times \boldsymbol{r}_v(u_0, v_0)$ *as the* `normal vector`.

The same surface can have different representations. The graph $z = f(x, y)$ is a level set $F(x, y, z) = 0$ with $F(x, y, z) = f(x, y) - z$ or $F(x, y, z) = z - f(x, y)$. The same graph can also be written in the parametric form $\boldsymbol{r}(u, v) = u\,\hat{\boldsymbol{\imath}} + v\,\hat{\boldsymbol{\kappa}} + f(u, v)\,\hat{\boldsymbol{\kappa}}$. Similarly, the vector $\boldsymbol{r} = x(u, v)\,\hat{\boldsymbol{\imath}} + y(u, v)\,\hat{\boldsymbol{\jmath}} + z(u, v)\,\hat{\boldsymbol{\kappa}}$ defines a parametric representation of the level set $F(x, y, z) = c$ if and only if $F(x(u, v), y(u, v), z(u, v)) = c$ for all (u, v). In particular, one can write the level set in parametric form by solving the equation $F(x, y, z) = c$ for one of the variables x, y, z. FOR EXAMPLE, the hemisphere $x^2 + y^2 + z^2 = 4$, $x > 0$, can be written as $x = \sqrt{4 - y^2 - z^2}$. Taking $u = y$ and $v = z$, we find the equivalent parametric representation $\boldsymbol{r}(u, v) = \sqrt{4 - u^2 - v^2}\,\hat{\boldsymbol{\imath}} + u\,\hat{\boldsymbol{\jmath}} + v\,\hat{\boldsymbol{\kappa}}$, where $u^2 + v^2 \le 4$. The most convenient representation of the given surface depends on the particular problem.

EXERCISE 3.1.10.^{C} *Verify that the surface* $x = \sqrt{4 - y^2 - z^2}$ *can be represented as* $\boldsymbol{r}(u, v) = 2\cos v\,\hat{\boldsymbol{\imath}} + 2\cos u \sin v\,\hat{\boldsymbol{\jmath}} + 2\sin u \sin v\,\hat{\boldsymbol{\kappa}}$, $0 \le u < 2\pi$, $0 \le v \le \pi/2$.

EXAMPLE. Consider the function $f(x, y) = 2x^2 + 3y^2$. This function has continuous partial derivatives of every order everywhere in $\mathbb{R}^2$: $f_x = 4x$, $f_{xx} = 4$, $f_y = 6y$, $f_{yy} = 6$, and all other derivatives are equal to zero. Thus, $\boldsymbol{\nabla} f(x, y) = 4x\,\hat{\boldsymbol{\imath}} + 6y\,\hat{\boldsymbol{\jmath}}$. Consider the point $A = (1, 2)$. Then $\boldsymbol{\nabla} f(A) = 4\hat{\boldsymbol{\imath}} + 12\hat{\boldsymbol{\jmath}}$, so that the most rapid increase of f at A is in the direction of $\hat{\boldsymbol{\imath}} + 3\hat{\boldsymbol{\jmath}}$, with the rate $\|\boldsymbol{\nabla} f(A)\| = 4\sqrt{10}$. Similarly, the most rapid decrease of f at A is in the direction of $-\hat{\boldsymbol{\imath}} - 3\hat{\boldsymbol{\jmath}}$, with the rate $-\|\boldsymbol{\nabla} f(A)\| = -4\sqrt{10}$. The rate of change of f at A in the direction toward the origin is $-28/\sqrt{5}$ and is computed by (3.1.6) as follows: $B = (0, 0)$, $\overrightarrow{AB} = -\hat{\boldsymbol{\imath}} - 2\hat{\boldsymbol{\jmath}}$, $\hat{\boldsymbol{u}}_B = -(\hat{\boldsymbol{\imath}} + 2\hat{\boldsymbol{\jmath}})/\sqrt{5}$, $\boldsymbol{\nabla} f(A) \cdot \hat{\boldsymbol{u}}_B = -28/\sqrt{5}$.

Next, consider the surface $z = 2x^2 + 3y^2$. Writing $2x^2 + 3y^2 - z = 0$, we conclude that the vector $4\hat{\imath} + 12\hat{\jmath} - \hat{\kappa}$ defines the plane tangent to the surface at the point $P_0 = (1, 2, 14)$. The equation of the plane at P_0 is $4x + 12y - z = 14$: note that $x = 1$, $y = 2$, $z = 14$ must satisfy the equation, and you get the number 14 on the right by plugging in these values for x, y, z.

EXERCISE 3.1.11. [B] *Verify that* $\boldsymbol{r}(u, v) = u\,\hat{\imath} + v\,\hat{\jmath} + (2u^2 + 3v^2)\,\hat{\kappa}$ *and* $\boldsymbol{r}(u, v) = (v/\sqrt{2})\,\cos u\,\hat{\imath} + (v/\sqrt{3})\,\sin u\,\hat{\jmath} + v^2\,\hat{\kappa}$ *are two possible parametric representations of the surface* $z = 2x^2 + 3y^2$. *Verify that, for both representations, the vector* $\boldsymbol{r}_u \times \boldsymbol{r}_v$ *at the point* $(1, 2, 14)$ *is parallel to the vector* $4\,\hat{\imath} + 12\,\hat{\jmath} - \hat{\kappa}$.

Recall that P is a `critical point` of a scalar field f if either $\boldsymbol{\nabla} f(P_0) = \mathbf{0}$ or f is defined but not differentiable at P_0. If P is not a critical point of f, then $-\boldsymbol{\nabla} f(P)$ is the direction of the most rapid decrease of the function f at the point P. This observation suggests the following iterative method for finding the point of local minimum of f: given a reference point O, a starting point P_1, and a collection of positive numbers a_k, $k = 2, 3, \ldots$, define the points P_k recursively by

$$\overline{OP}_{k+1} = \overline{OP}_k - a_k \boldsymbol{\nabla} f(P_k). \tag{3.1.12}$$

Sometimes it is convenient to identify the point P_k with the position vector $\boldsymbol{r}_k = \overline{OP}_k$ and to write (3.1.12) in an equivalent form as

$$\boldsymbol{r}_{k+1} = \boldsymbol{r}_k - a_k \boldsymbol{\nabla} f(\boldsymbol{r}_k).$$

The new point P_{k+1} can be defined as long as $\boldsymbol{\nabla} f(P_k) \neq 0$. If the sequence of the points P_k, $k \geq 1$, converges as $k \to \infty$, then the limit P^* must be the point of local minimum of f; otherwise $\boldsymbol{\nabla} f(P^*) \neq \mathbf{0}$ and the sequence can be continued. As a result, $\boldsymbol{\nabla} f(P_k) \to \mathbf{0}$, and in practice, the computations are stopped as soon as $\|\boldsymbol{\nabla} f(P_k)\|$ is sufficiently small. Special choices of the numbers a_k can speed up the convergence. This method is known as the `method of steepest descent`. It was one of the first numerical methods for minimizing a function. The method can converge very slowly, but is easy to implement. There exist more sophisticated numerical methods for finding critical points.

EXERCISE 3.1.12. [A] *(a) Draw a picture illustrating (3.1.12). (b) Write a computer program that implements (3.1.12) to find the local minimum of the function* $f(x, y) = 2x^2 + 3y^2$. *Experiment with various starting points, various values of* a_k *and different stopping criteria (for example,*

$\|\nabla f(P_k)\| \leq 10^{-m}$ *for some positive integer* m*).*

3.1.2 *Integration and Differentiation*

Let us recall the definition of the **Riemann integral** from the one-variable calculus:

$$\int_a^b f(x)dx = \lim_{\max \triangle x_k \to 0} \sum_{k=1}^{N} f(x_k^*)\triangle x_k,$$

where $a = x_0 < \ldots < x_N = b$ is a partition of the interval $[a, b]$, $\triangle x_k = x_k - x_{k-1}$, and x_k^* is a point in the interval $[x_{k-1}, x_k]$. By definition, the function f is Riemann-integrable if the limit exists and does not depend on the particular sequence of partitions or on the choice of the points x_k^*. The same definition is extendable to real-valued functions f defined on a set G other than an interval, as long as there exists a function $\mathfrak{m}$ that measures the sizes of the subsets of the set G; we will not discuss the deep question of measurability. Using the measure $\mathfrak{m}$, we define

$$\int_G f\, d\mathfrak{m} = \lim_{\max \mathfrak{m}(G_k) \to 0} \sum_{k=1}^{N} f(P_k)\mathfrak{m}(G_k), \tag{3.1.13}$$

where $G_1, \ldots, G_N$ are mutually disjoint ($G_k \bigcap G_m = \varnothing$) sets such that their union $\bigcup_{k=1}^{N} G_k$ contains the set G and every set G_k is **measurable** (the size $\mathfrak{m}(G_k)$ of G_k is defined) and has a non-empty intersection with G ($G_k \bigcap G \neq \varnothing$); P_k is a point in $G_k \bigcap G$. *Note that definition (3.1.13) is not tied to any coordinate system.*

We will integrate functions defined on intervals, curves, planar regions, surfaces, and solid regions:

Set	Measure $\mathfrak{m}$	$d\mathfrak{m}$
Interval	Length	dx
Curve	Arc length	$d\mathfrak{s}$
Planar region	Area	dA
Surface	Surface Area	$d\sigma$
Solid region	Volume	dV

The limit in (3.1.13) must be independent of the particular choice of the sets G_k and the points P_k, which, in general, puts certain restrictions on both the function f (*integrability*) and the set G (*measurability*). If f is integrable over a measurable set G, then the **average value** of f over G

is, *by definition*,

$$\overline{f}_G = \frac{1}{\mathfrak{m}(G)} \int_G f \, d\mathfrak{m}. \tag{3.1.14}$$

The idea of the Riemann integral was introduce by the German mathematician GEORG FRIEDRICH BERNHARD RIEMANN (1826–1866) in 1854 in his `Habilitation` dissertation (Habilitation is one step higher than Ph.D. and is the terminal degree in several European countries).

We now summarize the main facts about line, area, surface, and volume integrals.

LINE, OR PATH, INTEGRALS. The set G in (3.1.13) is a curve C defined by the vector-valued function $\boldsymbol{r} = \boldsymbol{r}(t)$, $a < t < b$. The set is measurable and has the line element $d\mathfrak{s}$ if C is piece-wise smooth; see pages 28–29. The `line integral of the first kind` $\int_C f \, d\mathfrak{s}$ is defined for a continuous scalar field f by the equality

$$\int_C f \, d\mathfrak{s} = \int_a^b f(\boldsymbol{r}(t)) \, \|\boldsymbol{r}'(t)\| \, dt, \tag{3.1.15}$$

where $f(\boldsymbol{r}(t))$ is the value of the scalar field f at the point on the curve with position vector $\boldsymbol{r}(t)$. If $f(P)$ represents the linear density of the curve at the point P, then $\int_C f \, d\mathfrak{s}$ is the mass of the curve.

The `line integral of the second kind` $\int_C \boldsymbol{F} \cdot d\boldsymbol{r}$ is defined for a continuous vector field $\boldsymbol{F}$ by the equality

$$\int_C \boldsymbol{F} \cdot d\boldsymbol{r} = \int_a^b \boldsymbol{F}(\boldsymbol{r}(t)) \cdot \boldsymbol{r}'(t) \, dt, \tag{3.1.16}$$

where $\boldsymbol{F}(\boldsymbol{r}(t))$ is the value of the vector field $\boldsymbol{F}$ at the point on the curve with position vector $\boldsymbol{r}(t)$. For a simple closed curve C, the corresponding line integral of the second kind is denoted by $\oint_C \boldsymbol{F} \cdot d\boldsymbol{r}$ and is called the `circulation` of $\boldsymbol{F}$ along C. The line integral of the second kind can be reduced to the line integral of the first kind:

$$\int_C \boldsymbol{F} \cdot d\boldsymbol{r} = \int_C \boldsymbol{F} \cdot \widehat{\boldsymbol{u}} \, d\mathfrak{s}, \tag{3.1.17}$$

where $\widehat{\boldsymbol{u}}$ is the unit tangent vector to the curve. The direction of $\widehat{\boldsymbol{u}}$ defines the orientation of C; the integral reverses sign if the orientation of C is reversed. In cartesian coordinates, $\boldsymbol{F} = F_1 \,\hat{\boldsymbol{\imath}} + F_2 \,\hat{\boldsymbol{\jmath}} + F_3 \,\hat{\boldsymbol{\kappa}}$, and $\int_C \boldsymbol{F} \cdot d\boldsymbol{r}$ is often written as $\int_C F_1 dx + F_2 dy + F_3 dz$.

We say that $\boldsymbol{F}$ is a `potential` vector field with potential f if $\boldsymbol{F} = \boldsymbol{\nabla} f$ for some scalar field f. If $\boldsymbol{F} = \boldsymbol{\nabla} f$, then, by (3.1.9),

$$\boldsymbol{F}(\boldsymbol{r}(t)) \cdot \boldsymbol{r}'(t) = \frac{d}{dt} f(\boldsymbol{r}(t)),$$

and the value of the integral (3.1.16) depends only on the starting and ending points $\boldsymbol{r}(a)$, $\boldsymbol{r}(b)$ of the curve and not on the curve itself:

$$\int_{\mathcal{C}} \boldsymbol{\nabla} f \cdot d\boldsymbol{r} = f\big(\boldsymbol{r}(b)\big) - f\big(\boldsymbol{r}(a)\big). \tag{3.1.18}$$

In particular, $\oint_{\mathcal{C}} \boldsymbol{\nabla} f \cdot d\boldsymbol{r} = 0$ for every simple closed curve $\mathcal{C}$.

The alternative name for a potential vector field is `conservative`, which comes from the physical interpretation of the line integral. If $\boldsymbol{F}$ is a force acting on a point mass moving along $\mathcal{C}$, then $\int_{\mathcal{C}} \boldsymbol{F} \cdot d\boldsymbol{r}$ is the work done by the force along the curve $\mathcal{C}$. By the Second Law of Newton, $\boldsymbol{F}(\boldsymbol{r}(t)) = m\,\boldsymbol{r}''(t)$, so that $\boldsymbol{F}(\boldsymbol{r}(t)) \cdot \boldsymbol{r}'(t) = (1/2)md\|\boldsymbol{r}'(t)\|^2/dt = d\mathcal{E}_K(t)/dt$, where $\mathcal{E}_K$ is the kinetic energy; verify this. For a potential force field $\boldsymbol{F}$, we write $\boldsymbol{F} = -\boldsymbol{\nabla} V$, where V is a scalar function, called the `potential energy`. Then

$$\int_{\mathcal{C}} \boldsymbol{F} \cdot d\boldsymbol{r} = \mathcal{E}_K(b) - \mathcal{E}_K(a) = V(\boldsymbol{r}(a)) - V(\boldsymbol{r}(b)),$$

so that $\mathcal{E}_K(a) + V(\boldsymbol{r}(a)) = \mathcal{E}_K(b) + V(\boldsymbol{r}(b))$, which means that the total energy $\mathcal{E}_K + V$ is *conserved* as the point mass moves along $\mathcal{C}$.

In general, we say that a vector field $\boldsymbol{F}$ has the `path independence property` in a domain G if $\oint_{\mathcal{C}} \boldsymbol{F} \cdot d\boldsymbol{r} = 0$ for every simple closed piece-wise smooth curve $\mathcal{C}$ in G. The following result should be familiar from a course in multi-variable calculus.

Theorem 3.1.1 *A continuous vector field has the path independence property in a domain G if and only if the field has a potential.*

EXERCISE 3.1.13. ^{C} *Prove the above theorem. Hint: in one direction, use (3.1.18). In another direction, verify that, as stated, the path independence property implies that, for every two points A, B in G the value of the integral $\int_{\mathcal{C}} \boldsymbol{F} \cdot d\boldsymbol{r}$ does not depend on the particular choice of the curve $\mathcal{C}$ in G connecting the points A and B. Then fix a point A in G and, for each point P in G define $f(P) = \int_{\mathcal{C}} \boldsymbol{F} \cdot d\boldsymbol{r}$, where $\mathcal{C}$ is a simple piece-wise smooth curve starting at A and ending at P. Path independence makes f a well-defined function of P. Verify that $\boldsymbol{\nabla} f = \boldsymbol{F}$.*

THE AREA INTEGRAL. We now take the set G in (3.1.13) to be a subset of the plane $\mathbb{R}^2$. The set is measurable, and has area A, if the boundary of G is a continuous, piece-wise smooth simple closed curve. In area integrals, we use the differential form of the **area element** dA. In cartesian coordinates, the area element is $dA = dxdy$. In the **generalized polar coordinates** $x = ar\cos\theta$, $y = br\sin\theta$, the area element is $dA = ab\, rdr\, d\theta$; the usual polar coordinates correspond to $a = b = 1$. The area integral is evaluated by reduction to a suitable iterated integral. The area integral is also known as the *double integral* and is often written as $\iint_G f\, dA$. The integral is defined if f is a continuous scalar field in G. If f is the area density of G, then $\iint_G f\, dA$ is the mass of G. Computation of double integrals is one of the key skills that must be acquired in the study of multi-variable calculus.

EXERCISE 3.1.14. *(a)^B Let G be the parallelogram with vertices $(0,0), (3,0), (4,1), (1,1)$, and f, a continuous function. Write the limits in the iterated integrals below:*

$$\iint\limits_G f\, dA = \int \int f dx\, dy = \int \int f dy\, dx = \int \int f r dr\, d\theta = \int \int f d\theta\, r dr.$$

Hint: some integrals must be split into several. (b)^C Verify that the area of the ellipse $(x^2/a^2) + (y^2/b^2) \le 1$ with semi-axis a and b is πab. Hint: use generalized polar coordinates.

THE SURFACE INTEGRAL. We now take the set G in (3.1.13) to be a surface S defined by the vector-valued function $\boldsymbol{r} = \boldsymbol{r}(u, v)$ of two variables, where (u, v) belong to some measurable planar set G_1. We say that the surface is **piece-wise smooth** if the function $\boldsymbol{r}$ is continuous on G_1 and the set G_1 can be split into finitely many non-overlapping pieces so that, on each piece, the vector field $\boldsymbol{r}_u \times \boldsymbol{r}_v$, representing the normal vectors to the surface, is continuous and is not equal to zero. We say that the surface is **orientable** if it is piece-wise smooth and has two sides (the intuitive idea of a side is enough for our discussion). A piece-wise smooth surface S is measurable and its area $\mathfrak{m}(S)$ is defined as

$$\mathfrak{m}(S) = \iint\limits_{G_1} \|\boldsymbol{r}_u \times \boldsymbol{r}_v\|\, dA(u, v), \tag{3.1.19}$$

where $dA(u, v)$ is the area element of G_1. We call

$$d\sigma = \|\boldsymbol{r}_u \times \boldsymbol{r}_v\|\, dA(u, v)$$

the **surface area element** of S.

EXERCISE 3.1.15. [B] *Justify (3.1.19). Hint. Verify that* $\|\boldsymbol{r}_u \times \boldsymbol{r}_v\|\triangle u \triangle v$ *is the area of the parallelogram in the tangent plane to* S *at the point* $\boldsymbol{r}(u,v)$, *with sides* $\triangle u$, $\triangle v$ *parallel to the vectors* $\boldsymbol{r}_u$, $\boldsymbol{r}_v$, *respectively. For small* $\triangle u$, $\triangle v$, *this area closely approximates a curved parallelogram-like surface element of* S *with vertices at* $\boldsymbol{r}(u,v)$, $\boldsymbol{r}(u+\triangle u, v)$, $\boldsymbol{r}(u, v+\triangle v)$, $\boldsymbol{r}(u+\triangle u, v+\triangle v)$. *Summing up over all such surface elements and passing to the limit as* $\triangle u \to 0, \triangle v \to 0$ *gives (3.1.19). For a slightly different derivation of (3.1.19), see Exercise 3.1.35(c) on page 149 below.*

The **surface integral of the first kind** $\iint\limits_S f\,d\sigma$ is defined for a continuous scalar field f by the equality

$$\iint\limits_S f\,d\sigma = \iint\limits_{G_1} f(\boldsymbol{r}(u,v))\,\|\boldsymbol{r}_u \times \boldsymbol{r}_v\|\,dA(u,v), \tag{3.1.20}$$

where $dA(u,v)$ is the area element. If $f(P)$ represents the areal density of the surface at point P, then $\iint\limits_S f\,d\sigma$ is the mass of the surface. The **surface integral of the second kind** $\iint\limits_S \boldsymbol{F}\cdot d\boldsymbol{\sigma}$ is defined for a continuous vector field $\boldsymbol{F}$ with values in $\mathbb{R}^3$ and an orientable surface S by the equality

$$\iint\limits_S \boldsymbol{F}\cdot d\boldsymbol{\sigma} = \iint\limits_{G_1} \boldsymbol{F}\big(\boldsymbol{r}(u,v)\big)\cdot\big(\boldsymbol{r}_u \times \boldsymbol{r}_v\big)dA(u,v); \tag{3.1.21}$$

the integral is also called the **flux** of $\boldsymbol{F}$ across S. A surface integral of the second kind can be reduced to a surface integral of the first kind:

$$\iint\limits_S \boldsymbol{F}\cdot d\boldsymbol{\sigma} = \iint\limits_S \boldsymbol{F}\cdot\widehat{\boldsymbol{n}}_S\,d\sigma, \tag{3.1.22}$$

where $\widehat{\boldsymbol{n}}_S$ is the unit normal vector to the surface. The direction of $\widehat{\boldsymbol{n}}_S$ defines the orientation of S; the integral reverses the sign if the orientation of S is reversed.

If $\boldsymbol{F} = \rho\boldsymbol{v}$, where $\boldsymbol{v}$ is the velocity of a fluid flow, and ρ is the density of the fluid, then $\iint\limits_S \boldsymbol{F}\cdot d\boldsymbol{\sigma}$ is the mass of the fluid passing across the surface S per unit time; verify that

$$\boldsymbol{F}\cdot\widehat{\boldsymbol{n}}_S\,d\sigma = \rho\,\boldsymbol{v}\cdot\widehat{\boldsymbol{n}}_S\,d\sigma$$

is the mass of fluid passing across the surface area element $d\sigma$ per unit time. On page 154 we derive the *equation of continuity* that governs such flows.

THE VOLUME INTEGRAL. We now take the set G in (3.1.13) to be a subset of the space $\mathbb{R}^3$. The set is measurable and has volume V if the boundary of G is a piece-wise smooth surface. In volume integrals, we use the differential form of the **volume element** dV. In cartesian coordinates, $dV = dxdydz$. In **generalized cylindrical coordinates** $x = ar\cos\theta,\ y = br\sin\theta, z = z,\ dV = ab\ rdr\,d\theta\,dz$; the usual cylindrical coordinates correspond to $a = b = 1$. In **generalized spherical coordinates** $x = ar\cos\theta\sin\varphi,\ y = br\sin\theta\sin\varphi,\ z = cr\cos\varphi,\ dV = abc\,r^2\sin\varphi\,dr\,d\theta\,d\varphi$; the usual spherical coordinates correspond to $a = b = c = 1$. The volume integral is evaluated by reduction to three iterated integrals; this is why the volume integral is also known as the *triple integral* and is often written as $\iiint\limits_G f\,dV$. The integral is defined if f is a continuous scalar field in G. If f is the volume density of G, then $\iiint\limits_G f\,dV$ is the mass of G.

EXERCISE 3.1.16.[B] *Find the volume of the ellipsoid* $(x^2/a^2) + (y^2/b^2) + (z^2/c^2) \leq 1$. *Hint: use generalized spherical coordinates.*

Let us summarize the main facts related to the orientation of domains and their boundaries. For a domain G in the $(\hat{\imath}, \hat{\jmath})$ plane, we denote by ∂G the boundary of G and assume that this boundary consists of finitely many simple closed piece-wise smooth curves. The **positive orientation** of ∂G means that the domain G stays on your left as you walk around the boundary in the direction of the orientation of the corresponding curve. An equivalent mathematical description is as follows. At all but finitely many points of ∂G, there exists the unit tangent vector $\hat{\boldsymbol{u}}$; the direction of $\hat{\boldsymbol{u}}$ defines the orientation. At every point of ∂G where $\hat{\boldsymbol{u}}$ exists, consider the unit vector $\hat{\boldsymbol{n}}$ that is perpendicular to $\hat{\boldsymbol{u}}$ and points *outside of* G. Draw a picture and convince yourself that the orientation of ∂G is positive if and only if $\hat{\boldsymbol{n}} \times \hat{\boldsymbol{u}}$ has the same direction as $\hat{\imath} \times \hat{\jmath}$.

For a domain G in $\mathbb{R}^3$, we denote by ∂G the boundary of G and assume that this boundary consists of finitely many piece-wise smooth simple closed surfaces. The **positive orientation** of ∂G means that, when it exists, the normal vector to ∂G points *outside* of G.

For a piece-wise smooth surface S that is not closed, we denote by ∂S the boundary of the surface and assume that this boundary consists of finitely many simple closed piece-wise smooth curves. We say that the `orientations of` S `and` ∂S `agree` if S stays on your left as you walk around the boundary in the direction of the orientation of the corresponding curve with your head in the direction of the normal vector to S. If this is possible, then S is orientable. An equivalent mathematical description is as follows. Assume that the orientation of S is fixed and given by the unit vectors $\widehat{\boldsymbol{n}}_S$ that exist at most points of S. Then a small piece of S near ∂S looks like a small piece of a domain in the $(\hat{\imath}, \hat{\jmath})$ plane, with $\widehat{\boldsymbol{u}}_S$ defining the direction of $\hat{\imath} \times \hat{\jmath}$. Then the positive orientation of ∂S means positive orientation of the boundary of this two-dimensional domain as described earlier.

A famous example of a smooth non-orientable surface is the `Möbius strip`, first discovered by the German mathematician AUGUST FERDINAND MÖBIUS (1790–1868). You can easily make this surface by twisting a long narrow strip of paper and taping the ends. Verify, for example, by drawing a line along the strip, that the surface has only one side

We will now use integrals to provide *coordinate-free* definitions for the *divergence* and *curl* of a vector field. A traditional presentation starts with the definitions in the cartesian coordinates. Then, after proving Gauss's and Stokes's Theorems, one can use the results to show that the definitions do not actually depend on the coordinate system. We take a different approach and start with the definitions that are not connected to any coordinate system.

We begin with the DIVERGENCE. Divergence makes a scalar field out of a vector field and is an example of an *operator*. This operator arises naturally in fluid flow models. Let $\boldsymbol{v} = \boldsymbol{v}(P)$ be the velocity at point P in a moving fluid, and let $\rho = \rho(P)$ be the density of the fluid at P. Consider a small surface of area $\triangle\sigma$ with unit normal vector $\widehat{\boldsymbol{u}}$. The rate of flow per unit time, or flux, of fluid across $\triangle\sigma$ is $\rho\boldsymbol{v} \cdot \widehat{\boldsymbol{u}}\,\triangle\sigma$. For a region G bounded by a smooth simple closed surface ∂G, the total flux across ∂G is $\Phi(G) = \iint\limits_{G} \rho\boldsymbol{v} \cdot \widehat{\boldsymbol{u}}\, d\sigma$. This flux is equal the rate of change in time of the mass of fluid contained in G. The quantity $\Phi(G)/\mathfrak{m}(G)$, where $\mathfrak{m}(G)$ is the volume of G, is the average flux per unit volume in G; in the limit $\mathfrak{m}(G) \to G$, that is, as the region G shrinks to a point, we get an important local characteristic of the flow. This naturally leads to the definition of divergence.

Let $\boldsymbol{F}$ be a continuous vector field with values in $\mathbb{R}^3$. Then the divergence of $\boldsymbol{F}$ at the point P is the number $\operatorname{div}\boldsymbol{F}(P)$ so that

$$\operatorname{div}\boldsymbol{F}(P) = \lim_{\mathfrak{m}(G)\to 0} \frac{1}{\mathfrak{m}(G)} \iint\limits_{\partial G} \boldsymbol{F}\cdot d\boldsymbol{\sigma}, \tag{3.1.23}$$

where $G = G(P)$ is a measurable solid region that contains the point P, and $\mathfrak{m}(G)$ is the volume of G. This definition assumes that the limit exists and does not depend on the particular choice of the regions G. Interpreting $\boldsymbol{F}$ as the velocity field of fluid, the right-hand side of expression (3.1.23) determines the density of sources ($\operatorname{div}\boldsymbol{F}(P) > 0$) or sinks ($\operatorname{div}\boldsymbol{F}(P) < 0$) at the point P. For an electric field, sources and sinks correspond to positive and negative charges, respectively. The vector field $\boldsymbol{F}$ is called solenoidal in a domain G if $\operatorname{div}\boldsymbol{F} = 0$ everywhere in G. The reason for this terminology is that, in the theory of magnetism, there are no analogs of positive and negative electric charges. Therefore, a magnetic field has no sources or sinks and its divergence is zero everywhere. On the other hand, a magnetic field is often produced by a solenoid — a coil of wire connected to an electric source.

EXERCISE 3.1.17.^{B} *Let $\boldsymbol{F} = F_1\,\hat{\imath} + F_2\,\hat{\jmath} + F_3\,\hat{\kappa}$ and assume that the functions F_1, F_2, F_3 have continuous first-order partial derivatives. Show that*

$$\operatorname{div}\boldsymbol{F} = F_{1x} + F_{2y} + F_{3z}, \tag{3.1.24}$$

where, as usual, $F_{1x} = \partial F_1/\partial x$, etc. Hint: let $P = (x_0, y_0, z_0)$ and consider a cube G with center at P, faces parallel to the coordinate planes, and side a. Then $\mathfrak{m}(G) = a^3$. The total flux through the two faces that are parallel to the $(\hat{\jmath}, \hat{\kappa})$ plane is then approximately

$$\big(F_1(x_0 + a/2, y_0, z_0) - F_1(x_0 - a/2, y_0, z_0)\big)a^2; \tag{3.1.25}$$

draw a picture and keep in mind the orientation of the boundary. Then consider the faces parallel to the other two coordinate planes. The total flux out of G is the sum of the resulting three fluxes. After dividing by the volume of G and passing to the limit, (3.1.25) results in $F_{1x}(x_0, y_0, z_0)$, and F_{2y}, F_{3z} come form the fluxes across the other two pairs of faces.

Next, we discuss the CURL. As a motivation, we again consider a vector field $\boldsymbol{v}$ representing the velocity of moving fluid or gas. The flow of fluid or gas can have vortices: think of the juice in a blender or hurricanes, tornadoes or large-scale circulation of atmospheric winds due to Coriolis force. Vortices are caused by circular motion of the particles in the flow, and, for a smooth simple closed planar curve C, the circulation $\Psi = \oint_C \boldsymbol{v}\cdot d\boldsymbol{r}$ measures the amount of vorticity. If S is the planar region enclosed by C, then $\Psi/\mathfrak{m}(S)$ is the average vorticity in the region, where $\mathfrak{m}(S)$ is the area of S. In the limit $\mathfrak{m}(S)\to 0$ we get another important local characteristic of the flow. This naturally leads to the definition of the curl.

Let $\boldsymbol{F}$ be a continuous vector field with values in $\mathbb{R}^3$. Fix a point P and a unit vector $\widehat{\boldsymbol{n}}$ and consider the plane passing through P and having $\widehat{\boldsymbol{n}}$ as the normal vector. Let C be a piece-wise smooth simple closed curve in the plane, and S, the part of the plane enclosed by the curve. We assume that $P\in S$, $P\notin C$, and the orientation of C is counterclockwise as seen from the top of $\widehat{\boldsymbol{n}}$. By definition, the `curl` of $\boldsymbol{F}$ at the point P is the vector $\operatorname{curl}\boldsymbol{F}(P)$ so that, for all C and $\widehat{\boldsymbol{u}}$,

$$\operatorname{curl}\boldsymbol{F}(P)\cdot\widehat{\boldsymbol{n}} = \lim_{\mathfrak{m}(S)\to 0}\frac{1}{\mathfrak{m}(S)}\oint_C \boldsymbol{F}\cdot d\boldsymbol{r}, \tag{3.1.26}$$

where $\mathfrak{m}(S)$ is the area of S. This definition assumes that the limit exists for every unit vector $\widehat{\boldsymbol{n}}$, and, for each vector $\widehat{\boldsymbol{n}}$, the limit does not depend on the particular choice of the curves C.

EXERCISE 3.1.18.^{B} *Let $\boldsymbol{F} = F_1\,\hat{\boldsymbol{\imath}} + F_2\,\hat{\boldsymbol{\jmath}} + F_3\,\hat{\boldsymbol{\kappa}}$ and assume that the functions F_1, F_2, F_3 have continuous first-order partial derivatives. Show that* $\operatorname{curl}\boldsymbol{F}$ *can be computed as the vector given by the symbolic determinant*

$$\operatorname{curl}\boldsymbol{F} = \begin{vmatrix} \hat{\boldsymbol{\imath}} & \hat{\boldsymbol{\jmath}} & \hat{\boldsymbol{\kappa}} \\ \frac{\partial}{\partial x} & \frac{\partial}{\partial y} & \frac{\partial}{\partial z} \\ F_1 & F_2 & F_3 \end{vmatrix} = (F_{3y}-F_{2z})\,\hat{\boldsymbol{\imath}} + (F_{1z}-F_{3x})\,\hat{\boldsymbol{\jmath}} + (F_{2x}-F_{1y})\,\hat{\boldsymbol{\kappa}}. \tag{3.1.27}$$

Hint: with $P=(x_0,y_0,z_0)$, take the plane through P parallel to one of the coordinate planes and let C be the boundary of the square in that plane with center at P and side a so that $\mathfrak{m}(S)=a^2$. Taking the plane so that $\widehat{\boldsymbol{n}}=\hat{\boldsymbol{\imath}}$, we find that

$$\oint_C \boldsymbol{F}\cdot d\boldsymbol{r} \approx a\big(F_3(x_0,y_0+a/2,z_0) - F_3(x_0,y_0-a/2,z_0)\big) + a\big(F_2(x_0,y_0,z_0-a/2) - F_2(x_0,y_0,z_0+a/2)\big);$$

draw the picture and keep track of the orientations. Dividing by a^2 and passing to the limit $a\to 0$ gives the $\hat{\boldsymbol{\imath}}$ component of $\operatorname{curl}\boldsymbol{F}$. *The other two components are computed similarly.*

Relations (3.1.24) and (3.1.27) suggest the introduction of the *vector differential operator*

$$\boldsymbol{\nabla} = \frac{\partial}{\partial x}\,\hat{\boldsymbol{\imath}} + \frac{\partial}{\partial y}\,\hat{\boldsymbol{\jmath}} + \frac{\partial}{\partial z}\,\hat{\boldsymbol{\kappa}} \tag{3.1.28}$$

and the symbolic vector representations of the divergence and curl:

$$\operatorname{div} \boldsymbol{F} = \boldsymbol{\nabla}\cdot\boldsymbol{F},\ \operatorname{curl}\boldsymbol{F} = \boldsymbol{\nabla}\times\boldsymbol{F}. \tag{3.1.29}$$

We can also interpret $\operatorname{grad} f$ as a multiplication of the vector $\boldsymbol{\nabla}$ by the scalar f. Then it is easy to remember the following identities:

$$\begin{aligned}
&\operatorname{div}(\operatorname{curl}\boldsymbol{F}) = 0:\ \boldsymbol{\nabla}\cdot(\boldsymbol{\nabla}\times\boldsymbol{F}) = 0;\ \operatorname{curl}(\operatorname{grad} f) = \boldsymbol{0}:\ \boldsymbol{\nabla}\times(\boldsymbol{\nabla} f) = \boldsymbol{0};\\
&\boldsymbol{\nabla}\cdot(f\,\boldsymbol{F}) = \boldsymbol{\nabla} f\cdot\boldsymbol{F} + f\,\boldsymbol{\nabla}\cdot\boldsymbol{F},\quad \boldsymbol{\nabla}\times(f\,\boldsymbol{F}) = \boldsymbol{\nabla} f\times\boldsymbol{F} + f\,\boldsymbol{\nabla}\times\boldsymbol{F};\\
&\boldsymbol{\nabla}\cdot(\boldsymbol{F}\times\boldsymbol{G}) = \boldsymbol{G}\cdot(\boldsymbol{\nabla}\times\boldsymbol{F}) - \boldsymbol{F}\cdot(\boldsymbol{\nabla}\times\boldsymbol{G});\quad \boldsymbol{\nabla}\cdot(\boldsymbol{\nabla} f\times\boldsymbol{\nabla} g) = 0.
\end{aligned} \tag{3.1.30}$$

Not all formulas are nice and easy, though: the expressions for $\boldsymbol{\nabla}(\boldsymbol{F}\cdot\boldsymbol{G})$ and $\operatorname{curl}(\boldsymbol{F}\times\boldsymbol{G})$ are rather complicated.

EXERCISE 3.1.19.[A] *Recall that, for two vector fields $\boldsymbol{F}, \boldsymbol{G}$, we define the vector field $(\boldsymbol{F}\cdot\boldsymbol{\nabla})\boldsymbol{G}$; see (3.1.11) on page 126. Show that*

$$\begin{aligned}
(\boldsymbol{F}\cdot\boldsymbol{\nabla})\boldsymbol{G} = \frac{1}{2}\Big(&\operatorname{curl}(\boldsymbol{F}\times\boldsymbol{G}) + \operatorname{grad}(\boldsymbol{F}\cdot\boldsymbol{G}) - \boldsymbol{G}\operatorname{div}\boldsymbol{F} + \boldsymbol{F}\operatorname{div}\boldsymbol{G}\\
&- \boldsymbol{G}\times\operatorname{curl}\boldsymbol{F} - \boldsymbol{F}\times\operatorname{curl}\boldsymbol{G}\Big).
\end{aligned}$$

Use the result to derive the following formulas:

$$\begin{aligned}
\operatorname{curl}(\boldsymbol{F}\times\boldsymbol{G}) &= \boldsymbol{F}\operatorname{div}\boldsymbol{G} - \boldsymbol{G}\operatorname{div}\boldsymbol{F} + (\boldsymbol{G}\cdot\boldsymbol{\nabla})\boldsymbol{F} - (\boldsymbol{F}\cdot\boldsymbol{\nabla})\boldsymbol{G},\\
\operatorname{grad}(\boldsymbol{F}\cdot\boldsymbol{G}) &= \boldsymbol{F}\times\operatorname{curl}\boldsymbol{G} + \boldsymbol{G}\times\operatorname{curl}\boldsymbol{F} + (\boldsymbol{F}\cdot\boldsymbol{\nabla})\boldsymbol{G} + (\boldsymbol{G}\cdot\boldsymbol{\nabla})\boldsymbol{F}.
\end{aligned}$$

For a scalar field f, $\operatorname{div}(\operatorname{grad} f)$ is called the `Laplacian` of f and denoted by $\nabla^2 f$; an alternative notation $\boldsymbol{\Delta} f$ is also used.

EXERCISE 3.1.20.[C] *Verify that, in cartesian coordinates,*

$$\nabla^2 f = f_{xx} + f_{yy} + f_{zz}. \tag{3.1.31}$$

For a vector field $\boldsymbol{F}$, we define

$$\nabla^2\boldsymbol{F} = \operatorname{grad}(\operatorname{div}\boldsymbol{F}) - \operatorname{curl}(\operatorname{curl}\boldsymbol{F}). \tag{3.1.32}$$

EXERCISE 3.1.21. *(a)*[A] *Verify the equalities in (3.1.30) Hint: choose cartesian coordinates so that $\boldsymbol{F} = F_1\hat{\boldsymbol{\imath}}$. (b) Verify that if $\boldsymbol{F} = F_1\,\hat{\boldsymbol{\imath}} + F_2\,\hat{\boldsymbol{\jmath}} + F_3\,\hat{\boldsymbol{\kappa}}$,* *then* $\nabla^2 F = \nabla^2 F_1\,\hat{\imath} + \nabla^2 F_2\,\hat{\jmath} + \nabla^2 F_3\,\hat{\kappa}$. *(c)*^{C} *Verify that* $\boldsymbol{\nabla}^2(fg) = g\boldsymbol{\nabla}^2 f + 2\boldsymbol{\nabla} f \cdot \boldsymbol{\nabla} g + f\boldsymbol{\nabla}^2 g$.

> Given a reference point O, there is a natural vector field, denoted by $\boldsymbol{r}$, which maps every point P to the corresponding position vector $\overrightarrow{OP}$. The corresponding scalar field $r = \|\boldsymbol{r}\|$ is the distance from O to P.

The following exercise establishes a number of remarkable properties of the fields $\boldsymbol{r}$ and r.

EXERCISE 3.1.22.^{C} *(a) Verify that if* $g = g(t)$ *is a differentiable function, then*

$$\boldsymbol{\nabla} g(r) = g'(r)\,\widehat{\boldsymbol{r}} = \frac{g'(r)}{r}\,\boldsymbol{r}. \tag{3.1.33}$$

Hint: use cartesian coordinates so that $\boldsymbol{r} = x\,\hat{\imath} + y\,\hat{\jmath} + z\,\hat{\kappa}$, *and* $r = \sqrt{x^2 + y^2 + z^2}$. *(b) Verify that, in* $\mathbb{R}^n$, $\operatorname{div}\boldsymbol{r} = n$, $n = 2, 3$. *(c) Verify that, in* $\mathbb{R}^2$, $\nabla^2 \ln r = 0$ *for* $r \neq 0$. *(d) Verify that, in* $\mathbb{R}^3$, $\nabla^2 r^{-1} = 0$ *for* $r \neq 0$. *Hint: for (c) and (d), use (3.1.33) together with some of the identities in (3.1.30). (e) Let* $h = h(t)$ *be a differentiable function. Verify that* $\operatorname{curl}(h(r)\,\boldsymbol{r}) = \boldsymbol{0}$. *Hint: the fastest way is to conclude from (3.1.33) that* $h(r)\,\boldsymbol{r} = \boldsymbol{\nabla} H(r)$ *for a suitable function* H*; then use (3.1.30). (f) Let* $\boldsymbol{b}$ *be a constant vector. Verify that*

$$\operatorname{grad}(\boldsymbol{b} \cdot \boldsymbol{r}) = \boldsymbol{b}. \tag{3.1.34}$$

Hint: write the dot product in cartesian coordinates. (g) Let $\boldsymbol{F} = c\widehat{\boldsymbol{r}}/r^2$ *be a central inverse-square force field in* $\mathbb{R}^3$ *Prove that* $\int_C \boldsymbol{F} \cdot d\boldsymbol{r}$ *is independent of the path* C *joining any two points, as long as* C *does not pass through the origin. Hint: show that* $\boldsymbol{F}$ *is conservative by finding* $f = f(r)$ *so that* $\boldsymbol{F} = \boldsymbol{\nabla} f$.

EXERCISE 3.1.23.^{C} *Let* $\boldsymbol{F} = c\,\widehat{\boldsymbol{r}}/r^\alpha = (c/r^{\alpha+1})\,\boldsymbol{r}$, *where* c *and* α *are real numbers. Verify that the flux of* $\boldsymbol{F}$ *across the sphere* S *with center at the origin and radius* R *is equal to* $4\pi R^{2-\alpha} c$. *Hint: Use (3.1.22);* $\widehat{\boldsymbol{n}}_S = \widehat{\boldsymbol{r}}$*; on* S, $r = R$.

EXERCISE 3.1.24.^{B} *Let* B_R *be a ball with center at the origin and radius* R. *Verify that* $\lim_{R\to 0} \iiint_{B_R} r^{-p} dV = 0$ *for all* $p < 3$.

Our motivation of the definition of the curl suggests that curl is associated with rotation; in fact, some books denote the curl of $\boldsymbol{F}$ by $\operatorname{rot}\boldsymbol{F}$. Let us look closer at this connection with rotation. Recall that, for fixed vectors $\boldsymbol{\omega}$ and $\boldsymbol{r}_0$, the vector $\boldsymbol{\omega} \times \boldsymbol{r}_0$ is the velocity of the point with position vector $\boldsymbol{r}_0$, rotating with constant angular speed $\omega = \|\boldsymbol{\omega}\|$ about a fixed axis; the vector $\boldsymbol{\omega}$ is the angular velocity of the rotation, see page 48. The vector field $\boldsymbol{v} = \boldsymbol{\omega} \times \boldsymbol{r}$ therefore corresponds to the rotation of all points in space with the same angular velocity $\boldsymbol{\omega}$. Direct computations in cartesian coordinates show that $\operatorname{curl} \boldsymbol{v} = 2\boldsymbol{\omega}$, establishing the connection between the curl and rotation: if $\boldsymbol{F} = \boldsymbol{F}(P)$ is interpreted as the velocity field of a continuum of moving points, then $\operatorname{curl} \boldsymbol{F}$ measures the rotational component of this velocity; the field $\boldsymbol{F}$ is called **irrotational** in a domain G if $\operatorname{curl} \boldsymbol{F} = \boldsymbol{0}$ in G.

3.1.3 *Curvilinear Coordinate Systems*

Recall that a vector field $(P, \boldsymbol{F}(P))$ is a collection of vectors with different starting points, and, to study the vector field, we need a *local coordinate system*, that is, a set of basis vectors, at every point P. So far, we mostly worked with cartesian coordinates, where this local coordinate system is the parallel translation of $(\hat{\boldsymbol{\imath}}, \hat{\boldsymbol{\jmath}}, \hat{\boldsymbol{k}})$ to the corresponding point P. More generally, these local coordinate systems are not necessarily parallel translations of one another. For example, the polar coordinates r, θ in $\mathbb{R}^2$ give rise to the unit vectors $\widehat{\boldsymbol{r}}, \widehat{\boldsymbol{\theta}}$, see (1.3.22) on page 35 or Figure 3.1.1 below. At every point, $\widehat{\boldsymbol{r}} \perp \widehat{\boldsymbol{\theta}}$, but the corresponding vectors $\widehat{\boldsymbol{r}}$ have, in general, different directions at different points. For various applications, it is natural to use cylindrical or spherical coordinates, and then it becomes necessary to express all the operations of differentiation (gradient, divergence, curl, Laplacian) in these coordinate systems.

The objective of this section is to study general *orthogonal curvilinear coordinate systems*, where the basis vectors at every point are mutually orthogonal, but their direction changes from point to point. Our first step is to establish the general relation between the coordinates and their basis vectors. For notational convenience, we denote by $\mathfrak{X} = (x_1, x_2, x_3)$ the cartesian coordinates, with the correspondence $x_1 = x, x_2 = y, x_3 = z$. For non-cartesian coordinates $\mathcal{Q} = (q_1, q_2, q_3)$, each q_k is a given function of x_1, x_2, x_3: $q_k = \mathfrak{q}_k(x_1, x_2, x_3)$, $k = 1, 2, 3$. The correspondence between the $\mathfrak{X}$ and $\mathcal{Q}$ coordinates must be invertible, so that each x_k is a corresponding function of q_1, q_2, q_3: $x_k = \chi_k(q_1, q_2, q_3)$, $k = 1, 2, 3$. FOR EXAMPLE, in **cylindrical coordinates**, with $q_1 = r$, $q_2 = \theta$, $q_3 = z$, we have $r = \sqrt{x_1^2 + x_2^2}$, $\theta = \tan^{-1}(x_2/x_1)$, $z = x_3$, so that $x_1 = r\cos\theta$, $x_2 = r\sin\theta$, $x_3 = z$. This example also shows that there could be some points where the correspondence between the $\mathfrak{X}$ and $\mathcal{Q}$-coordinates is not invertible: the polar angle θ is not defined when $x = y = 0$. We call such points *special* and allow their existence as long as there are not too many of them; at this point, we rely on intuition to decide how many is "many" and will not go into the details.

EXERCISE 3.1.25.[B] *Let* $q_1 = r$*,* $q_2 = \theta$*,* $q_3 = \varphi$ *be the usual* **spherical coordinates** *so that* $x_1 = r\cos\theta\,\sin\varphi$*,* $x_2 = r\sin\theta\,\sin\varphi$*,* $x_3 = r\cos\varphi$*. Express* q_1, q_2, q_3 *as functions of* x_k*,* $k = 1, 2, 3$*. Find all special points.*

A **coordinate curve** in system $\mathcal{Q}$ is obtained by fixing two of the $\mathcal{Q}$ coordinates, and using the remaining one as a parameter. For every non-special point P with $\mathcal{Q}$ coordinates (c_1, c_2, c_3), there are three coordinate curves passing through P. For example, we can fix q_2, q_3 and vary q_1. The corresponding coordinate curve $\mathcal{C}_1$ is given in cartesian coordinates by $x_k = \chi(q_1, c_2, c_3)$, $k = 1, 2, 3$. The coordinate curves $\mathcal{C}_2$ and $\mathcal{C}_3$ are obtained in a similar way. The coordinate system $\mathcal{Q}$ is called **orthogonal** if the coordinate curves are smooth and are mutually orthogonal at every non-special point. The unit tangent vectors $\widehat{\boldsymbol{q}}_k$, $k = 1, 2, 3$ to these curves at P define the local orthonormal basis at P; we choose the orientation and the ordering of the curves so that the basis is right-handed: $\widehat{\boldsymbol{q}}_1 \times \widehat{\boldsymbol{q}}_2 = \widehat{\boldsymbol{q}}_3$. Figure 3.1.1 shows the coordinate curves and unit tangent vectors for the polar coordinates $x = r\cos\theta, y = r\sin\theta$.

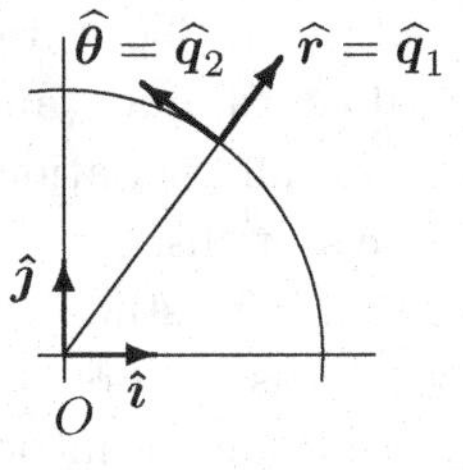

Fig. 3.1.1 Polar Coordinates

In the $\mathcal{Q}$ coordinates, the coordinate curves are also represented as intersections of pairs of surfaces: $\{(q_1, q_2, q_3) : q_n = c_n, q_m = c_m\}$, where $m, n = 1, 2, 3$, $m < n$, and c_m, c_n are real numbers. If we assume that, at time $t = 0$, each curve passes through the point P, then, in the $\mathcal{Q}$ coordinates,

- The q_1 coordinate curve with $q_2 = c_2$, $q_3 = c_3$, is the collection of points $(c_1 + t, c_2, c_3)$, $t \in \mathbb{R}$;
- The q_2 coordinate curve with $q_1 = c_1$, $q_3 = c_3$, is the collection of points

$(c_1, c_2 + t, c_3)$, $t \in \mathbb{R}$;

• The q_3 coordinate curve with $q_1 = c_1$, $q_2 = c_2$, is the collection of points $(c_1, c_2, c_3 + t)$, $t \in \mathbb{R}$.

Since, by assumption, $x_k = \chi_k(q_1, q_2, q_3)$, the equations of these curves in cartesian coordinates are

$$\begin{aligned} \boldsymbol{r}_1(t) &= \chi_1(c_1 + t, c_2, c_3)\,\hat{\boldsymbol{\imath}} + \chi_2(c_1 + t, c_2, c_3)\,\hat{\boldsymbol{\jmath}} + \chi_3(c_1 + t, c_2, c_3)\,\hat{\boldsymbol{\kappa}} \\ \boldsymbol{r}_2(t) &= \chi_1(c_1, c_2 + t, c_3)\,\hat{\boldsymbol{\imath}} + \chi_2(c_1, c_2 + t, c_3)\,\hat{\boldsymbol{\jmath}} + \chi_3(c_1, c_2 + t, c_3)\,\hat{\boldsymbol{\kappa}} \\ \boldsymbol{r}_3(t) &= \chi_1(c_1, c_2, c_3 + t)\,\hat{\boldsymbol{\imath}} + \chi_2(c_1, c_2, c_3 + t)\,\hat{\boldsymbol{\jmath}} + \chi_3(c_1, c_2, c_3 + t)\,\hat{\boldsymbol{\kappa}} \end{aligned} \tag{3.1.35}$$

We define the functions h_k, $k = 1, 2, 3$ so that the tangent vectors are

$$\boldsymbol{r}_k'(0) = h_k\,\widehat{\boldsymbol{q}}_k. \tag{3.1.36}$$

By (3.1.35),

$$h_k(P) = \|\boldsymbol{r}_k'(0)\| = \sqrt{\sum_{n=1}^{3} \left(\frac{\partial \chi_n}{\partial q_k}\right)^2}, \tag{3.1.37}$$

where the partial derivatives are evaluated at (c_1, c_2, c_3), the $\mathcal{Q}$ coordinates of the point P. The orthogonality condition means that

$$\boldsymbol{r}_m'(0) \cdot \boldsymbol{r}_n'(0) = \sum_{k=1}^{3} \frac{\partial \chi_k}{\partial q_m}\frac{\partial \chi_k}{\partial q_n} = 0,\ m \neq n. \tag{3.1.38}$$

The functions h_k defined in (3.1.36) will play the central role in many computations to follow. We will assume that, away from the special points, the functions $h_k = h_k(q_1, q_2, q_3)$ have continuous partial derivatives of every order.

EXERCISE 3.1.26.^{B} *(a) Verify the relations (3.1.37) and (3.1.38). (b) Verify that cylindrical and spherical coordinates are orthogonal. Hint: a picture is helpful.*

Our next goal is to study ARC LENGTH in $\mathcal{Q}$ coordinates. Let C be a curve represented in cartesian coordinates $\mathfrak{X}$ by the vector function $\boldsymbol{r}(t) = x_1(t)\,\hat{\boldsymbol{\imath}} + x_2(t)\hat{\boldsymbol{\jmath}} + x_3(t)\,\hat{\boldsymbol{\kappa}}$. We rewrite the expression (1.3.12) on page 29 for the line element as $(d\mathfrak{s}/dt)^2 = \sum_{k=1}^{3}(dx_k/dt)^2$. Since $x_k = \chi_k(q_1, q_2, q_3)$, we have $x_k(t) = \chi_k(q_1(t), q_2(t), q_3(t))$, where $q_k(t) = \mathfrak{q}_k(x_1(t), x_2(t), x_3(t))$, $k = 1, 2, 3$ By the chain rule, $dx_k/dt = \sum_{m=1}^{3}(\partial\chi_k/\partial q_m)\,(dq_m/dt)$, and then, using (3.1.36) and dropping $(dt)^2$, we find that the **line element** in $\mathcal{Q}$ coordinates is

$$(d\mathfrak{s})^2 = \sum_{k=1}^{3} h_k^2(q_1, q_2, q_3)\,(dq_k)^2. \tag{3.1.39}$$

EXERCISE 3.1.27.[B] *(a) Verify (3.1.39).*

(b) Verify that, for small a, the distance between the two nearby points with the $\mathcal{Q}$ coordinates (c_1, c_2, c_3) and $(c_1 + a, c_2, c_3)$ is approximately $h_1 a$.

(c) Verify that in cylindrical coordinates, $(d\mathfrak{s})^2 = (dr)^2 + r^2(d\theta)^2 + (dz)^2$.

(d) Verify that in spherical coordinates, $(d\mathfrak{s})^2 = (dr)^2 + r^2\sin^2\varphi\,(d\theta)^2 + r^2(d\varphi)^2$. Hint for (b) and (c): while direct substitution into (3.1.39) will work, getting closer to the basics produces an easier argument. For example, in spherical coordinates, if r and φ are fixed and θ is changing, the corresponding coordinate curve is a circle of radius $r\sin\varphi$ (draw a picture). Then a change of $\Delta\theta$ in θ produces the change $r\sin\varphi\,\Delta\theta$ in the arc length. Similar arguments work for the other two coordinate curves, and then it remains to use orthogonality. Note that, with this approach, you get the values of h_k without using (3.1.37).

(e) Verify that if the $\mathcal{Q}$ coordinate are not orthogonal, then

$$(d\mathfrak{s})^2 = \sum_{i,j=1}^{3} H_{ij}\,dq_i\,dq_j, \text{ where } H_{ij} = \sum_{k=1}^{3} \frac{\partial \chi_k}{\partial q_i}\frac{\partial \chi_k}{\partial q_j}. \tag{3.1.40}$$

We already used a four-dimensional analog of this relation in the study of general relativity; see (2.4.23) on page 107.

For future reference, let us summarize the values of the functions h_k for cylindrical and spherical coordinates:

	h_1	h_2	h_3
Cylindrical	1	r	1
Spherical	1	$r\sin\varphi$	r

We can now derive the expression for the GRADIENT in $\mathcal{Q}$ coordinates. The value of a function at a point does not depend on the coordinate system, but the formula for computing that value does: if $f = x_1x_2 + x_3$ in cartesian coordinates, the same function in spherical coordinates becomes $r^2\cos\theta\sin\theta\sin^2\varphi + r\cos\varphi$. Given an expression $f(q_1, q_2, q_3)$ for a function in $\mathcal{Q}$ coordinates, we define the partial derivatives $\partial f/\partial q_k$, $k = 1, 2, 3$ in the usual way; for example,

$$\frac{\partial f}{\partial q_1} = \lim_{t\to 0} \frac{f(q_1+t, q_2, q_3) - f(q_1, q_2, q_3)}{t}.$$

Unlike cartesian coordinates, the partial derivative $\partial f/\partial q_k$ is not equal to the directional derivative $D_{\widehat{\boldsymbol{q}}_k} f$. Instead, by (3.1.9), (3.1.35), and (3.1.36), we conclude that

$$\frac{\partial f}{\partial q_k} = \nabla f \cdot \boldsymbol{r}_k'(0) = h_k \nabla f \cdot \widehat{\boldsymbol{q}}_k = h_k D_{\widehat{\boldsymbol{q}}_k} f. \tag{3.1.41}$$

EXERCISE 3.1.28.^{B} *(a) Convince yourself that, in general, $\partial f/\partial q_k \neq D_{\widehat{\boldsymbol{q}}_k} f$. Hint: if P has $\mathcal{Q}$ coordinates (q_1, q_2, q_3), and P_t, the coordinates (q_1+t, q_2, q_3), then, in general, $\overrightarrow{OP_t} \neq \overrightarrow{OP} + t\,\widehat{\boldsymbol{q}}_1$, but the equality does hold in cartesian coordinates. (b) Verify (3.1.41).*

Since the representation $\nabla f = \sum_{k=1}^{3} (\nabla f \cdot \widehat{\boldsymbol{q}}_k)\, \widehat{\boldsymbol{q}}_k$ holds in every orthonormal coordinate system, equality (3.1.41) implies the following formula for computing the gradient in general orthogonal coordinates:

$$\nabla f = \frac{1}{h_1}\frac{\partial f}{\partial q_1}\widehat{\boldsymbol{q}}_1 + \frac{1}{h_2}\frac{\partial f}{\partial q_2}\widehat{\boldsymbol{q}}_2 + \frac{1}{h_3}\frac{\partial f}{\partial q_3}\widehat{\boldsymbol{q}}_3, \tag{3.1.42}$$

where the functions h_1, h_2, h_3 are defined in (3.1.37).

EXERCISE 3.1.29.^{B} *(a) Verify that*

$$\nabla f = \frac{\partial f}{\partial r}\widehat{\boldsymbol{r}} + \frac{1}{r}\frac{\partial f}{\partial \theta}\widehat{\boldsymbol{\theta}} + \frac{\partial f}{\partial z}\widehat{\boldsymbol{z}}$$

is the gradient in cylindrical coordinates. (b) Verify that

$$\nabla f = \frac{\partial f}{\partial r}\widehat{\boldsymbol{r}} + \frac{1}{r\sin\varphi}\frac{\partial f}{\partial \theta}\widehat{\boldsymbol{\theta}} + \frac{1}{r}\frac{\partial f}{\partial \varphi}\widehat{\boldsymbol{\varphi}}$$

is the gradient in spherical coordinates, and verify that, for $f(r,\theta,\varphi) = g(r)$, the result coincides with (3.1.33).

Next, we derive the expression for the DIVERGENCE in $\mathcal{Q}$ coordinates using the definition (3.1.23). Recall that, to compute the divergence in the cartesian coordinates, we consider a family of shrinking cubes with faces parallel to the coordinate planes. The same approach works in any coordinates by considering rectangular boxes whose sides are parallel to the *local* basis vectors $\widehat{\boldsymbol{q}}_1, \widehat{\boldsymbol{q}}_2, \widehat{\boldsymbol{q}}_3$. Let P be a point with $\mathcal{Q}$ coordinates (c_1, c_2, c_3), and, for sufficiently small $a > 0$, consider a rectangular box with vertices at the points with the $\mathcal{Q}$ coordinates $(c_1 \pm (a/2), c_2 \pm (a/2), c_3 \pm (a/2))$. The volume of this box is approximately $a^3|\boldsymbol{r}_1'(0) \cdot (\boldsymbol{r}_2'(0) \times \boldsymbol{r}_3'(0))| = a^3 h_1 h_2 h_3$; draw a picture and see equations (3.1.35). By (3.1.36) and (3.1.38), The basis vector $\widehat{\boldsymbol{q}}_k$ is normal to two of the faces so that the area of each of those faces is approximately $a^2 h_m h_n$, where $k \neq m \neq n$. Consider a continuously differentiable vector field $\boldsymbol{F}$ written in the $\mathcal{Q}$ coordinates as $\boldsymbol{F}(q_1, q_2, q_3) = F_1(q_1, q_2, q_3)\,\widehat{\boldsymbol{q}}_1 + F_2(q_1, q_2, q_3)\,\widehat{\boldsymbol{q}}_2 + F_3(q_1, q_2, q_3)\,\widehat{\boldsymbol{q}}_3$. Then the flux of this vector field through the pair of faces with the normal vector $\widehat{\boldsymbol{q}}_1$ is approximately

$$a^2\Big(h_2\big(c_1 + (a/2), c_2, c_3\big)h_3\big(c_1 + (a/2), c_2, c_3\big)F_1\big(c_1 + (a/2), c_2, c_3\big)$$
$$-h_2\big(c_1 - (a/2), c_2, c_3\big)h_3\big(c_1 - (a/2), c_2, c_3\big)F_1\big(c_1 - (a/2), c_2, c_3\big)\Big)$$
$$\approx a^3 \frac{\partial(h_2 h_3 F_1)}{\partial q_1},$$

with the approximation getting better as $a \to 0$. Similar expressions hold for the fluxes across the other two pairs of faces. Summing up all the fluxes, dividing by the volume of the box, and passing to the limit $a \to 0$, we get the formula for the divergence of $\boldsymbol{F}$ in $\mathcal{Q}$ coordinates:

$$\operatorname{div} \boldsymbol{F} = \frac{1}{h_1 h_2 h_3}\left(\frac{\partial(h_2 h_3 F_1)}{\partial q_1} + \frac{\partial(h_1 h_3 F_2)}{\partial q_2} + \frac{\partial(h_1 h_2 F_3)}{\partial q_3}\right). \tag{3.1.43}$$

EXERCISE 3.1.30.^{B} *(a) Verify (3.1.43). (b) Verify that, in cylindrical coordinates,*

$$\operatorname{div} \boldsymbol{F} = \frac{1}{r}\left(\frac{\partial(rF_1)}{\partial r} + \frac{\partial F_2}{\partial \theta} + \frac{\partial(rF_3)}{\partial z}\right).$$

(c) Verify that, in spherical coordinates,

$$\operatorname{div} \boldsymbol{F} = \frac{1}{r^2 \sin\varphi}\left(\frac{\partial(r^2 \sin\varphi\, F_1)}{\partial r} + \frac{\partial(rF_2)}{\partial \theta} + \frac{\partial(r \sin\varphi\, F_3)}{\partial \varphi}\right).$$

Recall that the Laplacian $\nabla^2 f$ of a scalar field f is defined in every coordinate system as $\nabla^2 f = \operatorname{div}(\operatorname{grad} f)$. Let $f = f(q_1, q_2, q_3)$ be a scalar field defined in an orthogonal coordinate system $\mathcal{Q}$.

EXERCISE 3.1.31. *(a)^B Using (3.1.42) and (3.1.43) verify that*

$$\nabla^2 f = \frac{1}{h_1 h_2 h_3}\left(\frac{\partial}{\partial q_1}\left(\frac{h_2 h_3}{h_1}\frac{\partial f}{\partial q_1}\right) + \frac{\partial}{\partial q_2}\left(\frac{h_1 h_3}{h_2}\frac{\partial f}{\partial q_2}\right) + \frac{\partial}{\partial q_3}\left(\frac{h_1 h_2}{h_3}\frac{\partial f}{\partial q_3}\right)\right).$$

(b)^C *Verify that, in cylindrical coordinates,*

$$\nabla^2 f = \frac{1}{r}\frac{\partial}{\partial r}\left(r\frac{\partial f}{\partial r}\right) + \frac{1}{r^2}\frac{\partial^2 f}{\partial \theta^2} + \frac{\partial^2 f}{\partial z^2}. \tag{3.1.44}$$

and use the result to provide an alternative proof that $\nabla^2 \ln r = 0$ *in* $\mathbb{R}^2$.
(c)^C *Verify that, in spherical coordinates,*

$$\nabla^2 f = \frac{1}{r^2}\left(\frac{\partial}{\partial r}r^2\frac{\partial f}{\partial r}\right) + \frac{1}{r^2\sin^2\varphi}\frac{\partial^2 f}{\partial \theta^2} + \frac{1}{r^2\sin\varphi}\frac{\partial}{\partial \varphi}\left(\sin\varphi\frac{\partial f}{\partial \varphi}\right). \tag{3.1.45}$$

and use the result to provide an alternative proof that $\nabla^2 r^{-1} = 0$ *in* $\mathbb{R}^3$.

Next, we derive the formula for the CURL using (3.1.26). Consider a continuously differentiable vector field $\boldsymbol{F}$ written in the $\mathcal{Q}$ coordinates as $\boldsymbol{F}(q_1, q_2, q_3) = F_1(q_1, q_2, q_3)\,\widehat{\boldsymbol{q}}_1 + F_2(q_1, q_2, q_3)\,\widehat{\boldsymbol{q}}_2 + F_3(q_1, q_2, q_3)\,\widehat{\boldsymbol{q}}_3$. Let us compute $\operatorname{curl}\boldsymbol{F}(P)\cdot\widehat{\boldsymbol{q}}_1$, where the point P has $\mathcal{Q}$-coordinates (c_1, c_2, c_3). For a sufficiently small $a > 0$, consider a rectangle spanned by the vectors $a\boldsymbol{r}'_k(0)$, $k = 2, 3$, so that P is at the center of the rectangle. The vertices of the rectangle have the $\mathcal{Q}$ coordinates $(c_1, c_2 \pm (a/2), c_3 \pm (a/2))$ and the area of this rectangle is approximately $a^2 h_2 h_3$; draw a picture and see equations (3.1.35). The line integral of $\boldsymbol{F}$ along the two sides parallel to $\widehat{\boldsymbol{q}}_2$ is approximately

$$\begin{aligned}
&a\Big(h_2\big(c_1, c_2, c_3 - (a/2)\big)F_2\big(c_1, c_2, c_3 - (a/2)\big)\\
&- h_2\big(c_1, c_2, c_3 + (a/2)\big)F_2\big(c_1, c_2, c_3 + (a/2)\big)\Big)\\
&\approx -a^2\frac{\partial(h_2F_2)}{\partial q_3},
\end{aligned}$$

with the quality of approximation improving as $a \to 0$. The line integral over the remaining two sides is approximately $a^2\partial(h_3F_3)/\partial q_2$. Dividing by the area of the rectangle and passing to the limit $a \to 0$, we find that

$$\operatorname{curl}\boldsymbol{F}(P)\cdot\widehat{\boldsymbol{q}}_1 = \frac{1}{h_1h_2h_3}\left(\frac{\partial(h_3F_3)}{\partial q_2} - \frac{\partial(h_2F_2)}{\partial q_3}\right)h_1$$

The other two components, $\operatorname{curl}\boldsymbol{F}(P)\cdot\widehat{\boldsymbol{q}}_2$ and $\operatorname{curl}\boldsymbol{F}(P)\cdot\widehat{\boldsymbol{q}}_3$ are computed similarly, and then we get the representation of the curl as a symbolic determinant similar to (3.1.27):

$$\operatorname{curl} \boldsymbol{F} = \frac{1}{h_1 h_2 h_3} \begin{vmatrix} h_1 \widehat{\boldsymbol{q}}_1 & h_2 \widehat{\boldsymbol{q}}_2 & h_3 \widehat{\boldsymbol{q}}_3 \\ \dfrac{\partial}{\partial q_1} & \dfrac{\partial}{\partial q_2} & \dfrac{\partial}{\partial q_3} \\ h_1 F_1 & h_2 F_2 & h_3 F_3 \end{vmatrix}. \tag{3.1.46}$$

EXERCISE 3.1.32. [A] *(a) Verify (3.1.46). (b) Use (3.1.46) in spherical coordinates to show that* $\operatorname{curl}(g(r)\,\boldsymbol{r}) = \boldsymbol{0}$ *in* $\mathbb{R}^3$.

We conclude this section with a brief discussion of INTEGRATION in curvilinear coordinates. As before, let $\mathfrak{X}$ be the cartesian coordinates x_1, x_2, x_3 and $\mathcal{Q}$, some other coordinates q_1, q_2, q_3 so that $x_k = \chi_k(q_1, q_2, q_3)$, $k = 1, 2, 3$; we no longer assume that the $\mathcal{Q}$ coordinates are orthogonal. At a non-special point P, consider the three vectors $\boldsymbol{r}_k'(0)$, $k = 1, 2, 3$, defined in (3.1.35), and assume that

$$\boldsymbol{r}_1'(0) \cdot (\boldsymbol{r}_2'(0) \times \boldsymbol{r}_3'(0)) \neq 0. \tag{3.1.47}$$

EXERCISE 3.1.33.[B] *Verify that*

$$\boldsymbol{r}_1'(0) \cdot (\boldsymbol{r}_2'(0) \times \boldsymbol{r}_3'(0)) = \begin{vmatrix} \dfrac{\partial \chi_1}{\partial q_1} & \dfrac{\partial \chi_2}{\partial q_1} & \dfrac{\partial \chi_3}{\partial q_1} \\ \dfrac{\partial \chi_1}{\partial q_2} & \dfrac{\partial \chi_2}{\partial q_2} & \dfrac{\partial \chi_3}{\partial q_2} \\ \dfrac{\partial \chi_1}{\partial q_3} & \dfrac{\partial \chi_2}{\partial q_3} & \dfrac{\partial \chi_3}{\partial q_3} \end{vmatrix}. \tag{3.1.48}$$

Hint: we have cartesian coordinates in (3.1.35).

The determinant in (3.1.48) is called the **Jacobian**, after the German mathematician CARL GUSTAV JACOB JACOBI (1804–1851), who, in 1841, published a paper containing a detailed study of such determinants. The Jacobian (3.1.48) is denoted either by $J = J(q_1, q_2, q_3)$ or by $\partial(x_1, x_2, x_3)/\partial(q_1, q_2, q_3)$; similarly, we either call it the Jacobian of the system of functions χ_1, χ_2, χ_3 or the Jacobian of the transformation from the $\mathfrak{X}$ coordinates to the $\mathcal{Q}$ coordinates. It was one of the results in the original paper by Jacobi that the change of variables is one-to-one if $J \neq 0$.

EXERCISE 3.1.34. *(a)*[A] *We know that, in orthogonal coordinates,*

$J(q_1, q_2, q_3) = h_1h_2h_3$. *Try to derive this equality directly by expanding the determinant in (3.1.48) and using the orthogonality conditions (3.1.38).* *(b)*^{C} *Verify that* $J(r, \theta, z) = r$ *(cylindrical coordinates) and* $J(r, \theta, \varphi) = r^2 \sin\varphi$ *(spherical coordinates).*

Let G be a measurable domain in $\mathbb{R}^3$ with cartesian coordinates $\mathfrak{X}$, and let $f = f(x_1, x_2, x_3)$ be a continuous scalar field on G. A change of coordinates $q_k = \mathfrak{q}_k(x_1, x_2, x_3)$, $k = 1, 2, 3$, maps every point (x_1, x_2, x_3) to a point (q_1, q_2, q_3), which is also in $\mathbb{R}^3$. Under this map,

- The domain G transforms into another domain $\tilde{G}$ (of course, there are certain conditions on the functions $\mathfrak{q}_k$ to ensure that the image of a domain is again a domain);
- A small box with sides a, b, c around a point $P = (x_1, x_2, x_3)$ becomes a small curvilinear box around the image of P; by (3.1.47), the volume of this box is approximately $abc|J(q_1, q_2, q_3)|$, with approximation getting better as $\max(a, b, c) \to 0$;
- The function f becomes $\tilde{f} = \tilde{f}(q_1, q_2, q_2)$ so that $\tilde{f}(q_1, q_2, q_3) = f(\chi_1(q_1, q_2, q_3), \chi_2(q_1, q_2, q_3), \chi_3(q_1, q_2, q_3))$.

Using the general definition of the Riemann integral (3.1.13), we get the following **change of variables formula for the volume integral**:

$$\iiint_G f(x_1, x_2, x_3)\, dV(\mathfrak{X}) = \iiint_{\tilde{G}} \tilde{f}(q_1, q_2, q_3)|J(q_1, q_2, q_3)|\, dV(\mathcal{Q}). \tag{3.1.49}$$

Notice that the change of coordinates from $\mathfrak{X}$ to $\mathcal{Q}$ is given by the functions $\mathfrak{q}_k$, $k = 1, 2, 3$, whereas, to compute the Jacobian and to find the function $\tilde{f}$, we need the functions χ_k defining the change in the opposite direction.

A result similar to (3.1.49) holds for area integrals, if we write $q_k = \mathfrak{q}_k(x_1, x_2)$, $k = 1, 2$, and $q_3 = x_3$.

EXERCISE 3.1.35.^{C} *(a) Verify (3.1.49). (b) Write the analogs of (3.1.47) and (3.1.49) in two dimensions. Hint:* $\boldsymbol{r}_1'(0) \times \boldsymbol{r}_2'(0) = J(q_1, q_2)\hat{\boldsymbol{k}}$. *(c) Interpret the vector parametric representation* $\boldsymbol{r} = \boldsymbol{r}(u, v)$, $(u, v) \in G$, *of a surface* S *as a map from* G *to* S *and derive the familiar formula for the surface area*

$$\mathfrak{m}(S) = \iint_G \|\boldsymbol{r}_u \times \boldsymbol{r}_v\| dA(u, v).$$

Hint: show that a small rectangle with sides a, b *in* G *becomes a curvilinear figure on* S *with area approximately equal to* $ab\|\boldsymbol{r}_u \times \boldsymbol{r}_v\|$. *Compare this with Exercise 3.1.15 on page 134.*

3.2 The Three Integral Theorems of Vector Analysis

The theorems of Green, Gauss, and Stokes are usually the high point of a multi-variable calculus class. We will use these theorems to derive the equations of electromagnetic theory.

3.2.1 *Green's Theorem*

Theorem 3.2.1 *In the space $\mathbb{R}^2$ with cartesian coordinates, consider a domain G whose boundary ∂G consists of finitely many simple closed piecewise smooth curves. Let $P = P(x, y)$, $Q = Q(x, y)$ be two functions having continuous first-order partial derivatives in an open domain containing G. Assume that the orientation of ∂G is positive (G stays to the left as you walk around ∂G). Then*

$$\oint_{\partial G} P dx + Q dy = \iint_G \left(\frac{\partial Q}{\partial x} - \frac{\partial P}{\partial y} \right) dA. \tag{3.2.1}$$

If the boundary of G consists of several pieces, then the integral on the left is the sum of the corresponding integrals over each piece.

This result is known as **Green's theorem**, after the English scientist George Green (1793–1841), who published it in 1828 as a part of an essay on electricity and magnetism. The essay was self-published for private distribution and did not receive much attention at the time. Green did all his major scientific work while managing the family mill and baking business, with about two years of elementary school as his only formal education. Later, he quit the family business and, in 1837, got an undergraduate degree from Cambridge University.

Exercise 3.2.1. [B] *Prove Green's theorem. Hint: Consider the vector field $\boldsymbol{F} = P\,\hat{\imath} + Q\,\hat{\jmath} + 0\,\hat{\kappa}$ and verify that $\boldsymbol{\nabla} \times \boldsymbol{F} = (Q_x - P_y)\,\hat{\kappa}$. Then break G into small pieces G_k and note that, by (3.1.26), $(Q_x - P_y)\mathfrak{m}(G_k) \approx \oint_{\partial G_k} P dx + Q dy$, where $\mathfrak{m}(G_k)$ is the area of G_k, and ∂G_k is the boundary of G_k with the suitable orientation. Now sum over all the small pieces and note that the line integrals over the interior pieces cancel one another. Finally, pass to the limit $\mathfrak{m}(G_k) \to 0$.*

An interesting application of (3.2.1) is computing areas by line integrals:

$$\mathfrak{m}(G) = \iint_G dA = \oint_{\partial G} x dy = -\oint_{\partial G} y dx = \frac{1}{2} \oint_{\partial G} x dy - y dx. \tag{3.2.2}$$

EXERCISE 3.2.2. ^{B} *(a) Verify all equalities in (3.2.2). (b) Verify that, in polar coordinates $x = r\cos\theta$, $y = r\sin\theta$, we have*

$$\mathfrak{m}(G) = \frac{1}{2}\oint_{\partial G} r^2\, d\theta.$$

Hint: verify that, in polar coordinates, $xdy - ydx = r^2 d\theta$.

FOR EXAMPLE, let us compute the area under one arc of the `cycloid`, the trajectory of a point on the rim of a wheel rolling without slippage along a straight line. If a is the radius of the wheel, and the wheel is rolling to the right so that, at time $t = 0$, the point on the rim is at the origin, then the vector parametric equation of the cycloid is

$$\boldsymbol{r}(t) = a(t - \sin t)\,\hat{\boldsymbol{\imath}} + a(1 - \cos t)\,\hat{\boldsymbol{\jmath}}. \tag{3.2.3}$$

One revolution of the wheel corresponds to $t = 2\pi$. The region G under one arc is bounded by the x-axis on the bottom and the arc on the top. By (3.2.2), keeping in mind the orientation of the boundary, the area of the region is $\mathfrak{m}(G) = -\oint_{\partial G} ydx = -\int_0^{2\pi} 0dx - a^2\int_{2\pi}^0 (1-\cos t)^2 dt = \boxed{3\pi a^2}$. On the arc, $x'(t) = a(1-\cos t)$, $y(t) = a(1-\cos t)$, and, of the three possible formulas for the area, the one involving only ydx results in the easiest expression to integrate.

EXERCISE 3.2.3. ^{B} *(a) Verify that (3.2.3) is indeed a vector parametric equation of the trajectory of the point on the rim of a rolling wheel. Hint: consider $0 < t < \pi$. Denote by A the point on the rim, A', the current point of contact between the wheel and the ground, and C, the center of the wheel. Then $\boldsymbol{r}(t) = \overrightarrow{OA'} + \overrightarrow{A'C} + \overrightarrow{CA}$, $|OA'| = at$, and the angle between $\overrightarrow{CA'}$ and $\overrightarrow{CA}$ is t; draw the picture. (b) Draw the picture of the cycloid and verify the integration.*

There is a hardware device, called `planimeter`, which implements the formula $\mathfrak{m}(G) = (1/2)\oint_{\partial G} xdy - ydx$ and measures the area of a planar figure by traversing the perimeter. The Swiss mathematician JACOB AMSLER (1823–1912) is credited with the invention of a mechanical planimeter in 1854. Today, both mechanical and electronic planimeters are in use.

3.2.2 *The Divergence Theorem of Gauss*

Depending on the source, the following result is known as `Gauss's theorem`, the `Gauss-Ostrogradsky theorem`, or the `divergence theorem`. This theorem can be motivated by the following physical argument. Let $\boldsymbol{v}$

and ρ be the velocity field and the density of a fluid in a region G. In our discussion of divergence, we saw that $\operatorname{div}(\rho\,\boldsymbol{v})$ is the time rate of change of the amount fluid in G per unit volume. Therefore, $\iiint\limits_G \operatorname{div}(\rho\,\boldsymbol{v})dV$ is the time rate of change of the amount of fluid in G, and must be equal to the total flux across the boundary ∂G of G:

$$\iiint\limits_G \operatorname{div}(\rho\,\boldsymbol{v})\,dV = \iint\limits_{\partial G} \rho\,\boldsymbol{v}\cdot d\boldsymbol{\sigma}$$

This is the essential meaning of the divergence theorem, which holds for every continuously differentiable vector field $\boldsymbol{F}$ as stated below.

Theorem 3.2.2 *Consider a bounded domain G in $\mathbb{R}^3$ whose boundary ∂G consists of finitely many piece-wise smooth orientable surfaces. Let $\boldsymbol{F}$ be a vector field so that, for every unit vector $\widehat{\boldsymbol{u}}$, the scalar field $\boldsymbol{F}\cdot\widehat{\boldsymbol{u}}$ has a continuous gradient in an open set containing G. Then*

$$\iint\limits_{\partial G} \boldsymbol{F}\cdot d\boldsymbol{\sigma} = \iiint\limits_G \operatorname{div}\boldsymbol{F}\,dV, \tag{3.2.4}$$

where the normal vector to ∂G is pointing outside of G; if the boundary of G consists of several pieces, then the integral on the left is the sum of the corresponding integrals over all those pieces.

The great German mathematician Carl Friedrich Gauss (1777–1855) discovered this result in 1813 while studying the laws of electrostatics. As with many other his discoveries, he was not in a hurry to publish it; the exact year is known from the detailed mathematical diary Gauss kept all his adult life. Based on the entries in that diary, many believe that Gauss could have advanced 19th century mathematics by another 50 years or more, had he published all his results promptly. The Russian mathematician Mikhail Vasil'evich Ostrogradsky (1801–1862) made an independent discovery of (3.2.4) and published it around 1830.

Exercise 3.2.4.[B] *(a) Prove Gauss's Theorem. Hint: the argument is the same as in the proof of Green's Theorem, but using (3.1.23). (b) Let $\boldsymbol{F} = \boldsymbol{r}/r^3 = \widehat{\boldsymbol{r}}/r^2$ (see page 140). Let S be a piece-wise smooth closed surface so that the origin O of the frame is not on the surface. Show that*

$$\iint\limits_S \boldsymbol{F}\cdot d\boldsymbol{\sigma} = \begin{cases} 4\pi, & S \text{ encloses the origin,} \\ 0, & S \text{ does not enclose the origin.} \end{cases} \tag{3.2.5}$$

Hint: Note that $\operatorname{div}\boldsymbol{F} = \nabla^2(1/r) = 0$ *if* $r \neq 0$. *If* S *encloses the origin, then there exists a small sphere centered at the origin and lying inside the domain enclosed by* S. *Show that the flux through* S *is equal to the flux through the sphere; then see Exercise 3.1.23 on page 140. (c) Let* ∂G *be an orientable piece-wise smooth surface enclosing a domain* G, *and let* $\boldsymbol{F}$ *be a vector field, continuously differentiable in an open set containing* G. *Show that*

$$\iint_{\partial G} \boldsymbol{F} \times \widehat{\boldsymbol{n}}_G \, d\sigma = - \iiint_G \operatorname{curl}\boldsymbol{F} \, dV, \tag{3.2.6}$$

where $\widehat{\boldsymbol{n}}_G$ *is the outside normal vector to* ∂G. *Hint: the left- and right-hand sides of (3.2.6) are vectors; denote them by* $\boldsymbol{v}$ *and* $\boldsymbol{w}$, *respectively. The claim is that* $\boldsymbol{v}\cdot\boldsymbol{b} = \boldsymbol{w}\cdot\boldsymbol{b}$ *for every constant vector* $\boldsymbol{b}$; *in cartesian coordinates, it is enough to verify this equality for only three vectors,* $\hat{\boldsymbol{\imath}}, \hat{\boldsymbol{\jmath}}, \hat{\boldsymbol{\kappa}}$ *instead of every* $\boldsymbol{b}$. *By the properties of the scalar triple product,* $\boldsymbol{b}\cdot(\boldsymbol{F}\times\widehat{\boldsymbol{n}}_G) = -(\boldsymbol{F}\times\boldsymbol{b})\cdot\widehat{\boldsymbol{n}}_G$. *By Gauss's Theorem,*

$$\iint_{\partial G} (\boldsymbol{F}\times\boldsymbol{b})\cdot\widehat{\boldsymbol{n}}_G \, d\sigma = \iiint_G \operatorname{div}(\boldsymbol{F}\times\boldsymbol{b}) \, dV,$$

and by (3.1.30), $\operatorname{div}(\boldsymbol{F}\times\boldsymbol{b}) = \boldsymbol{b}\cdot\operatorname{curl}\boldsymbol{F}$.

We will now use Gauss's Theorem to derive the basic equation of fluid flow, called the EQUATION OF CONTINUITY. Assume that the space $\mathbb{R}^3$ is filled with a continuum of point masses (particles). This continuum can represent, for example, liquid, gas, or electric charges; accordingly, "mass" represents the appropriate characteristic of that continuum, such as the mass of matter or electric charge. Denote by $\boldsymbol{v} = \boldsymbol{v}(P,t)$ the velocity of the point mass at the point P and time t, and by $\rho = \rho(P,t)$, the density of the point masses at the point P and time t. By definition,

$$\rho(P,t) = \lim_{\triangle V \to 0} \frac{\triangle M}{\triangle V},$$

where $\triangle V$ is the volume of a region containing the point P, $\triangle M$ is the mass of that region, and limit is assumed to exist at every point P. We also allow the existence of sources and sinks, that is, in some areas of the space, the particles can be injected into the space (source) or removed from the space (sink). The density of the sources and sinks is described by a continuous scalar field $\nu = \nu(P,t)$ so that the net mass per unit time, injected into and removed from G due to sources and sinks is $\iiint_G \rho\,\nu\,dV$. We also assume that the functions $\rho, \nu, \boldsymbol{v}$ are continuously differentiable as many time as necessary. Since the total mass M of the particles in G also changes due to the flux across ∂G, we have

$$\frac{dM}{dt} = \iiint\limits_{G} \rho\,\nu\, dV - \iint\limits_{\partial G} \rho\, \boldsymbol{v} \cdot d\boldsymbol{\sigma}; \tag{3.2.7}$$

recall that the positive direction of the flux is out of G, which reduces the mass. On the other hand, $M = \iiint\limits_{G} \rho\, dV$. Assuming that we can differentiate under the integral, $dM/dt = \iiint\limits_{G} (\partial\rho/\partial t)\, dV$. If the functions ρ and $\boldsymbol{v}$ are continuously differentiable, then, by Gauss's Theorem (3.2.4), $\iint\limits_{\partial G} \rho\, \boldsymbol{v} \cdot d\boldsymbol{\sigma} = \iiint\limits_{G} \operatorname{div}(\rho\, \boldsymbol{v})\, dV$. As a result, (3.2.7) implies

$$\iiint\limits_{G} \left(\frac{\partial \rho}{\partial t} + \operatorname{div}(\rho\, \boldsymbol{v}) - \rho\nu \right) dV = 0.$$

Given a point in space, this equality holds for every domain G containing that point, and we therefore conclude that the **equation of continuity**

$$\frac{\partial \rho}{\partial t} + \operatorname{div}(\rho\, \boldsymbol{v}) = \rho\nu \tag{3.2.8}$$

holds at every point of the space. Conversely, after integration, (3.2.8) implies (3.2.7).

EXERCISE 3.2.5.^{C} *(a) Convince yourself that (3.2.7) implies (3.2.8). Hint: assume that the left-hand side of (3.2.8) is bigger than $\nu\rho$ at one point. By continuity, the inequality will hold in some neighborhood of that point. Integrate over that neighborhood and get a contradiction. (b) Verify that (3.2.8) is equivalent to*

$$\frac{d\rho}{dt} + \rho \operatorname{div} \boldsymbol{v} = \nu\rho. \tag{3.2.9}$$

Hint: Consider a trajectory $\boldsymbol{r} = \boldsymbol{r}(t)$ of a moving particle so that $\boldsymbol{v} = \boldsymbol{r}'$ and $\rho = \rho(\boldsymbol{r}(t), t)$. Then $d\rho/dt = \boldsymbol{\nabla}\rho \cdot \boldsymbol{v} + \partial\rho/\partial t$. On the other hand, by (3.1.30), $\operatorname{div}(\rho\, \boldsymbol{v}) = \boldsymbol{\nabla}\rho \cdot \boldsymbol{v} + \rho \operatorname{div} \boldsymbol{v}$.

A continuum is called **incompressible** if the density ρ does not change in time and space: $\rho(t, x) = const$. For an incompressible continuum, equation (3.2.8) becomes $\operatorname{div} \boldsymbol{v} = \nu$; the velocity $\boldsymbol{v}$ of an incompressible continuum without sources or sinks satisfies $\operatorname{div} \boldsymbol{v} = 0$.

We can also state Gauss's Theorem in a two dimensional domain G. Let the functions P, Q and the domain G satisfy the conditions of Green's

Theorem. Define the vector field $\boldsymbol{F} = P\,\hat{\boldsymbol{\imath}} + Q\,\hat{\boldsymbol{\jmath}} + 0\,\hat{\boldsymbol{\kappa}}$ and consider the *outside* unit normal vector $\widehat{\boldsymbol{n}}_G$ to ∂G; since the unit tangent vector $\widehat{\boldsymbol{u}}_G$ to ∂G exists at all but finitely many points of ∂G, so does $\widehat{\boldsymbol{n}}_G$, being, by definition, a unit vector perpendicular to $\widehat{\boldsymbol{u}}_G$. Then, by Gauss's Theorem,

$$\oint_{\partial G} \boldsymbol{F}\cdot\widehat{\boldsymbol{n}}_G\, d\mathfrak{s} = \iint\limits_{G} \operatorname{div} \boldsymbol{F} dA. \tag{3.2.10}$$

EXERCISE 3.2.6.[C] *Derive (3.2.10) from Green's Theorem. Hint: if* $\boldsymbol{r}(t) = x(t)\,\hat{\boldsymbol{\imath}} + y(t)\,\hat{\boldsymbol{\jmath}}$ *is the parametrization of* ∂G*, then* $\boldsymbol{r}' = x'\,\hat{\boldsymbol{\imath}} + y'\,\hat{\boldsymbol{\jmath}}$ *is tangent to* ∂G*; the vector* $y'\,\hat{\boldsymbol{\imath}} - x'\hat{\boldsymbol{\jmath}}$ *is therefore perpendicular to* $\boldsymbol{r}\,'$ *and points outside (verify this). Then* $\boldsymbol{F}\cdot\widehat{\boldsymbol{n}}_G\, d\mathfrak{s} = -Qdx + Pdy$*. Note that* $\operatorname{div}\boldsymbol{F} = P_x + Q_y$*.*

3.2.3 *Stokes's Theorem*

The following result is known as `Stokes's Theorem`, after the English mathematician GEORGE GABRIEL STOKES (1819–1903).

Theorem 3.2.3 *Let S be a piece-wise smooth orientable surface whose boundary ∂S consists of finitely many simple closed piece-wise smooth curves so that the orientations of S and ∂S agree (see page 135). Let $\boldsymbol{F}$ be a vector field, continuously differentiable in a domain containing S. Then*

$$\oint_{\partial S} \boldsymbol{F}\cdot d\boldsymbol{r} = \iint\limits_{S} \operatorname{curl} \boldsymbol{F}\cdot d\boldsymbol{\sigma}; \tag{3.2.11}$$

if the boundary of S consists of several closed curves, then the integral on the left is the sum of the corresponding integrals over all those curves.

According to some accounts, the first statement of this theorem appeared *in a letter to Stokes*, written in 1850 by another English scientist, Sir WILLIAM THOMSON (1824–1907), also known as LORD KELVIN.

EXERCISE 3.2.7. *(a)*[C] *Verify that Stokes's Theorem implies Green's theorem. (b)*[B] *Prove Stokes's Theorem (the arguments are similar to the proof of Greens Theorem).*

The following result should be familiar from a course in multi-variable calculus; Stokes's Theorem provides a key component for the proof.

Theorem 3.2.4 *Let $\boldsymbol{F}$ be a continuous vector field in a simply connected domain G, and assume that, for every unit vector $\widehat{\boldsymbol{n}}$, the scalar field $\boldsymbol{F}\cdot\widehat{\boldsymbol{n}}$*

has a continuous gradient in G. Then the following three conditions are equivalent:
(i) The field $\boldsymbol{F}$ has the path independence property: $\oint_{C} \boldsymbol{F} \cdot d\boldsymbol{r} = 0$ for every simple closed piece-wise smooth curve C in G.
(ii) The field $\boldsymbol{F}$ has a potential: there exists a function f so that $\boldsymbol{F} = \boldsymbol{\nabla} f$ everywhere in G.
(iii) The field $\boldsymbol{F}$ is irrotational: $\operatorname{curl} \boldsymbol{F} = \boldsymbol{0}$ *everywhere in G.*

EXERCISE 3.2.8. *(a)^C Prove the theorem. Hint: for (i) $\Rightarrow$ (ii), see Theorem 3.1.1. For (ii) $\Rightarrow$ (iii), use (3.1.30). For (iii) $\Rightarrow$ (i) use Stokes's Theorem. (b)^B Verify that the vector field $\boldsymbol{F}(x,y,z) = (-y/(x^2+y^2))\,\hat{\imath} + (x/(x^2+y^2))\,\hat{\jmath} + 0\,\hat{\kappa}$ is irrotational in the domain $\{(x,y,z) : x^2+y^2 > 0\}$ (direct computation) but does not have the path independence property there (integrate over a unit circle in the $(\hat{\imath}, \hat{\jmath})$ plane, centered at the origin). Explain why the result is consistent with the above theorem (note that the function is not defined anywhere on the z axis).*

Assume that the vector field $\boldsymbol{F}$ satisfies the conditions of Stoke's Theorem everywhere in $\mathbb{R}^3$. Then, by (3.2.11), the value of $\iint\limits_{S} \operatorname{curl} \boldsymbol{F} \cdot d\boldsymbol{\sigma}$ does not depend on the particular surface S, as long as the boundary of S is the given simple closed piece-wise smooth curve C. This conclusion is consistent with Gauss's Theorem. Indeed, let S_1 and S_2 be two orientable piece-wise smooth surfaces with the same boundary C and no other points in common, and denote by G the region enclosed by S_1 and S_2 (draw a picture). The boundary ∂G of G is the union of S_1 and S_2. If the vector field $\boldsymbol{F}$ is sufficiently smooth so that $\operatorname{curl} \boldsymbol{F}$ satisfies the conditions of Gauss's Theorem, then, by (3.2.11), $\iint\limits_{\partial G} \operatorname{curl} \boldsymbol{F} \cdot d\boldsymbol{\sigma} = \iiint\limits_{G} \operatorname{div}(\operatorname{curl} \boldsymbol{F}) dV = 0$, where the last equality follows from (3.1.30) on page 139, and therefore $\iint\limits_{S_1} \operatorname{curl} \boldsymbol{F} \cdot d\boldsymbol{\sigma} = \iint\limits_{S_2} \operatorname{curl} \boldsymbol{F} \cdot d\boldsymbol{\sigma}$, if the orientations of both S_1 and S_2 agree with the orientation of C.

By now we have seen enough connections between Green's, Gauss's, and Stokes's Theorems to suspect that there should be a way to unify the three into one formula. Such a formula does indeed exist:

$$\int_{\partial M} \omega = \int_{M} d\omega, \tag{3.2.12}$$

where ∂M is the boundary of the set M, and $d\omega$ is a certain derivative of the expression ω being integrated on the left-hand side. Making a precise meaning of this formula and demonstrating that (3.2.1), (3.2.4), and (3.2.11) all follow as particular cases requires a higher level of abstraction and is done in differential geometry, where (3.2.12) is known as Stokes's Theorem. The details are beyond the scope of this book and can be found in most books on differential geometry, for example, in Section 8.2 of the book *Differentiable Manifolds: A First Course* by L. Conlon, 1993.

The following exercise provides yet another application of Stokes's Theorem. One of the results, namely, (3.2.14), will be used later in the analysis of the magnetic dipole.

EXERCISE 3.2.9. ^{C} *Let S be piece-wise smooth orientable surface and its boundary ∂S, a simple closed piece-wise smooth curve. Assume that the orientations of S and ∂S agree. Denote by $\widehat{\boldsymbol{n}}$ the field of unit normal vectors on S, and by $\widehat{\boldsymbol{u}}$, the field of unit tangent vectors on ∂S. (a) Show that, for every continuously differentiable scalar function f,*

$$\iint_S \operatorname{grad} f \times \widehat{\boldsymbol{n}}\, d\sigma = -\oint_{\partial S} f\, \widehat{\boldsymbol{u}}\, ds. \tag{3.2.13}$$

Hint: the left- and right-hand sides of (3.2.13) are vectors; denote them by $\boldsymbol{v}$ and $\boldsymbol{w}$, respectively. The claim is that $\boldsymbol{v}\cdot\boldsymbol{b} = \boldsymbol{w}\cdot\boldsymbol{b}$ for every constant vector $\boldsymbol{b}$; in the cartesian coordinates, it is enough to verify this equality for only three vectors, $\hat{\imath}, \hat{\jmath}, \hat{k}$ instead of $\boldsymbol{b}$. Note that, by Stokes's Theorem, $\boldsymbol{w} = \iint_S \big(\operatorname{curl}(f\boldsymbol{b})\big)\cdot\widehat{\boldsymbol{n}}\, d\sigma$. By (3.1.30), $\operatorname{curl}(f\boldsymbol{b}) = \operatorname{grad} f \times \boldsymbol{b}$. Then use the properties of the scalar triple product. (b) Taking $f = \boldsymbol{r}_0\cdot\boldsymbol{r}$ in (3.2.13), where $\boldsymbol{r}_0$ is a constant vector and $\boldsymbol{r}$ is the vector field defined on page 140, and using (3.1.34), show that

$$\oint_{\partial S} (\boldsymbol{r}_0\cdot\boldsymbol{r})\, \widehat{\boldsymbol{u}}\, ds = -\boldsymbol{r}_0 \times \iint_S \widehat{\boldsymbol{n}}\, d\sigma. \tag{3.2.14}$$

3.2.4 *Laplace's and Poisson's Equations*

This section is a very brief introduction to potential theory and harmonic functions. Gauss's Theorem will be the main tool in the investigation.

A scalar function f is called `harmonic` in a domain G if f is twice continuously differentiable in G and $\nabla^2 f = 0$. Recall that $\nabla^2 f = \operatorname{div}(\operatorname{grad} f)$, so that, in cartesian coordinates, $\nabla^2 f = f_{xx} + f_{yy}$ in $\mathbb{R}^2$ and $\nabla^2 f = f_{xx} + f_{yy} + f_{zz}$ in $\mathbb{R}^3$. Similar expression exist in every $\mathbb{R}^n$, $n > 3$. While we restrict our discussion to $\mathbb{R}^3$, all the results of this section are valid in every $\mathbb{R}^n$, $n \geq 2$.

The equation

$$\nabla^2 f = g, \tag{3.2.15}$$

where g is a known continuous function, is called **Poisson's equation**, after the French mathematician SIMÉON DENIS POISSON (1781–1840). The particular case of the Poisson equation,

$$\nabla^2 f = 0, \tag{3.2.16}$$

is important enough to have its own name, and is called **Laplace's equation**, after the French mathematician PIERRE-SIMON LAPLACE (1749–1827). **Potential theory** is the branch of mathematics studying these two equations; the large number of problems in both mathematics and physics that are reduced to either (3.2.15) or (3.2.16) justifies the allocation of a whole branch of mathematics to the study of just two equations. This section provides the most basic introduction to potential theory. While the results we discuss are true in both two and three dimensions, our presentation will be in $\mathbb{R}^3$, both for the sake of concreteness, and because in $\mathbb{R}^2$ complex numbers provide a much more efficient method to study Laplace's and Poisson's equations; see Theorem 4.2.6 on page 205 below.

Let f be a continuously differentiable function in a domain G in $\mathbb{R}^3$, and S, a closed orientable piece-wise smooth surface. Denote by $\widehat{\boldsymbol{n}}_S = \widehat{\boldsymbol{n}}_S(P)$ the outside unit normal vector to S at the point P. The **normal derivative** of f at P is, by definition,

$$\frac{df}{dn}(P) = \boldsymbol{\nabla} f(P) \cdot \widehat{\boldsymbol{n}}_S(P); \text{ where } \boldsymbol{\nabla} f = \operatorname{grad} f. \tag{3.2.17}$$

EXERCISE 3.2.10.^{C} *Assume that f is a harmonic function in a domain G and S, a closed orientable piece-wise smooth surface in G. Assume that the domain enclosed by S is a subset of G. (a) Show that*

$$\iint_S \frac{df}{dn}\, d\sigma = 0. \tag{3.2.18}$$

Hint: Apply Gauss's Theorem to $\boldsymbol{F} = \boldsymbol{\nabla} f$. (b) Show that (3.2.18) can fail if the domain enclosed by S contains points that are not in G. Hint: take $f = r^{-1}$ and G, the unit ball around the origin with the center removed.

The following result is the analog of the integration-by-parts formula from the one-variable calculus.

Theorem 3.2.5 *Let G be a domain in $\mathbb{R}^3$. Assume that the boundary ∂G of G consists of finitely many closed piece-wise smooth orientable surfaces. As usual, we choose the orientation of ∂G so that every unit normal vector to ∂G points outside of G. Let f, g be two scalar fields defined in an open set G_1 containing G.*
(a) If f is continuously differentiable in G_1 and g is twice continuously differentiable, then

$$\iiint_G (\nabla^2 g)\, f\, dV = -\iiint_G \nabla f \cdot \nabla g\, dV + \iint_{\partial G} f\, \frac{dg}{dn}\, d\sigma. \qquad (3.2.19)$$

(b) If both f and g are twice continuously differentiable in G_1, then

$$\iiint_G (f\nabla^2 g - g\nabla^2 f)\, dV = \iint_{\partial G} \left(f\frac{dg}{dn} - g\frac{df}{dn} \right) d\sigma. \qquad (3.2.20)$$

EXERCISE 3.2.11.[C] *(a) Prove (3.2.19). Hint: apply Gauss's Theorem to the vector field $f\nabla g$. (b) Prove (3.2.20). Hint: in (3.2.19), switch f and g, then take the difference of the two resulting identities.*

Equalities (3.2.19) and (3.2.20) are known as **Green's first and second formulas** (or identities).

EXERCISE 3.2.12[C] *State and prove the two-dimensional versions of (3.2.18), (3.2.19), and (3.2.20). Hint: you just have to replace triple integrals with double, and surface integrals with line. For example, (3.2.19) becomes*

$$\iint_G f\nabla^2 g\, dA = -\iint_G \nabla f \cdot \nabla g\, dA + \oint_{\partial G} f\, \frac{dg}{dn}\, ds. \qquad (3.2.21)$$

Laplace's and Poisson's equations are examples of *partial differential equations*. While there exist many different methods for finding explicit solutions of *ordinary differential equations*, most partial differential equations are not explicitly solvable. As a result, theorems that ensure existence and uniqueness of solutions of such equations are very useful. Below, we discuss uniqueness of solution for the Poisson equation.

Theorem 3.2.6 *Let S be a closed orientable piece-wise smooth surface. Consider two functions f_1, f_2 that are continuously differentiable in an open set containing S and are harmonic everywhere in the domain G_S enclosed by S.*

(a) If $f_1 = f_2$ on S, then $f_1 = f_2$ everywhere in G_S.

(b) If $df_1/dn = df_2/dn$ on S then there exists a real number c so that $f_1 - f_2 = c$ everywhere in G_S.

EXERCISE 3.2.13.^{C} *Prove Theorem 3.2.6. Hint: consider the function $f = f_1 - f_2$ and apply (3.2.19) with $g = f$ to conclude that $\boldsymbol{\nabla} f = \mathbf{0}$ everywhere in G_S. Therefore f must be constant (Exercise 3.1.3 on page 125); in part (a), $f = 0$ on S, so the constant must be zero.*

Let G be a bounded domain whose boundary ∂G is a closed orientable piece-wise smooth surface. Let g be a continuous function in G, and h, a function continuous in some domain containing ∂G. The `Dirichlet`, or `first`, boundary value problem for the Poisson equation (3.2.15) in G is to find a function f that is twice continuously differentiable in an open set containing G, satisfies equation (3.2.15) everywhere in G, and $f = h$ on ∂G. The `Neumann`, or `second`, boundary value problem for the Poisson equation is to find a function f that is twice continuously differentiable in an open set containing G, satisfies equation (3.2.15) everywhere in G, and $df/dn = h$ on ∂G. With some additional constructions, a formulation of both problems is possible by considering the function f only in the closure of G, that is, without using an open set containing G

EXERCISE 3.2.14. ^{C} *(a) Show that the Dirichlet problem for the Poisson equation can have at most one solution. Hint: use the first part of Theorem 3.2.6; the difference of two solutions is a harmonic function. (b) Show that every two solutions of the Neumann problem differ by a constant. Hint: use the second part of Theorem 3.2.6. (c) Show that the Neumann problem for the Laplace equation has no solutions unless $\iint_{\partial G} h\, d\sigma = 0$. Hint: use (3.2.18).*

The difference between the Dirichlet and Neumann problems is the `boundary conditions` imposed on the function f: in the first case, we prescribe the values on the boundary of G to the function itself, and in the second case, to the normal derivative of the function. A combination of the two is possible and is called the `Robin` boundary value problem, when we prescribe the values on ∂G to the expression $af + (df/dn)$, where a is a known function defined on ∂G. Finally, there is an `oblique derivative problem`, when the values on ∂G are prescribed to the expression $\boldsymbol{\nabla} f \cdot \widehat{\boldsymbol{u}}$, where $\widehat{\boldsymbol{u}}$ is a continuous unit vector field on ∂G. We only mention these boundary conditions to illustrate the numerous problems that can be studied for the Poisson and Laplace equations.

While the Dirichlet problem is named after the German mathematician J.P.G.L. Dirichlet, who did pioneering work in potential theory, the origins of the other two names, Neumann and Robin, are less clear. Most probably, the Neumann problem is named after the German mathematician CARL GOTTFRIED NEUMANN (1832–1925) for his contributions to potential theory. His father FRANZ ERNST NEUMANN (1798–1895) worked in mathematical physics and could have contributed to the subject as well.

We conclude this section by establishing the *Dirichlet principle* for the Poisson equation. As before, let G be a bounded domain whose boundary ∂G is a closed orientable piece-wise smooth surface. Let g be a continuous function on G, and h, a function continuous on some domain containing ∂G. We assume that all functions are real-valued.

> Denote by $\mathcal{U}$ the set of functions so that every function in $\mathcal{U}$ is twice continuously differentiable in an open set containing G and is equal to h on ∂G.

For every $u \in \mathcal{U}$, define the number

$$I(u) = \iiint\limits_G \left(\frac{1}{2}\|\boldsymbol{\nabla} u\|^2 + gu\right) dV.$$

In mathematics, a rule that assigns a number to a function is called a `functional`; thus, I is a functional defined on the set $\mathcal{U}$.

Theorem 3.2.7 `(Dirichlet principle)` *If f is a function in $\mathcal{U}$ and $\nabla^2 f = g$ in G, then*

$$I(f) = \min_{u\in\mathcal{U}} I(u). \tag{3.2.22}$$

Conversely, if the function f in $\mathcal{U}$ satisfies (3.2.22), then $\nabla^2 f = g$ in G.

Proof. Assume that f is a function in $\mathcal{U}$ and $\nabla^2 f = g$ in G. Then

$$0 = \iiint\limits_G (\nabla^2 f - g)(f-u)dV = \iiint\limits_G \nabla^2 f\,(f-u)dV - \iiint\limits_G g\,(f-u)dV \tag{3.2.23}$$

for every $u \in \mathcal{U}$. Being elements of the set $\mathcal{U}$, both f and u are equal to h on ∂G. Therefore, $f - u = 0$ on ∂G, and we find from (3.2.19) that

$$\iiint\limits_G \nabla^2 f\,(f-u)dV = -\iiint\limits_G \|\boldsymbol{\nabla} f\|^2\,dV + \iiint\limits_G \boldsymbol{\nabla} f\cdot\boldsymbol{\nabla} u\,dV. \tag{3.2.24}$$

Combining (3.2.23) and (3.2.24) yields

$$\iiint_G \left(\|\boldsymbol{\nabla} f\|^2 + fg\right) dV = \iiint_G \left(\boldsymbol{\nabla} f \cdot \boldsymbol{\nabla} u + gu\right) dV. \tag{3.2.25}$$

By the definition of the dot product, $|\boldsymbol{\nabla} f \cdot \boldsymbol{\nabla} u| \leq \|\boldsymbol{\nabla} f\|\,\|\boldsymbol{\nabla} u\|$. Also, for every two non-negative numbers a, b, $0 \leq (a-b)^2 = a^2 + b^2 - 2ab$, which means that $ab \leq (a^2+b^2)/2$. Therefore, (3.2.25) implies

$$\iiint_G \left(\|\boldsymbol{\nabla} f\|^2 + fg\right) dV \leq \frac{1}{2} \iiint_G \|\boldsymbol{\nabla} f\|^2 dV + \iiint_G \left(\frac{1}{2}\|\boldsymbol{\nabla} u\|^2 + gu\right) dV.$$

Hence, $I(f) \leq I(u)$, which is equivalent to (3.2.22).

Conversely, assume that f satisfies (3.2.22). Take any function v that is twice continuously differentiable in G and is equal to zero outside of an open set contained in G. For this function v define the function $F = F(t)$ by $F(t) = I(f + tv)$. By assumption, $t = 0$ is a local minimum of F. Direct computations show that $F = F(t)$ is differentiable and

$$F'(0) = \iiint_G \left(\boldsymbol{\nabla} f \cdot \boldsymbol{\nabla} v + gv\right) dV. \tag{3.2.26}$$

From the one-variable calculus we know that $F'(0)$ must be equal to zero. Applying (3.2.19) and keeping in mind that, by assumption, $v = 0$ on ∂G, we conclude that

$$\iiint_G \left(-\nabla^2 f + g\right) v \, dV = 0. \tag{3.2.27}$$

Since the function v is arbitrary, this implies $\nabla^2 f = g$ in G and completes the proof of the theorem. □

EXERCISE 3.2.15.^{B} *(a) Verify (3.2.26). Hint: replace u in the definition of I with $f + tv$ and expanding the squares; note that, by construction, $f + tv \in \mathcal{U}$ for every v and t. (b) Verify that (3.2.27) indeed implies $\nabla^2 f = g$ in G. Hint: assume that $\nabla^2 f > g$ at some point P in G. Argue that the same inequality must then hold in some open ball centered at P and lying completely in G. Then take a function v that is equal to one in that ball — there is still some space between the boundary of the ball and the boundary of G where v can come down to zero — and come to a contradiction with (3.2.27).*

Theorem 3.2.7 holds also in $\mathbb{R}^2$, and, in fact, in every $\mathbb{R}^n$, $n \geq 2$.

3.3 Maxwell's Equations and Electromagnetic Theory

Electromagnetic theory, or electromagnetism, is the branch of physics that studies electric and magnetic fields. Numerous experiments have shown that these two fields are physically related. Maxwell's equations provide a mathematical model that establishes the precise relation between the electric and magnetic fields. The original publication in 1864 by the Scottish mathematician and physicist JAMES CLERK MAXWELL (1831–1879) contained 20 equations in 20 unknowns. Later, in the 1880s, J. W. Gibbs and O. Heaviside put the equations in a more compact vector form, which we will present below. On the basic physical level, Maxwell's equations generalize a number of experimental facts; there are some physicists these days who believe that one cannot *derive* Maxwell's equations, although Maxwell himself would probably disagree. On the mathematical level, these equations admit numerous derivations, for different models and on various levels of abstraction, and we discuss some of these derivations below.

If an equation describes a physical law, the particular form of the equation often depends on the units used to measure the corresponding physical quantities. In our discussion of electromagnetism, we will use the `International System of Units` (SI). Some of the units in SI are Ampere (A) for electric current, Coulomb (C) for electric charge, meter (m) for length, Newton (N) for force, second (s) for time.

3.3.1 *Maxwell's Equations in Vacuum*

In vacuum, the electric and magnetic fields are described mathematically by the vectors $\boldsymbol{E}$ and $\boldsymbol{B}$, respectively. A possible interpretation of these vectors is through the force $\boldsymbol{F}$ exerted by the fields on a point charge q moving with velocity $\boldsymbol{v}$:

$$\boldsymbol{F} = q(\boldsymbol{E} + \boldsymbol{v} \times \boldsymbol{B}). \tag{3.3.1}$$

The charges generating the fields can be stationary or moving; moving charges constitute an electric current. Denote by ρ the density per unit volume of free stationary charges, and by $\boldsymbol{J}$, the density, per unit area, of electric currents. In general, $\boldsymbol{E}$, $\boldsymbol{B}$, ρ, and $\boldsymbol{J}$ are time-dependent. Maxwell's equations establish the connection between $\boldsymbol{E}$, $\boldsymbol{B}$, ρ, and $\boldsymbol{J}$. We begin by stating the equation, and then show how to derive them by combining the basic empirical laws of electricity and magnetism with the theorems of

Gauss and Stokes from vector analysis.

In the International System of Units (SI), `Maxwell's equations` are

$$\operatorname{div} \boldsymbol{E} = \frac{\rho}{\varepsilon_0}; \tag{3.3.2}$$

$$\operatorname{div} \boldsymbol{B} = 0; \tag{3.3.3}$$

$$\operatorname{curl} \boldsymbol{E} = -\frac{\partial \boldsymbol{B}}{\partial t}; \tag{3.3.4}$$

$$\operatorname{curl} \boldsymbol{B} = \mu_0 \boldsymbol{J} + \mu_0 \varepsilon_0 \frac{\partial \boldsymbol{E}}{\partial t}. \tag{3.3.5}$$

The positive numbers $\mu_0 = 4\pi \cdot 10^{-7}$ N/A^2, $\varepsilon_0 = 8.85 \cdot 10^{-12}$ C/(N·m^2) are called the `magnetic permeability` and `electrical permittivity` of the free space, respectively; the relation

$$c^2 = \frac{1}{\mu_0 \varepsilon_0} \tag{3.3.6}$$

holds, where c is the speed of light in vacuum.

The starting point in the derivation of (3.3.2) is `Coulomb's Law`, discovered experimentally in 1785 by the French physicist CHARLES AUGUSTIN DE COULOMB (1736–1806). Originally stated as the inverse-square law for the force between two charges, the result also provides the electric field $\boldsymbol{E}$ produced by a point charge q: if q is located at the point O, then, for every point $P \neq O$,

$$\boldsymbol{E}(P) = \frac{q}{4\pi\varepsilon_0} \frac{\overrightarrow{OP}}{|OP|^3} = \frac{q}{4\pi\varepsilon_0} \frac{\widehat{\boldsymbol{r}}}{r^2}. \tag{3.3.7}$$

Let S be closed piece-wise smooth surface enclosing the point O. Then $\boldsymbol{E}(P)$ is defined at all points P on S. According to (3.2.5), page 152, $\iint\limits_S \boldsymbol{E} \cdot d\boldsymbol{\sigma} = q/\varepsilon_0$. By linearity, if there are finitely many point charges $q_1, \ldots, q_n$ inside S, and $\boldsymbol{E}$ is the total electric field produced by these charges, then

$$\iint\limits_S \boldsymbol{E} \cdot d\boldsymbol{\sigma} = (1/\varepsilon_0) \sum_{k=1}^{n} q_k. \tag{3.3.8}$$

Now assume that there is a continuum of charges in the domain G enclosed by S, and a small region G_B around a point $B \in G$ has the approximate charge $\rho(B)\mathfrak{m}(G_B)$, where ρ is a continuous density function and $\mathfrak{m}(G_B)$ is the volume of G_B; the smaller the region G_B, the better this approximation of the charge. Then (3.3.8) becomes

$$\iint_S \boldsymbol{E} \cdot d\boldsymbol{\sigma} = \frac{1}{\varepsilon_0} \iiint_G \rho\, dV, \tag{3.3.9}$$

and equation (3.3.2) follows from Gauss's Theorem (see Theorem 3.2.2).

EXERCISE 3.3.1.^{C} *Verify that (3.3.9) indeed implies (3.3.2).*

Equation (3.3.9) is important in its own right and is known as `Gauss's Law of Electric Flux`.

Equation (3.3.3) is the analog of (3.3.2) for the magnetic field and reflects the experimental fact that there are no single magnetic charges, or monopoles.

The starting point in the derivation of (3.3.4) is `Faraday's Law`, discovered experimentally in 1831 by the English physicist and chemist MICHAEL FARADAY (1791–1867): a change in time of a magnetic field $\boldsymbol{B}$ induces an electric field $\boldsymbol{E}$. This law is the underlying principle of electricity generation: the electric field $\boldsymbol{E}$ produces an electric current in a conductor. Mathematically, if S is a piece-wise smooth orientable surface with a closed piece-wise smooth boundary ∂S so that the orientations of S and ∂S agree, then

$$\oint_{\partial S} \boldsymbol{E} \cdot d\boldsymbol{r} = -\iint_S \frac{\partial \boldsymbol{B}}{\partial t} \cdot d\boldsymbol{\sigma}. \tag{3.3.10}$$

Equation (3.3.4) now follows from Stoke's Theorem (Theorem 3.2.3). Note that $\oint_{\partial S} \boldsymbol{E} \cdot d\boldsymbol{r} = V$ is the voltage drop along ∂S. For a coil without a ferromagnetic core, electric current I through the coil produces magnetic field $\boldsymbol{B}$ whose flux is proportional to I: $-\iint_S \boldsymbol{B} \cdot d\boldsymbol{\sigma} = LI$, where L is the inductance of the coil. Then (3.3.10) becomes $V = L dI/dt$, which is a fundamental relation in the theory of electrical circuits.

EXERCISE 3.3.2.^{C} *Verify that (3.3.10) indeed implies (3.3.4).*

The starting point in the derivation of (3.3.5) is an experimental fact that a steady electric current produces a magnetic field. In 1820, the Danish physicist HANS CHRISTIAN ØRSTED (1777–1851) observed that a steady electric current deflects a compass needle. The French scientist ANDRÉ-MARIE AMPÈRE (1775–1836) learned about this discovery on September 11, 1820; one week later, he presented a paper to the French Academy with a detailed explantation and generalizations of this phenomenon. In 1827, he gave a mathematical formulation, connecting the steady electric current with the induced magnetic field. In modern terms, the formulation is as follows Let S be a piece-wise smooth orientable surface with a piece-wise smooth boundary ∂S so that the orientations of S and ∂S agree. If $\boldsymbol{J}$ is the density, per unit area, of the *stationary* (that is, time-independent) electric current, then the induced magnetic field $\boldsymbol{B}$ satisfies $\oint_{\partial S} \boldsymbol{B}\cdot d\boldsymbol{r} = \mu_0 \iint_S \boldsymbol{J}\cdot d\boldsymbol{\sigma}$; see also Exercise 3.3.11 on page 170 below. By Stokes's Theorem, this implies `Ampère's Law`

$$\operatorname{curl} \boldsymbol{B} = \mu_0 \boldsymbol{J}; \tag{3.3.11}$$

Maxwell used the equation of continuity (3.2.8) and his first equation (3.3.2) to extend (3.3.11) to *time-varying* currents in the form (3.3.5). Recall that $\operatorname{div}(\operatorname{curl} \boldsymbol{F}) = 0$ for every twice continuously differentiable vector field $\boldsymbol{F}$. Then, assuming $\boldsymbol{B}$ is twice continuously differentiable, we get from (3.3.11) that $\operatorname{div} \boldsymbol{J} = 0$. On the other hand, if ρ is the density, per unit volume, of the charges moving with velocity $\boldsymbol{v}$, then $\boldsymbol{J} = \rho\boldsymbol{v}$, and, by the equation of continuity with no sources and sinks,

$$\frac{\partial \rho}{\partial t} + \operatorname{div} \boldsymbol{J} = 0, \tag{3.3.12}$$

or $\operatorname{div} \boldsymbol{J} = -\partial\rho/\partial t$. For a stationary current, $\partial\rho/\partial t = 0$, which is consistent with $\operatorname{div}(\operatorname{curl} \boldsymbol{B}) = 0$. For time-varying currents, we use (3.3.2) to find $\partial\rho/\partial t = \varepsilon_0 \operatorname{div}(\partial \boldsymbol{E}/\partial t)$ so that $\operatorname{div}(\boldsymbol{J} + \varepsilon_0 \partial \boldsymbol{E}/\partial t) = 0$. To ensure the equality $\operatorname{div}(\operatorname{curl} \boldsymbol{B}) = 0$, we therefore replace (3.3.11) with (3.3.5).

EXERCISE 3.3.3.^{B} *Go over the above arguments and verify that the divergence of the right-hand side of (3.3.5) is indeed equal to zero.*

Equations (3.3.4) and (3.3.5) express the observed interdependence between $\boldsymbol{B}$ and $\boldsymbol{E}$, leading to the single theory of electromagnetism. This is a striking example of the power of mathematical modelling in describing physical phenomena. A successful model can also predict new physical phenomena, and we will see later how Maxwell's equations lead to the prediction of electromagnetic waves (see page 348 below).

We conclude this section by establishing the connection between a stationary electromagnetic field and the Poisson equation. We start with the stationary electric field. Recall that $\operatorname{grad}(1/r) = -\boldsymbol{r}/r^3$. Equation (3.3.7) therefore implies that the electric field $\boldsymbol{E}$ produced by a single point charge is potential: $\boldsymbol{E}(P) = -\boldsymbol{\nabla} U(P)$, where $U(P) = q/(4\pi\varepsilon_0 |OP|)$, $P \neq O$. By linearity, the electric field at the point P, produced by n point charges is

$$\boldsymbol{E}(P) = -\boldsymbol{\nabla} U(P), \quad U(P) = \sum_{k=1}^{n} \frac{q_k}{4\pi\varepsilon_0} \frac{1}{\|\boldsymbol{r}_i - \boldsymbol{r}_P\|}, \tag{3.3.13}$$

where $\boldsymbol{r}_k = \overrightarrow{OO_k}$, $k = 1, \ldots, n$, are the position vectors of the point charges, $\boldsymbol{r}_P = \overrightarrow{OP}$ is the position vector of P, and we assume $P \neq O_k$ for all k (draw a picture).

EXERCISE 3.3.4.^{C} *Verify (3.3.13).*

Now assume that there is a continuum of charges in $\mathbb{R}^3$ with density ρ so that the charge of a small region G_Q around a point $Q \in \mathbb{R}^3$ is approximately $\rho(Q)\mathfrak{m}(G_Q)$, where ρ is a twice continuously differentiable function in $\mathbb{R}^3$ and $\mathfrak{m}(G_Q)$ is the volume of G_Q; the approximation becomes better as the volume of G_Q becomes smaller. We assume that $\rho = 0$ outside of some bounded domain. This continuum of charges produces an electric field $\boldsymbol{E}$ at every point $P \in \mathbb{R}^3$. Using (3.3.13) and the result of Exercise 3.1.24,

$$\boldsymbol{E}(P) = -\boldsymbol{\nabla} U(P), \quad U(P) = \frac{1}{4\pi\varepsilon_0} \iiint_{\mathbb{R}^3} \frac{\rho(Q)}{|PQ|} \, dV, \tag{3.3.14}$$

with integration over the points $Q \in \mathbb{R}^3$ where $\rho(Q)$ is not zero.

EXERCISE 3.3.5.^{C} *Verify (3.3.14). Hint: by assumption on ρ, the integral in (3.3.14) is actually over a bounded domain.*

Equalities (3.3.13) and (3.3.14) show that the stationary electric field $\boldsymbol{E}$, produced either by finitely many point charges or by a continuum of distributed charges, is a potential vector field: $\boldsymbol{E} = -\boldsymbol{\nabla} U$. The solutions of the vector differential equation $\dot{\boldsymbol{r}}(t) = \boldsymbol{E}(\boldsymbol{r}(t))$ for all possible initial conditions are called `lines of force`, and the surfaces $\{P \in \mathbb{R}^3 : U(P) = c\}$, for all possible values of c, are called `equipotential surfaces`.

EXERCISE 3.3.6.^{B} *(a) Show that lines of forces are orthogonal to the equipotential surfaces. Hint: $\boldsymbol{\nabla} U$ gives the normal direction to the surface, and, because $\boldsymbol{E} = -\boldsymbol{\nabla} U$, also given the tangential direction for the line of force. (b) Verify that the lines of force for a point charge are straight lines and the equipotential surfaces are spheres with the charge at the center. Hint: for a suitable $f = f(t)$ and every unit vector $\widehat{\boldsymbol{u}}$, the function $\boldsymbol{r}(t) = \widehat{\boldsymbol{u}}\, f(t)$ satisfies $\dot{\boldsymbol{r}}(t) = c\boldsymbol{r}(t)/\|\boldsymbol{r}(t)\|^3$.*

By (3.3.2), $\operatorname{div} \boldsymbol{E} = \rho/\varepsilon_0$. The first equality in (3.3.14) then implies

$$\nabla^2 U = -\frac{\rho}{\varepsilon_0}, \tag{3.3.15}$$

that is, *the potential of the electric field satisfies the Poisson equation* (3.3.15) in $\mathbb{R}^3$. Our arguments also suggest that the second equality (3.3.14) gives a solution of (3.3.15); see Problem 4.2 on page 427 for the precise statement and proof of the corresponding result.

EXERCISE 3.3.7. *(a)*^{C} *Verify that (3.3.15) follows from (3.3.2). (b)*^{B} *Taking for granted that the function U defined in (3.3.14) is twice continuously differentiable in $\mathbb{R}^3$, use (3.3.15) to show that U is a harmonic function outside of some bounded domain. Hint: recall that we assume that $\rho = 0$ outside of some bounded domain.*

The analogs of (3.3.2), (3.3.14), and (3.3.15) exist for every *central, inverse-square* field, such as the gravitational field.

EXERCISE 3.3.8.^{C} *Let $\boldsymbol{E} = \boldsymbol{E}(P)$, $P \in \mathbb{R}^3$, be the* `gravitational field intensity vector`*, so that $\boldsymbol{F} = m\,\boldsymbol{E}(P)$ is the force acting on a point mass m placed at the point P. Verify that*

$$\boldsymbol{E} = -\boldsymbol{\nabla} V \quad \text{and} \quad \nabla^2 V = 4\pi \mathrm{G}\rho, \tag{3.3.16}$$

where G *is the universal gravitational constant (see page 45) and ρ is the density, per unit volume, of the mass producing the gravitational field. Hint: for a point mass M at the origin, $\boldsymbol{E}(P) = -MGr^{-2}\,\widehat{\boldsymbol{r}} = MG\,\boldsymbol{\nabla}(1/r)$. Then repeat the arguments that lead from (3.3.7) to (3.3.9) and from (3.3.13) to (3.3.15).*

Let us now consider the magnetic field. By Maxwell's equation (3.3.4), the vector field $\boldsymbol{B}$ is solenoidal, that is, has zero divergence. The condition $\operatorname{div} \boldsymbol{B} = 0$ is satisfied if

$$\boldsymbol{B} = \operatorname{curl} \boldsymbol{A} \tag{3.3.17}$$

for some vector field $\boldsymbol{A}$; if we can find $\boldsymbol{A}$, then we can find $\boldsymbol{B}$. This field $\boldsymbol{A}$ is called a `vector potential` of $\boldsymbol{B}$. Note that, if it exists, $\boldsymbol{A}$ is not unique: for every sufficiently smooth scalar field f, $\boldsymbol{A}_1 = \boldsymbol{A} + \operatorname{grad} f$ is also a vector potential of $\boldsymbol{B}$, because $\operatorname{curl}(\operatorname{grad} f) = \boldsymbol{0}$. On the other hand, $\operatorname{div} \boldsymbol{A}_1 = \operatorname{div} \boldsymbol{A} + \nabla^2 f$; with a suitable choice of f we can ensure that the vector potential satisfies

$$\operatorname{div} \boldsymbol{A} = 0. \tag{3.3.18}$$

This choice of $\boldsymbol{A}$ corresponds to `Coulomb's gauge`; for more about gauging see page 351.

Assume that the current density $\boldsymbol{J}$ and the corresponding electric field $\boldsymbol{E}$ do not depend on time. Then Maxwell's equation (3.3.5) becomes the original Ampère's Law (3.3.11).

EXERCISE 3.3.9.[B] *Verify that (3.3.11), (3.3.17), and (3.3.18) imply*

$$\nabla^2 \boldsymbol{A} = -\mu_0 \boldsymbol{J}. \tag{3.3.19}$$

Hint: see (3.1.32) on page 139.

Let us assume that the vector field $\boldsymbol{J}$ is twice continuously differentiable and is equal to $\mathbf{0}$ outside of a bounded domain. By analogy with (3.3.14) and (3.3.15) (see also Problem 4.2 on page 427), we write the solution of the vector Poisson equation (3.3.19) as

$$\boldsymbol{A}(P) = \frac{\mu_0}{4\pi} \iiint\limits_{\mathbb{R}^3} \frac{\boldsymbol{J}(Q)}{|PQ|}\, dV, \tag{3.3.20}$$

with integration over the points $Q \in \mathbb{R}^3$ where $\boldsymbol{J}$ is not zero.

EXERCISE 3.3.10.[C] *Assuming that differentiation under the integral sign is justified, verify that (3.3.17) and (3.3.20) imply*

$$\boldsymbol{B}(P) = \frac{\mu_0}{4\pi} \iiint\limits_{\mathbb{R}^3} \frac{\boldsymbol{J}(Q) \times \overrightarrow{PQ}}{|PQ|^3}\, dV. \tag{3.3.21}$$

Hint: Note that, for the purpose of differentiating the expression under the integral in (3.3.20), the point Q is fixed and the point P is variable. Accordingly, place the origin at the point Q; then you need to compute $\operatorname{curl}(\boldsymbol{J}/r)$, *where $\boldsymbol{J}$ is a constant vector. Recall that $\boldsymbol{\nabla}(1/r) = -\boldsymbol{r}/r^3$ and use a suitable formula from the collection (3.1.30) on page 139.*

Formula (3.3.21) is the three-dimensional version of the `Biot-Savart Law`, discovered experimentally in 1820 by the French physicists JEAN BAPTISTE BIOT (1774–1862) and FELIX SAVART (1791–1841); in its original form, the law states that the magnetic field produced by a constant current I in an infinitesimally thin straight wire of infinite length is

$$\boldsymbol{B}(P) = \frac{\mu_0 I}{2\pi} \frac{\widehat{\boldsymbol{u}} \times \overrightarrow{QP}}{|PQ|^2}, \tag{3.3.22}$$

where $\widehat{\boldsymbol{u}}$ is the unit vector along the wire in the direction of the current, Q is the point on the wire closest to P, and I is the current measured in amperes. In particular, the magnitude of $\boldsymbol{B}$ is inversely proportional to the distance from the wire.

EXERCISE 3.3.11. *(a)*^{B} *Draw a picture illustrating (3.3.22) and derive (3.3.22) from (3.3.21). Hint: consider a uniform cylindrical wire of small radius.* *(b)*^{C} *Let C be a circle of radius R in the plane perpendicular to the wire and center the wire. Show that (3.3.22) implies $\oint_C \boldsymbol{B}\cdot d\boldsymbol{r} = \mu_0 I$.* *(c)*^{B} *Show that the result of part (b) is true for every simple closed piece-wise smooth curve enclosing the wire.*

EXERCISE 3.3.12.^{A} *Let C be a simple piece-wise smooth curve representing an infinitesimally thin conducting wire with current I (measured in amperes); the value of the current does not depend on time but can be different at different points of the wire. Assume that orientation of C is in the direction of the current. Use (3.3.21) to show that the resulting magnetic field $\boldsymbol{B}$ and the corresponding vector potential $\boldsymbol{A}$ at a point P not in C are*

$$\boldsymbol{B}(P) = -\frac{\mu_0}{4\pi}\int_C \frac{I(Q)\,(\overrightarrow{QP}\times\widehat{\boldsymbol{u}})}{|PQ|^3}\,d\mathfrak{s},\quad \boldsymbol{A}(P) = \frac{\mu_0}{4\pi}\int_C \frac{I(Q)\,\widehat{\boldsymbol{u}}}{|PQ|}\,d\mathfrak{s}, \tag{3.3.23}$$

where Q is a (varying) point on C and $\widehat{\boldsymbol{u}}$ is the unit tangent vector to C. Hint: assume that the wire has a small radius a and is bent as C. Then $\pi a^2\boldsymbol{J} = I\widehat{\boldsymbol{u}}$ and $\iiint(\cdots)\,dV = \int(\cdots)\,\pi a^2\,d\mathfrak{s}$.

3.3.2 *The Electric and Magnetic Dipoles*

As an example illustrating some of the general discussions from the previous section, we will consider electric and magnetic dipoles. In the following section, these dipoles will help us to study the electromagnetic field in material media.

An `electric dipole` is a pair of point charges q and $-q$, $q>0$, with equal magnitude and opposite signs, placed at a fixed distance from each other. Let O be a reference point in $\mathbb{R}^3$, and let the position vectors of the charges $q, -q$ be $\boldsymbol{r}_1$ and $\boldsymbol{r}_2$, respectively. By (3.3.13), the potential U of the resulting electric field at the point with position vector $\boldsymbol{r}$ is

$$U(\boldsymbol{r}) = \frac{q}{4\pi\varepsilon_0}\left(\frac{1}{\|\boldsymbol{r}-\boldsymbol{r}_1\|} - \frac{1}{\|\boldsymbol{r}-\boldsymbol{r}_2\|}\right). \tag{3.3.24}$$

Let $\boldsymbol{r}_0 = (\boldsymbol{r}_1+\boldsymbol{r}_2)/2$ be the position vector of the midpoint of the line segment connecting q and $-q$. In what follows, we will find the approximate value of U at the points that are far away from both q and $-q$, that is, when $\|\boldsymbol{r}-\boldsymbol{r}_0\|$ is much bigger than the distance $\|\boldsymbol{r}_1-\boldsymbol{r}_2\|$ between the point charges. In Problem 4.3, page 429, we discuss the lines of force for the dipole, that is, the solutions of the differential equation $\dot{\boldsymbol{r}}(t)=-\boldsymbol{\nabla}U(\boldsymbol{r})$.

To simplify the computations, define $\mathfrak{r}=\boldsymbol{r}-\boldsymbol{r}_0$ and $\boldsymbol{r}_{12}=\boldsymbol{r}_1-\boldsymbol{r}_2$. Then

$$U(\boldsymbol{r})=\frac{q}{4\pi\varepsilon_0}\left(\frac{1}{\|\mathfrak{r}-(1/2)\boldsymbol{r}_{12}\|}-\frac{1}{\|\mathfrak{r}+(1/2)\boldsymbol{r}_{12}\|}\right). \tag{3.3.25}$$

EXERCISE 3.3.13.[C] *Draw a picture and verify (3.3.25).*

For every two vectors $\boldsymbol{u},\boldsymbol{v}$, we have $\|\boldsymbol{u}-\boldsymbol{v}\|^2=(\boldsymbol{u}-\boldsymbol{v})\cdot(\boldsymbol{u}-\boldsymbol{v})=\|\boldsymbol{u}\|^2+\|\boldsymbol{v}\|^2-2\,\boldsymbol{u}\cdot\boldsymbol{v}$. If $\|\boldsymbol{u}\|\gg\|\boldsymbol{v}\|$ (that is, $\|\boldsymbol{u}\|$ is much larger than $\|\boldsymbol{v}\|$), then $\|\boldsymbol{u}-\boldsymbol{v}\|^2\approx\|\boldsymbol{u}\|^2(1-(2\,\boldsymbol{u}\cdot\boldsymbol{v}/\|\boldsymbol{u}\|^2))$. Applying the linear approximation to the function $f(x)=(1-x)^{-1/2}$ at the point $x=0$, we find that $f(x)\approx 1+(x/2)$ and

$$\frac{1}{\|\boldsymbol{u}-\boldsymbol{v}\|}\approx\frac{1}{\|\boldsymbol{u}\|}\left(1+\frac{\boldsymbol{u}\cdot\boldsymbol{v}}{\|\boldsymbol{u}\|^2}\right),\quad \|\boldsymbol{u}\|\gg\|\boldsymbol{v}\|. \tag{3.3.26}$$

For the exact expansion of $1/\|\boldsymbol{u}-\boldsymbol{v}\|$ in the powers of $\|\boldsymbol{v}\|/\|\boldsymbol{u}\|$, see Problem 4.4, page 430. Applying this approximation to (3.3.25) with $\boldsymbol{u}=\mathfrak{r}=\boldsymbol{r}-\boldsymbol{r}_0$ and $\boldsymbol{v}=\pm\boldsymbol{r}_{12}/2$, we find

$$U(\boldsymbol{r})\approx\frac{q}{4\pi\varepsilon_0}\,\frac{(\boldsymbol{r}-\boldsymbol{r}_0)\cdot\boldsymbol{r}_{12}}{\|\boldsymbol{r}-\boldsymbol{r}_0\|^3},\quad \|\boldsymbol{r}-\boldsymbol{r}_0\|\gg\|\boldsymbol{r}_{12}\|. \tag{3.3.27}$$

EXERCISE 3.3.14.[B] *(a) Verify (3.3.27) and show that the approximation error is of order $(\|\boldsymbol{r}_1-\boldsymbol{r}_2\|/\|\boldsymbol{r}-\boldsymbol{r}_0\|)^2$. (b) Using the relation $\boldsymbol{E}=-\boldsymbol{\nabla}U$, verify that*

$$\boldsymbol{E}(\boldsymbol{r})\approx\frac{q}{4\pi\varepsilon_0}\left(\frac{3(\mathfrak{r}\cdot\boldsymbol{r}_{12})\mathfrak{r}}{\|\mathfrak{r}\|^5}-\frac{\boldsymbol{r}_{12}}{\|\mathfrak{r}\|^3}\right),\quad \|\mathfrak{r}\|\gg\|\boldsymbol{r}_{12}\|, \tag{3.3.28}$$

(i) by computing the gradient of the approximate value U from (3.3.27); (ii) by computing the gradient of the exact value of U from (3.3.25) and then approximating the result.

The vector $\boldsymbol{d}=q\boldsymbol{r}_{12}$, pointing from the negative charge to the positive charge, is called the `dipole moment` of the electric dipole. Taking the limit as $\boldsymbol{r}_{12}\to\boldsymbol{0}$ and $q\to\infty$ so that the vector $\boldsymbol{d}$ stays constant, we get the `point electric dipole`. For the point electric dipole at the origin, we have the potential and the electric field at every point with position vector $\boldsymbol{r}$ as follows:

$$U(\boldsymbol{r}) = -\frac{1}{4\pi\varepsilon_0}\frac{\boldsymbol{d}\cdot\boldsymbol{r}}{r^3},\quad \boldsymbol{E}(\boldsymbol{r}) = \frac{1}{4\pi\varepsilon_0}\left(\frac{3(\boldsymbol{d}\cdot\boldsymbol{r})\,\boldsymbol{r}}{r^5} - \frac{\boldsymbol{d}}{r^3}\right). \tag{3.3.29}$$

EXERCISE 3.3.15. *(a)*[B] *Verify that* $\nabla^2(\boldsymbol{d}\cdot\boldsymbol{r}/r^3) = 0,\ r \neq 0$, *and explain how the result is connected with Maxwell's equation (3.3.2). Hint: use (3.1.33) on page 140, as well as suitable relations from the collection (3.1.30) on page 139; note that* $\operatorname{grad}(\boldsymbol{d}\cdot\boldsymbol{r}) = \boldsymbol{d}$. *(b)*[A] *Consider a point electric dipole with moment* $\mathbf{d}$ *in an external electric field* $\mathbf{E}$ *and assume the electric field is uniform, that is, has the same magnitude and direction at every point in space:* $\boldsymbol{E}(P) = \boldsymbol{E}$, *where* $\boldsymbol{E}$ *is a constant vector. Show that the total force acting on the dipole by the field is zero and the torque is* $\boldsymbol{T} = \boldsymbol{d}\times\boldsymbol{E}$. *Thus, the external field will tend to align the dipole with the field. Draw the picture illustrating the* stable *orientation of the dipole relative to the field.*

MAGNETIC DIPOLE. Since currents produce magnetic fields, we define a `magnetic dipole` as an infinitesimally thin closed circular wire in the shape of a simple smooth closed curve $\mathcal{C}$, carrying a constant current I. By (3.3.23) on page 170, the vector potential characterizing the magnetic field of this current is

$$\boldsymbol{A}(P) = \frac{\mu_0 I}{4\pi}\oint_{\mathcal{C}} \frac{\widehat{\boldsymbol{u}}}{|PQ|}\,d\mathfrak{s}, \tag{3.3.30}$$

where $\widehat{\boldsymbol{u}}$ is the unit tangent vector to $\mathcal{C}$ in the direction of the current, and Q is a (varying) point on $\mathcal{C}$. Denote by O the center of the circle $\mathcal{C}$. Then $|PQ| = \|\overrightarrow{OP} - \overrightarrow{OQ}\|$, and $\overrightarrow{OQ}$ is perpendicular to the tangent vector at the point Q. As with the electric dipole, we want to get an approximation of $\boldsymbol{A}$ at the point P that is far away from $\mathcal{C}$, that is, when $|OP|$ is much larger than the radius $|OQ|$ of the circle (draw a picture!) Using (3.3.26), we find

$$\frac{1}{|PQ|} \approx \frac{1}{|OP|}\left(1 + \frac{\overrightarrow{OP}\cdot\overrightarrow{OQ}}{|OP|^2}\right).$$

Then (3.3.30) becomes

$$\boldsymbol{A}(P) \approx \frac{\mu_0 I}{4\pi|OP|}\left(\oint_{\mathcal{C}} \widehat{\boldsymbol{u}}\,d\mathfrak{s} + \oint_{\mathcal{C}} \frac{\overrightarrow{OP}\cdot\overrightarrow{OQ}}{|OP|^2}\,\widehat{\boldsymbol{u}}\,d\mathfrak{s}\right) \tag{3.3.31}$$

Denote by $\widehat{\boldsymbol{n}}$ the unit vector in the direction of $\overrightarrow{OQ} \times \widehat{\boldsymbol{u}}$, and by a, the radius of the circle.

EXERCISE 3.3.16.[B] *(a) Verify that the direction of $\widehat{\boldsymbol{n}}$ is the same for every point Q of the circle. How should a small bar magnet be placed to produce the same magnetic field as the magnetic dipole? Hint: draw a picture; if C is a circle in the $(\widehat{\boldsymbol{\imath}},\ \widehat{\boldsymbol{\jmath}})$ plane and is oriented counterclockwise, then $\widehat{\boldsymbol{n}} = \widehat{\boldsymbol{k}}$. (b) Verify that (3.3.31) can be written as*

$$\boldsymbol{A}(P) \approx \frac{\mu_0}{4\pi}\frac{\pi a^2 I\,\widehat{\boldsymbol{n}} \times \overrightarrow{OP}}{|OP|^3}. \tag{3.3.32}$$

Hint: to show that $\oint_C \widehat{\boldsymbol{u}}\, d\mathfrak{s} = \mathbf{0}$, use cartesian coordinates so that C is a circle in the $(\widehat{\boldsymbol{\imath}},\ \widehat{\boldsymbol{\jmath}})$ plane and is oriented counterclockwise. Then $\widehat{\boldsymbol{u}} = -\sin t\,\widehat{\boldsymbol{\imath}} + \cos t\,\widehat{\boldsymbol{\jmath}}$ and $d\mathfrak{s} = a dt$, $0 < t < 2\pi$. For the second integral, use (3.2.14) on page 157 with $\boldsymbol{r}_0 = \overrightarrow{OP}$, $\boldsymbol{r} = \overrightarrow{OQ}$, and S, the disk enclosed by C.

Similar to the electric dipole, we define the `dipole moment` of the magnetic dipole as the vector $\boldsymbol{m} = \pi a^2 I\,\widehat{\boldsymbol{n}}$. In the limit $a \to 0$ and $I \to \infty$ so that the vector $\boldsymbol{m}$ stays the same, we get the `point magnetic dipole`. If a point magnetic dipole is placed at the origin, then the corresponding vector potential is

$$\boldsymbol{A}(\boldsymbol{r}) = \frac{\mu_0}{4\pi}\frac{\boldsymbol{m} \times \boldsymbol{r}}{r^3}. \tag{3.3.33}$$

EXERCISE 3.3.17.[B] *Find the magnetic field $\boldsymbol{B}$ of the point magnetic dipole.*

3.3.3 *Maxwell's Equations in Material Media*

In conductors, such as metals and electrolytes, there are charged particles that can move when an external electric field $\boldsymbol{E}$ is applied. In metals, these particles are free electrons, and in electrolytes, ions. By Ohm's Law, the resulting current density $\boldsymbol{J}$ is proportional to the external field $\boldsymbol{E}$:

$$\boldsymbol{J} = \sigma\,\boldsymbol{E}, \tag{3.3.34}$$

where σ is the conductivity of the material. Similar to other laws of electricity and magnetism, relation (3.3.34) is a mathematical expression of empirical observations and was originally published in 1827 by the German physicist GEORG SIMON OHM (1789–1854). Today, (3.3.34) is known as the *microscopic* form of `Ohm's Law`, as opposite to the more familiar *macro-*

scopic, or averaged-out, version $V = IR$, where V is the voltage measured in volts, I is the current in amperes, and R is the resistance in ohms.

If an insulated conductor is placed in an external electric field, then the induced electric current redistributes the charged particles on the surface of the conductor so that, in equilibrium, the total electric field inside the conductor is zero: if there were non-zero field inside the conductor, then, by Ohm's Law, there would be a non-zero current, which is impossible for an isolated conductor in equilibrium. The force lines of the electric field are therefore normal to the surface of the conductor; otherwise, the tangential component would produce a surface current. There is no charge accumulations inside the conductor either, as such accumulations would produce electric current. Thus, a closed conducting surface isolates the interior from the exterior electric field. For that reason, delicate microchips and electronic equipment are often stored and shipped in aluminum foil or other conducting cover.

In dielectrics, such as glass, plastics, and rubber, there is little or no free motion of charged particles when an external electric field is applied. Instead, the external field penetrates the material, causing polarization of molecules in the substance and producing numerous point electric dipoles. In biological materials, cell membranes are good insulators against ion flux, and ions form dipoles near the surface of cell membranes. Protein molecules form dipoles by ionization of the amino-acid chains.

Denote by $\boldsymbol{P}$ the resulting density of dipole moments per unit volume in the dielectric material. Recall that a point dipole has zero net charge; direct computations show that the flux of the corresponding electric field $\boldsymbol{E}$ (see (3.3.29) through *every* closed piece-wise smooth surface is zero, as long as the dipole is not on the surface. Accordingly, instead of looking at the flux of $\boldsymbol{E}$, we will look at the flux of $\boldsymbol{P}$. Note that $\boldsymbol{P}$ has the dimension of charge per unit area, and so the divergence $\operatorname{div} \boldsymbol{P}$ of $\boldsymbol{P}$ has the dimension of charge per unit volume. Recall that the direction of the dipole moment is from the negative charge to the positive. It is then natural to define the scalar field ρ_b so that $\rho_b = -\operatorname{div} \boldsymbol{P}$. With this definition, if G is any part of the dielectric, then $\iiint\limits_G (\operatorname{div} \boldsymbol{P} - \rho_b)\, dV = 0$, which is consistent with the idea that there is no movement of charges inside the dielectric. The scalar field ρ_b is called the **density of bound charges**. It is also possible that some *free charges* are present inside the dielectric, with density per unit volume ρ_f. By Gauss's Law of Electric Flux (3.3.9), a region in free space with density $\rho_b + \rho_f$ of charges per unit volume produces the electric field $\boldsymbol{E}$ so that $\operatorname{div}\boldsymbol{E} = (\rho_b + \rho_f)/\varepsilon_0$. In the dielectric material, we have $\rho_b = -\operatorname{div}\boldsymbol{P}$ so that $\operatorname{div}(\varepsilon_0\boldsymbol{E} + \boldsymbol{P}) = \rho_f$. It is then natural to define the vector $\boldsymbol{D} = \varepsilon_0\boldsymbol{E} + \boldsymbol{P}$, called the `electric displacement`. We get the following modification of (3.3.2):

$$\operatorname{div}\boldsymbol{D} = \rho_f. \tag{3.3.35}$$

Dielectric material is called linear if $\boldsymbol{P}$ is proportional to the external field $\boldsymbol{E}$:

$$\boldsymbol{P} = \varepsilon_0\chi_e\boldsymbol{E}, \quad \boldsymbol{D} = \varepsilon_0\boldsymbol{E} + \varepsilon_0\chi_e\boldsymbol{E} = \varepsilon\boldsymbol{E}, \tag{3.3.36}$$

where χ_e is called the `susceptibility` and $\varepsilon = \varepsilon_0(1+\chi_e)$, the `permittivity` of the dielectric material; $\kappa = 1+\chi_e$ is called the `dielectric constant` or `relative permittivity`. In general, ε (as well as χ_e and κ) is a `tensor field` and the value of ε depends on both the location and the direction. For time-dependent $\boldsymbol{E}$, the value of ε can also depend on the frequency of $\boldsymbol{E}$. In homogeneous isotropic materials, ε is a constant number so that (3.3.35) becomes $\operatorname{div}\boldsymbol{E} = \rho_f/\varepsilon$.

AMPERE'S LAW FOR MAGNETIZED MATERIALS. In material media, there exists an internal magnetic field created by the motion of electrons around the atoms. At distances much larger than the size of an atom, the magnetic field produced by an electron moving around an atom is well approximated by the field of a point magnetic dipole. These magnetic dipoles affect the overall magnetic field in the material, just as the bound charges in dielectrics affect the electric field.

Denote by $\boldsymbol{M}$ the density, per unit volume, of the point magnetic dipoles in the material; this density is called `magnetization`. According to (3.3.33), the overall vector potential $\boldsymbol{A}$ of the resulting magnetic field is

$$\boldsymbol{A}(P) = \frac{\mu_0}{4\pi}\iiint\limits_G \frac{\boldsymbol{M}(Q)\times\overrightarrow{QP}}{|QP|^3}\,dV, \tag{3.3.37}$$

where G is the region in space occupied by the material, Q is a point in G and P is the point in $\mathbb{R}^3$ at which the magnetic field is computed. We now treat P as a fixed point and Q, as variable. Then

$$\operatorname{grad}\left(\frac{1}{|PQ|}\right) = \frac{\overrightarrow{QP}}{|QP|^3} \tag{3.3.38}$$

and (3.3.37) becomes

$$\boldsymbol{A}(P) = \frac{\mu_0}{4\pi} \iiint_G \boldsymbol{M}(Q) \times \operatorname{grad}\left(\frac{1}{|PQ|}\right) dV. \tag{3.3.39}$$

EXERCISE 3.3.18.[B] *Verify (3.3.38). Hint: use (3.1.33) on page 140, taking P as the origin of the frame.*

By (3.1.30) on page 139,

$$\boldsymbol{M}(Q) \times \operatorname{grad}\left(\frac{1}{|PQ|}\right) = \frac{1}{|PQ|} \operatorname{curl} \boldsymbol{M}(Q) - \operatorname{curl}\left(\frac{\boldsymbol{M}(Q)}{|PQ|}\right), \tag{3.3.40}$$

and then we use relation (3.2.6) on page 153 to write (3.3.39) as

$$\boldsymbol{A}(P) = \frac{\mu_0}{4\pi}\left(\iiint_G \frac{1}{|PQ|} \operatorname{curl} \boldsymbol{M}(Q)\, dV + \iint_{\partial G} \frac{1}{|PQ|} \boldsymbol{M}(Q) \times \widehat{\boldsymbol{n}}_G\, d\sigma\right), \tag{3.3.41}$$

where ∂G is the boundary of G, which we assume to be piece-wise smooth, closed, and orientable, and $\widehat{\boldsymbol{n}}_G$ is the outside unit normal to ∂G. We call $\boldsymbol{J}_b = \operatorname{curl} \boldsymbol{M}$ the **bound current**, and $\boldsymbol{K}_b = \boldsymbol{M} \times \widehat{\boldsymbol{n}}_G$, the **surface density of the bound current**. Recall that a magnetic dipole is modelled by a circular movement of electric charges, and the vector $\boldsymbol{J}_b$ can be interpreted as the total current produced by the charge movement in all the point magnetic dipoles in G.

EXERCISE 3.3.19. *(a)*[C] *Verify that (3.3.41) indeed follows from (3.3.39), (3.3.40), and (3.2.6). (b)*[B] *Find the magnetic field corresponding to the potential (3.3.41). Hint: recall that the magnetic field is* curl $\boldsymbol{A}$*; in the resulting computations, you now treat P as a variable and Q as fixed.*

Suppose that the material has non-zero conductivity and there is a steady current $\boldsymbol{J}_e$ flowing in the material due to an external source of electric field, for example, two electrodes embedded in the material and connected to an external battery. The total current density in the material is then $\boldsymbol{J} = \boldsymbol{J}_b + \boldsymbol{J}_e = \operatorname{curl} \boldsymbol{M} + \boldsymbol{J}_e$. We assume the the system is in equilibrium so that both $\boldsymbol{J}_b$ and $\boldsymbol{J}_e$ do not depend on time. By the original Ampère's Law (3.3.11) on page 166, the resulting magnetic field $\boldsymbol{B}$ satisfies $\operatorname{curl} \boldsymbol{B} = \mu_0(\operatorname{curl} \boldsymbol{M} + \boldsymbol{J}_e)$ or $\operatorname{curl}\big(\boldsymbol{B}/\mu_0 - \boldsymbol{M}\big) = \boldsymbol{J}_e$. We therefore define the net magnetic field $\boldsymbol{H}$, also called the **magnetic field strength**, as

$$\boldsymbol{H} = \boldsymbol{B}/\mu_0 - \boldsymbol{M}, \tag{3.3.42}$$

so that Ampère's Law in material media becomes

$$\operatorname{curl}\boldsymbol{H} = \boldsymbol{J}_e. \tag{3.3.43}$$

EXERCISE 3.3.20. [B] *Verify that, in material media, Maxwell's equation (3.3.5) becomes*

$$\operatorname{curl}\boldsymbol{H} = \boldsymbol{J}_e + \frac{\partial \boldsymbol{D}}{\partial t}. \tag{3.3.44}$$

Hint: use the same arguments as in the derivation of (3.3.5).

In **paramagnetic** materials, the magnetic dipoles are aligned so that the induced magnetic field $\boldsymbol{B}$ is stronger than in vacuum; the vector $\boldsymbol{M}$ has, on average, the same direction as $\boldsymbol{H}$: $\iiint_G (\boldsymbol{M}\cdot\boldsymbol{H})\,dV > 0$. In **diamagnetic** materials, the magnetic dipoles are aligned so that the induced magnetic field $\boldsymbol{B}$ is weaker than in vacuum; the directions of $\boldsymbol{M}$ and $\boldsymbol{H}$ are, on the average, opposite: $\iiint_G (\boldsymbol{M}\cdot\boldsymbol{H})\,dV < 0$.

A material is called linear if magnetization $\boldsymbol{M}$ depends linearly on $\boldsymbol{H}$, that is, $\boldsymbol{M} = \chi_m \boldsymbol{H}$, where χ_m is called the **magnetic susceptibility** of the material. This linear relation holds for most materials when $\|\boldsymbol{H}\|$ is sufficiently small. By (3.3.42),

$$\boldsymbol{B} = \mu\boldsymbol{H}, \text{ where } \mu = \mu_0(1+\chi_m)$$

is called the **permeability** of the material. In vacuum, $\chi_m = 0$ and $\mu = \mu_0$. Similar to (electric) permittivity, permeability is, in general, a *tensor field* and the value of μ depends on both the location and the direction. For time-varying magnetic fields, the value of μ can also depend on the frequency of the field. The value of μ usually depends also on temperature and density. In *linear homogeneous isotropic materials*, μ is a number and (3.3.43) becomes

$$\boldsymbol{\nabla}\times\boldsymbol{B} = \mu\boldsymbol{J}_e, \tag{3.3.45}$$

which is the same as (3.3.11), but with μ instead of μ_0.

EXERCISE 3.3.21. [C] *Assuming the frequencies of* $\boldsymbol{B}$ *and* $\boldsymbol{E}$ *are constant, verify that, in a linear homogeneous isotropic material, equation (3.3.44) becomes*

$$\operatorname{curl}\boldsymbol{B} = \mu\boldsymbol{J}_e + \mu\varepsilon\frac{\partial \boldsymbol{E}}{\partial t}. \tag{3.3.46}$$

Hint: use (3.3.36).

For most linear materials, $|\chi_m|$ is close to zero. In *linear homogeneous isotropic paramagnetic* materials, $\chi_m > 0$ and $\mu > \mu_0$. For example, aluminum has $\chi_m \approx 2 \times 10^{-5}$. In *linear homogeneous isotropic diamagnetic* materials, $\chi_m < 0$ and $\mu < \mu_0$. For example, both copper and water have $\chi_m \approx -10^{-5}$ (we write $\approx$ rather than $=$ because the precise values are sensitive to the temperature of the materials and the frequency of the field; the above values correspond to room temperature and a time-homogeneous field).

The main examples of *nonlinear materials* are **ferromagnetics** and **superconductors**. In ferromagnetics, the induced magnetic field is much stronger than in vacuum. In superconductors the induced magnetic field is essentially zero.

Let us summarize **Maxwell's equations** in material media:

$$\operatorname{div} \boldsymbol{D} = \rho_f; \tag{3.3.47}$$

$$\operatorname{div} \boldsymbol{B} = 0; \tag{3.3.48}$$

$$\operatorname{curl} \boldsymbol{E} = -\frac{\partial \boldsymbol{B}}{\partial t}; \tag{3.3.49}$$

$$\operatorname{curl} \boldsymbol{H} = \boldsymbol{J}_e + \frac{\partial \boldsymbol{D}}{\partial t}, \tag{3.3.50}$$

where ρ_f is the density, per unit volume, of the free electric charges, and $\boldsymbol{J}_e$ is the density, per unit area, of the externally produced electric currents. These equations describe the electromagnetic field at distances much larger than atomic size (atomic size is usually taken as 10^{-10} meters.) In Section 6.3.4, we further investigate Maxwell's equations for some simple models, both in vacuum and in material media.

At sub-atomic distances, of the order of $e^2/(4\pi\varepsilon m_e c^2) \approx 2.8 \cdot 10^{-17} m$, Maxwell's equations are no longer applicable, and **quantum electrodynamics** takes over. A quantum-theoretic analog of Maxwell's equations was suggested in 1954 by two American physicists, CHEN-NING YANG (b. 1922) and ROBERT L. MILLS (1927–1999). Mathematical analysis of these **Yang-Mills equations** is mostly an open problem and is outside the scope of our discussion. In fact, this analysis is literally a million-dollar question, being one of the seven **Millennium Problems** announced by the Clay Mathematics Institute in 2000; for details, see the book *The Millennium Problems* by K. Devlin, 2002.

Chapter 4

Elements of Complex Analysis

4.1 The Algebra of Complex Numbers

4.1.1 *Basic Definitions*

Complex numbers occur naturally in several areas of physics and engineering, for example, in the study of Fourier series and transforms and related applications to signal processing and wave propagation. They also appear in the mathematical models of quantum mechanics. Still, the original motivation to introduce complex numbers was the study of roots of polynomials.

A `polynomial` $p = p(x)$ in the `variable` x is an expression

$$a_n x^n + a_{n-1} x^{n-1} + \cdots a_1 x + a_0.$$

The numbers a_j are called the `coefficients` of the polynomial; if $a_n \neq 0$, then n is called `the degree` of the polynomial. A `polynomial equation` is $p(x) = 0$, and a root of p is, by definition, a solution of this equation.

So far, our underlying assumption was that the reader has some basic familiarity with the construction of the real numbers. At this point, though, we will go back to the foundations of the theory of real numbers. In the late 1800's, German mathematicians GEORG FERDINAND LUDWIG PHILIPP CANTOR (1845–1918) and JULIUS WILHELM RICHARD DEDEKIND (1831–1916) put the construction of the real number system on a precise mathematical foundation by combining, in a rather sophisticated way, set theory and analysis. A modern approach that we will outline next, is more algebraic and leads to the complex numbers in a natural way. The remark

God made the integers and all the rest is the work of man,

attributed to the German mathematician LEOPOLD KRONECKER (1823–1891), suggests the set $\mathbb{N} = \{1, 2, 3, \ldots\}$ of positive integers as the starting point of the construction.

The set $\mathbb{N}$ is naturally equipped with two binary operations, *addition* $(a, b) \mapsto a + b$, and *multiplication* $(a, b) \mapsto ab$. These operations are **associative**:

$$(a + b) + c = a + (b + c), (ab)c = a(bc),$$

and **distributive**:

$$a(b + c) = ab + ac.$$

To solve the linear equation $x + a = b$ for every $a, b \in \mathbb{N}$ the set of positive integers is extended to the set $\mathbb{Z} = \{0, \pm 1, \pm 2, \ldots\}$ of all integers by introducing the special number 0 so that $a + 0 = a$ for all a and adjoining to every non-zero $a \in \mathbb{N}$ the *additive inverse* $-a$ so that $a + (-a) = 0$.

To solve the linear equation $ax = b$ for every $a, b \in \mathbb{Z}$, the set of integers is extended to the set $\mathbb{Q}$ of *rational* numbers, that is, expressions of the form $p/q \equiv pq^{-1}$, where $p, q \in \mathbb{Z}$ and $q \neq 0$. For every non-zero $p/q \in \mathbb{Q}$, the element q/p satisfies $(p/q)(q/p) = 1$ and is called **the multiplicative inverse** of p/q.

To solve the quadratic equation $x^2 - 2 = 0$ and other general polynomial equations with coefficients in $\mathbb{Q}$, the set of rational numbers is extended by creating and including the *algebraic irrational* numbers, that is, the numbers such as $\sqrt{2}$ or $\sqrt[3]{4}$, that can be roots of polynomial equations with coefficients in $\mathbb{Q}$. The result is the *algebraic closure* of $\mathbb{Q}$.

It turns out there are other irrational real numbers that are not algebraic, that is, are not roots of any polynomial with rational coefficients. These irrational numbers are called **transcendental**. To put it differently, the ordered set of algebraic numbers contains many holes, and these holes are filled with transcendental numbers. It was only in 1844 that the French mathematician JOSEPH LIOUVILLE (1809–1882) showed the existence of such numbers by explicitly constructing a few. Later, it was proved that the two familiar numbers, π and e, are also transcendental. Together, the algebraic and transcendental numbers make up the **real** numbers. The rigorous definition of a real number is necessary to put calculus on a sound basis.

Cantor *defined* a real number as the limit of a sequence of rational numbers. The operations of addition, subtraction, multiplication, and division are introduced on the set of real numbers in an obvious way as the limits of the corresponding sequences; the non-trivial part is the proof that the definitions do not depend on the approximating sequences. The set of real numbers is denoted by $\mathbb{R}$. Geometrically, we represent $\mathbb{R}$ as points on a one-dimensional *continuum*, known as the *real line*. The construction of Cantor ensures that the set of real numbers, also known as the real line, does not contain any holes.

Finally, to solve equation $x^2 + 1 = 0$ and other similar polynomial equations with real coefficients but without real solutions, the set of real numbers is extended by adjoining to $\mathbb{R}$ an element i (also denoted by j in some engineering books) such that $i^2 = -1$. This element is called the `imaginary unit`. The resulting extension is denoted by $\mathbb{C}$ and is the collection of the expressions $x + iy$, where $x, y \in \mathbb{R}$. The elements of $\mathbb{C}$ are called `complex numbers`. As far as solving polynomial equations, we are all set now: the `Fundamental Theorem of Algebra` states that every polynomial with coefficients in $\mathbb{C}$ has at least one root in $\mathbb{C}$; in other words, the set $\mathbb{C}$ is *algebraically closed*.

If $z = x + iy \in \mathbb{C}$, and $x, y \in \mathbb{R}$, then x, denoted by $\Re z$, is called `the real part of` z. Also, $y = \Im z$ is called `the imaginary part of` z, and $\bar{z} = a - bi$, the `complex conjugate` of z. By definition,

$$(a+bi)+(c+di) = (a+c)+(b+d)i, \ (a+bi)(c+di) = (ac-bd)+(ad+bc)i,$$

where the multiplication rule follows naturally from the distributive law and the equality $i^2 = -1$.

EXERCISE 4.1.1.^{C} *(a) Verify that* $\Re z = (z + \bar{z})/2$, $\Im z = (z - \bar{z})/(2i)$.
(b) Verify that the complex conjugate of the sum, difference, product, and ratio of two complex numbers is equal to the sum, difference, product, and ratio, respectively, of the corresponding complex conjugates: for example, $\overline{z_1 z_2} = \bar{z}_1 \bar{z}_2$. *(c) Verify that if* $P = P(z)$ *is a polynomial with* real *coefficients, then* $\overline{P(z)} = P(\bar{z})$, *and therefore the non-real roots of this polynomial come in complex conjugate pairs. (d) Conclude that a polynomial of odd degree and with real coefficients has at least one real root.*

Complex numbers appear naturally in computations involving square roots of negative numbers. Some records indicate that such computations can be traced to the Greek mathematician and inventor HERON OF ALEXANDRIA (c.10 – c.70 AD). In 1545, the Italian mathematician GEROLAMO CARDANO (1501–1576) published the general solutions to the cubic (degree three) and quartic (degree four) equations. The publication boosted the interest in the complex numbers, because the corresponding formulas required manipulations with square roots of negative numbers, even when the final result was a real number; for one example of this kind, see Exercise 4.1.5 on page 185 below. It still took some time to get used to the new concept, and to develop the corresponding theory. In fact, the term "imaginary" in connection with the complex numbers was introduced around 1630 by Descartes, who intended the term to be derogatory. In 1777, Euler suggested the symbol i for $\sqrt{-1}$, and the complex numbers started to get the respect they deserve. The foundations of modern complex analysis were laid during the first half of the 19th century. By the end of the 19th century, it was already impossible to imagine mathematics without complex numbers. Part of the reason could be that, as the French mathematician JACQUES SALOMON HADAMARD (1865–1963) put it, *the shortest path between two truths in the real domain passes through the complex domain.* In the following sections, we will see plenty of examples illustrating this statement.

Since solving polynomial equations was the main motivation for the introduction of complex numbers, let us say a bit more about these equations. Given a polynomial, the objective is to find an *algebraic formula* for the roots, that is, an expression involving a *finite* number of additions, subtractions, multiplications, divisions, and root extractions, performed on the coefficients of the polynomial. The formula $x = (-b \pm \sqrt{b^2 - 4ac})/(2a)$ for the roots of the quadratic equation $ax^2 + bx + c = 0$ was apparently known to ancient Babylonians some 4000 years ago. The formulas of Cardano for equations of degree three and four are much more complicated but still algebraic. Ever since the discovery of those formulas, various mathematicians tried to extend the results to equations of degree five or higher, until, around 1820, the Norwegian mathematician NIELS HENRIK ABEL (1802–1829) proved the non-existence of such algebraic representations for the solutions of a general fifth-degree equation. In 1829, the French mathematician EVARISTE GALOIS (1811–1832) resolved the issue completely by proving the non-existence of an algebraic formula for the solution of a general polynomial equation of degree five or higher, and also describing all the equations for which such a formula does exist. Note that both Abel and Galois were under 20 years of age when they made their discoveries.

The solutions of a polynomial equation can exist even without an algebraic formula to compute them; the fundamental theorem of algebra ensures that a polynomial of degree n and with coefficients in $\mathbb{C}$ has exactly n roots in $\mathbb{C}$, and there are many ways to represent the roots using *infinitely* many operations of addition, subtraction, multiplication, and division, performed on the coefficients. Such representations lead to various *numerical* methods of solving the polynomial equations, but these topics fall outside the scope of our discussions. For more on the history of complex numbers, see the book *An Imaginary Tale: The Story of* $\sqrt{-1}$ by P. J. Nahin, 1998.

4.1.2 *The Complex Plane*

The field $\mathbb{C}$ can be represented geometrically as a set of points in the plane, with the real part along the horizontal axis, and the imaginary part along the vertical axis. This representation identifies a complex number $z = x+iy$ with either the point (x, y) or the vector $\boldsymbol{r} = x\,\hat{\imath} + y\,\hat{\jmath}$ in the Euclidean space $\mathbb{R}^2$; see Figure 4.1.1. The `upper` (`lower`) half-plane contains complex numbers with positive (negative) *imaginary* part. Similarly, the `right` or `left` half-plane refers to complex numbers with positive or negative real parts, respectively.

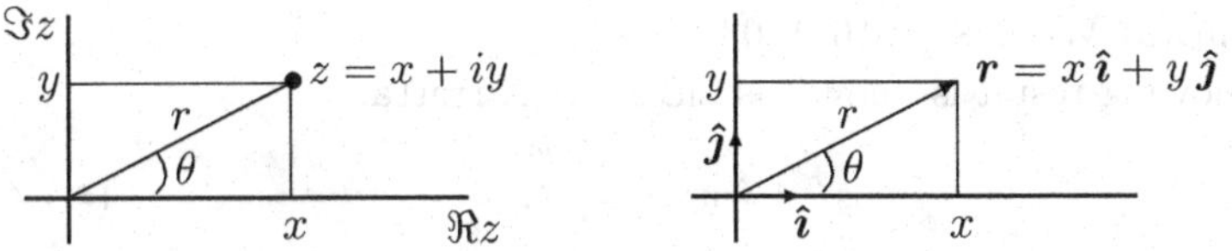

Fig. 4.1.1 Complex plane and $\mathbb{R}^2$

In polar coordinates,

$$z = r(\cos\theta + i\sin\theta), \tag{4.1.1}$$

where $r = \sqrt{x^2+y^2} = |z|$ is the `modulus` or `absolute value` of z, and $\theta = \arg(z)$ is the `argument` of z. Similar to the polar angle at the origin, the argument is not defined for $z = 0$.

Note that if θ is the argument of z, so is $\theta + 2\pi k$ for every integer k. Accordingly, the `principal value of the argument` $\mathrm{Arg}(z)$ is defined as the value of $\arg z$ in the interval $(-\pi, \pi]$.

EXERCISE 4.1.2.^{B} *Verify the following formula for the principal value of the argument:*

$$\mathrm{Arg}(z) = \begin{cases} \tan^{-1}(y/x), & x \geq 0,\ y \neq 0; \\ \pi + \tan^{-1}(y/x), & x < 0,\ y \geq 0; \\ -\pi + \tan^{-1}(y/x), & x < 0,\ y < 0. \end{cases} \tag{4.1.2}$$

EXERCISE 4.1.3.^{C} *Using the suitable trigonometric identities, verify that if* $z_1 = r_1(\cos\theta_1 + i\sin\theta_1)$ *and* $z_2 = r_2(\cos\theta_2 + i\sin\theta_2)$, *then*

$$\begin{aligned} z_1 z_2 &= r_1 r_2(\cos(\theta_1+\theta_2) + i\sin(\theta_1+\theta_2)),\\ \frac{z_1}{z_2} &= \frac{r_1}{r_2}\Big(\cos(\theta_1-\theta_2) + i\sin(\theta_1-\theta_2)\Big). \end{aligned} \tag{4.1.3}$$

Some sources call the left picture in Figure 4.1.1 the `Argand diagram`, in honor of the Swiss-born non-professional mathematician JEAN-ROBERT ARGAND (1768–1822), who published the idea of the geometric interpretation of complex numbers in 1806, while managing a bookstore in Paris. Mathematical formulas are neither copyrightable nor patentable, and the names of those formulas are often assigned in an unpredictable way. Argand's diagram is one such example: in 1685, when complex number were much less popular, a similar idea appeared in a book by the English mathematician JOHN WALLIS (1616–1703).

The following result is known as `Euler's formula`:

$$\cos\theta + i\sin\theta = e^{i\theta}. \tag{4.1.4}$$

In this case, the name is true to the fact: the result was first published by L. Euler in 1748. At this point, we will take (4.1.4) for granted and only mention that (4.1.4) is consistent with (4.1.3); later on, we will prove (4.1.4) using power series. A particular case of (4.1.4), $e^{i\pi}+1=0$, collects the five most important numbers in mathematics, $0, 1, e, \pi, i$, in one simple equality.

An immediate consequence of (4.1.4) is that multiplication of a complex number z by $e^{i\theta}$ is equivalent to a rotation of z in the complex plane by θ radians counterclockwise. Note also that $e^{i\theta} = e^{i(\theta+2\pi)}$. As a result, a time-varying periodic quantity A is conveniently represented as $A(t) = A_0 e^{i\omega t}$, where A_0 is the amplitude, and ω is the angular frequency (so that $2\pi/\omega$ is the period). Below, we will use such representations in the analysis of electrical circuits and planar electromagnetic waves.

EXERCISE 4.1.4.^{C} *Let* $z_1 = x_1 + iy_1$, $z_2 = x_2 + iy_2$ *be two complex numbers, and* $\boldsymbol{r}_1 = x_1\,\hat{\boldsymbol{\imath}} + y_1\,\hat{\boldsymbol{\jmath}}$, $\boldsymbol{r}_2 = x_2\,\hat{\boldsymbol{\imath}} + y_2\,\hat{\boldsymbol{\jmath}}$, *the corresponding vectors. (a) Verify that* $\overline{z}_1 z_2 = (\boldsymbol{r}_1 \cdot \boldsymbol{r}_2) + i\big((\boldsymbol{r}_1 \times \boldsymbol{r}_2)\cdot\hat{\boldsymbol{\kappa}}\big)$, *where, as usual,* $\hat{\boldsymbol{\kappa}} = \hat{\boldsymbol{\imath}} \times \hat{\boldsymbol{\jmath}}$. *(b) Express the angle between the vectors* $\boldsymbol{r}_1$ *and* $\boldsymbol{r}_2$ *in terms of* $Arg(z_1)$ *and* $Arg(z_2)$. *Hint: draw a picture; the angle is always between* 0 *and* π.

One application of (4.1.4) is computing `roots of complex numbers`.

Writing

$$z = |z|e^{(\text{Arg}(z)+2\pi k)i},$$

we find

$$z^{1/m} = |z|^{1/m}\, e^{(\text{Arg}(z)/m+2\pi k/m)i}, \; k = 0, \dots, m-1. \tag{4.1.5}$$

Thus, the m-th roots of a complex number have exactly m distinct values. On the complex plane, these values are at the vertices of a regular m gon. FOR EXAMPLE, taking $z = 1$ and $m = 3$ so that $\text{Arg}(z) = 0$, we find the three values of $\sqrt[3]{1}$: $1, (-1 + i\sqrt{3})/2, (-1 - i\sqrt{3})/2$.

EXERCISE 4.1.5. *(a)^C Find all the values of $\sqrt[3]{8i}$ and $\sqrt[6]{64i}$ and draw them in the complex plane. (b)^A Find all the values of $\sqrt{1 + i\sqrt{3}} + \sqrt{1 - i\sqrt{3}}$. Hint: one of them is $\sqrt{6}$: simply square the expression.*

Another application of (4.1.4) is deriving certain trigonometric identities. By taking the n-th power on both sides, we get `de Moivre's formula`:

$$(\cos\theta + i\sin\theta)^n = \cos n\theta + i\sin n\theta; \tag{4.1.6}$$

of course, the French mathematician ABRAHAM DE MOIVRE (1667–1754), who published (4.1.6) in 1722, did not use (4.1.4) in his derivations.

By equating the real and imaginary parts of (4.1.6), we get the expressions for $\cos n\theta$ and $\sin n\theta$ in terms of the products of $\sin\theta$ and $\cos\theta$. FOR EXAMPLE, $n = 2$ yields the familiar results: $\cos 2\theta = cos^2\theta - \sin^2\theta$, $\sin 2\theta = 2\sin\theta\cos\theta$.

EXERCISE 4.1.6.^{B} *Find similar expressions for* $\cos 3\theta$ *and* $\sin 3\theta$.

Yet another application of (4.1.4) is illustrated by the following exercise.

EXERCISE 4.1.7.^{C} *Evaluate the indefinite integral $\int e^{-3x}\cos 2x\, dx$ without integration by parts. Hint: note that the integrand is $\Re(e^{(-3+2i)x})$. Integrate the complex exponential as if it were real, and them compute the real part of the result.*

We conclude this section with a few definitions related to sets in the complex plane; these definitions are identical to those on page 121. As with ordinary points in the plane, $|z_1 - z_2|$ is the distance between z_1 and z_2, and the set $\{z : |z - z_0| < r\}$ is an `open disk` with center at z_0 and radius $r > 0$; similarly, $\{z : |z - z_0| \le r\}$ is a `closed disk`. A `neighborhood` of a point z is an open disk centered at z. A point in a set is called `interior` if there exists a neighborhood of the point that lies entirely in the set. FOR EXAMPLE, the center of the disk (open or closed) is an interior point of the disk.

A point P is called a **boundary** point of a set if every neighborhood of the point contains at least one point that is not in the set, and at least one point that belongs to the set and is different from P. FOR EXAMPLE, the boundary points of the set $\{z : |z - z_0| < r\}$ are exactly the points of the circle $\{z : |z - z_0| = r\}$.

A point P is called an **isolated** point of a set if there exists a neighborhood of P in which P is the only point belonging to the set. FOR EXAMPLE, the set $\{z : z = a + ib\}$ in the complex plane, where a, b are real integers, consists entirely of isolated points.

A set is called

- **Bounded**, if it lies entirely inside an open disk of sufficiently large radius.
- **Closed**, if it contains all its boundary points.
- **Connected**, if every two points in the set can be connected with a continuous curve lying completely in the set.
- **Open**, if every point belongs to the set together with some neighborhood. In other words, all points of an open set are interior points.
- **Domain**, if it is open and connected.
- **Simply connected**, if it has no holes. More precisely, consider a simple, closed, continuous curve that lies entirely in the set (see page 25); such a curve encloses a domain (recall that we take for granted the *Jordan curve theorem, see page 123*). The set is simply connected if this domain lies entirely in the set.

The **closure of a set** is the set together with all its boundary points. The **complement** of a set are all the points that are not in the set.

FOR EXAMPLE, the closure of an open disk $\{z : |z| < 1\}$ is the closed disk $\{z : |z| \leq 1\}$, and the complement of that open disk is the set $\{z : |z| \geq 1\}$. Every disk, open or closed, is both connected and simply connected, while the set $\{z : 0 < |z| < 1\}$ is open, connected, but not simply connected because the point $z = 0$ is missing from the set.

EXERCISE 4.1.8.^{C} *Give an example of a set in $\mathbb{C}$ that is not bounded, is neither open nor closed, is not connected, but is simply connected.*

4.1.3 *Applications to Analysis of AC Circuits*

In this section, we continue to use i as the notation for the imaginary unit. The basic components of an alternating current (AC) circuit are the resistor R, the capacitor C, and the inductor L; see Figure 4.1.2. The main facts about electric circuits are summarized in Section 8.4 in Appendix. If A is a quantity, such as current or voltage, changing periodically in time with period $T = 1/\nu$, then we represent this quantity as $A(t) = A_0 e^{i(\omega t+\phi_0)}$, where ϕ_0 is the initial phase, and ω is the underlying angular frequency, related to the usual frequency ν by $\omega = 2\pi\nu$ (a household outlet in the US has $\nu = 60$ Hz or 60 cycles per second; in Europe, the standard is $f = 50$ Hz). *Since taking the real part of $A(t)$ brings us back to physical reality and can be done at any moment, we will work only with complex currents and voltages.*

Denote by I_Y and V_Y the current through and the voltage across the element Y, respectively, with Y being a resistor R, a capacitor C, or an inductor L. We assume that all the elements are linear, so that

- by Ohm's Law, see page 173, $I_R = V_R/R$;
- by the definition of the capacitance, $C = q_C/V_C$, where $q_C(t) = \int_0^t I_C(s)ds$ is the charge; thus, $I_C = C\, dV_C/dt$;
- by Faraday's Law, see page 165, $V_L = LdI_L(t)/dt$.

EXERCISE 4.1.9. ^{C} *Taking $I(t) = I_0 e^{i\omega t+i\phi_0}$, verify that (a) The current through the resistor is in phase with the voltage; (b) The current though the capacitor is ahead of, or leads, the voltage by the phase $\pi/2$; (c) The current through the inductor is behind, or lags, the voltage by the phase $\pi/2$. Hint: $i = e^{i\pi/2}$.*

Consider the series circuit on the left-hand side of Figure 4.1.2, with $E(t) = E_0 e^{i(\omega t+\phi_0)}$. All the elements of the circuit have the same current $I(t)$ passing through them; we take $I(t) = I_0 e^{i\omega t}$. Then we have $V_R(t) = I(t)R$ (the voltage across R is in phase with the current), $V_C(t) = (1/\omega C)I(t)e^{-i\pi/2}$ (the voltage across C is behind the current by $\pi/2$), and $V_L(t) = L\omega I(t)e^{i\pi/2}$ (the voltage across L is ahead of the current by $\pi/2$). The vector diagram corresponds to time $t = 0$; for $t > 0$ the diagram rotates counterclockwise with angular speed $\omega = 2\pi\nu$; linear frequency $\nu = 60$ Hz, corresponds to 60 full turns per second. Since

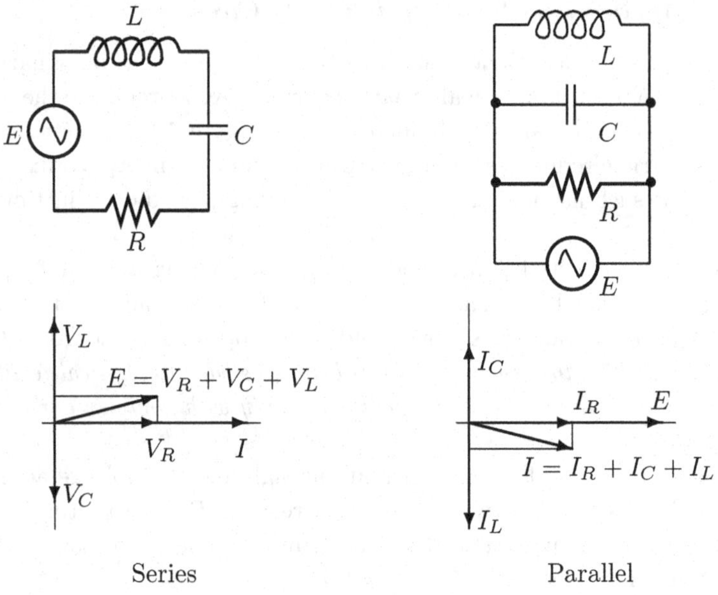

Fig. 4.1.2 *RCL* Circuits

$V_R + V_C + V_L = E$, we have $E(t) = (R + 1/(i\omega C) + i\omega L)I(t)$, and therefore

$$I_0 = \frac{E_0}{\sqrt{R^2 + \left(\omega^2 L^2 - \frac{1}{\omega^2 C^2}\right)^2}}, \quad \tan\phi_0 = \frac{\omega L - \frac{1}{\omega C}}{R}. \tag{4.1.7}$$

EXERCISE 4.1.10.[B] *(a) Verify (4.1.7). Hint: Write* $(R + 1/(i\omega C) + i\omega L)$ *in the complex exponential form; keep in mind that* $1/i = -i$. *(b) Verify that, for every input* $E(t)$, *the current* $I(t)$ *in the series circuit satisfies*

$$LI''(t) + RI'(t) + I(t)/C = E'(t). \tag{4.1.8}$$

Substitute $E(t) = E_0 e^{i(\omega t + \phi_0)}$ *and* $I(t) = I_0 e^{i\omega t}$ *to recover (4.1.7). Hint: if* q *is the charge, then* $E = E_L + E_R + E_C = Lq'' + Rq' + q/C$; $I = q'$.

For fixed E_0, the largest value of $I_0 = E_0/R$ is achieved at the *resonance frequency* $\omega_0 = 1/\sqrt{LC}$; the corresponding phase shift ϕ_0 at this resonance frequency is zero. According to (4.1.8), a series circuit is a damped **harmonic oscillator**, with **damping** proportional to R. The "ideal" series circuit with $R = 0$ is a pure harmonic oscillator and has infinite current at the resonance frequency $\omega_0 = 1/\sqrt{LC}$.

Now let us consider the parallel RCL circuit on the right-hand side of Figure 4.1.2. This time, the voltage $E(t)$ is the same across all components; we take $E(t) = E_0 e^{i\omega t}$. For the currents, we have $I_R(t) = E(t)/R$ (in phase with E), $I_C(t) = E(t)\omega C e^{i\pi/2}$ (ahead of E by $\pi/2$), $I_L(t) = (E(t)/(\omega L))e^{-i\pi/2}$ (behind E by $\pi/2$). The vector diagram corresponds to $t = 0$; for $t > 0$, the diagram rotates counterclockwise with angular speed ω. The total current is $I = I_R + I_C + I_L = (1/R + i\omega C + 1/(i\omega L))E(t)$. Taking $I(t) = I_0 e^{i(\omega t+\phi_0)}$, we conclude that

$$I_0 = \sqrt{\frac{1}{R^2} + \left(\omega C - \frac{1}{\omega L}\right)^2}\, E_0, \ \tan\phi_0 = R(\omega C - 1/(\omega L)). \qquad (4.1.9)$$

EXERCISE 4.1.11. [B] *(a) Verify (4.1.9). (b) Verify that, for every input voltage $E(t)$, the current $I(t)$ in the parallel circuit satisfies*

$$CE''(t) + E'(t)/R + E(t)/L = I'(t). \qquad (4.1.10)$$

Substitute $E(t) = E_0 e^{i(\omega t)}$ and $I(t) = I_0 e^{i(\omega t+\phi_0)}$ to recover (4.1.9).

Unlike the series circuit, the resonance frequency $\omega_0 = 1/\sqrt{LC}$ now corresponds to the *smallest* absolute value $I_0 = E_0/R$ of the current for given E_0; the corresponding phase shift ϕ_0 at the resonance frequency is zero. According to (4.1.10), a parallel circuit is a damped `harmonic oscillator` with damping proportional to $1/R$. The "ideal" parallel circuit with $R = \infty$ is a pure harmonic oscillator and has zero total current at the resonance frequency $\omega_0 = 1/\sqrt{LC}$.

The above analysis also demonstrates that the effective resistance, called `reactance`, of the capacitor and the inductor is, respectively, $X_C = 1/(i\omega C)$ and $X_L = i\omega L$. The usual laws for series or parallel connection apply: the total effective resistance Z, known as the `complex impedance`, satisfies $Z = R + X_C + X_L$ in the series circuit and $1/Z = (1/R) + (1/X_C) + (1/X_L)$ in the parallel circuit.

EXERCISE 4.1.12.[B] *Sketch the graph of $|Z|$ as a function of ω for (i) the series circuit; (ii) the parallel circuit.*

4.2 Functions of a Complex Variable

4.2.1 *Continuity and Differentiability*

The study of functions of a complex variable is called `complex analysis`. As in the usual calculus, we define a `complex function`, that is, function $f = f(z)$ of a complex variable z as a rule that assigns to every complex number z *at most one* complex number $f(z)$. This definition is especially important to keep in mind: rules that assign *several* values to the same complex number z also appear in complex analysis and are called `multi-valued functions`.

A polynomial function $p = p(x) = \sum_{k=0}^{n} a_k x^k$ of a real variable x easily extends to the complex plane by replacing x with z and allowing the coefficients a_k to be complex numbers. Similarly, a `rational function`, that is, a ratio of two polynomial functions, extends to the complex plane except at the points where the denominator vanishes. For other functions of the real variable, the replacement of x with z is not as straightforward, and a special *theory of the functions of a complex variable* must be developed.

Definition 4.1 A function $f = f(z)$ is `continuous` at z_0 if f is defined in some neighborhood of z_0 and

$$\lim_{|z|\to 0} f(z_0 + z) = f(z_0).$$

A function $f = f(z)$ is `differentiable` at z_0 if f is defined in some neighborhood of z_0 and there exists a complex number, denoted by $f'(z_0)$, so that

$$\lim_{|z|\to 0} \frac{f(z_0 + z) - f(z_0)}{z} = f'(z_0).$$

EXERCISE 4.2.1.^{C} *Verify that a function differentiable at z_0 is continuous at z_0.*

EXERCISE 4.2.2.^{C} *Which of the following functions are differentiable at zero: $f(z) = z^2, f(z) = \Im(z), f(z) = |z|, f(z) = |z|^2$?*

EXERCISE 4.2.3.^{C} *Verify that if $f(z) = z^k$ for a positive integer k, then $f'(z) = kz^{k-1}$.*

Definition 4.2 A function is called `analytic at a point` if it is differentiable in some neighborhood of the point. A function is called `analytic in a domain` if it is analytic at every point of the domain. In general, we say that a function is **analytic** if it is analytic *somewhere.* An **entire function** is a function that is analytic everywhere in the complex plane.

Remark 4.1 *Sometimes, the word* holomorphic *is used instead of* analytic.

EXERCISE 4.2.4.[C] *(a) Convince yourself that a function is analytic at a point if and only if the function is analytic in some neighborhood of the point. Hint: a neighborhood of a point is an open set; see page 186. (b) Verify that a polynomial is an entire function. (c) Verify that a* **rational function** $f(z) = P(z)/Q(z)$, *where* P, Q *are polynomials, is analytic everywhere except at the roots of* $Q(z)$.

4.2.2 *Cauchy-Riemann Equations*

Let $f = f(z)$, $z = x + iy$, be a function of a complex variable. Being a complex number itself, $f(z)$ can be written as

$$f(z) = u(x, y) + iv(x, y) \tag{4.2.1}$$

for some real functions u, v of two real variables x, y. FOR EXAMPLE, if $f(z) = z^2$, then $f(z) = (x+iy)^2 = (x^2-y^2)+2ixy$, so that $u(x, y) = x^2-y^2$ and $v(x, y) = 2xy$.

EXERCISE 4.2.5.[C] *Find the functions* u, v *if* $f(z) = z^3$.

The objective of this section is to investigate the connection between differentiability of the complex function f and differentiability of the functions u, v. The motivation for this investigation comes from the following exercise, showing that differentiability of a function of a complex variable requires more than mere differentiability of the real and imaginary parts.

EXERCISE 4.2.6.[C] *(a) Show that the function* $f(z) = u(x, y) + iv(x, y)$ *is continuous at the point* $z_0 = x_0 + iy_0$, *in the sense of Definition 4.1, if and only if both functions* u, v *are continuous at* (x_0, y_0), *as real functions of two variables. (b) Show that if the function* $f = f(z)$ *is differentiable, then the corresponding functions* u, v *are also differentiable, as functions of* x *and* y. *(c) Show that the function* $f(z) = x$ *is not differentiable anywhere, in the sense of Definition 4.1.*

To understand what is going on, let us assume that $f = f(z)$ is differentiable at z_0 so that, according to Definition 4.1, we have $f(z_0 + z) - f(z_0) = zf'(z_0) + \varepsilon(z)$, where $|\varepsilon(z)|/|z| \to 0$, $|z| \to 0$. Writing

$\varepsilon(z) = \varepsilon_1(x,y) + i\varepsilon_2(x,y)$, $f'(z_0) = A + iB$, $r = |z|$, we find that, for all x, y sufficiently close to zero,

$$\begin{aligned}\big(u(x_0+x, y_0+y) - u(x_0,y_0)\big) + i\big(v(x_0+x, y_0+y) - v(x_0,y_0)\big)\\ = (x+iy)(A+iB) + \varepsilon_1(x,y) + i\varepsilon_2(x,y),\end{aligned}$$

where $|\varepsilon_k(x,y)/r| \to 0$, $r \to 0$, for $k = 1, 2$.

EXERCISE 4.2.7. ^{C} *By comparing the real and imaginary parts in the last equality, convince yourself that the functions u, v are differentiable at (x_0, y_0) and the following equalities hold:*

$$A = \frac{\partial u}{\partial x}(x_0,y_0) = \frac{\partial v}{\partial y}(x_0,y_0), \quad B = -\frac{\partial u}{\partial y}(x_0,y_0) = \frac{\partial v}{\partial x}(x_0,y_0).$$

We now state the main result of this section.

Theorem 4.2.1 *A function $f = f(z)$ is differentiable at the point $z_0 = x_0 + iy_0$ if and only if the functions $u = \Re f$, $v = \Im f$ are differentiable at the point (x_0, y_0) and the equalities*

$$\frac{\partial u}{\partial x} = \frac{\partial v}{\partial y}, \quad \frac{\partial u}{\partial y} = -\frac{\partial v}{\partial x} \tag{4.2.2}$$

hold at the point (x_0, y_0).

Equalities (4.2.2) are known as the **Cauchy-Riemann equations**. Riemann introduce many key concepts of complex analysis in 1851 in his Ph.D. dissertation; his advisor was Gauss. And, as we mentioned earlier in connection with the Cauchy-Bunyakovky-Schwartz inequality, one should not be surprised to see the name of Cauchy attached to an important result.

EXERCISE 4.2.8. *(a)^B Prove the above theorem. Hint: You already have the proof in one direction, and you reverse the arguments to get the other direction. (b)^C Verify that the derivative $f'(z)$ can be written in one of the four equivalent ways:*

$$f'(z) = \frac{\partial u}{\partial x} + i\frac{\partial v}{\partial x} = \frac{\partial v}{\partial y} - i\frac{\partial u}{\partial y} = \frac{\partial u}{\partial x} - i\frac{\partial u}{\partial y} = \frac{\partial v}{\partial y} + i\frac{\partial v}{\partial x}.$$

(c)^B Verify that if the function f is analytic in the domain G, then each of the following implies that f is constant in G: (i) $f'(z) = 0$ in G; (ii) Either $\Re f$ or $\Im f$ is constant in G; (iii) $|f(z)|$ is constant in G.

We will see later that if a function $f = f(z)$ is differentiable (analytic) in a domain, then it is infinitely differentiable there. Then the functions u

and v have continuous partial derivatives of every order. Recall that the alternative notation for the partial derivative $\partial u/\partial x$ is u_x, and when the second-order partial derivatives are continuous, we have $u_{xy} = u_{yx}$. Then from (4.2.2) we find $u_{xx} = v_{yx}$ and $v_{xy} = -u_{yy}$, so that $u_{xx} + u_{yy} = 0$. Similarly, $v_{xx} + v_{yy} = 0$, that is, both u and v are *harmonic functions.* Two harmonic functions that satisfy (4.2.2) are called `conjugate`. We conclude that *a function $f = f(z)$ is analytic in a domain G if and only if the real and imaginary parts of f are conjugate harmonic functions.* This remarkable connection between analytic and harmonic functions leads to numerous applications of complex analysis in problems such as the study of two-dimensional electrostatic fields and two-dimensional flows of heat and fluids. We discuss mathematical foundations of these applications in Problem 5.3 on page 433. The lack of a three-dimensional analog of complex numbers is one of the reasons why the corresponding problems in three dimensions are much more difficult.

If u and v are conjugate harmonic functions, then, because of (4.2.2), one of the functions uniquely determines the other up to an additive constant. Similarly, either the real part u or the imaginary part v specify the corresponding analytic function $f = u + iv$ uniquely up to an additive constant. FOR EXAMPLE, if $u(x, y) = xy$, then, by the first equality in (4.2.2), $u_x = y = v_y$, so that $v(x, y) = y^2/2 + h(x)$ and hence $v_x = h'(x)$. By the second equality in (4.2.2), $v_x = -u_y = -x$, that is, $h'(x) = -x$ and $v(x, y) = (y^2 - x^2)/2 + c$, where c is a real number. Note that $v_{xx} + v_{yy} = -1 + 1 = 0$, as it should; this is a useful computation to ensure that the computations leading to the formula for v are correct. With $z = x + iy$, the corresponding function $f(z)$ is recovered from the equality

$$f(z) = u\big((z + \bar{z})/2, (z - \bar{z})/2i\big) + iv\big((z + \bar{z})/2, (z - \bar{z})/2i\big),$$

which in this case results in $f(z) = -iz^2/2 + ic$, where c is a real number. This answer is easy to check: $f(z) = -iz^2/2 + ic = xy + i(y^2 - x^2)/2 = u(x, y) + iv(x, y)$.

EXERCISE 4.2.9.^{C} *Find all functions $f = f(z)$ that are differentiable for all z and have $\Im f(z) = x^2 - xy - y^2$. Check that your answer is correct.*

If we write $z = r(\cos\theta + i\sin\theta)$ in polar coordinates, then $f(z) = u(r, \theta) + iv(r, \theta)$.

EXERCISE 4.2.10.^{B} *(a) Verify that in polar coordinates equations (4.2.2) imply*

$$u_r = r^{-1} v_\theta, \quad v_r = -r^{-1} u_\theta, \tag{4.2.3}$$

and conversely, (4.2.3) imply (4.2.2). (b) Verify that each of the following functions satisfies (4.2.3): (i) $f(z) = r^\alpha e^{i\alpha\theta}$, *$\alpha$ a real number; (ii)* $f(z) = \ln r + i\theta$

The usual rules of differentiation (for the sum, difference, product, ratio, and composition (chain rule) of two functions, and for the inverse of a function) hold for the functions of complex variable just as for the functions of real variable. If the function $f = f(x)$ is differentiable, the chances are good that the corresponding function $f = f(z)$ is analytic, and to compute the derivative of $f(z)$, you treat z the same way as you would treat x. The functions $f(z)$ that contain $\bar{z}$, $\Re z$, $\Im z$, $\arg(z)$, and $|z|$ require special attention because they do not have clear analogs in the real domain. Analyticity of such functions must be studied using the Cauchy-Rieman equations. FOR EXAMPLE, the function $f(z) = \bar{z}$ is not analytic anywhere, because for this function $u_x = 1$, $v_y = -1$, and so $u_x \neq v_y$.

EXERCISE 4.2.11.[C] *Check whether the following functions f are analytic. If the function is analytic, find the derivative. (a)* $f(z) = (\bar{z})^2$, *(b)* $f(z) = \Im z / \Re z$, *(c)* $f(z) = \Re z^3 - i \Im z^3$, *(d)* $f(z) = z^2/(1 - z^2)$.

4.2.3 *The Integral Theorem and Formula of Cauchy*

In this section, we study integrals of analytic functions and establish two results from which many of the properties of the analytic functions follow.

We start with integration in the complex plain. Consider a curve C in $\mathbb{R}^2$ defined by the vector-valued function $\boldsymbol{r}(t) = x(t)\,\hat{\boldsymbol{\imath}} + y(t)\,\hat{\boldsymbol{\jmath}}$, $a \leq t \leq b$. Equivalently, we can define this curve using a complex-valued function $z = z(t)$, $t \in [a, b]$, by setting $z(t) = x(t) + iy(t)$. As usual, we write $\dot{z}(t) = \dot{x}(t) + i\dot{y}(t)$, provided the derivatives of x and y exist.

Assume that the curve C is *piece-wise smooth*, that is, consists of finitely many smooth pieces; see page 28 for details. Let $f = f(z)$ be a function, continuous in some domain containing the curve C. We define the integral $\int_C f(z)dz$ of f along C by

$$\int_C f(z)dz = \int_a^b f(z(t))\dot{z}(t)dt. \tag{4.2.4}$$

In what follows, *we consider only piece-wise smooth curves.* The line integral over a closed curve $\mathcal{C}$ (that is, a curve for which $\boldsymbol{r}(a) = \boldsymbol{r}(b)$) is often denoted by $\oint_{\mathcal{C}}$. The letter z is not the only possible notation for the variable of integration; in particular, we will often use the Greek letter ζ when z is being used for other purposes.

EXERCISE 4.2.12.^{C} *(a) Writing $f(z) = u(x,y) + iv(x,y)$, verify that*

$$\int_{\mathcal{C}} f(z)dz = \int_{\mathcal{C}} (udx - vdy) + i\int_{\mathcal{C}} (vdx + udy). \tag{4.2.5}$$

In particular, by (3.1.17) on page 131, change of orientation of the curve $\mathcal{C}$ reverses the sign of $\int_{\mathcal{C}} f(z)dz$. (b) Verify that the length of the curve can be written as $\int_{\mathcal{C}} |dz|$.

EXERCISE 4.2.13.^{C} *Let $f(z) = (z - z_0)^n$, where z_0 is a fixed complex number and n, an integer: $n = 0, \pm 1, \pm 2, \ldots$. Let $\mathcal{C}$ be a circle with radius ρ, center at z_0, and orientation counterclockwise. Verify that $\oint_{\mathcal{C}} f(z)dz = 2\pi i$ if $n = -1$ and $\int_{\mathcal{C}} f(z)dz = 0$ otherwise. Hint: $z(t) = z_0 + \rho e^{it}$, $0 \le t < 2\pi$.*

In other words, you show in Exercise 4.2.13 that

$$\oint_{|z-z_0|=\rho} \frac{1}{(z-z_0)^n} dz = \begin{cases} 2\pi i, & n = 1, \\ 0, & n = 0, -1, \pm 2, \pm 3, \ldots . \end{cases} \tag{4.2.6}$$

Recall that, for a real-valued continuous function $h = h(x)$, we have $|\int_a^b h(x)dx| \le (b-a)\max_{x\in[a,b]} |h(x)|$, and the easiest way to prove this inequality is to apply the triangle inequality to the approximation of the integral according to the rectangular rule and then pass to the limit. We will now derive a similar inequality for the complex integral (4.2.4).

By the left-point rectangular rule, we have

$$\int_a^b f(z(t))\dot{z}(t)dt = \lim_{\max|t_{k+1}-t_k|\to 0} \sum_{k=0}^{n-1} f(z(t_k))\dot{z}(t_k)(t_{k+1} - t_k), \tag{4.2.7}$$

where $a = t_0 < t_1 < \ldots < t_n = b$. Note that $|\dot{z}(t_k)| = \|\dot{\boldsymbol{r}}(t_k)\|$ where $\boldsymbol{r} = \boldsymbol{r}(t)$ is the vector function that defines the curve, and we saw on page 29 that

$$\lim_{\max|t_{k+1}-t_k|\to 0} \sum_{k=0}^{n-1} \|\dot{\boldsymbol{r}}(t_k)\|(t_{k+1} - t_k) = \int_a^b \|\dot{\boldsymbol{r}}(t)\|dt = L_{\mathcal{C}}(a,b),$$

the length of the curve $\mathcal{C}$. We then apply the triangle inequality on the right-hand side of (4.2.7) and denote by $\max_{z\in\mathcal{C}} |f(z)|$ the maximal value of $|f(z)|$ on the curve C. The result is the following inequality for the integral (4.2.4):

$$\left|\int_C f(z)dz\right| \leq L_C(a,b) \max_{z\in C} |f(z)|. \tag{4.2.8}$$

Note that $\max_{z\in C}|f(z)| = \max_{a\leq t\leq b}|f(z(t))|$; since the functions $f = f(z)$ and $z = z(t)$ are both continuous and the interval $[a,b]$ is closed and bounded, a theorem from the one-variable calculus ensures that the maximal value indeed exists. If you remember that the curve C is only *piece-wise* smooth, and insist on complete rigor, apply the above argument to each smooth piece of the curve separately and then add the results.

We will now look more closely at the line integrals along closed curves. Recall (page 25) that a curve C, defined by a vector-valued function $\boldsymbol{r} = \boldsymbol{r}(t)$, $a \leq t \leq b$, is called *simple closed* if the equality $\boldsymbol{r}(t_1) = \boldsymbol{r}(t_2)$ holds for $t_1 = a$, $t_2 = b$ and for no other $t_1, t_2 \in [a,b]$. By default, the orientation of such a curve is *counterclockwise*: as you walk along the curve, the domain enclosed by the curve stays on your *left*.

EXERCISE 4.2.14.^{C} *Let f be a function, analytic in a domain G, and let C be a simple, closed, piece-wise smooth curve in G so that the domain enclosed by C lies entirely in G. Assuming that the derivative f' of f is continuous in G, show that $\oint_C f(z)dz = 0$. Hint: use Green's formula to evaluate the line integrals in (4.2.5), then use the Cauchy-Riemann equations (4.2.2).*

As with line integrals of real-valued vector functions, we say that the function $f = f(z)$ has the `path independence property in a domain` G of the complex plane if f is continuous in G and $\oint_C f(z)dz = 0$ for every simple closed curve C in G (recall that we always assume that C is piece-wise smooth).

EXERCISE 4.2.15.^{C} *Show that if the function f has the path independence property in G and C_1, C_2 are two curves in G with a common starting point and with a common ending point, then $\int_{C_1} f(z)dz = \int_{C_2} f(z)dz$. Hint: make a closed curve by combining C_1 and C_2.*

EXERCISE 4.2.16.^{C} *Verify that the function $f(z) = 1$ has the path independence property in every domain, and therefore $\int_{C(z_1,z_2)} dz = z_2 - z_1$, where $C(z_1, z_2)$ is a curve that starts at z_1 and ends at z_2. Hint: use (4.2.5) and the result about path independence from vector analysis.*

We will show next that a continuous function with the path independence property in a simply connected domain has an *anti-derivative* there.

Theorem 4.2.2 *Assume that a continuous function $f = f(z)$ has the path independence property in a domain G, and assume that the domain G is* simply connected *(see page 186). Then there exists an analytic function $F = F(z)$ such that $F'(z) = f(z)$ in G.*

Proof. Define the function $F(z) = \int_{C(z_0,z)} f(\zeta)d\zeta$, where $C(z_0, z)$ is a curve in G, starting at a fixed point $z_0 \in G$ and ending at $z \in G$. By Exercise 4.2.15 this function is well defined, because the integral depends only on the points z_0, z and not on the particular curve. Define the number $A = A(z, \triangle z)$ by

$$A = \frac{F(z+\triangle z) - F(z)}{\triangle z} - f(z). \tag{4.2.9}$$

We need to show that $\lim_{\triangle z \to 0} A = 0$ for every $z \in G$, which is equivalent to proving that $F'(z) = f(z)$. The result will also imply that F is analytic in G.

Let $C = C(z, z+\triangle z)$ be any curve that starts at z and ends at $z+\triangle z$. Using the result of Exercise 4.2.16, we have $f(z) = f(z)(1/\triangle z)\int_C d\zeta = (1/\triangle z)\int_C f(z)d\zeta$, because $f(z)$ is constant on C. Therefore,

$$A = \frac{1}{\triangle z}\int_{C(z,z+\triangle z)} (f(\zeta) - f(z))d\zeta,$$

and, by (4.2.8), $|A| \leq (L_C/|\triangle z|)\max_{\zeta\in C}|f(\zeta) - f(z)|$. Since we are free to choose the curve $C(z, z+\triangle z)$, let it be the line segment from the point z to the point $z+\triangle z$. Then $L_C = |\triangle z|$. Since f is continuous at z, for every $\varepsilon > 0$, there exists a $\delta > 0$ so that $|\triangle z| < \delta$ implies $|f(\zeta) - f(z)| < \varepsilon$ for all $\zeta \in C$. As a result, for those $\triangle z$, $|A| < \varepsilon$, which completes the proof. □

We will now state the first main result of this section, the Integral Theorem of Cauchy, which says that the functions having the path independence property in a simply connected domain are exactly the analytic functions.

Theorem 4.2.3 **(The Integral Theorem of Cauchy)** *A continuous function f has the path independence property in a simply connected domain G if and only if f is analytic in G.*

In Exercise 4.2.14, you already proved this theorem in one direction (analyticity implies path independence) under an additional assumption that f' is continuous; this is exactly what Cauchy did in 1825. This is not the best proof, because the definition of an analytic function requires only the existence of the derivative.

In 1900, the French mathematician ÉDOUARD GOURSAT (1858–1936) produced a proof that does not rely on the continuity of f'. He published the result in 1900 in the very first issue of the Transactions of the American Mathematical Society (Volume 1, No. 1, pages 14–16; the paper is in French, though). Goursat's proof is too technical to discuss here and is more suitable for a graduate-level course in complex analysis.

The converse statement (path independence implies analyticity) was proved by the Italian mathematician GIACINTO MORERA (1856–1909). We almost have the proof of this result too: with Theorem 4.2.2 at our disposal, all we need is the infinite differentiability of analytic functions, which we will establish later in this section.

Remark 4.2 *Assume that the function f is continuous in the closure of G (that is, f is defined on the boundary of G and, for every z_0 on the boundary of G, $f(z) \to f(z_0)$ as $z \to z_0$, $z \in G$.) If f is analytic in G and G is simply connected and bounded with a piece-wise smooth boundary C_G, then $\oint_{C_G} f(z)dz = 0$. Indeed, continuity of f implies that, for every $\varepsilon > 0$, there exists a simple, closed, piece-wise smooth path C_ε lying inside of G such that $|\oint_{C_G} f(z)dz - \oint_{C_\varepsilon} f(z)dz| < \varepsilon$. By the Integral Theorem of Cauchy, $\oint_{C_\varepsilon} f(z)dz = 0$.*

The Integral Theorem of Cauchy says nothing about domains that are not simply connected; Exercise 4.2.13, in which G is the disk with the center removed, shows that anything can happen in those domains. If the holes in G are sufficiently nice, then we can be more specific.

Theorem 4.2.4 *Assume that the boundary of G consists of $n+1$ closed, simple, piece-wise smooth curves $C_0, \dots, C_n$ so that $C_1, \dots, C_n$ do not have points in common and are all inside the domain enclosed by C_0. Assume that f is a function analytic in G and is continuous in the closure of G. Then*

$$\oint_{C_0} f(z)dz + \sum_{k=1}^{n} \oint_{C_k} f(z)dz = 0, \tag{4.2.10}$$

where the curve C_0 is oriented counterclockwise, and all other curves, clockwise, so that, if you walk in the direction of the orientation, the domain G is always on the left.

Proof. Remember that the change of the orientation reverses the sign of the line integral; see Exercise 4.2.12, page 195. Connect the curves C_k, $k = 1, \ldots, n$ to C_0 with smooth curves (for example, line segments), and apply the Integral Theorem of Cauchy to the resulting simply connected domain (draw a picture). Then note that the integrals over the connecting curves vanish (you get two for each, with opposite signs due to opposite orientations), and you are left with (4.2.10). □

EXERCISE 4.2.17.^{C} *Use (4.2.10) to show that if f is analytic in G and C_1, C_2 are two simple closed curves in G with the same orientation (both clockwise or both counterclockwise) so that C_2 is in the domain enclosed by C_1, then $\oint_{C_1} f(z)dz = \oint_{C_2} f(z)dz$. Note that G does not have to be simply connected. Hint: consider the domain G_1 that lies in between C_1 and C_2.*

We now use the "analyticity implies path independence" part of the Integral Theorem of Cauchy to establish the `Integral Formula of Cauchy`.

Theorem 4.2.5 **(Integral Formula of Cauchy)** *Let $f = f(z)$ be an analytic function in a simply connected domain G, and C, a simple closed curve in G, oriented counterclockwise. Then the equality*

$$f(z) = \frac{1}{2\pi i} \oint_C \frac{f(\zeta)}{\zeta - z} d\zeta \tag{4.2.11}$$

holds for every point z inside the domain bounded by C.

Proof. *Step 1.* Fix the curve C and the point z. Since f is analytic in G, the function $\tilde{f}(\zeta) = f(\zeta)/(\zeta - z)$ is analytic in the domain G with the point z removed. Let C_ρ be a circle with center at z and radius ρ small enough so that C_ρ lies completely inside the domain bounded by C. We orient the circle counterclockwise and use the result of Exercise 4.2.17 to conclude that $\oint_C \tilde{f}(\zeta)d\zeta = \oint_{C_\rho} \tilde{f}(\zeta)d\zeta$.

Step 2. By (4.2.6), $f(z) = (1/2\pi i) \int_{C_\rho} \big(f(z)/(\zeta - z)\big)d\zeta$ (keep in mind that the variable of integration is ζ, whereas z is fixed). Combining this with the result of Step 1, we find:

$$\frac{1}{2\pi i} \oint_C \frac{f(\zeta)}{\zeta - z} d\zeta - f(z) = \frac{1}{2\pi i} \oint_{C_\rho} \frac{f(\zeta) - f(z)}{\zeta - z} d\zeta. \tag{4.2.12}$$

Step 3. Note that the left-hand side of (4.2.12) does not depend on ρ.

To complete the proof of (4.2.11), it is therefore enough to show that

$$\lim_{\rho\to 0}\frac{1}{2\pi i}\oint_{\mathcal{C}_\rho}\frac{f(\zeta)-f(z)}{\zeta - z}d\zeta = 0,$$

and the argument is very similar to the proof of Theorem 4.2.2; see the discussion following equation (4.2.9). Indeed, the continuity of f implies that, for every $\varepsilon > 0$, we can find $\rho > 0$ so that $|f(z) - f(\zeta)| < \varepsilon$ as long as $|\zeta - z| \le \rho$. Using inequality (4.2.8) and keeping in mind that $L_{\mathcal{C}_\rho} = 2\pi\rho$ and $|\zeta - z| = \rho$ when $\zeta \in \mathcal{C}_\rho$, we find:

$$\left|\frac{1}{2\pi i}\oint_{\mathcal{C}_\rho}\frac{f(\zeta)-f(z)}{\zeta - z}d\zeta\right| < \varepsilon, \tag{4.2.13}$$

which completes the proof. □

Similar to Remark 4.2, if the domain G is bounded and has a piece-wise smooth boundary $\mathcal{C}_G$, and the function f is analytic in G and is continuous in the closure of G, then, for every $z \in G$, we have

$$f(z) = \frac{1}{2\pi i}\oint_{\mathcal{C}_G}\frac{f(\zeta)}{\zeta - z}d\zeta \tag{4.2.14}$$

EXERCISE 4.2.18. *(a)^B Fill in the details leading to inequality (4.2.13). (b)^C What is the value of the integral on the right-hand side of (4.2.11) if the point z does not belong to the closure of the domain enclosed by $\mathcal{C}$. Hint: 0 by the Integral Theorem of Cauchy.*

EXERCISE 4.2.19. ^{A} *(a) Use the Integral Formula of Cauchy to prove the* `mean-value property` *of the analytic functions: if f is analytic in a domain G (not necessarily simply connected), z_0 is a point in G, and $\{z : |z - z_0| \le \rho\}$ is a closed disk lying completely inside G, then*

$$f(z_0) = \frac{1}{2\pi}\int_0^{2\pi} f\left(z_0 + \rho e^{it}\right)dt. \tag{4.2.15}$$

Hint: Use (4.2.14) in the closed disk; $\zeta = z_0 + \rho e^{it}$. (b) In (4.2.15, can you replace the average over the circle with the average over the disk? Hint: yes.

An almost direct consequence of representation (4.2.11) is that *an analytic function is differentiable infinitely many times.* More precisely, if $f = f(z)$ is analytic in a domain G (not necessarily simply connected), the

n-th order derivative $f^{(n)}$ of f exists at every point in G and

$$f^{(n)}(z) = \frac{n!}{2\pi i} \oint_{|z-\zeta|=\rho} \frac{f(\zeta)}{(\zeta - z)^{n+1}} d\zeta, \tag{4.2.16}$$

as long as the closed disk $\{z : |z - \zeta| \le \rho\}$ lies entirely in G; recall that $0! = 1$ and, for a positive integer k, $k! = 1 \cdot 2 \cdot \ldots \cdot k$. Using the result of Exercise 4.2.17, we can replace the circle of radius ρ in (4.2.16) with any simple, closed, piece-wise smooth curve, as long as the domain bounded by that curve contains the point z and lies entirely in G.

An informal way to derive (4.2.16) from (4.2.11) is to differentiate (4.2.11) n times, bring the derivative inside the integral, and observe that $d^n(\zeta - z)^{-1}/dz^n = n!(\zeta - z)^{-n-1}$. A rigorous argument could go by induction, with both the basis and induction step computations similar to the proof of (4.2.11). This proof is rather lengthy and does not introduce anything new to our discussion; we leave the details to the interested reader. Later, we will discuss an alternative rigorous derivation of (4.2.16) using power series; see Exercise 4.3.8(c) on page 211 below.

The consequences of (4.2.16) are far-reaching indeed, as demonstrated by the following results. The best part is, you can now easily prove all of them yourself.

EXERCISE 4.2.20. *(a)^B Complete the proof of the Cauchy Integral Theorem by showing that a continuous function that has the path-independence property in a simply connected domain is analytic there. Hint: by Theorem 4.2.2, there exists an analytic function F so that $F'(z) = f(z)$ for all z in G. By (4.2.16), all derivatives of F are continuous in G, and you have $f'(z) = F''(z)$.* *(b)^C Use (4.2.16) to prove* `Cauchy's Inequality`*: If $|f(\zeta)| \le M$ when $|\zeta - z| = \rho$, then*

$$|f^{(n)}(z)| \le \frac{Mn!}{\rho^n}. \tag{4.2.17}$$

Hint: use (4.2.8). Also note that since the function f is continuous, such a number M always exists. *(c)^B Now use (4.2.17) to prove* `Liouville's Theorem`*: A bounded entire function is constant. (That is, if a function is analytic everywhere in the complex plain and is bounded, then the function must be constant.) Hint: Taking $n = 1$ and ρ arbitrarily large, you conclude that $f'(z) = 0$ for all z. Then recall part (c) of Exercise 4.2.8.* *(d)^B Finally, use Liouville's Theorem to prove the* `Fundamental Theorem of Algebra`*: every polynomial of degree $n \ge 1$ with complex coefficients has at least one* *root. Hint: if the polynomial $P = P(z)$ has no roots, then $1/P(z)$ is a bounded entire function, hence constant — a contradiction.*

The Liouville theorem is due to the same Joseph Liouville who discovered transcendental numbers; he has one more famous theorem, related to Hamiltonian mechanics. The first proof of the fundamental theorem of algebra appeared in 1799 in Gauss's Doctoral dissertation; as with other important theorems, there had been numerous unsuccessful attempts at the proof prior to that.

EXERCISE 4.2.21.[A] *Let us say that a function $f = f(z)$ is analytic at the point $z = \infty$ if and only if $h(z) = f(1/z)$ is analytic at $z = 0$. Prove that if an entire function is analytic at $z = \infty$, then the function is everywhere constant.*

4.2.4 *Conformal Mappings*

No discussion of analytic functions is complete without mentioning conformal mappings.

As the name suggests, a conformal mapping is a mapping that preserves (local) *form*, or, more precisely, angles. We will see that an analytic function with non-zero derivative defines a conformal mapping. Before giving the precise definitions, let us recall how to compute the angle between two curves in $\mathbb{R}^2$.

As a set of points, a curve in $\mathbb{R}^2$ is defined in one of the two ways: (a) by a vector-valued function $\boldsymbol{r}(t) = x(t)\,\hat{\imath} + y(t)\,\hat{\jmath}$; (b) as a *level set* of a function $F = F(x, y)$, that is, a set $\{(x, y) : F(x, y) = \text{const.}\}$. If $\boldsymbol{r}_1 = \boldsymbol{r}_1(t)$ and $\boldsymbol{r}_2 = \boldsymbol{r}_2(s)$ define two *smooth* curves and $\boldsymbol{r}_1(t_0) = \boldsymbol{r}_2(s_0)$, then the angle θ between the curves at the point of intersection is defined by $\cos\theta = |\widehat{\boldsymbol{u}}_1(t_0) \cdot \widehat{\boldsymbol{u}}_2(s_0)|$: it is either the angle between the unit tangent vectors or π minus that angle, whichever is smaller. If the *differentiable* functions $F = F(x, y)$ and $G = G(x, y)$ define two curves so that $F(x_0, y_0) = G(x_0, y_0)$ and the point (x_0, y_0) is not critical for F and G, then the angle θ between the curves at the point of intersection satisfies

$$\cos\theta = \frac{|\boldsymbol{\nabla} F(x_0, y_0) \cdot \boldsymbol{\nabla} G(x_0, y_0)|}{\|\boldsymbol{\nabla} F(x_0, y_0)\|\, \|\boldsymbol{\nabla} G(x_0, y_0)\|}.$$

EXERCISE 4.2.22.[B] *Show that conjugate harmonic functions have orthogonal level sets. In other words, let u, v be two differentiable functions satisfying the Cauchy-Riemann equations (4.2.2) on page 192, and C_u, C_v, two curves* *corresponding to some level sets of u and v. Show that, at every point of intersection, the angle between C_u and C_v is $\pi/2$. Hint: use (4.2.2) on page 192.*

In the complex plane, the vector representation of a smooth curve is equivalent to defining a continuously differentiable complex-valued function $z = z(t)$ of a real variable t, so that $\dot{z}(t) \neq 0$; the tangent vector to the curve is represented by the function $\dot{z}(t)$. Using the properties of complex numbers, we conclude that if $z_1 = z_1(t)$ and $z_2 = z_2(s)$ represent two smooth curves and $z_1(t_0) = z_2(s_0)$, then, according to Exercise 4.1.4 on page 184, the angle θ between the curves is

$$\theta = \min\left(|\mathrm{Arg}(\dot{z}_1(t_0)) - \mathrm{Arg}(\dot{z}_2(s_0))|, \pi - |\mathrm{Arg}(\dot{z}_1(t_0)) - \mathrm{Arg}(\dot{z}_2(s_0))|\right). \tag{4.2.18}$$

We will now see that an analytic function $f = f(z)$ with non-zero derivative does not change the angle between two curves. Indeed, let $z_1 = z_1(t)$ and $z_2 = z_2(s)$ represent two smooth curves. The function f maps these curves to $w_1(t) = f(z_1(t))$ and $w_2(s) = f(z_2(s))$. By the chain rule, $\dot{w}_1(t) = f'(z_1(t))\dot{z}_1(t)$, $\dot{w}_2(s) = f'(z_2(s))\dot{z}_2(s)$. If $z^* = z_1(t_0) = z_2(s_0)$ is the point of intersection of the original curves, then $f(z^*)$ is the point of intersection of the images of these curves under f. By Exercise 4.1.4, the argument of the product of two non-zero complex numbers is equal to the sum of the arguments, so that, under the mapping f, all smooth curves that pass through the point z^* are mapped onto curves $w(t) = f(z(t))$, which are turned by the same angle $\mathrm{Arg}(f'(z^*))$. Then relation (4.2.18) implies that *the angle between the original curves at the point z^* is the same as the angle between the images at the point $f(z^*)$*. Note that the above calculations do not go through if $f'(z^*) = 0$, because in that case the argument of $f'(z^*)$ is not defined.

EXERCISE 4.2.23.^{C} *The derivative of the analytic function $f(z) = z^2$ is equal to zero when $z = 0$. By considering two lines passing through the origin, show that this function doubles angles between curves at $z = 0$. Hint: a line through the origin is defined by $Arg(z) = const$, and $Arg(z^2) = 2Arg(z)$.*

Recall that, to every mapping from $\mathbb{R}^2$ to $\mathbb{R}^2$, we associate the `Jacobian`, the function describing how the areas change locally at every point (see page 148 for a three-dimensional version). If written $f(z) = u(x, y) + iv(x, y)$, every function of a complex variable defines a mapping from $\mathbb{R}^2$ to $\mathbb{R}^2$ by sending a point (x, y) to the point $(u(x, y), v(x, y))$.

EXERCISE 4.2.24. [B] *Using the Cauchy-Riemann equations (4.2.2), verify that the Jacobian of the mapping defined by an analytic function $f = f(z)$ is equal to $|f'(z)|^2$.*

Beside the conservation of angles, another important property of the mapping defined by an analytic function with non-zero derivative is the `uniform scaling` at every point, that is, the linear dimensions near every point z are changed by the same factor $|f'(z)|$ near $w = f(z)$ in *all directions.* This property is the consequence of the definition of the derivative: if $w_0 = f(z_0)$, $w = f(z)$, $\Delta z = z - z_0$, $\Delta w = w - w_0$, then, for $|\Delta z|$ close to zero, we have $(\Delta w/\Delta z) \approx f'(z_0)$ or $|\Delta w| \approx |f'(z_0)|\,|\Delta z|$, and the direction from z_0 to z or from w_0 to w does not matter.

By definition, a mapping is called `conformal` at a point if it preserves angles and has uniform scaling at that point. We just saw that an *analytic function defines a conformal mapping at all points where the derivative of the function is non-zero.*

EXERCISE 4.2.25. [B] *Let $f = f(z)$ be an analytic function in a domain G and $f'(z) \neq 0$ in G. (a) Denoting by $\tilde{G}$ the image of G under f, verify that the area of $\tilde{G}$ is $\iint_G |f'(z)|^2 dA$. (b) Denoting by $\tilde{C}$ the image under f of a piece-wise smooth curve C in G, show that the length of $\tilde{C}$ is $\int_C |f'(z)|\,|dz|$.*

Figuring out how a particular function f transforms a certain domain or a curve is usually a matter of straightforward computations. In doing these calculations, one should keep in mind that, while a conformal mapping preserves the *local geometry*, the *global geometry* can change quite dramatically. FOR EXAMPLE, let us see what the mapping $f(z) = 1/z$, which is conformal everywhere except $z = 0$, does to the family of circles $|z - ic|^2 = c^2$, where $c > 0$ is a real number. It is convenient to consider two different complex planes: where the function f is defined, and where the function f takes its values. Since $z = x + iy$ denotes the generic complex number in the complex plane where f is defined, it is convenient to introduce a different letter, w, to denote the generic complex number in the complex plane where f takes its values. The equation of the circle is $x^2 + y^2 - 2cy = 0$. The relation between z and w is $w = 1/z = (x - iy)/(x^2 + y^2)$. When z is on the circle, we have $x^2 + y^2 = 2cy$, and then $w = x/(2cy) - i/(2c)$. In other words, if a point z is on the circle $|z - ic|^2 = c^2$, then the point $w = 1/z$ satisfies $\Im w = -1/(2c)$; you should convince yourself that, as we take different points z on the circle, we can get all possible values of the real part of w. Since the collection of all w with the same imaginary part is a line parallel to the real axis, we conclude that the function $f(z) = 1/z$ maps a circle $|z - ic|^2 = c^2$ to the line $\{w : \Im(w) = -1/(2c)\}$.

EXERCISE 4.2.26.^{C} *(a) Verify that the family of circles $|z - c|^2 = c^2$, $c > 0$, $c \neq 0$ is orthogonal to the family of circles $|z - ic|^2 = c^2$, $c \in \mathbb{R}$, $c > 0$ (draw a picture). (b) Verify that the function $f(z) = 1/z$ maps a circle $|z - c|^2 = c^2$ to the line line $\Re w = 1/(2c)$. Again, draw a picture and convince yourself that all the right angles stayed right. (c) What happens if we allow $c < 0$?*

For more examples of conformal mappings, see Problem 5.4, page 434.

The following theorem is one of the main tools in the application of complex analysis to the study of Laplace's equation in two dimensions.

Theorem 4.2.6 *Let G and $\widetilde{G}$ be two domains in $\mathbb{R}^2$ and let $f(z) = u(x, y) + iv(x, y)$ be a conformal mapping of G onto $\widetilde{G}$. If $\widetilde{U} = \widetilde{U}(\xi, \eta)$ is a harmonic function in $\widetilde{G}$, then the function $U(x, y) = \widetilde{U}(u(x, y), v(x, y))$ is a harmonic function in G.*

The domain $\widetilde{G}$ is often either the upper half-plane or the unit disk with center at the origin. Note the direction of the mapping: for example, to find a harmonic function in a domain G given a harmonic function in the unit disk, we need a conformal mapping *of the domain G onto the unit disk.*

EXERCISE 4.2.27.^{B} *(a) Prove Theorem 4.2.6 in two ways: (i) by interpreting U as the real part of analytic function $F(f(z))$, where F is an analytic function with real part $\widetilde{U}$; (b) by showing, with the help of the Cauchy-Riemann equations, that*

$$U_{xx} + U_{yy} = \left(\widetilde{U}_{\xi\xi} + \widetilde{U}_{\eta\eta}\right)\left(u_x^2 + u_y^2\right).$$

(b) Suppose you can solve the Dirichlet problem $\nabla^2 U = 0$, $U|_{\partial D} = g$, for every continuous function g when D is the unit disk. Let G be a bounded domain with a smooth boundary ∂G, and $f : G \to D$, a conformal mapping of G onto D. How to solve the Dirichlet problem $\nabla^2 V = 0$, $V|_{\partial G} = h$, for a given continuous function h?

4.3 Power Series and Analytic Functions

4.3.1 *Series of Complex Numbers*

A series of complex numbers $c_k, k \geq 0$ is the infinite sum $c_0 + c_1 + c_2 + \ldots = \sum_{k\geq 0} c_k$. The sum $\sum_{k=1}^{n} c_k$, $n \geq 1$, is called the n-th partial sum of the series. The series is called **convergent** if the sequence of its partial sums converges, that is, if there exists a complex number C such that $\lim_{n\to\infty} |C - \sum_{k=0}^{n} c_k| = 0$. The series is called **absolutely convergent** if the series $\sum_{k\geq 0} |c_k|$ converges. A series that converges but does not converge absolutely is said to converge **conditionally**. Note that the sequence $\sum_{k=1}^{n} |c_k|$, $n \geq 1$, of the partial sums of the series $\sum_{k\geq 0} |c_k|$ is non-decreasing and therefore converges if and only if $\sum_{k=1}^{n} |c_k| \leq C$ for some number C independent of n. As a result, we often indicate the convergence of $\sum_{k\geq 0} |c_k|$ by writing $\sum_{k\geq 0} |c_k| < \infty$.

EXERCISE 4.3.1.^{C} *Show that (a) the condition* $\lim_{n\to\infty} |c_n| \to 0$ *is necessary but not sufficient for convergence of the series* $\sum_{k\geq 0} c_k$*; (b) absolute convergence implies convergence, but not conversely.*

Recall that, for a sequence of real numbers a_n, $n \geq 0$, the **upper limit** is defined by

$$\limsup_{n} a_n = \lim_{n\to\infty} \sup_{k\geq n} a_k,$$

where sup means the least upper bound. Since the sequence $A_n = \sup_{k\geq n} a_k$, $n \geq 0$, is non-increasing, the upper limit either exists or is equal to $+\infty$. For a convergent sequence, the upper limit is equal to the limit of the sequence. Similarly, the **lower limit** $\liminf_n a_n = \lim_{n\to\infty} \inf_{k\geq n} a_k$, where inf is the greatest lower bound, is a limit of a non-decreasing sequence; this limit is either $-\infty$ or a finite number, and, for a convergent sequence, is equal to the limit of the sequence.

EXERCISE 4.3.2. *Verify that, for every sequence* $\{a_n,\ n \geq 0\}$ *of real numbers, (a) the sequence* $\{A_n,\ n \geq 0\}$*, defined by* $A_n = \sup_{k\geq n} a_k$ *is non-increasing, that is,* $A_{n+1} \leq A_n$ *for all* $n \geq 0$*; (b) the sequence* $\{B_n,\ n \geq 0\}$*, defined by* $B_n = \inf_{k\geq n} a_k$ *is non-decreasing, that is,* $B_{n+1} \geq B_n$ *for all* $n \geq 0$. *Hint. Use the following argument: if you remove a number from a finite collection, then the largest of the remaining numbers will be as large as or smaller than the largest number in the original collection; the smallest number will be as small as or larger.*

The following result is known as the `ratio test` for convergence.

Theorem 4.3.1 *Define*

$$K = \limsup_n \frac{|c_{n+1}|}{|c_n|}, \quad L = \liminf_n \frac{|c_{n+1}|}{|c_n|}.$$

Then the series $\sum_{k\geq 0} c_k$

- *Converges absolutely if* $K < 1$;
- *Diverges if* $L > 1$;
- *Can either converge or diverge, if* $K \geq 1$ *or* $L \leq 1$.

Proof. If $K < 1$, then, by the definition of lim sup, there exists a $q \in (K, 1)$ and a positive integer N so that $|c_{n+1}| \leq q|c_n|$ for all $n \geq N$. Then $|c_{N+k}| \leq q^k |c_N|$, $k \geq 1$, and

$$\sum_{k\geq 0} |c_k| \leq \sum_{k=0}^{N-1} |c_k| + |c_N| \sum_{k\geq 1} q^k = \sum_{k=0}^{N-1} |c_k| + \frac{|c_N|}{1-q} < \infty.$$

If $L > 1$, then, by the definition of lim inf, there exists a $q \in (1, L)$ and a positive integer N so that $|c_{n+1}| \geq q|c_n|$ for all $n \geq N$. Therefore, $\lim_{n\to\infty} |c_n| \neq 0$, and the series diverges.

The following exercise completes the proof of the theorem. □

EXERCISE 4.3.3. [B] *Construct three sequences* $\sum_{k\geq 0} c_k^{(j)}$, $j = 1, 2, 3$, *so that* $\sum_{k\geq 0} c_k^{(1)}$ *converges absolutely,* $\sum_{k\geq 0} c_k^{(2)}$ *converges conditionally, and* $\sum_{k\geq 0} c_k^{(3)}$ *diverges, while the ratio test in all three cases gives* $K = 2$ *and* $L = 0$.

The following result is known as the `root test`.

Theorem 4.3.2 *Define*

$$M = \limsup_n |c_n|^{1/n}.$$

Then the series $\sum_{k\geq 0} c_k$

- *Converges absolutely if* $M < 1$;
- *Diverges if* $M > 1$;
- *Can either converge or diverge, if* $M = 1$.

Proof. If $M < 1$, then, by the definition of lim sup, there exists a $q \in (M, 1)$ and a positive integer N so that $|c_n|^{1/n} \le q$, that is, $|c_n| \le q^n$ for all $n \ge N$. Then

$$\sum_{k\ge 0} |c_k| \le \sum_{k=0}^{N-1} |c_k| + \sum_{k\ge N} q^k = \sum_{k=0}^{N-1} |c_k| + \frac{q^N}{1-q} < \infty.$$

If $M > 1$, then, by the definition of lim sup, there exists a $q \in (1, K)$ so that $|c_n| \ge q^n$ for infinitely many n. Then $\lim_{n\to\infty} |c_n| \ne 0$ and the series diverges.

The following exercise completes the proof of the theorem. □

EXERCISE 4.3.4. ^{B} *Construct three sequences* $\sum_{k\ge 0} c_k^{(j)}$, $j = 1, 2, 3$, *so that* $\sum_{k\ge 0} c_k^{(1)}$ *converges absolutely,* $\sum_{k\ge 0} c_k^{(2)}$ *converges conditionally, and* $\sum_{k\ge 0} c_k^{(3)}$ *diverges, while the root test in all three cases gives* $M = 1$.

4.3.2 *Convergence of Power Series*

In this section, we will see that analytic functions are exactly the functions that can be represented by convergent power series. We start with the general properties of power series.

A **power series** around (or at) point $z_0 \in \mathbb{C}$ is a series

$$\sum_{k\ge 0} a_k (z - z_0)^k = a_0 + a_1(z - z_0) + a_2(z - z_0)^2 + \cdots, \tag{4.3.1}$$

where $a_k, k \ge 0$, are complex numbers. The following result about the convergence of power series is usually attributed to Cauchy and Hadamard.

Theorem 4.3.3 *For the power series (4.3.1), define the number R by*

$$R = \frac{1}{\limsup_n |a_n|^{1/n}}, \tag{4.3.2}$$

with the convention $1/+\infty = 0$ and $1/0 = +\infty$. If $R = 0$, then (4.3.1) converges only for $z = z_0$. If $R = +\infty$, then (4.3.1) converges for all $z \in \mathbb{C}$. If $0 < R < +\infty$, then (4.3.1) converges absolutely for $|z - z_0| < R$ and diverges for $|z - z_0| > R$.

An alternative representation for R is

$$R = \limsup_n \frac{|a_n|}{|a_{n+1}|} \tag{4.3.3}$$

Proof. Define $\tilde{z} = z - z_0$. For $\tilde{z} \neq 0$, apply the root test to the resulting series (4.3.1):

$$K = \limsup_n |a_n \tilde{z}^n|^{1/n} = |\tilde{z}| \limsup_n |a_n|^{1/n} = |\tilde{z}|/R.$$

If $R = 0$, then $K = +\infty$ for all $\tilde{z} \neq 0$, which means that the series diverges for $\tilde{z} \neq 0$. If $R = +\infty$, then $K = 0$ for all $\tilde{z}$, which means that the series converges for all $\tilde{z}$. If $0 < R < +\infty$, then $K < 1$ for $|\tilde{z}| < R$ and $K > 1$ for $|\tilde{z}| > R$, so that the result again follows from the root test.

Representation (4.3.3) follows in the same way from the ratio test. □

EXERCISE 4.3.5.^{C} *Verify representation (4.3.3).*

Definition 4.3 The number R introduced in Theorem 4.3.3 is called the **radius of convergence** of the power series (4.3.1), and the set $\{z : |z - z_0| < R\}$ is called the **disk of convergence**.

EXERCISE 4.3.6. ^{C} *(a) Verify that the power series* $\sum_{k \geq 0} a_k z^k$ *and* $\sum_{k \geq 0} k a_k z^k$ *have the same radius of convergence. (b) Give an example of a power series that converges at one point on the boundary of the disk of convergence, and diverges at another point. Explain why the convergence in this example is necessarily conditional (in other words, explain why the absolute convergence at one point on the boundary implies absolute convergence at all points of the boundary).*

In practice, it is more convenient to compute the radius of convergence of a given series by directly applying the ratio test or the root test, rather than by formulas (4.3.2) or (4.3.3). FOR EXAMPLE, to find the radius of convergence of the power series

$$\sum_{n \geq 1} \frac{(-1)^n z^{3n+1} n!}{(2n)^n} \tag{4.3.4}$$

we write $c_n = \frac{(-1)^n z^{3n+1} n!}{(2n)^n}$ and

$$\frac{|c_{n+1}|}{|c_n|} = \frac{(n+1)(2n)^n |z|^3}{(2(n+1))^{n+1}} = \frac{|z|^3}{2}(1 + n^{-1})^{-n} \to |z|^3/(2e), \text{ as } n \to \infty,$$

where we used $\lim_{n \to \infty}(1 + n^{-1})^n = e$. The ratio test guarantees convergence if $|z|^3/(2e) < 1$ or $|z| < \sqrt[3]{2e}$. Therefore, the radius of convergence is $\sqrt[3]{2e}$. Direct application of either (4.3.2) or (4.3.3) is difficult because many of the coefficients a_n in the series (4.3.4) are equal to zero.

EXERCISE 4.3.7.^{C} *Find the radius of convergence of the power series*

$$\sum_{k\geq 1}(5+(-1)^n)^n \frac{z^{2n+1}(4n)!}{(3n)^n}.$$

Let us now state and prove the main result of this section.

Theorem 4.3.4 *A function $f = f(z)$ is analytic at the point z_0 if and only if there exists a power series $\sum_{k\geq 0} a_k(z-z_0)^k$ with some radius of convergence $R > 0$ so that $f(z) = \sum_{k\geq 0} a_k(z-z_0)^k$ for all $|z-z_0| < R$.*

Proof. *Step 1.* Let us show that an analytic function can be written as a convergent power series. Recall that, by definition, $f = f(z)$ is analytic at z_0 if and only if f has a derivative in some neighborhood of z_0. Therefore, there exists an $R > 0$ so that f is analytic in the open disk $G_R = \{z : |z-z_0| < R\}$. Take $z \in G_R$ and find a number ρ so that $|z-z_0| < \rho < R$. Then the circle C_ρ with center at z_0 and radius ρ encloses z and lies entirely in G_R (draw a picture). By the Cauchy Integral Formula,

$$f(z) = \frac{1}{2\pi i}\oint_{C_\rho} \frac{f(\zeta)}{\zeta - z}\, d\zeta. \tag{4.3.5}$$

Next, we note that $|z-z_0| < |\zeta - z_0| = \rho$ for $\zeta \in C_\rho$ and use the formula for the geometric series to write

$$\frac{1}{\zeta - z} = \frac{1}{(\zeta - z_0)\left(1 - \frac{z-z_0}{\zeta - z_0}\right)} = \sum_{k\geq 0}\frac{(z-z_0)^k}{(\zeta - z_0)^{k+1}}. \tag{4.3.6}$$

We now plug the result into (4.3.5) and find

$$f(z) = \frac{1}{2\pi i}\oint_{C_\rho}\left(\sum_{k\geq 0}\frac{f(\zeta)}{(\zeta - z_0)^{k+1}}(z-z_0)^k\right) d\zeta. \tag{4.3.7}$$

Finally, let us assume for the moment that we can integrate term-by-term in (4.3.7), that is, first do the integration and then, summation; this is certainly true for sums with a finite number of terms, but must be justified for (4.3.7), and we will do the justification later. Then the term-by-term integration results in the equality

$$f(z) = \sum_{k\geq 0} a_k(z-z_0)^k, \tag{4.3.8}$$

where

$$a_k = \frac{1}{2\pi i}\oint_{C_\rho} \frac{f(\zeta)}{(\zeta - z_0)^{k+1}}\, d\zeta. \tag{4.3.9}$$

By (4.3.6), the radius of convergence of the power series in (4.3.8) is at least ρ. This shows that an analytic function can be represented by a convergent power series.

Step 2. Let us show that a convergent power series defines a continuous function. Define $g(z) = \sum_{k\geq 0} a_k(z - z_0)^k$, $|z - z_0| < R$. We will show that the function g is continuous for $|z - z_0| < R$, that is, for every z_1 in the disk of convergence and for every $\varepsilon > 0$, there exits a $\delta > 0$ so that $|g(z_1) - g(z)| < \varepsilon$ as long as $|z_1 - z| < \delta$ and $|z - z_0| < R$; the interested reader can then use similar arguments to prove that g is analytic. To prove the continuity of g, fix a z_1 with $|z_1 - z_0| < R$ and an $\varepsilon > 0$. Next, let us define $r = (R + |z_1 - z_0|)/2$ and find N so that $\sum_{k\geq N+1} |a_k|\, r^k < \varepsilon/4$. Such an N exists because $r < R$ and the power series converges absolutely inside the disk of convergence. Now consider $g_N(z) = \sum_{k=1}^{N} a_k z^k$. Being a polynomial, $g_N(z)$ is a continuous function at z_1, and therefore there exists a $\delta_1 > 0$ so that $|g_N(z_1) - g(z)| < \varepsilon/2$ as long as $|z - z_1| < \delta_1$. Finally, define δ as the minimum of δ_1 and $(R - |z_1 - z_0|)/4$. By construction, if $|z_1 - z| < \delta$, then $|z_0 - z| < r$ (draw a picture). Now let us look at the value of $|g(z_1) - g(z)|$ for $|z - z_1| < \delta$. By the triangle inequality,

$$|g(z_1) - g(z)| \leq |g_N(z_1) - g_N(z)| + \sum_{k\geq N+1} |a_k|\, |z_1 - z_0|^k + \sum_{k\geq N+1} |a_k|\, |z - z_0|^k; \tag{4.3.10}$$

by the choice of N and δ, $|g_N(z_1) - g_N(z)| < \varepsilon/2$, and, by the choice of r,

$$\sum_{k\geq N+1} |a_k|\, |z_1 - z_0|^k + \sum_{k\geq N+1} |a_k|\, |z - z_0|^k \leq 2 \sum_{k\geq N+1} |a_k|\, r^k < \varepsilon/2.$$

As a result, $|g(z_1) - g(z)| < \varepsilon$, which proves the continuity of g.

The following exercise completes the proof of the theorem. □

EXERCISE 4.3.8.^{A} *(a) Use the same arguments as in Step 2 above to show that* $g(z) = \sum_{k\geq 0} a_k(z - z_0)^k$ *is differentiable for* $|z - z_0| < R$ *and* $g'(z) = \sum_{k\geq 0} k a_k (z - z_0)^{k-1}$. *Thus, g is indeed an analytic function for* $|z - z_0| < R$. *Hints: (i) you can differentiate the polynomials; (ii) by Exercise 4.3.6, the power series* $\sum_{k\geq 0} k a_k z^{k-1}$ *has the radius of convergence equal to R. (b) Use the same arguments to justify the switch of summation and integration in (4.3.7). Hints: (i) you can do this switching when the number of terms is finite. (ii)*

The value of $|z - z_0|/|z - \zeta|$ *is constant for all* ζ *on* C_ρ *and is less than one, which allows you to make the sum* $\sum_{k \geq N}$ *arbitrarily small. (c) Show that the n-th derivative* $g^{(n)}$ *of* $g(z) = \sum_{k \geq 0} a_k (z - z_0)^k$ *exists inside the disk of convergence and satisfies* $g^{(n)}(z_0) = n! a_n$. *Then use (4.3.8) and (4.3.9) to give the rigorous proof of (4.2.16) on page 201. Hint: part (a) of this exercise shows that you can differentiate the power series term-by-term as many times as you want.*

Corollary 4.1 **Uniqueness of power series.** *(a) Assume that the two power series* $\sum_{k=0}^{\infty} a_k (z - z_0)^k$, $\sum_{k=0}^{\infty} b_k (z - z_0)^k$ *converge in the some neighborhood of* z_0 *and* $\sum_{k=0}^{\infty} a_k (z - z_0)^k = \sum_{k=0}^{\infty} b_k (z - z_0)^k$ *for all* z *in that neighborhood. Then* $a_k = b_k$ *for all* $k \geq 0$.

(b) If f *is analytic at* z_0, *then*

$$f(z) = f(z_0) + \sum_{k=1}^{\infty} \frac{f^{(k)}(z_0)}{n!} (z - z_0)^k \tag{4.3.11}$$

for all z *in some neighborhood of* z_0. *Therefore, if a function has a power series representation at a point* z_0, *this power series is necessarily (4.3.11).*

EXERCISE 4.3.9.[A] *Prove both parts of Corollary 4.1.*

As in the ordinary calculus, the series in (4.3.11) is called the **Taylor series** for f at z_0; when $z_0 = 0$, we also call it the **Maclaurin series**. It appears, though, that the original idea to represent real functions by a power series belongs to neither Taylor nor Maclaurin and can be traced back to Newton. The English mathematician BROOK TAYLOR (1685–1731) and the Scottish mathematician COLIN MACLAURIN (1698–1746) were the first to make this idea clear enough and spread it around, and thus got their names attached to this power series representation, even in the complex domain.

The methods for finding the power series expansion of a particular complex function are the same as for the real functions. With these methods, you never compute the derivatives of the function. One of the key relations is the sum of the geometric series:

$$\frac{1}{1-z} = \sum_{k \geq 0} z^k, \; |z| < 1. \tag{4.3.12}$$

FOR EXAMPLE, let us find the Taylor series for $f(z) = 1/z^2$ at point $z_0 = 1$. We have $f(z) = -g'(z)$, where $g(z) = 1/z$. Now, $1/z = 1/(1 + (z - 1)) = \sum_{k \geq 0} (-1)^k (z - 1)^k$, where we used (4.3.12) with $-(z - 1)$ instead of z. Differentiating term-by-term the power series for g gives $f(z) = \sum_{k\geq 1}(-1)^{k+1}k(z-1)^{k-1} = \sum_{k\geq 0}(-1)^k(k+1)(z-1)^k$, where in the last equality we changed the summation index to start from zero. The series converges for $|z-1| < 1$.

EXERCISE 4.3.10. ^{C} *Find the power series expansions of the following functions at the given points, and find the radius of convergence: (a) $f(z) = z^2+z+1$ at $z_0 = i$. Hint: put $z = i+(z-i)$ and simplify. Alternatively, put $w = z-i$ so that $z = w+i$, and find the expansion of the resulting function in powers of w; then replace w with $(z-i)$. (b) $f(z) = 1/(z^2+2z+2)$ at $z_0 = 0$ Hint: use partial fractions: $z^2+2z+2 = (z+1)^2+1 = (z+1-i)(z+1+i)$, $f(z) = A/(z+1-i) + B/(z+1+i)$.*

Given an analytic function, you usually do not need the explicit form of its Taylor series to find the `radius of convergence` of the series. Indeed, by Theorem 4.3.4, the radius must be the distance from z_0 to the closest point at which f is not analytic (think about it...). FOR EXAMPLE, the function $f(z) = 1/(z^2+2z+2)$ is not analytic only at points $z = 1 \pm i$, and both points are $\sqrt{2}$ away from the origin. As a result, the Maclaurin series for f has the radius of convergence equal to $\sqrt{2}$.

EXERCISE 4.3.11. ^{C} *Consider the function $f(z) = \frac{1}{(z+2)(z-5)}$ at the point $z_0 = 1+2i$. (a) Without computing the series expansion, find the radius of convergence of the Taylor series of f at z_0. (b) Find the series expansion and verify that it has the same radius of convergence as you computed in part (a).*

4.3.3 *The Exponential Function*

For $z \in \mathbb{C}$, we *define*

$$e^z = \sum_{k\geq 0}\frac{z^k}{k!}, \ \sin z = \sum_{k\geq 0}\frac{(-1)^k z^{2k+1}}{(2k+1)!}, \ \cos z = \sum_{k\geq 0}\frac{(-1)^k z^{2k}}{(2k)!}. \qquad (4.3.13)$$

EXERCISE 4.3.12. *(a)^C Verify that all three series in (4.3.13) converge for every $z \in \mathbb{C}$. (b)^B Verify the two main properties of the exponential function:*

$$e^{z_1+z_2} = e^{z_1}e^{z_2}, \ (e^z)' = e^z,$$

`Euler's formula`

$$e^{i\theta} = \cos\theta + i\sin\theta, \tag{4.3.14}$$

and the relations

$$\sin z = \frac{e^{iz} - e^{-iz}}{2i}, \ \cos z = \frac{e^{iz} + e^{-iz}}{2}. \tag{4.3.15}$$

We use the Euler formula to evaluate the complex exponentials: it follows from (4.3.14) that, for $z = x + iy$,

$$e^z = e^x(\cos y + i\sin y), \ |e^z| = e^x. \tag{4.3.16}$$

Relation (4.3.16) shows that, as a function of complex variable, e^z can take *all* complex values except zero and is a periodic function with period $2\pi i$. Similarly, (4.3.15) implies that $\sin z$ and $\cos z$ can take *all* complex values. In particular, the familiar inequalities $e^x > 0$, $|\sin x| \leq 1$, $|\cos x| \leq 1$ that are true for real x, no longer hold in the complex domain.

EXERCISE 4.3.13.^{B} *(a) Verify that the mapping defined by the exponential function $f(z) = e^z$ is conformal everywhere in the complex plane. (b)Find the image of the set $\{z : \Re z > 0, \ 0 < \Im z < \pi\}$ under this mapping.*

Two other related functions are the `hyperbolic sine` sinh and `hyperbolic cosine` cosh:

$$\sinh z = \frac{e^z - e^{-z}}{2}, \ \cosh z = \frac{e^z + e^{-z}}{2}. \tag{4.3.17}$$

EXERCISE 4.3.14.^{B} *(a) Verify that*

$$\sinh z = \sum_{k\geq 0} \frac{z^{2k+1}}{(2k+1)!}, \ \cosh z = \sum_{k\geq 0} \frac{z^{2k}}{(2k)!}.$$

(b) Verify that

$$\sin(x+iy) = \sin x \cosh y + i\cos x \sinh y, \quad \sinh(z) = -i\sin(iz),$$
$$\cos(x+iy) = \cos x \cosh y - i\sin x \sinh y, \quad \cosh(z) = \cos(iz).$$

The `natural logarithm` $\ln z$ of $z \neq 0$ is defined as the number whose exponential is equal to z. If $a = \ln z$, then, because of (4.3.14), $a+2\pi i$ is also a natural logarithm of z. In other words, $\ln z$ has infinitely many values, and, since $z = |z|e^{i\arg(z)}$, we have $\ln z = \ln|z| + i\arg(z)$. The `principal value` of the natural logarithm is defined by $\operatorname{Ln} z = \ln|z| + i\operatorname{Arg}(z)$, which is specified by the condition $\text{Ln}1 = 0$. More generally, by fixing the value of $\ln z$ at one point $z_0 \neq 0$, we define a `branch` of the natural logarithm. For example, condition $\ln 1 = 2\pi i$ results in a different branch $f = f(z)$ of $\ln z$, such that $f(i) = (5\pi/2)i$. The values of different branches at the same point vary by an integer multiple of $2\pi i$.

EXERCISE 4.3.15.[C] *(a) Use the Cauchy-Riemann equations in polar coordinates to verify that the derivative of* $\text{Ln}\, z$ *(and, in fact, of every branch of the natural logarithm) is* $1/z$. *(b) Verify that* $-\text{Ln}(1-z) = \sum_{k\geq 1} z^k/k$. *(If everything else fails, just integrate term-by-term the expansion of* $1/(1-z)$.*)*

Using the exponential and the natural logarithm, we define the `complex power` of a complex number:

$$z^w = e^{w \ln z}. \tag{4.3.18}$$

While the natural logarithm $\ln z$ has infinitely many different values, the power can have one, finitely many, or infinitely many values, depending on w. Quite surprisingly, a complex power of a complex number can be real:

$$i^i = e^{i(\pi i/2 + 2\pi k i)} = e^{-\pi/2 - 2\pi k},\ k = 0, \pm 1, \pm 2, \ldots,$$

which was first noticed by Euler in 1746. By selecting a branch of the natural logarithm, we select the corresponding branch of the complex power. For example $i^i = e^{-\pi/2}$ corresponds to the principal value of the natural logarithm. What is the value of i^i if we take $\ln 1 = -2\pi i$?

EXERCISE 4.3.16. *(a)*[C] *Verify that if* m *is an integer, and* $w = 1/m$, *then the above definition of* z^w *is consistent with (4.1.5), page 185. (b)*[B] *For what complex numbers* w *does the expression* z^w *have finitely many different values? Hint: recall that the exponential function is periodic with period* $2\pi i$. *(c)*[A] *Assume that in (4.3.18) we take the principal value of the natural logarithm. Verify that, for every complex number* w, *the corresponding function* $f(z) = z^w$ *is analytic at every point* $z \neq 0$, *and* $f'(z) = wz^{w-1}$. *Convince yourself that this is true for every branch of the complex power.*

4.4 Singularities of Complex Functions

4.4.1 *Laurent Series*

We know that a function analytic at a point z_0 can be written as a Taylor series that converges to the values of the function in some neighborhood of z_0. The proof of this series representation essentially relies on the fact that a neighborhood of a point is a simply connected set. It turns out that a somewhat similar expansion exists in domains that are not simply connected. This expansion is known as the *Laurent series*, the study of which is the main goal of this section. The French mathematician PIERRE ALPHONSE LAURENT (1813–1854) published the result in 1843.

Before we state the result, let us make a simple yet important observation that will be an essential part of many computations to follow.

Recall that the geometric series formula

$$\frac{1}{1-z} = \sum_{k=0}^{\infty} z^k \tag{4.4.1}$$

is true for $|z| < 1$. On the other hand, writing

$$\frac{1}{1-z} = -\frac{1}{z\left(1-\frac{1}{z}\right)}$$

and replacing z with $1/z$ in (4.4.1), we can write

$$\frac{1}{1-z} = -\sum_{k=0}^{\infty} \frac{1}{z^{k+1}},$$

which is now true for $|z| > 1$. Therefore, we have two representations for $1/(1-z)$, one, for small $|z|$, and the other, for large.

We will now state and prove the main result of this section. Recall that the Taylor series is written in an open disk $\{z : |z - z_0| < R\}$, which is the basic example of a simply connected domain; we allow $R = \infty$ to include the whole plane. Similarly, the basic domain that is not simply connected is an **annulus**, a set of the type $\{z : R_1 < |z - z_0| < R\}$, where z_0 is a fixed complex number and, for consistency, we allow $R_1 = 0$ and/or $R = \infty$; the Latin words *ānulus* and *ānus* mean "a ring". The Laurent series expansion is written in an annulus.

Theorem 4.4.1 Laurent series expansion. *If a function $f = f(z)$ is analytic in the annulus $G = \{z : R_1 < |z - z_0| < R\}$, then, for all $z \in G$,*

$$\begin{aligned} f(z) &= \sum_{k=-\infty}^{\infty} c_k (z - z_0)^k \\ &= \ldots + \frac{c_{-2}}{(z-z_0)^2} + \frac{c_{-1}}{(z-z_0)} + c_0 + c_1(z-z_0) + c_2(z-z_0)^2 + \ldots, \end{aligned} \tag{4.4.2}$$

where

$$c_k = \frac{1}{2\pi i} \oint_{C_\rho} \frac{f(\zeta)}{(\zeta - z_0)^{k+1}} d\zeta, \quad k = 0, \pm 1, \pm 2, \ldots, \tag{4.4.3}$$

and C_ρ is a circle with center at z_0 and radius ρ so that $R_1 < \rho < |z - z_0|$; the circle is oriented counterclockwise. Representation (4.4.2) is `unique`*: if $f(z) = \sum_{k=-\infty}^{\infty} c_k(z - z_0)^k = \sum_{k=-\infty}^{\infty} a_k(z - z_0)^k$ in G, then $a_k = c_k$ for all k.*

Proof. We present the main steps of the proof. The details are in the exercise below.

Step 1. Fix the point z. Let C_r be a circle with center z_0 and radius r so that $\rho < r < R$ and $|z - z_0| < r$ (draw a picture). Combining the proof of Theorem 4.2.4, page 198, with the Integral Formula of Cauchy, we conclude that

$$f(z) = \frac{1}{2\pi i} \oint_{C_r} \frac{f(\zeta)}{\zeta - z} d\zeta - \frac{1}{2\pi i} \oint_{C_\rho} \frac{f(\zeta)}{\zeta - z} d\zeta, \tag{4.4.4}$$

with both C_r and C_ρ oriented counterclockwise.

Step 2. We already know that

$$\frac{1}{2\pi i} \oint_{C_r} \frac{f(\zeta)}{\zeta - z} d\zeta = \sum_{k=0}^{\infty} c_k(z - z_0)^k, \; c_k = \frac{1}{2\pi i} \oint_{C_r} \frac{f(\zeta)}{(\zeta - z_0)^{k+1}} d\zeta; \tag{4.4.5}$$

see the proof of Theorem 4.3.4, page 210.

Step 3. By Exercise 4.2.17, page 199, we conclude that

$$\frac{1}{2\pi i} \oint_{C_r} \frac{f(\zeta)}{(\zeta - z_0)^{k+1}} d\zeta = \frac{1}{2\pi i} \oint_{C_\rho} \frac{f(\zeta)}{(\zeta - z_0)^{k+1}} d\zeta = c_k.$$

Step 4. Let us show that

$$\frac{1}{2\pi i} \oint_{C_\rho} \frac{f(\zeta)}{\zeta - z} d\zeta = \sum_{k=1}^{\infty} \frac{c_{-k}}{(z - z_0)^k}, \quad c_{-k} = \frac{1}{2\pi i} \oint_{C_\rho} f(\zeta)(\zeta - z_0)^{k-1} d\zeta. \tag{4.4.6}$$

Once again, in (4.4.6) we have *positive* integer k.

To establish (4.4.6), we write

$$-\frac{1}{\zeta - z} = \frac{1}{z - z_0 - (\zeta - z_0)} = \frac{1}{(z - z_0)\left(1 - \frac{\zeta - z_0}{z - z_0}\right)}$$

$$= \sum_{k=0}^{\infty} \frac{(\zeta - z_0)^k}{(z - z_0)^{k+1}} = \sum_{k=1}^{\infty} \frac{(\zeta - z_0)^{k-1}}{(z - z_0)^k}.$$

Then

$$-\frac{f(\zeta)}{\zeta - z} = \sum_{k=1}^{\infty} \frac{f(\zeta)(\zeta - z_0)^{k-1}}{(z - z_0)^k}, \tag{4.4.7}$$

and it remains to integrate this equality.

Step 5. We combine the results of the above steps to get both (4.4.2) and (4.4.3). Note that *any* simple, closed, piece-wise smooth curve can be used instead of C_ρ in (4.4.3), as long as the curve is completely inside G, the closed disk $\{z : |z - z_0| \leq R_1\}$ is inside the domain enclosed by the curve, and the point z is outside that domain.

Step 6. To prove the uniqueness, we multiply the equality $\sum_{k=-\infty}^{\infty} c_k(z - z_0)^k = \sum_{k=-\infty}^{\infty} a_k(z - z_0)^k$ by $(z - z_0)^{-m-1}$ for some integer m and integrate both sums term-by-term over the circle C_ρ. By (4.2.6), page 195, all integrals become zero except for those corresponding to $k = m$, and we get $c_m = a_m$. □

EXERCISE 4.4.1. ^{A} *(a) Verify (4.4.4). Hint: connect the circles C_r and C_ρ with a line segment that does not pass through the point z and write the Integral Theorem of Cauchy in the resulting simply connected domain. Then simplify the result, keeping in mind that you integrate along the line segment twice, but in opposite directions, and that the orientation of C_ρ is clockwise. (b) Justify the term-by-term integration in (4.4.7). Hint: note that, for $\zeta \in C_\rho$, we have $|(\zeta - z_0)/(z - z_0)| = \rho/|z - z_0| < 1$. Then use the same argument as in the proof of Theorem 4.3.4, page 210. (c) Fill in the details in the proof of the uniqueness of the expansion. Hint: once again, the key step is justifying the term-by-term integration, and once again, you use the same arguments as in the previous similar cases.*

Note that if, in the above theorem, the function is analytic for all z satisfying $|z - z_0| = R_1$, then we can decrease R_1. Similarly, we can increase R if f is analytic for all z satisfying $|z - z_0| = R$ (make sure you understand this). In other words, with no loss of generality, we will always assume that the annulus $R_1 < |z - z_0| < R$ is *maximal*, that is, each of the sets $|z - z_0| = R_1$ and $|z - z_0| = R$ (assuming $R < \infty$) contains at least one point where the function f is not analytic. There are several types of such points.

Definition 4.4 A point $z_0 \in \mathbb{C}$ is called `an isolated singular point` or `an isolated singularity` of a function f if the function f is not analytic at z_0 and there exists a $\delta > 0$ so that the function f is analytic in the region $\{z : 0 < |z - z_0| < \delta\}$.

An isolated singular point z_0 is called

- `removable`, if the function can be defined at z_0 so that the result is an analytic function at z_0.
- `a pole of order` k, if k is the smallest value of the positive integer power n with the property that function $(z - z_0)^n f(z)$ has a removable singularity at z_0.
- `an essential singularity`, if it is neither a removable singularity nor a pole.

A pole of order 1 is called `simple`.

Without going into the details, let us mention that the point $z = 0$ is *not* an isolated singularity of $f(z) = z^{1/2}$ in the sense of the above definition, but rather a `branching point` of order two. The reason is that the square root $\sqrt{z}$ has two different values in every neighborhood of $z = 0$. As a result, in any neighborhood of zero, there is no *unique* number assigned to $z^{1/2}$ and $f(z) = z^{1/2}$ is *not a function* in the sense of our definition. Similarly, $z = 0$ is a branching point of order three for the function $f(z) = z^{-1/3}$ (the fact that f is unbounded near $z = 0$ is not as important as the three different values of $\sqrt[3]{z}$ in every neighborhood of $z = 0$), and $z = 0$ is a branching point of infinite order for the function $f(z) = \ln z$. The study of branching points and the related topics (multi-valued analytic functions, Riemann surfaces, etc.) is beyond the scope of our discussions.

Note that the closed disk $\{z : |z - z_0| \leq R_1\}$ can contain several points where f is not analytic, and z_0 is not necessarily one of them. In the special case of the Laurent series with $R_1 = 0$, z_0 *is* an isolated singular point of the function f, with no other singular points in the domain $\{z : 0 < |z - z_0| < R\}$ for some $R > 0$. The corresponding Laurent series is called the `expansion of` f `at (or around) the isolated singular point` z_0. This expansion has two distinct parts: the `regular part`, consisting of the terms with non-negative k, $\sum_{k \geq 0} c_k (z - z_0)^k$, and the `principal part`, consisting of the terms with the negative values of k, $\sum_{k<0} c_k (z - z_0)^k$. As the names suggest, the regular part is a function that is analytic at z_0 (this follows from Theorem 4.3.4), and the principal part determines the type of the singularity at z_0 (this follows from the exercise below).

EXERCISE 4.4.2.^{C} *(a) Let z_0 be an isolated singular point of the function $f = f(z)$, and consider the corresponding Laurent series $\sum_{k=-\infty}^{\infty} c_k(z-z_0)^k$ converging for $0 < |z - z_0| < R$. Verify that z_0 is*
(i) a removable singularity if and only if $c_k = 0$ for all $k < 0$ Hint: this pretty much follows from Theorem 4.3.4. Equivalently, z_0 is a removable singularity if and only if there exists a $\delta > 0$ so that the function $f(z)$ is analytic and bounded for $0 < |z - z_0| < \delta$.
(ii) a pole of order N if and only $c_{-N} \neq 0$ and $c_k = 0$ for all $k \leq -(N+1)$ Hint: for the proof in one direction, multiply the Laurent series by $(z - z_0)^N$; for the proof in another direction, multiply f by $(z - z_0)^N$ and use Theorem 4.3.4.
(iii) an essential singularity if and only if $c_k \neq 0$ for infinitely many $k < 0$. Hint: by definition, the essential singularity is the only remaining option.

(b) Verify that the point $z_0 = 0$ is
(i) a removable singularity for the functions $f(z) = \sin z/z$, $f(z) = (e^z - 1)^2/(1 - \cos z)$ and $f(z) = (z\cos z - \sin z)/(z \sin z)$;
(ii) a second-order pole for the function $f(z) = (1 - \cos z)/(e^z - 1)^4$;
(iii) an essential singularity for the function $f(z) = e^{1/z}$;
(iv) not an isolated singularity for the function $f(z) = 1/\sin(1/z)$.

As with the Taylor series, we usually do not use the formula (4.4.3) to find the coefficients of the Laurent series, and use other methods instead. FOR EXAMPLE, consider the function

$$f(z) = \sin\left(\frac{z}{z-1}\right).$$

This function is analytic everywhere except at the point $z_0 = 1$. Let us find the Laurent series for f at z_0. We have $z/(z-1) = (z-1+1)/(z-1) = 1 + (z-1)^{-1}$ and so

$$f(z) = \sin(1 + (z-1)^{-1}) = \sin 1\, \cos((z-1)^{-1}) + \sin((z-1)^{-1})\cos 1,$$

where we use the formula for the sine of the sum. Then we use the standard Taylor expansions for the sine and cosine to conclude that

$$f(z) = \sin 1 \sum_{k\geq 0} \frac{(-1)^k}{(z-1)^{2k}(2k)!} + \cos 1 \sum_{k\geq 0} \frac{(-1)^k}{(z-1)^{2k+1}(2k+1)!};$$

one could write this as a single series, but it will not add anything essential to the final answer. We therefore conclude that $z_0 = 1$ is an essential singularity of f.

EXERCISE 4.4.3.[C] *Find the Laurent series for the function*

$$f(z) = \cos\left(\frac{2z+5}{z+2}\right)$$

around the point $z_0 = -2$. *What is the type of the singularity of* f *at* z_0*?*

When writing the Laurent series expansion, one should pay attention to the domain in which this expansion should hold. Recall that the expansion is written in an annulus $\{z : R_1 < |z - z_0| < R\}$, where z_0 is a fixed complex number, and none of the singular points of f should be inside this annulus. The expansion in such an annulus has the form $\sum_{k=-\infty}^{\infty} c_k(z-z_0)^k$. *The point* z_0 *does not have to be a singular point of* f*, but at least one singular point of* f *must be in the set* $\{z : |z - z_0| = R_1\}$*, and, unless* R *is infinite, at least one singular point of* f *must be in the set* $\{z : |z - z_0| = R\}$ (otherwise, we are able to expand the annulus). The Laurent series that converges in the disk $\{z : |z - z_0| < R\}$ is the same as the Taylor series at z_0. As a result, *the same function can have different expansions in different domains*, even when the point z_0 is the same: recall that $(1-z)^{-1} = \sum_{k\geq 0} z^k$ for $|z| < 1$ and $(1-z)^{-1} = \sum_{k\geq 0} z^{-k-1}$ for $|z| > 1$.

AS A DIFFERENT EXAMPLE, consider the function

$$f(z) = \frac{1}{2z - z^2} = \frac{1}{2z} + \frac{1}{2(2-z)}. \tag{4.4.8}$$

The Laurent series of this function at $z_0 = 0$ is $f(z) = (2z)^{-1} + \sum_{k\geq 0} 2^{-k-2} z^k$ (check it), and z_0 is a simple pole (a pole of order one). The expansion is true for $0 < |z| < 2$. Similarly, the Laurent series in the domain $\{z : 2 < |z-2|\}$ is computed as follows: $1/z = (z-2+2)^{-1} = (z-2)^{-1}(1 + 2/(z-2))^{-1}$, and so

$$f(z) = -\frac{1}{2(z-2)} + \sum_{k\geq 0} \frac{2^k}{(z-2)^{k+1}} = \frac{1}{2(z-2)} + \sum_{k\geq 1} \frac{2^k}{(z-2)^{k+1}}. \tag{4.4.9}$$

EXERCISE 4.4.4.[C] *Find the Laurent series of the function* f *from (4.4.8) in the domain* $\{z : 0 < |z-2| < 2\}$.

It follows from (4.4.8) that $z_0 = 2$ is a pole of order one; this is also what you should conclude from the previous exercise. On the other hand, the expansion in (4.4.9) contains infinitely many negative powers of $(z-2)$. Should we conclude from (4.4.9) that $z_0 = 2$ is an essential singularity of f, and how do we reconcile these seemingly contradictory conclusions? After looking more closely at (4.4.8), we realize that (4.4.8) is *not* a Laurent series of f at $z_0 = 2$; the Laurent series at the point $z_0 = 2$ must converge when $0 < |z - 2| < R$ for some R, and that is the series you compute in Exercise 4.4.4.

Finally, let us emphasize once again that a Laurent series does *not* have to be around a singular point. We go back to the equality $(1 - z)^{-1} = \sum_{k \geq 0} z^{-1-k}$, which is true for $|z| > 1$. Even though we have infinitely many negative powers of z, zero is not an essential singularity of $f(z) = 1/(1-z)$, because, for $f(z) = 1/(1 - z)$, the point $z_0 = 0$ is not a singularity at all!

EXERCISE 4.4.5.^{C} *Find the expansion of the function from (4.4.8) in the domain* $\{z : 1 < |z + 1| < 3\}$. *Hint: use geometric series; your answer should be* $\sum_{k=-\infty}^{\infty} a_k (z + 1)^k$ *for suitable numbers* a_k.

EXERCISE 4.4.6.^{A} *We say that the point* $z = \infty$ *is an isolated singular point of the function* $f = f(z)$ *if and only if the point* $z = 0$ *is an isolated singular point of the function* $h(z) = f(1/z)$. *Show that* $z = \infty$ *is (a) a removable singularity of a rational function* $P(z)/Q(z)$ *if the degree of* P *is less than or equal to the degree of* Q*; (b) A pole of order* n *for a polynomial of degree* n*; (c) an essential singularity for* $f(z) = \sin z$.

4.4.2 *Residue Integration*

Let z_0 be an isolated singular point of the function $f = f(z)$ and let

$$f(z) = \sum_{k=-\infty}^{\infty} c_k (z - z_0)^k \tag{4.4.10}$$

be the corresponding Laurent series expansion of f at z_0. The coefficient c_{-1} in this expansion is called the **residue** of the function $f = f(z)$ at the point z_0 and is denoted by $\operatorname{Res}_{z=z_0} f(z)$. As we saw in the previous section, there could be several different expansions of f in powers of $(z - z_0)$; the expansion we use in (4.4.10) is *around* the point z_0 and must converge when $0 < |z - z_0| < R$ for some R.

The Latin word *residuus* means "left behind," and we will see next that c_{-1} is the only coefficient in the expansion (4.4.10) that contributes to the integral of f along a closed curve around z_0.

Recall that

$$\oint_{C(z_0)} \frac{dz}{(z - z_0)^n} = \begin{cases} 2\pi i, & n = 1, \\ 0, & n = 0, -1, \pm 2, \pm 3, \dots, \end{cases} \tag{4.4.11}$$

where $C(z_0)$ is a simple, closed, piece-wise smooth curve enclosing the point z_0 and oriented counterclockwise; see Exercises 4.2.13 and 4.2.17. Term-by-term integration of (4.4.10) over the curve $C(z_0)$ then results in

$$\oint_{C(z_0)} f(z)dz = 2\pi i c_{-1} = 2\pi i \operatorname*{Res}_{z=z_0} f(z). \tag{4.4.12}$$

Note also that we get (4.4.12) after setting $k = -1$ in formula (4.4.3) for the coefficients c_k (remember that, for our integration purposes, the curve $C(z_0)$ can be replaced with a circle centered at z_0). Consequently, c_{-1} is the only coefficient in the Laurent series expansion contributing to the integral over a closed curve around the singular point.

The more general result is as follows.

Theorem 4.4.2 *Let G be a simply connected domain. Assume that the function $f = f(z)$ is analytic at all points in G* except *finitely many points $z_1, \ldots, z_n$. If a simple, closed, piece-wise smooth curve C in G encloses the points $z_1, \ldots, z_n$ and is oriented counterclockwise, then*

$$\oint_C f(z)dz = 2\pi i \sum_{k=1}^{n} \operatorname*{Res}_{z=z_k} f(z). \tag{4.4.13}$$

EXERCISE 4.4.7. *(a)^B Prove the above theorem. Hint: surround each z_k with a small circle that stays inside the domain bounded by C, then apply Theorem 4.2.4, page 198, in the (not simply connected) domain bounded by C and the n circles around $z_1, \ldots, z_n$. (b)^C Explain why both the Cauchy Integral Formula (4.2.11), page 199, and formula (4.2.16), page 201, are particular cases of (4.4.13). Hint: for (4.2.16), write the Taylor expansion and integrate term-by-term.*

The idea of the residue integration is to find the residues of a function without computing any integrals, and then use the above theorem to evaluate the integrals of the function over different closed curves. Let us mention that if the curve does not enclose any singular points of the function, then the integral along the curve is zero by the Integral Theorem of Cauchy. If the curve passes through a singular point, then, in general, the integral along such a curve is not defined (remember that, to define the integral, we require the function to be continuous at all points on the curve).

COMPUTING THE RESIDUES. The residue is a certain coefficient in the Laurent series expansion at an isolated singular point, and there are three types of isolated singular points: removable singularity, pole, and essential singularity. Computation of the residue at a removable singularity or a pole does not require the explicit knowledge of the Laurent series expansion.

If z_0 is a removable singularity, then, according to Exercise 4.4.2, $c_{-1} = 0$ and $\operatorname*{Res}_{z=z_0} f(z) = 0$.

If z_0 is a simple pole of f, then

$$\operatorname*{Res}_{z=z_0} f(z) = \lim_{z=z_0} \big((z-z_0)f(z)\big), \tag{4.4.14}$$

because the Laurent series around a simple pole is

$$f(z) = \frac{c_{-1}}{z-z_0} + \sum_{k=0}^{\infty} c_n(z-z_0)^k.$$

An immediate consequence of (4.4.14) is the following result.

EXERCISE 4.4.8.[C] *Assume that* $f(z) = h(z)/g(z)$, *where* h, g *are analytic at* z_0, $h(z_0) \neq 0$, $g(z_0) = 0$, $g'(z_0) \neq 0$. *Show that*

$$\operatorname*{Res}_{z=z_0} f(z) = \frac{h(z_0)}{g'(z_0)}. \tag{4.4.15}$$

Hint: use (4.4.14) and note that $g'(z_0) = \lim_{z\to z_0}(g(z)-g(z_0))/(z-z_0) = \lim_{z\to z_0} g(z)/(z-z_0)$.

Usually, formula (4.4.15) is easier to use than (4.4.14). FOR EXAMPLE, if $f(z) = (z+5)/(z^3-z)$, then

$$\operatorname*{Res}_{z=-1} f(z) = \left.\frac{z+5}{3z^2-1}\right|_{z=1} = 6/2 = 3.$$

If z_0 is a pole of order $N > 1$, then the function $(z-z_0)^N f(z)$ has a removable singularity at z_0 and

$$(z-z_0)^N f(z) = \sum_{k=-N}^{\infty} c_k(z-z_0)^{k+N} = c_{-N} + c_{-N+1}(z-z_0)$$
$$+ \ldots + c_{-2}(z-z_0)^{N-2} + c_{-1}(z-z_0)^{N-1} + \ldots$$

We differentiate this equality $N-1$ times with respect to z, so that all the terms on the right that are before $c_{-1}(z-z_0)^{N-1}$ disappear, while the term $c_{-1}(z-z_0)^{N-1}$ becomes $(N-1)!\,c_{-1}$. We then set $z = z_0$ so that all the terms after $(N-1)!\,c_{-1}$ disappear as well. As a result,

$$c_{-1} = \operatorname*{Res}_{z=z_0} f(z) = \frac{1}{(N-1)!} \lim_{z \to z_0} \left(\frac{d^{N-1}}{dz^{N-1}} \left((z - z_0)^N f(z) \right) \right). \qquad (4.4.16)$$

In particular, if $f(z) = g(z)(z - z_0)^{-N}$, where the function g is analytic at z_0 and $g(z_0) \neq 0$, then

$$\operatorname*{Res}_{z=z_0} f(z) = \frac{1}{(N-1)!}\, g^{(N-1)}(z_0). \qquad (4.4.17)$$

For example, if $f(z) = (z^6 + 3z^4 + 2z - 1)/(z-1)^3$, then $N = 3$ and

$$\operatorname*{Res}_{z=1} f(z) = \frac{1}{2}(6 \cdot 5z^4 + 3 \cdot 4 \cdot 3z^2)|_{z=1} = 33.$$

If $z = z_0$ is an essential singularity, then the computation of the residue usually requires the computation of the corresponding terms in the Laurent series. For example, if $f(z) = z^2 \sin(1/z)$, then $f(z) = z^2\big(1/z - 1/(6z^3) + 1/(120z^5) - \ldots\big)$ and so $\operatorname*{Res}_{z=0} f(z) = -1/6$.

Exercise 4.4.9.^{C} *Taking $C = \{z : |z - 5i| = 5\}$, verify that*

$$\oint_C \frac{z}{e^z + 1}\, dz = 8\pi^2.$$

Hint: how many poles are enclosed by C?

One of the most elegant applications of residue integration is evaluation of *real* integrals. In what follows, we describe several classes of real integrals that can be evaluated using residues.

The easiest class is integrals involving rational expressions (sums, differences, products, and ratios) of $\sin\varphi$ and $\cos\varphi$, integrated from 0 to 2π:

$$\int_0^{2\pi} H(\cos\varphi, \sin\varphi)\, d\varphi.$$

Such integrals are immediately reduced to complex integrals over the unit circle. Indeed, writing $z = e^{i\varphi}$, we have z going around the unit circle counterclockwise; $dz = ie^{i\varphi} d\varphi$, so that $d\varphi = dz/(iz)$; also, by the Euler formula, $\cos\varphi = (z + z^{-1})/2$, $\sin\varphi = (z - z^{-1})/(2i)$. As a result,

$$\int_0^{2\pi} H(\cos\varphi, \sin\varphi)\, d\varphi = \int_{|z|=1} H\left(\frac{z^2+1}{2z}, \frac{z^2-1}{2iz} \right) \frac{dz}{iz},$$

assuming that the integral on the left is well-defined. Then integration is reduced to computing the residues inside the unit disk for the modified integrand; the computations can be rather brutal, even for seemingly simple functions H.

FOR EXAMPLE, let us evaluate the integral

$$I = \int_0^{2\pi} \frac{2+\cos\varphi}{4+3\sin\varphi}\, d\varphi.$$

Step 1. We write $z = e^{i\varphi}$, $\cos\varphi = (z^2+1)/(2z)$, $\sin\varphi = (z^2-1)/(2iz)$, $d\varphi = dz/(iz)$, and so

$$I = \oint_{|z|=1} \frac{z^2+4z+1}{z(3z^2+8iz-3)}\, dz.$$

Step 2. The function $f(z) = (z^2+4z+1)/(z(3z^2+8iz-3))$ we are integrating has three simple poles: $z_0 = 0$, z_1, and z_2, where $z_{1,2}$ are the solutions of $3z^2+8iz-3=0$. By the quadratic formula, $z_{1,2} = (-4i \pm \sqrt{-16+9})/3 = i(-4\pm\sqrt{7})/3$; only $z_1 = i(-4+\sqrt{7})/3$ has $|z_1| < 1$ and therefore lies inside the unit circle. As a result,

$$I = 2\pi i \left(\operatorname*{Res}_{z=0} f(z) + \operatorname*{Res}_{z=z_1} f(z) \right).$$

Step 3. We compute the residue at zero by formula (4.4.14):

$$\operatorname*{Res}_{z=z_0} f(z) = (zf(z))|_{z=0} = -\frac{1}{3}.$$

We compute the residue at z_1 by formula (4.4.15). It is more convenient to compute the derivative of the denominator by the product rule: $\big(z(3z^2+8iz-3)\big)' = (3z^2+8iz-3) + z(6z+8i)$, because the first term on the right vanishes at z_1. As a result,

$$\operatorname*{Res}_{z=z_1} f(z) = \frac{z_1^2+4z_1+1}{2z_1(3z_1+4i)}, \text{ where } z_1 = \frac{-4+\sqrt{7}}{3}i.$$

We simplify the expression as follows:

$$z_1^2 + 4z_1 + 1 = \frac{-(16 - 8\sqrt{7} + 7) + i12(-4 + \sqrt{7}) + 9}{9}$$
$$= \frac{-14 + 8\sqrt{7}}{9} + 4i\frac{(-4 + \sqrt{7})}{3};$$
$$2z_1(3z_1 + 4i) = -2\sqrt{7}\left(\frac{-4 + \sqrt{7}}{3}\right) = \frac{-14 + 8\sqrt{7}}{3}.$$

As a result,

$$\operatorname*{Res}_{z=z_1} f(z) = \frac{1}{3} - 2i\frac{1}{\sqrt{7}}.$$

Step 4. We now get the final answer:

$$I = 2\pi i\left(-\frac{1}{3} + \frac{1}{3} - 2i\frac{1}{\sqrt{7}}\right) = \boxed{\frac{4\pi}{\sqrt{7}}}.$$

EXERCISE 4.4.10.^{C} *Without using a computer, verify that*

$$\int_0^{2\pi} \frac{1 + 2\sin\varphi}{5 + 4\cos\varphi} d\varphi = \frac{2\pi}{3}.$$

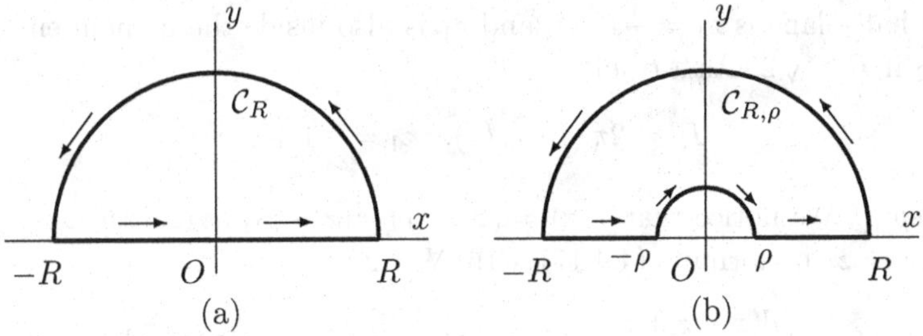

Fig. 4.4.1 Computing Real Integrals

Next, we consider integrals of the type

$$\int_{-\infty}^{\infty} \frac{P(x)}{Q(x)}\,dx, \tag{4.4.18}$$

where P, Q are polynomials with real coefficients and no common roots, the polynomial Q has no real roots, and the degree of Q is at least two units higher than the degree of P (to ensure the convergence of the integral). In particular, the degree of Q must be *even*; think about it.

Such integrals are evaluated by integrating the function $P(z)/Q(z)$ over the curve $\mathcal{C}_R$ on Figure 4.4.1(a), and then passing to the limit $R \to \infty$. The integral over $\mathcal{C}_R$ is determined by the residues of $P(z)/Q(z)$ at the roots of Q that have positive imaginary part, and the difference of degrees of P and Q ensures that the integral over the semi-circle tends to zero as R increases. If R is sufficiently large, then the curve $\mathcal{C}_R$ encloses all the roots of Q with positive real part.

Compared to the integration of the trigonometric expressions, we have more complicated theoretical considerations, such as passing to the limit as $R \to \infty$, but often much easier computations of the residues.

FOR EXAMPLE, consider the integral

$$I = \int_{-\infty}^{\infty} \frac{dx}{(2+2x+x^2)^2}.$$

Step 1. We consider the complex integral

$$I_R = \oint_{\mathcal{C}_R} \frac{dz}{(2+2z+z^2)^2},$$

where the curve $\mathcal{C}_R$ is from Figure 4.4.1(a), and R is sufficiently large.

Step 2. The function $f(z) = (2+2z+z^2)^{-2}$ we are integrating has two second-order poles $z_{1,2}$ at the roots of the polynomial $z^2+2z+2 = (z+1)^2+1$, and these roots are $z_1 = -1+i$, $z_2 = -1-i$. The root in the upper half-plane is $z_1 = -1+i$, and z_1 is also inside the domain enclosed by C_R if $R > \sqrt{2}$. As a result,

$$I_R = 2\pi i \operatorname*{Res}_{z=z_1} f(z), \quad z_1 = -1+i.$$

Step 3. We notice that $f(z) = (z-z_1)^{-2}(z-z_2)^{-2}$ and then find the residue at z_1 by formula (4.4.17) with $N=2$:

$$\operatorname*{Res}_{z=z_1} f(z) = \left.\frac{d(z-z_2)^{-2}}{dz}\right|_{z=z_1} = -2(z_1-z_2)^{-3} = -2(2i)^{-3} = -i/4;$$

recall that $i^{-1} = -i$. Accordingly, $I_R = \pi/2$ for all $R > \sqrt{2}$.

Step 4. We have $I_R = I_{R,r} + I_{R,c}$, where $I_{R,r}$ is the integral over the real axis from $-R$ to R, and $I_{R,h}$ is the integral over the circular arc. Then $I = \lim_{R\to\infty} I_{R,r}$. As for $I_{R,c}$, we note that $|z| = R$ on the arc, and, for $R > 10$, we have $|z^2+2z+2| > |z|^2 - 2|z| - 2 = R^2 - 2R - 2 > R(R-4) > R^2/2$. Since the length of the arc is πR, inequality (4.2.8) on page 196 implies

$$|I_{R,c}| \le \frac{4\pi R}{R^4} \to 0, \ R \to \infty.$$

As a result,

$$I = \lim_{R\to\infty} I_R = \boxed{\frac{\pi}{2}}.$$

EXERCISE 4.4.11.[C] *(a) Without using a computer, verify that*

$$\int_{-\infty}^{\infty} \frac{dx}{(4+2x+x^2)^3} = \frac{\pi\sqrt{3}}{72}.$$

(b) Convince yourself that a computation of (4.4.18) using a mirror image of the curve C_R in the lower half-plane leads to the same final answer. Hint: the reason is that the roots of Q come in complex conjugate pairs.

The same method applies to the computation of integrals

$$\int_{-\infty}^{\infty} \cos(ax)\,\frac{P(x)}{Q(x)}\,dx \text{ and } \int_{-\infty}^{\infty} \sin(ax)\,\frac{P(x)}{Q(x)}\,dx, \tag{4.4.19}$$

where a is a real number, P, Q are polynomials without common roots, the degree of Q is at least two units bigger than the degree of P, and Q has no real roots. Assuming that $a > 0$, you evaluate the integral

$$\int_{-\infty}^{\infty} e^{iax}\,\frac{P(x)}{Q(x)}\,dx \tag{4.4.20}$$

using the steps described above, and then take the real or imaginary part of the result. The integral in (4.4.20) can exist even when the degree of Q is only one unit bigger than the degree of P, see Problem 5.8 on page 436.

EXERCISE 4.4.12.[C] *(a) Evaluation of the integrals in (4.4.19) will not go through if, instead of e^{iaz} you try to work directly with $\cos az$ or $\sin az$. What goes wrong? Hint: look carefully at $\cos az$ and $\sin az$ when $|z| = R$. (b) How should you change the integral in (4.4.20) if you want to evaluate (4.4.19) with $a > 0$ by integrating over a semi-circle with $\Im z < 0$? Hint: you should go with e^{-iax}. (c) Without using a computer, verify that*

$$\int_{-\infty}^{\infty} \frac{\cos x}{1+x^2}\,dx = \frac{\pi}{e}.$$

Then ask your computer algebra system to evaluate this integral.

Some real integrals are evaluated using residues and integration over the curve $\mathcal{C}_{R,\rho}$ in Figure 4.4.1(b). FOR EXAMPLE, let us evaluate the integral

$$I = \int_0^\infty \frac{\cos(3x) - \cos x}{x^2}\, dx. \tag{4.4.21}$$

Step 0. EXERCISE 4.4.13. ^{C} *Verify that the integral converges. Hint:* $\int_\delta^\infty dx/x^2$ *for every* $\delta > 0$*, and, for* x *near* 0*,* $\cos(ax) = 1 - a^2x^2/2 + \ldots$.

Step 1. By the Integral Theorem of Cauchy, we have

$$\oint_{\mathcal{C}_{R,\rho}} \frac{e^{3iz} - e^{iz}}{z^2}\, dz = 0.$$

Step 2. The integrals over the parts of $\mathcal{C}_{R,\rho}$ on the real axis result in

$$\int_{-R}^{-\rho} \frac{e^{3ix} - e^{ix}}{x^2}\, dx + \int_\rho^R \frac{e^{3ix} - e^{ix}}{x^2}\, dx$$
$$= \int_\rho^R \left(\frac{e^{3ix} + e^{-3ix}}{x^2} - \frac{e^{ix} + e^{-ix}}{x^2} \right) dx \to 2I,\ \rho \to 0,\ R \to \infty.$$

Step 3. The integral over the semi-circle of radius R tends to zero as $R \to \infty$. All you need to notice is that, for $z = R(\cos\theta + i \sin\theta)$, we have $|e^{iaz}| = e^{-aR\sin\theta} \leq 1$.

Step 4. The integral over the semi-circle of radius ρ tends to -2π as $\rho \to 0$. Indeed, the Taylor expansion shows that, near $z = 0$, we have $z^{-2}(e^{3iz} - e^{iz}) = 2iz^{-1} + g(z)$, where $g(z)$ is analytic at $z = 0$. The integral of g will then tend to zero, and the integral of $2iz^{-1}$, once you take into account only half of the circle, the clock-wise orientation and the factors of i, produces 2π.

Step 5. By combining the results of the above steps, we conclude that

$$\int_0^\infty \frac{\cos(3x) - \cos x}{x^2}\, dx = -\pi.$$

EXERCISE 4.4.14. *(a)^A Provide the details in the steps 3–5 above. (b)^C Is it possible to evaluate (4.4.21) by integrating* $(\cos(3z) - \cos z)/z^2$ *over the curve in Figure 4.4.1(a)?*

With all these new techniques of integration, we should never forget the basic rules related to symmetry. First and foremost, *the integral of an odd function* (that is, a function $f = f(x)$ satisfying $f(x) = -f(-x)$) *over a symmetric interval is zero.* FOR EXAMPLE,

$$\int_{-\infty}^{\infty} \frac{\sin(23x)}{1+2x^6+5x^8}\,dx = 0,$$

even though some computer algebra systems do not recognize this. Second, if the original integral is not over the region we want, but otherwise is of the suitable type, we can try and use symmetry to extend the integration to the large interval. One basic example is $\int_0^{\infty} F(x)dx = (1/2)\int_{-\infty}^{\infty} F(x)dx$, if the function F is even, that is, $F(x) = F(-x)$. Still, as (4.4.21) shows, sometimes it is better to keep the original interval. More subtle symmetry can happen with trigonometric functions. FOR EXAMPLE, verify that

$$\int_0^{\pi} \frac{1-\cos\varphi}{5+3\cos\varphi}\,d\varphi = \frac{1}{2}\int_0^{2\pi} \frac{1-\cos\varphi}{5+3\cos\varphi}\,d\varphi = \frac{\pi}{3}.$$

There exist many other classes of real integrals that are evaluated using residues, but most of the examples include roots and natural logarithms. Evaluation of such integrals relies on the theory of multi-valued analytic functions and is discussed in a special Complex Analysis course.

4.4.3 *Power Series and Ordinary Differential Equations*

Many problems in mathematics, physics, engineering, and other sciences are reduced to a linear second-order ordinary differential equation (ODE)

$$A(x)y''(x) + B(x)y'(x) + C(x)y(x) = 0. \tag{4.4.22}$$

Even though the solution of such equations is usually not expressed in terms of elementary functions, a rather comprehensive theory exists describing various properties of the solution. This theory is based on power series and was developed in the second half of the 19th century by the German mathematicians LAZARUS FUCHS (1833–1902) and GEORG FROBENIUS (1849–1917). In what follows, we will outlines the main ideas of the theory; we will need the results later in the study of partial differential equations.

We assume that the functions A, B, C can be extended to the complex plane, and, instead of (4.4.22), we consider the equation in the complex domain

$$w''(z) + p(z)w'(z) + q(z)w(z) = 0 \tag{4.4.23}$$

for the unknown function w of the complex variable z, where $p(z) = B(z)/A(z)$, $q(z) = C(z)/A(z)$.

To proceed, we need the following two equalities:

$$\sum_{k=N}^{\infty} a_k(z-z_0)^k = \sum_{n=0}^{\infty} a_{n+N}(z-z_0)^{n+N}, \ N \geq 1, \tag{4.4.24}$$

and

$$\left(\sum_{k=0}^{\infty} a_k(z-z_0)^k\right)\left(\sum_{k=0}^{\infty} b_k(z-z_0)^k\right) = \sum_{k=0}^{\infty}\left(\sum_{m=0}^{k} a_m b_{k-m}\right)(z-z_0)^k. \tag{4.4.25}$$

EXERCISE 4.4.15.^{C} *(a) Verify (4.4.24). Hint: set* $n = k-N$, *so that* $k = n+N$. *(b) Verify (4.4.25). Hint: to get the idea, look at the first few terms by writing* $(a_0 + a_1(z-z_0) + a_2(z-z_0^2) + \ldots)(b_0 + b_1(z-z_0) + b_2(z-z_0^2) + \ldots)$ *and then multiplying through.*

If the functions p, q are analytic in a neighborhood of a point z_0, then the point z_0 is called **regular** for equation (4.4.23), and it is natural to expect that all solutions of (4.4.23) are also analytic functions in some neighborhood of z_0. Indeed, writing

$$p(z) = \sum_{k=0}^{\infty} p_k(z-z_0)^k, \quad q(z) = \sum_{k=0}^{\infty} q_k(z-z_0)^k, \quad w(z) = \sum_{k=0}^{\infty} w_k(z-z_0)^k$$

and substituting into (4.4.23), we find:

$$\begin{aligned} &w''(z) + p(z)w'(z) + q(z)w(z) \\ &= \sum_{k=2}^{\infty} k(k-1)w_k(z-z_0)^{k-2} + \left(\sum_{k=0}^{\infty} p_k(z-z_0)^k\right)\left(\sum_{k=1}^{\infty} kw_k(z-z_0)^{k-1}\right) \\ &+ \left(\sum_{k=0}^{\infty} q_k(z-z_0)^k\right)\left(\sum_{k=0}^{\infty} w_k(z-z_0)^k\right) = \sum_{k=0}^{\infty}\Bigg((k+1)(k+2)w_{k+2} \\ &+ \sum_{m=1}^{k+1} mw_m p_{k+1-m} + \sum_{m=0}^{k} w_m q_{k-m}\Bigg)(z-z_0)^k = 0, \end{aligned} \tag{4.4.26}$$

where the last equality follows from (4.4.24) and (4.4.25).

EXERCISE 4.4.16.^{B} *Verify (4.4.26).*

By Corollary 4.1 on page 212, we conclude that the coefficients w_k are computed recursively by

$$w_{k+2} = -\frac{1}{(k+1)(k+2)}\left(\sum_{m=1}^{k+1} m w_m p_{k+1-m} + \sum_{m=0}^{k} w_m q_{k-m}\right) \qquad (4.4.27)$$

for $k = 0, 1, 2, \ldots$. The values of w_k, $k \geq 2$, are uniquely determined by the given initial conditions $w_0 = w(z_0)$ and $w_1 = w'(z_0)$. These computations suggest that the following statement is true.

Theorem 4.4.3 *Assume that the functions $p = p(z)$ and $q = q(z)$ are analytic in some domain G of the complex plain, and $z_0 \in G$. Then, given w_0 and w_1, equation (4.4.23) with the initial conditions $w(z_0) = w_0$ and $w'(z_0) = w_1$ has a unique solution $w = w(z)$, and the function $w = w(z)$ is analytic in the domain G.*

The proof of this theorem is beyond the scope of our discussion; for proofs and extensions of this and many other results in this section, an interested reader can consult, for example, Chapter 4 of the book *Theory of Ordinary Differential Equations* by E. A. Coddington and N. Levinson, 1955.

EXERCISE 4.4.17.^{C} *Write the expressions for w_2, w_3, and w_4 without using the $\sum$ sign.*

Next, we will study the solutions of equation (4.4.23) near a **singular point**, that is, a point where at least one of the functions p, q is not analytic. As the original equation (4.4.22) suggests, singular points often correspond to zeroes of the function $A = A(z)$. A singular point z_0 is called a **regular singular point** of equation (4.4.23) if the functions $(z - z_0)p(z)$ and $(z - z_0)^2 q(z)$ are both analytic at z_0. In other words, the point z_0 is a regular singular point of (4.4.23) if and only if $p(z) = B(z)/(z - z_0)$ and $q(z) = C(z)/(z - z_0)^2$ for some functions B, C that are analytic at z_0.

Accordingly, we now consider the equation

$$(z - z_0)^2 w''(z) + (z - z_0)B(z)w'(z) + C(z)w(z) = 0, \qquad (4.4.28)$$

and assume that the point z_0 is a regular singular point of this equation. Being a linear second-order equation, (4.4.28) has the general solution

$$w(z) = A_1 W_1(z) + A_2 W_2(z), \qquad (4.4.29)$$

where A_1, A_2 are arbitrary complex numbers, and W_1, W_2 are two *linearly independent solutions* of (4.4.28); see Exercise 8.2.1, page 455. In what follows, we will compute W_1 and W_2 using power series. While in the regular case both W_1 and W_2 are analytic at z_0, for equation (4.4.28) we usually have at most one of the functions W_1, W_2 analytic at the point z_0.

We start by looking for a solution of (4.4.28) in the form

$$w(z) = \sum_{k=0}^{\infty} w_k (z - z_0)^{k+\mu}, \tag{4.4.30}$$

where μ and w_k, $k \geq 0$, are unknown *complex* numbers and $w_0 \neq 0$. In other words, we choose μ so that the function $(z - z_0)^{-\mu} w(z)$ is analytic and non-vanishing at $z = z_0$.

Similar to (4.4.26), we write $B(z) = \sum_{k=0}^{\infty} b_k (z - z_0)^k$, $C(z) = \sum_{k=0}^{\infty} c_k (z - z_0)^k$, and substitute into (4.4.28) to conclude that

$$\begin{aligned} &(z - z_0)^{\mu} \sum_{k=0}^{\infty} \Big((k + \mu)(k + \mu - 1) w_k \\ &+ \sum_{m=0}^{k} \big((m + \mu) b_{k-m} + c_{k-m} \big) w_m \Big) (z - z_0)^k = 0. \end{aligned} \tag{4.4.31}$$

EXERCISE 4.4.18.^{C} *Verify (4.4.31).*

Similar to (4.4.27), we get a recursive system to find w_k:

$$(k + \mu)(k + \mu - 1) w_k + \sum_{m=0}^{k} \big((m + \mu) b_{k-m} + c_{k-m} \big) w_m = 0, \tag{4.4.32}$$

$k = 0, 1, 2, \ldots$. For $k = 0$, (4.4.32) yields $\big(\mu(\mu - 1) + \mu b_0 + c_0\big) w_0 = 0$, or, since we assumed that $w_0 \neq 0$,

$$\mu^2 + (b_0 - 1)\mu + c_0 = 0. \tag{4.4.33}$$

Equation (4.4.33) is called the `indicial equation` of the differential equation (4.4.28). We will see that the *indicial* equation *indicates* the general solution of (4.4.28) by providing the roots μ. Equation (4.4.33) has two solutions μ_1, μ_2, and there are two main possibilities to consider: (a) $\mu_1 - \mu_2$ is not an integer; (b) $\mu_1 - \mu_2$ is an integer; this includes the possible double root $\mu_1 = \mu_2 = (1 - b_0)/2$. The reason for this distinction is that, for $k \geq 1$, equation (4.4.32) is $\big((\mu + k)^2 + (b_0 - 1)(\mu + k) + c_0\big) w_k = F_k(w_0, \ldots, w_{k-1})$ for some function F_k. As a result, if $(\mu + k)^2 + (b_0 - 1)(\mu + k) + c_0 \neq 0$ for every $k \geq 1$, then, starting with $w_0 = 1$ and two different values of μ, we can get two different sets of the coefficients w_k, and the two linearly independent solutions W_1, W_2, both in the form (4.4.30). If we have two identical values of μ or if $(\mu + N)^2 + (b_0 - 1)(\mu + N) + c_0 = 0$ for some integer $N \geq 1$, then only one of the functions W_1, W_2 will be of the form (4.4.30), and extra effort is necessary to find the other function.

IF THE ROOTS μ_1, μ_2 OF (4.4.33) DO NOT DIFFER BY AN INTEGER, then the two linearly independent solutions W_1, W_2 of (4.4.28) are

$$W_n(z) = (z - z_0)^{\mu_n} \sum_{k=0}^{\infty} w_{k,n}(z - z_0)^k, \; n = 1, 2, \tag{4.4.34}$$

where $w_{k,n}$, $k \geq 1$, $n = 1, 2$, are determined recursively from (4.4.32) with $w_{0,n} = 1$. Note that this is the case when b_0, c_0 are real and μ_1, μ_2 are complex so that $\mu_1 = \overline{\mu_2}$.

IF THE ROOTS μ_1, μ_2 OF (4.4.33) SATISFY $\mu_1 - \mu_2 = N \geq 0$, where N is an integer, then one solution of (4.4.28) is

$$W_1(z) = (z - z_0)^{\mu_1} \sum_{k=0}^{\infty} w_k(z - z_0)^k, \tag{4.4.35}$$

where μ_1 is the larger root of the indicial equation (4.4.33) and $w_k, k \geq 1$, are determined from (4.4.32) with $\mu = \mu_1$ and $w_0 = 1$. To find W_2, we use `Liouville's formula`:

$$\begin{vmatrix} W_1(z) & W_2(z) \\ W_1'(z) & W_2'(z) \end{vmatrix} = \exp\left(-\int_{\mathcal{C}(z_1,z)} \frac{B(z)}{z - z_0} dz \right), \tag{4.4.36}$$

where the left-hand side is a two-by-two determinant, z_1 is a fixed point in the neighborhood of z_0, and and $\mathcal{C}(z_1, z)$ is a piece-wise smooth path from z_1 to z so that z_0 is not on $\mathcal{C}(z_1, z)$. For the proof of this formula, see an ODE textbook, such as *Theory of Ordinary Differential Equations* by E. A. Coddington, and N. Levinson, 1955. By assumption, the function $B(z)/(z - z_0)$ is analytic away from z_0, and so the value of the integral does not depend on the particular path. After some computations, (4.4.36) yields

$$W_2(z) = W_1(z) \int_{\mathcal{C}(z_2,z)} H(z)dz, \; H(z) = \frac{1}{W_1^2(z)} \exp\left(- \int_{\mathcal{C}(z_1,z)} \frac{B(z)}{z - z_0} dz \right). \tag{4.4.37}$$

EXERCISE 4.4.19.^{A} *Verify (4.4.37). Hint: the determinant in (4.4.36) can be written as the square of* $W_1(z)$ *times the derivative of* $W_2(z)/W_1(z)$; $\mathcal{C}(z_2, z)$ *is a path similar to* $\mathcal{C}(z_1, z)$, *starting at* z_2.

By assumption, $B(z)/(z-z_0) = b_0/(z-z_0) + b_1 + b_2(z-z_0) + \ldots$, so that

$$\int_{\mathcal{C}(z_1,z)} \frac{B(z)}{z-z_0}\, dz = B_0 \ln(z-z_0) + \varphi(z),$$

where $\ln z$ is some branch of the natural logarithm, and the function φ is analytic at z_0. Recalling that $W_1(z) = (z-z_0)^{\mu_1}(1+w_1 z+\ldots)$, we conclude that
$H(z) = (z-z_0)^{-(2\mu_1+b_0)}\psi(z)$, where the function ψ is analytic at z_0 and $\psi(z_0) \neq 0$. By assumption, $\mu_1 - \mu_2 = N$, while equation (4.4.33) implies $\mu_1 + \mu_2 = 1 - b_0$. Therefore, $2\mu_1 + b_0 = 1 + N$. Writing

$$\psi(z) = \sum_{k=0}^{\infty} \psi_k (z-z_0)^k \tag{4.4.38}$$

and integrating (4.4.37), we conclude that

$$W_2(z) = (z-z_0)^{\mu_2} h(z) + \psi_N\, W_1(z) \ln(z-z_0), \tag{4.4.39}$$

where the function h is analytic at z_0, and ψ_N is the coefficient of z^N in the expansion (4.4.38). Note that

- $\psi_0 \neq 0$, while, for $N > 0$, $\psi_N = 0$ is a possibility.
- If $N > 0$, then $h(z_0) = \psi_0 \neq 0$.

In particular, *if the indicial equation (4.4.33) has a double root, then one of the solutions of (4.4.28) always has a logarithmic term.*

EXERCISE 4.4.20. *(a)*^{A} *Verify (4.4.39).*
Hint: $H(z) = \sum_{k=0}^{\infty} \psi_k (z-z_0)^{k-N-1}$, *and* ψ_N *is the coefficient of* $(z-z_0)^{-1}$.

(b)^C *Consider* `Bessel's differential equation`

$$z^2 w''(z) + z w'(z) + (z^2 - q) w(z) = 0, \tag{4.4.40}$$

where q *is complex number; the equation is named after the German astronomer and mathematician* FRIEDRICH WILHELM BESSEL *(1784–1846). Verify that this equation has a solution that does not have a singularity at* $z_0 = 0$ *if and only if* $q = N^2$ *for some non-negative integer* N, *and, if it exists, this non-singular solution is unique up to a constant multiple.*

Hint: the inidicial equation is $\mu^2 - q = 0$; if $\mu = s + it$ for some real s, t, then $z^\mu = z^s z^{it} = z^s e^{it \ln z} = z^s(\cos(t \ln z) + i \sin(t \ln z))$, see page 215. Then use (4.4.34), (4.4.35), and (4.4.39).

EXERCISE 4.4.21.^{C} *Let $w = w(z)$ be a solution of (4.4.23). Show that the function $\varphi(z) = w(1/z)$ is a solution of*

$$\varphi''(z) + (2z^{-1} - z^{-2}p(1/z))\varphi'(z) + z^{-4}q(1/z)\varphi(z) = 0. \tag{4.4.41}$$

By definition, the point $z_0 = \infty$ is a **regular singular point** of equation (4.4.23) if and only if the point $z_0 = 0$ is a regular singular point of equation (4.4.41).

EXERCISE 4.4.22.^{B} *Verify that equation*

$$z(1-z)w''(z) + [c - (a+b+1)z]\, w'(z) - ab\, w(z) = 0, \;\; a, b, c \in \mathbb{C}, \tag{4.4.42}$$

has exactly three regular singular points at $z_0 = 0, 1, \infty$.

Equation (4.4.42) is called the **hypergeometric differential equation**, and has a special significance: *for many second-order linear ordinary differential equations with at most three regular singular points, including, if necessary, $z_0 = \infty$, and no other singular points,* the solution is expressed in terms of the solutions of (4.4.42).

We now list some other particular equations of the type (4.4.23) or (4.4.28); all of them arise in many mathematical, physical, and engineering problems, and we will encounter some of the equations later in our discussion of partial differential equations:

$$(1 - z^2)w''(z) + (\mu + \nu z)\, w'(z) + \lambda\, w(z) = 0, \tag{4.4.43}$$

$$z^2 w''(z) + \nu z\, w'(z) + \lambda\, w(z) = 0, \tag{4.4.44}$$

$$w''(z) + \nu z w'(z) + \lambda\, w(z) = 0, \tag{4.4.45}$$

$$w''(z) - (z^2 + \lambda)w(z) = 0, \tag{4.4.46}$$

$$z\, w''(z) + (\nu - z)w'(z) + \lambda\, w(z), \tag{4.4.47}$$

$$w''(z) + \nu z\, w(z) = 0, \tag{4.4.48}$$

$$w''(z) + w'(z) + (z + \lambda)w(z) = 0, \tag{4.4.49}$$

$$w''(z) + (\lambda - z^{2n})w(z) = 0, \tag{4.4.50}$$

$$w''(z) + (\nu \cos(z) + \lambda)w(z) = 0, \tag{4.4.51}$$

$$(z^2-a^2)(z^2-b^2)w''(z)+z(2z^2-b^2-c^2)w'(z) \\ -\big((n(n+1)z^2-(b^2+c^2)p)\big)w(z)=0, \quad (4.4.52)$$

$$w''(z)+w'(z)+(\mu z^{-1}+\nu z^{-2})w(z)=0, \quad (4.4.53)$$

$$w''(z)-\left(\frac{\mu}{\sin^2(az)}+\frac{\nu}{\cos^2(az)}+\lambda\right)w(z)=0, \quad (4.4.54)$$

$$w''(z)-\left(\frac{\mu}{\sinh^2(az)}+\frac{\nu}{\cosh^2(az)}+\lambda\right)w(z)=0. \quad (4.4.55)$$

In these equations, n is a non-negative integer, a, b, c, p are real numbers, μ, ν, λ are complex numbers. For each equation, the reader is encouraged to do the following: (a) find all singular points (the point $z_0 = \infty$ must always be investigated), (b) find the first few terms in the expansion of the two linearly independent solutions, around $z_0 = 0$ and around all other finite regular singular points, if any.

For certain values of the parameters, the above equations can have especially interesting properties. FOR EXAMPLE, consider (4.4.47) with positive integer $\nu = m+1$ and non-negative integer $\lambda = n$; $m, n = 0, 1, 2, \ldots$. Writing $w(z) = \sum_{k=0}^{\infty} w_k z^k$ and substituting into the equation, we find that

$$w_{k+1} = \big((k-n)/(k+1)(k+m+1)\big)w_k$$

(verify this!). Then $w_{n+1} = 0$ and therefore $w_k = 0$ for all $k > n$. In other words, a solution of (4.4.47) is a POLYNOMIAL. This remarkable fact certainly deserves special recognition, and (4.4.47) with $\nu = m+1$ and $\lambda = n$ is called `Laguerre's differential equation`, after the French mathematician EDMOND LAGUERRE (1834–1894); the polynomial solution of (4.4.47) with $\nu = 1$, $\lambda = n$, and $w_0 = 1$ is called `Laguerre's polynomial` of degree n and denoted by $L_n(z)$.

EXERCISE 4.4.23.^{C} *Verify that if $L_n(z)$ satisfies $zw'' + (1-z)w' + nw = 0$, then the k-th derivative $L_n^{(k)}(z)$ of $L_n(z)$, $k \leq n$, satisfies* $zw'' + (k+1-z)w' + (n-k)w = 0$.

EXERCISE 4.4.24. *(a)^C Find the Laguerre polynomials L_n for $n = 0, 1, 2, 3, 4$. (b)^B The Laguerre differential equation has another solution with a singularity at $z_0 = 0$. Find the type of this singularity for different m and n. (c)^A Verify that equation (4.4.43) has a polynomial solution for the following values of μ, ν, and λ:*
(i) $\mu = 0$, $\nu = -1$, $\lambda = n^2$: `Chebyshev's differential equation` *of the first kind, after the Russian mathematician* PAFNUTI L'VOVICH CHEBY-

SHEV *(1821–1894), see also Problem 6.3 on page 439;*
(ii) $\mu = 0$, $\nu = -2$, $\lambda = n(n+1)$*:* `Legendre's differential equation`, *after the French mathematician* ADRIEN MARIE LEGENDRE *(1752–1833);*
(iii) $\mu = 0$, $\nu = -3$, $\lambda = n(n+2)$*:* `Chebyshev's differential equation` *of the second kind;*
(iv) $\mu = b - a$, $\nu = -(a+b+2)$, $\lambda = n(n+a+b+1)$*:* `Jacobi's differential equation`, *after C. G. J. Jacobi. Note that this includes the previous three equations as particular cases.*
(d)[A] *Verify that equation (4.4.45) with* $\nu = -2$ *and* $\lambda = 2n$*, known as* `Hermite's differential equation`, *after the French mathematician* CHARLES HERMITE *(1822–1901), has a polynomial solution.*

Some equations have several names. For example, (4.4.47) with general complex ν and λ is known as the `confluent hypergeometric differential equation`, and, with real ν and real negative λ, as `Kummer's differential equation`, after the German mathematician ERNST EDUARD KUMMER (1810–1893). By the general terminology, an *XYZ function* or an *XYZ polynomial* is a certain solution of the *XYZ equation.* The book *Orthogonal Polynomials* by G. Szegö, re-published by the AMS in 2003, and the *Handbook of Differential Equations* by D. Zwillinger, 1997, provide more information on the subject. *Keep in mind that the notations and terminology can vary from source to source.*

To conclude this section, we note that the power series method can work for linear equations of any order. FOR EXAMPLE, consider the equation $w'''(z) = zw(z)$, with initial conditions $w(0) = w'(0) = 0$, $w''(0) = 12$. The reader is encouraged to verify that $w(z) = 6z^2 + z^6/20 + z^{10}/14400 + \ldots$. How will the answer change if $w'(1) = 5$ rather than $w'(0) = 0$?

For certain NONLINEAR EQUATIONS, the power series method can still be used, because the product of two power series is again a power series. Of course, there is no longer any hope of getting nice recursive relations of the type (4.4.27). The following analog of Theorem 4.4.3 holds; see Theorem 6 in Section 11, Chapter 3, of the book *Ordinary Differential Equations* by G. Birkhoff and G.-C. Rota, 1969.

Theorem 4.4.4 *If* $F = F(w, z)$ *is a function of two complex variables, analytic at* (w_0, z_0)*, then there exists a neighborhood of the point* z_0 *in which the solution* $w = w(z)$ *of the initial value problem* $w'(z) = F(w(z), z)$*,* $w(z_0) = z_0$*, exists, is unique, and is an analytic function of* z*.*

Even if the function F is analytic everywhere, the solution of $w'(z) =$

$F(w(z), z)$ may fail to be analytic everywhere. FOR EXAMPLE, the solution of the equation $w'(z) = w^2$, $w(0) = 1$, is $w(z) = 1/(1 - z)$. The reader is encouraged to derive this representation of w using power series.

Chapter 5

Elements of Fourier Analysis

5.1 Fourier Series

In mathematics, we often represent general functions and other objects using elementary building blocks; the number of these blocks can be finite or infinite. For example, we write vectors in $\mathbb{R}^3$ as a linear combinations of three unit basis vectors $\hat{\boldsymbol{\imath}}$, $\hat{\boldsymbol{\jmath}}$, $\hat{\boldsymbol{k}}$, and we use Taylor series to represent analytic functions as an infinite linear combination of powers of $(z - z_0)$.

Similar to powers, sines and cosines can serve as elementary building blocks of functions, and the Fourier series in sines and cosines provides the corresponding representation for many functions. This representation has both theoretical and practical benefits: we will see that, on a bounded interval, the class of functions that can be represented by a Fourier series is much larger than the class of functions that can be represented by a Taylor series, which leads to numerous applications in signal processing, communications, and other ares.

While the representation of certain functions using sines and cosines was known to many eighteenth-century mathematicians, it was the French mathematician JEAN-BAPTISTE JOSEPH FOURIER (1768–1830) who, in the early 1800s, developed a general method for solving partial differential equations using what we now know as Fourier series and Fourier transforms. The practical importance of this ground-breaking method more than outweighed the lack of rigor on the part of Fourier. The first rigorous result about the convergence of Fourier series appeared only in 1828 and was due to Dirichlet.

5.1.1 *Fourier Coefficients*

A **trigonometric polynomial** of degree N on the interval $[-\pi, \pi]$ is an expression $P_N(x) = a_0 + \sum_{k=1}^{N}(a_k \cos kx + b_k \sin kx)$, where a_k, $k = 0, \ldots, N$, and b_k, $k = 1, \ldots, N$, are real numbers. Using *complex numbers*, we can both simplify and generalize this expression as follows. Recall that, by the Euler formula (4.3.14), page 214, $\cos kx = (e^{ikx} + e^{-ikx})/2$, $\sin kx = (e^{ikx} - e^{-ikx})/(2i)$. Then the expression for the trigonometric polynomial becomes $P_N(x) = \sum_{k=-N}^{N} c_k e^{ikx}$, where $c_0 = a_0$, $c_k = (a_k - ib_k)/2$ for $k > 0$, and $c_k = (a_k + ib_k)/2$ for $k < 0$. It is therefore natural to consider trigonometric polynomials of the form

$$P_N(x) = \sum_{k=-N}^{N} C_k e^{ikx} \tag{5.1.1}$$

with complex coefficients C_k. Thus, P_N becomes a *complex-valued* function of a *real variable* x.

The following **orthogonality relation** will be essential in many computations to follow.

EXERCISE 5.1.1.^{C} *Verify that*

$$\int_{-\pi}^{\pi} e^{ikx}\, dx = \begin{cases} 2\pi, & k = 0 \\ 0, & k = \pm 1, \pm 2, \ldots. \end{cases} \tag{5.1.2}$$

EXERCISE 5.1.2.^{C} *(a) Verify that all values of $P_N(x)$ are real if and only if $C_{-k} = \overline{C_k}$ for all k; recall that $\overline{C_k}$ is the complex conjugate of C_k. Hint: we already proved the "if" part; for the "only if" part, write $P(x) = \overline{P(x)}$, multiply by e^{ikx} and integrate from $-\pi$ to π, using (5.1.2). (b) Verify that*

$$\int_{-\pi}^{\pi} |P_N(x)|^2 dx = 2\pi \sum_{k=-N}^{N} |C_k|^2. \tag{5.1.3}$$

Hint: use (5.1.2) and the equality $|P_N(x)|^2 = P_N(x)\overline{P_N(x)}$; remember that the complex conjugate of e^{ikx} is e^{-ikx}.

In what follows, we show that the Fourier coefficients of the function f are the coefficients of the trigonometric polynomial that is the best mean-square approximation of f.

Let f be a *reasonably good function* defined on $[-\pi, \pi]$, for example, bounded and (Riemann) integrable; it can take complex values. How should we choose the coefficients C_k of P_N so that P_N is the *best* approximation of f? Of course, with many different ways to measure the quality of the approximation, we must specify what "best" means in our case, and relative simplicity of equality (5.1.3) suggests that we use the *mean-square error*. In other words, we want to find the numbers C_k so that the value of $\int_{-\pi}^{\pi} |f(x) - P_N(x)|^2 dx$ is as small as possible.

To find the corresponding numbers C_k, we write

$$\int_{-\pi}^{\pi} |f(x) - P_N(x)|^2 dx = \int_{-\pi}^{\pi} |f(x)|^2 dx \\ - \int_{-\pi}^{\pi} \left(f(x)\overline{P_N(x)} + P_N(x)\overline{f(x)} \right) dx + \int_{-\pi}^{\pi} |P_N(x)|^2 dx,$$

and define, for $k = -N, \ldots, N$, the numbers

$$c_k(f) = \frac{1}{2\pi} \int_{-\pi}^{\pi} f(x) e^{-ikx} dx; \tag{5.1.4}$$

the numbers $c_k(f)$ are called the `Fourier coefficients` of f. Then algebraic manipulations show that

$$\int_{-\pi}^{\pi} |f(x) - P_N(x)|^2 dx = \int_{-\pi}^{\pi} |f(x)|^2 dx - 2\pi \sum_{k=-N}^{N} |c_k(f)|^2 \\ + 2\pi \sum_{k=-N}^{N} |c_k(f) - C_k|^2. \tag{5.1.5}$$

EXERCISE 5.1.3.^{C} *Verify (5.1.5). Hint: you work, back and forth, with the equality $|z-w|^2 = |z|^2 + |w|^2 - z\overline{w} - w\overline{z}$; keep in mind that, for complex numbers, $(z-w)^2 \neq |z-w|^2$.*

An immediate consequence of (5.1.5) is that $\int_{-\pi}^{\pi} |f(x) - P_N(x)|^2 dx$ is minimal when the coefficients of P_N are the Fourier coefficients of f: $C_k = c_k(f)$ for all $k = -N, \ldots, N$, with $c_k(f)$ defined in (5.1.4). Accordingly, we define a special trigonometric polynomial,

$$S_{f,N}(x) = \sum_{k=-N}^{N} c_k(f) e^{ikx},$$

and call it the N-th `partial sum` of the Fourier series for f. Using (5.1.5) with $C_k = c_k(f)$ and $P_N = S_{f,N}$, we get

$$\int_{-\pi}^{\pi} |f(x) - S_{N,f}(x)|^2 dx = \int_{-\pi}^{\pi} |f(x)|^2 dx - 2\pi \sum_{k=-N}^{N} |c_k(f)|^2. \quad (5.1.6)$$

Since the left-hand side of this equality is always non-negative, we conclude that, for *every* N, $2\pi \sum_{k=-N}^{N} |c_k(f)|^2 \leq \int_{-\pi}^{\pi} |f(x)|^2 dx$. Thus, if we start with a bounded integrable function f and, for each $k = 0, \pm 1, \pm 2, \ldots$, define the numbers $c_k(f)$ according to (5.1.4), then the series $\sum_{k=-\infty}^{\infty} |c_k(f)|^2$ converges and, in fact,

$$\sum_{k=-\infty}^{\infty} |c_k(f)|^2 \leq \frac{1}{2\pi} \int_{-\pi}^{\pi} |f(x)|^2 dx. \quad (5.1.7)$$

Inequality (5.1.7) implies that the Fourier coefficients of f tend to zero:

$$\lim_{|k| \to \infty} |c_k(f)| = 0. \quad (5.1.8)$$

Inequality (5.1.7) is (a particular case of) `Bessel's inequality`, named after W. F. Bessel. Equality (5.1.8) is (a particular case of) the `Riemann-Lebesgue Theorem`. As we saw earlier, the Ph.D. dissertation of Riemann was a major contribution to complex analysis. His other dissertation (`Habilitation`), was a major contribution to Fourier analysis. Written in 1854, it also introduced what we now know as the Riemann integral. The French mathematician Henri Léon Lebesque (1875–1941) developed a generalization of the Riemann integral, which allowed him to generalize Riemann's results about Fourier series. What we now know as the Lebesgue integral was introduced by Lebesgue in 1902 in his Ph.D. dissertation. Both (5.1.7) and (5.1.8) are extendable to the Lebesgue integral, but this is beyond the scope of our discussion.

The next natural question is whether we can have the *equality* in (5.1.7), and the answer, for *sufficiently nice functions* f, is positive: if f is bounded and integrable on $[-\pi, \pi]$, then

$$\sum_{k=-\infty}^{\infty} |c_k(f)|^2 = \frac{1}{2\pi} \int_{-\pi}^{\pi} |f(x)|^2 dx. \quad (5.1.9)$$

This equality is known as `Parseval's identity`, after the French mathematician Marc-Antoine Parseval des Chênes (1755–1836), even though his original result was not directly connected with Fourier series. The proof of (5.1.9) is not at all trivial and is too technical to discuss here; we will take the result for granted. Below (see page 256), we discuss the physical interpretation of Parseval's identity in the context of signal processing. With this interpretation, (5.1.9) means conservation of energy.

Definition 5.1 The `Fourier series` S_f of a bounded, Riemann integrable function $f = f(x)$ on the interval $[-\pi, \pi]$ is

$$S_f(x) = \sum_{k=-\infty}^{\infty} c_k(f)e^{ikx}, \text{ where } c_k(f) = \frac{1}{2\pi}\int_{-\pi}^{\pi} f(x)e^{-ikx}dx. \quad (5.1.10)$$

At this point, we do not know whether $S_f(x) = f(x)$; in fact, it is not even clear in what sense the infinite sum in (5.1.10) is defined (keep in mind that convergence of $\sum_{k=-\infty}^{\infty} |c_k(f)|^2$ *does not imply* convergence of $\sum_{k=-\infty}^{\infty} |c_k(f)|$). What is certainly true is that, because of the Parseval identity, the Fourier series converges to f `in the mean square`: because of (5.1.6) and (5.1.9),

$$\lim_{N\to\infty} \int_{-\pi}^{\pi} |f(x) - S_{f,N}(x)|^2 dx = 0.$$

Unfortunately, this `mean-square convergence` has nothing to do with the convergence of $S_{f,N}(x)$ to $f(x)$ for *individual* values of x. Indeed, one can construct a sequence of functions $g_1, g_2, \ldots$ on $(0,1)$ so that $\lim_{N\to\infty} \int_0^1 |g_N(x)|^2 dx = 0$ but $\lim_{N\to\infty} g_N(x)$ does not exists for *all* $x \in (0,1)$, see Problem 6.1, page 437.

EXERCISE 5.1.4.^{B} *Show that the Fourier series of f is unique: if a_k, $k \geq 1$, is a collection of complex numbers with the property* $\lim_{N\to\infty} \int_{-\pi}^{\pi} |f(x) - \sum_{k=-N}^{N} a_k e^{ikx}|^2 dx = 0$, *then $a_k = c_k(f)$ for all k. Hint: use (5.1.5).*

For some functions, the equality $S_f(x) = f(x)$ is easy to prove. Indeed, if f is a trigonometric polynomial of degree N, then $S_{f,N}(x) = f(x)$ for all x, because in this case $c_k(f) = 0$ for $|k| > N$. Another example is discussed in the following exercise.

EXERCISE 5.1.5. ^{B} *Let $F = F(z)$ be a function, analytic in an annulus $G = \{z : r_1 < |z| < r_2\}$, where $0 \leq r_1 < 1$, $r_2 > 1$, and define the function $f(\varphi) = F(e^{i\varphi})$. Show that if $F(z) = \sum_{k=-\infty}^{\infty} c_k z^k$ is the* `Laurent series expansion` *of F in G, then $c_k = c_k(f)$ for all k and $S_f(\varphi) = \sum_{k=-\infty}^{\infty} c_k e^{ik\varphi}$ for all $\varphi \in [0, 2\pi]$, and so in this case we have $f(\varphi) = S_f(\varphi)$. Hint: use formula (4.4.3) on page 217 for the coefficients of the Laurent series, with $z_0 = 0$ and $\rho = 1$. After changing the variable of integration $\zeta = e^{i\varphi}$ in (4.4.3), you will get 5.1.4.*

From the practical point of view, neither trigonometric polynomials nor analytic functions are very interesting to expand in a Fourier series, but fortunately the class of functions that can be represented by a Fourier series is much larger. We will see in the next section that, if one can draw the graph of the function $f = f(x)$, then $S_f(x)$ is well-defined for all x and $S_f(x) = f(x)$ for all x where f is continuous. In other words, the equality $S_f(x) = f(x)$ holds for most reasonable functions f. Note that the graphs that can be drawn must be smooth at all but finitely many points, simply because of the finite thickness of the line.

We conclude this section with an alternative form of the Fourier series for the REAL-VALUED FUNCTIONS. For such functions, it is natural to write

$$S_f(x) = a_0 + \sum_{k=1}^{\infty}(a_k \cos kx + b_k \sin kx). \tag{5.1.11}$$

EXERCISE 5.1.6.^{C} *(a) Verify that, in (5.1.11), we have*

$$\begin{gathered} a_0 = \frac{1}{2\pi}\int_{-\pi}^{\pi} f(x)\,dx,\ a_k = \frac{1}{\pi}\int_{-\pi}^{\pi} f(x)\cos kx\,dx,\ k = 1,2,\ldots; \\ b_k = \frac{1}{\pi}\int_{-\pi}^{\pi} f(x)\sin kx\,dx,\ k = 1,2,\ldots. \end{gathered} \tag{5.1.12}$$

(b) Verify that, for real-valued functions, Parseval's identity (5.1.9) becomes

$$2a_0^2 + \sum_{k=1}^{\infty}(a_k^2 + b_k^2) = \frac{1}{\pi}\int_{-\pi}^{\pi} f^2(x)dx, \tag{5.1.13}$$

with a_k, b_k from (5.1.12).

EXERCISE 5.1.7.^{C} *Verify the following* `orthogonality relations` *for the trigonometric functions, with integer $m, n \geq 1$:*

$$\begin{gathered} \int_{-\pi}^{\pi} \cos nx \cos mx\,dx = \int_{-\pi}^{\pi} \sin nx \sin mx\,dx = \begin{cases} \pi, & m = n, \\ 0, & m \neq n; \end{cases} \\ \int_{-\pi}^{\pi} \cos nx \sin mx\,dx = 0. \end{gathered} \tag{5.1.14}$$

Hint: you have at least three options: (1) trigonometric identities; (2) sines and cosines as complex exponentials; (3) equality $e^{imx}e^{-inx} = (\cos mx + i\sin mx)(\cos nx - i\sin nx)$ followed by (5.1.2) on page 242. Treat the cases $m = n$ and $m \neq n$ separately.

5.1.2 *Point-wise and Uniform Convergence*

Using formulas (5.1.4) or (5.1.12), we can compute the Fourier series for specific functions f defined on $[-\pi, \pi]$. Still, the computations will be much more efficient, and also make much more sense, once we understand when, and in what sense, $S_f = f$.

The aim of the present section is to develop the theory that justifies all the computations we will perform in the following section, and this development requires more than the usual number of theorems and proofs; those who do not like that can move on to the next section: after all, people have been computing the Fourier series well before the necessary theory was developed. On the other hand, those who want a more detailed account of the uniform convergence should consult a book such as *Principles of Mathematical Analysis* by W. Rudin, 1976, or *Introduction to Analysis* by A. Mattuck, 1998.

But why should one study the point-wise convergence of the Fourier series? Indeed, Fourier himself believed that the equality $S_f(x) = f(x)$ was always true, although he did not provide any proofs. It was only in 1828 that Dirichlet published the first rigorous result about the point-wise convergence of Fourier series (something along the lines of Theorem 5.1.5 below), which put certain restrictions on the function f. After this result, Dirichlet and many others started to believe that the equality $S_f(x) = f(x)$ should hold for all *continuous* functions f; then, in 1873, PAUL DAVID GUSTAV DU BOIS-REYMOND (1831–1889) constructed a continuous function for which the Fourier series diverges at one point, causing many to suspect that there could be continuous functions with an *everywhere divergent* Fourier series. In the early 1920s, A. N. Kolmogorov, who was not yet 20 year old, and who had just changed his major from history to mathematics, constructed a function for which the Fourier series diverges everywhere, but the function is not continuous. The question was finally settled in 1964, when the Swedish mathematician LENNART CARLESON (b. 1928) proved that the Fourier series of a continuous function converges *almost everywhere*. It is a mathematically precise statement; see his paper *On convergence and growth of partial sums of Fourier series* in *Acta Mathematica*, Vol. 116 (1966), pages 137–157. The proof of this result is considered by many to be one of the hardest in all of analysis.

To summarize, the question of point-wise convergence of Fourier series is very nontrivial, and our discussion below provides only the most basic ideas. We will use the complex version of the Fourier series (5.1.10).

For fixed x, the Fourier series $S_f(x)$ is a numerical series; see Section 4.3.1, page 206. Note that $|c_k(f)e^{ikx}| = |c_k(f)|$, so the Fourier series will converge absolutely for all x if $\sum_{k=-\infty}^{\infty} |c_k(f)| < \infty$. In fact, the series in this case also converges *uniformly*, which is a special type of convergence we define next.

Definition 5.2 A sequence of functions $f_1, f_2, \dots$ on an interval I of the real line `converges uniformly` to the function f if, for every $\varepsilon > 0$, there exists an $m \geq 1$ such that, for all $n > m$ and *all x in the interval I*, we have $|f_n(x) - f(x)| < \varepsilon$.

A series $\sum_{k=1}^{\infty} f_k(x)$ converges uniformly if the sequence of partial sums $\sum_{k=1}^{n} f_k(x)$ converges uniformly; for the series $\sum_{k=-\infty}^{\infty} f_k(x)$, we consider partial sums of the form $\sum_{k=-n}^{n} f_k(x)$.

The same definition applies to functions of a complex variable, defined in a complex domain instead of a real interval.

Recall that a sequence of functions $f_1, f_2, \dots$ converges to f at every point of the interval if for every $\varepsilon > 0$, and *for every x in the interval,* there exists an $m \geq 1$ such that, for all $n > m$ we have $|f_n(x) - f(x)| < \varepsilon$. The difference from uniform convergence is therefore in the possible dependence of m on x. In other words, a uniformly convergent sequence converges point-wise, but a point-wise convergent sequence does not need to converge uniformly. FOR EXAMPLE, the sequence $f_n(x) = x/n$ converges uniformly to zero on every bounded interval, while the sequence $f_n(x) = x^n$ converges to zero on $[0, 1)$ point-wise, but not uniformly; the same sequence $f_n(x) = x^n$ does converge uniformly on $[0, a]$ for every $a < 1$.

EXERCISE 5.1.8.[B] *Verify that if a Fourier series converges uniformly on $[-\pi, \pi]$, then the series converges uniformly on $\mathbb{R}$. Hint: use periodicity.*

One reason for considering uniform convergence is that the limit of a uniformly convergent sequence inherits many properties of the individual functions in the sequence.

Theorem 5.1.1 *Assume that a sequence of functions $f_1, f_2, \dots$ on a closed bounded interval I of the real line converges uniformly to some function f.*

(1) If each f_n is continuous on I, then f is continuous on I and $\lim_{n\to\infty} \int_I f_n(x)dx = \int_I f(x)dx$.

(2) If each f_n is differentiable on I and each f_n' is continuous on I, and the sequence of derivatives $f_1', f_2', \dots$ converges uniformly on I to some

function g, then f is also differentiable on I and $f' = g$.

We essentially proved this theorem in the special case of power series, see Theorem 4.3.4, page 210; an interested reader can easily adjust the arguments for the general case. The key in the proof is that, by taking n sufficiently large, we can make the difference $|f_n(x) - f(x)|$ as small as we want *for all* $x \in I$ *at once.*

EXERCISE 5.1.9.[B] *(a) State the analog of the above theorem for a series of functions. (b) Show that if the sequence* $f_1, f_2, \ldots$ *of continuous functions converges uniformly to* f *on the closed bounded interval* $[a, b]$, *then, for all sufficiently large* n, *the graph of* f_n *resembles the graph of* f. *In other words, show that* $\lim_{n\to\infty} \max_{x\in[a,b]} |f_n(x) - f(x)| = 0$. *Hint: let* $\max_{x\in[a,b]} |f_n(x) - f(x)| = |f_n(x_n) - f(x_n)|$.

The main test for uniform convergence for a *series of functions* is known as `Weierstrass's M-test`, in honor of the German mathematician KARL THEODOR WILHELM WEIERSTRASS (1815-1897), who was one of the founders of modern analysis (both real and complex).

Theorem 5.1.2 WEIERSTRASS'S M-TEST. *If* $|f_n(x)| \le M_n$ *for all* x *in closed bounded interval* I, *and if the series of numbers* $\sum_{n=1}^{\infty} M_n$ *converges, then the series* $\sum_{n=1}^{\infty} f_n(x)$ *converges uniformly in* I.

EXERCISE 5.1.10.[C] *(a) Show that if* $|f_n(x)| \le a_n$ *for all* x *in the interval, and the sequence* $a_1, a_2, \ldots$ *converges to zero, then the sequence* $f_1, f_2, \ldots$ *converges to zero uniformly. (b) By considering the sequence* $\sum_{k=n}^{\infty} f_k$, $n = 1, 2, \ldots$, *prove the Weierstrass M-test. (c) Convince yourself that if* $\sum_{k=-\infty}^{\infty} |c_k(f)| < \infty$, *then the Fourier series converges uniformly on* $\mathbb{R}$ *and* S_f *is a continuous function on the whole real line. (d) Consider the power series* $\sum_{k=0}^{\infty} a_k(z - z_0)^k$ *with the radius of convergence* $R > 0$. *By taking* $M_n = |a_n|r^n$, $r < R$, *convince yourself that the power series converges uniformly inside the closed disk* $\{z : |z - z_0| \le r\}$.

We can now establish a rather general result about the convergence of Fourier series.

Theorem 5.1.3 *Assume that the function* f *is continuous on* $[-\pi, \pi]$, $f(\pi) = f(-\pi)$, *and the Fourier series* S_f *converges uniformly on* $[-\pi, \pi]$. *Then* $S_f(x) = f(x)$ *for all* $x \in [-\pi, \pi]$.

Proof. Let us pass to the limit $N \to \infty$ in equality (5.1.6) on page 244. By assumption, the sequence $\{S_{f,N},\ N \ge 1\}$ converges uniformly to S_f. Then, by Theorem 5.1.1 we have

$$\lim_{N\to\infty}\int_{-\pi}^{\pi}|f(x)-S_{N,f}(x)|^2dx=\int_{-\pi}^{\pi}|f(x)-S_f(x)|^2dx. \qquad (5.1.15)$$

On the other hand, (5.1.6) and Parseval's identity (5.1.9) imply that the left-hand side of (5.1.15) is equal to zero. As a result, $\int_{-\pi}^{\pi}|f(x)-S_f(x)|^2dx=0$, and since both f and S_f are continuous, we conclude that $f(x)=S_f(x)$ for all $x\in[-\pi,\pi]$. □

Note that the function S_f necessarily satisfies $S_f(\pi)=S_f(-\pi)$, because $e^{ik\pi}=e^{-ik\pi}$ for all k. In fact, S_f is a function with period 2π. As a result, the statement of the theorem cannot hold without the assumption $f(\pi)=f(-\pi)$. This assumption implies that the function f can be extended to the whole real line so that the extension has period 2π and is continuous everywhere.

Sometimes, it is possible to establish the equality $S_f(x)=f(x)$ without computing the coefficients $c_k(f)$. We can easily prove the following result.

Theorem 5.1.4 *Assume that $f(\pi)=f(-\pi)$ and the function f is differentiable on $(-\pi,\pi)$ and f' is bounded and Riemann integrable. Then the series $\sum_{k=-\infty}^{\infty}|c_k(f)|$ converges and $S_f(x)=f(x)$ for all $x\in[-\pi,\pi]$.*

Proof. Denote by d_k the Fourier coefficients of f'. Then we integrate by parts to find

$$d_k=\int_{-\pi}^{\pi}f'(x)e^{-ikx}dx=f(x)e^{ikx}\Big|_{x=-\pi}^{x=\pi}+ik\int_{-\pi}^{\pi}f(x)e^{-ikx}dx.$$

By assumption, the first term on the right-hand side of the last equality is equal to zero, because $f(-\pi)e^{-i\pi k}=f(\pi)e^{i\pi k}$. As a result, $|d_k|=|kc_k(f)|$. Then the Cauchy-Schwartz inequality (1.2.14) on page 17 implies

$$\sum_{k=-\infty}^{\infty}|c_k(f)|=|c_0(f)|+\sum_{\substack{k=-\infty\\k\neq 0}}^{\infty}|d_k/k|\le\left(\sum_{k=-\infty}^{\infty}|d_k|^2\right)\left(\sum_{\substack{k=-\infty\\k\neq 0}}^{\infty}k^{-2}\right).$$

Since $\sum_{k=-\infty}^{\infty}|d_k|^2$ converges by the Bessel inequality (5.1.7), and

$$\sum_{\substack{k=-\infty\\k\neq 0}}^{\infty}k^{-2}=2\sum_{k=1}^{\infty}k^{-2}<\infty,$$

we conclude that $\sum_{k=-\infty}^{\infty} |c_k(f)| < \infty$. This implies uniform convergence of the Fourier series, and, together with Theorem 5.1.3, completes the proof. $\square$

To proceed, let us recall the definitions of the one-sided limits of a function at a point: the limit from the right $f(a^+)$, also denoted by $\lim_{x\to a^+} f(x)$, is defined by

$$f(a^+) = \lim_{\varepsilon\to 0, \varepsilon>0} f(a+\varepsilon).$$

Similarly, the limit from the left $f(a^-)$, also denoted by $\lim_{x\to a^-} f(x)$, is defined by

$$f(a^-) = \lim_{\varepsilon\to 0, \varepsilon>0} f(a-\varepsilon).$$

If one can physically draw the graph of a function, then the function must be piece-wise smooth: it is impossible to draw infinitely many cusps or discontinuities just because of the finite thickness of the line. In particular, $f(x^+)$ and $f(x^-)$ must exist at every point. These are the functions we encounter in every-day life. The functions can have jump-discontinuities, that is, points where $f(x^+) \neq f(x^-)$. The following result shows that we can represent such functions using a Fourier series. IT IS ALSO THE MAIN RESULT OF THIS SECTION.

Theorem 5.1.5 *Let f be a function with period 2π and assume that, on every interval of length 2π, the function has a continuous derivative except at a finite number of points. Also assume that there exists a number M so that $|f'(x)| \leq M$ for all x where f' exists. Then*

$$S_f(x) = \begin{cases} f(x), & \text{if } f \text{ is continuous at } x \\ \big(f(x^+) + f(x^-)\big)/2, & \text{if } f \text{ is not continuous at } x. \end{cases} \tag{5.1.16}$$

Moreover, if f is continuous everywhere, then the Fourier series converges to f uniformly on $\mathbb{R}$.

Note that conditions of the theorem imply that, on every bounded interval, there exist finitely many numbers $x_1 < x_2 < \ldots < x_n$ such that the function f is bounded and continuously differentiable on every interval (x_k, x_{k+1}); this, in particular, implies the existence of $f(x^+)$ and $f(x^-)$ for every x. The proof of this theorem is rather long; an interested reader will find the main steps in Problem 6.2 on page 437.

The next result shows that many convergent Fourier series can be integrated term-by-term even without uniform convergence, and the resulting integrated sequence will always converge uniformly.

Theorem 5.1.6 *Assume that the function f satisfies the conditions of Theorem 5.1.5 and, in addition, $\int_{-\pi}^{\pi} f(x)dx = 0$. Then the Fourier series for f can be integrated term-by-term, and the result converges uniformly. More precisely, if $F(x) = \int_0^x f(t)dt$, then the Fourier series S_F converges to F uniformly on $\mathbb{R}$ and*

$$S_F(x) = i \sum_{\substack{k=-\infty \\ k\neq 0}}^{\infty} \frac{c_k(f)}{k} - i \sum_{\substack{k=-\infty \\ k\neq 0}}^{\infty} \frac{c_k(f)}{k} e^{ikx},$$

where $c_k(f), k = \pm 1, \pm 2, \ldots$, are the Fourier coefficients of f.

Proof. We outline the main steps; the details are in the Exercise below.

Step 1. By assumption, $c_0(f) = 0$. Integrating by parts for $k \neq 0$, we find

$$c_k(f) = \int_{-\pi}^{\pi} f(x)e^{-ikx}dx = \left(f(x)\frac{e^{-ikx}}{-ik}\right)\Big|_{x=-\pi}^{x=\pi} - \int_{-\pi}^{\pi} f'(x)\frac{e^{-ikx}}{-ikx}\,dx.$$

Since both f and f' are bounded, there exists a positive number A so that $|c_k(f)| \leq A/|k|$ for all $k = \pm 1, \pm 2, \ldots$.

Step 2. Recall that, for $x < 0$, $\int_0^x f(t)dt = -\int_0^{-x} f(t)dt$. Denote by $C_k,\ k = 0, \pm 1, \pm 2, \ldots$ the Fourier coefficients of F. Then

$$C_k = \int_{-\pi}^{\pi} F(x)d\left(\frac{e^{-ikx}}{-ik}\right).$$

For $k \neq 0$, we integrate by parts and use $\int_{-\pi}^{\pi} f(x)dx = F(\pi) - F(-\pi) = 0$ to conclude that $C_k = -ic_k(f)/k$. In particular, $|C_k| \leq A/k^2$, and so the Fourier series of F converges uniformly by Weierstrass's M-test on page 249. Continuity of F implies $S_F(x) = \sum_{k=-\infty}^{\infty} C_k e^{ikx} = F(x)$ for all x. In particular, $S_F(0) = \sum_{k=-\infty}^{\infty} C_k = F(0) = 0$, and so

$$C_0 = -\sum_{\substack{k=-\infty \\ k\neq 0}}^{\infty} C_k = i \sum_{\substack{k=-\infty \\ k\neq 0}}^{\infty} \frac{c_k(f)}{k},$$

which completes the proof. □

EXERCISE 5.1.11.[B] *Fill in the details in the above proof. In particular, (a) Verify that condition* $\int_{-\pi}^{\pi} f(x)dx = 0$ *implies* $c_0(f) = 0$ *and also implies that the function* F *is periodic with period* 2π. *(b) Verify that* F *is continuous everywhere. Hint: for* $y > x$, $|F(y) - F(x)| \leq \int_x^y |f(t)|dt \leq A|x - y|$. *(c) Carry out the calculations in Step 2 to conclude that* $C_k = -ic_k(f)/k$. *(d) Verify that* $S_F(x)$ *is indeed the result of the term-by-term integration of* $S_f(x)$. *Hint: for* $k \neq 0$, $\int_0^x e^{ikt}dt = (e^{ikx} - 1)/(ik) = i(1 - e^{ikx})/k$.

Let us emphasize that the *construction* of the Fourier series S_f depends only on the behavior of the function f on the interval $(-\pi, \pi)$; it does not even matter how f is defined at the end points of the interval. The Fourier series S_f, if it converges, is a 2π-periodic function and represents the 2π-*periodic extension* of f rather than f itself. As a result, the *convergence* of the Fourier series depends on the behavior of this extension of f, and not on the behavior of original f. In particular, even if f is continuous on $[-\pi, \pi]$ and is infinitely differentiable on $(-\pi, \pi)$, the Fourier series will not converge uniformly unless $f(-\pi^+) = f(\pi^-)$, that is, unless the periodic extension of f is continuous. In general, the more derivatives the periodic extension of f has, the faster the Fourier series converges to f; the rate of convergence is determined by the rate at which the coefficients $c_k(f)$ tend to zero.

EXERCISE 5.1.12.[C] *(a) Convince yourself that if the periodic extension of* f *is not continuous, then the Fourier series cannot converge uniformly on* $[-\pi, \pi]$. *Hint: if it did converge uniformly, the result would be a continuous function, which it is not. (b) Assume that the periodic extension of* f *has* N *continuous derivatives. Show that there exists a positive number* A *so that, for all* k, $|c_k(f)| < A/(|k|+1)^N$. *Hint: integrate by parts* N *times and observe that the derivatives of the periodic extension, if exist, are also periodic.*

The non-uniform convergence of Fourier series can be visualized. Consider the 2π-periodic function $g = g(x)$ so that $g(x) = x$ for $|x| < \pi$, $g(\pm\pi) = 0$; this one is known as the sawtooth function. The function g is not continuous at the points $\pi + 2\pi n$ and therefore the Fourier series does not converge uniformly. At points π and $-\pi$, the Fourier series of g converges to $\big(g(\pi^+) + g(\pi^-)\big)/2 = 0$. Let us investigate what happens near the point π.

EXERCISE 5.1.13.[B] *(a) Verify that* $S_g(x) = 2\sum_{k=1}^{\infty}(-1)^{k+1}\frac{\sin kx}{k}$. *Hint: you have to integrate by parts.*

(b) Writing $S_{g,N}(x) = 2\sum_{k=1}^{N}(-1)^{k+1}\frac{\sin kx}{k}$, *use a computer algebra*

system to plot the graph of $S_{g,N}$ *on the interval* $[-\pi, \pi]$ *for* $N = 10, 50, 100$.
(c) Verify that $\lim_{N\to\infty} S_{g,N}\big(\pi(1-1/N)\big) = 2\int_0^{\pi}(\sin x/x)\,dx$.

(d) Show that $2\int_0^{\pi}(\sin x/x)\,dx > 1.17\pi$. *Hint: first try it analytically using the power series expansion of* $\sin x/x$ *at zero; if not successful, try a computer algebra system.*

In other words, even though $\lim_{N\to\infty} S_{g,N}(\pi - \varepsilon) = \pi - \varepsilon$ for every $\varepsilon \in (0, \pi]$, the maximal value of $S_{g,N}(x)$ on the interval $[0, \pi]$, being achieved at different points, does not converge to π, the supremum of g on the interval $[0, \pi]$; by Exercise 5.1.9, this could not have happened had $S_{g,N}$ converged uniformly.

In applications, we cannot sum infinitely many terms and approximate a function f with the trigonometric polynomial $S_{f,N}(x) = \sum_{k=-N}^{N} c_k(f)e^{ikx}$. If the periodic extension of the function f is not continuous, then the *maximal* value of the approximation error will not converge to zero. This effect was first observed in 1898 for the sawtooth function g by A. A. Michelson, of the Michelson-Morley experiment, who made a machine to recover the function from the Fourier coefficients. Since it was J. W. Gibbs who provided the mathematical explanation in 1899, the effect is now known as the `Gibbs phenomenon`.

EXERCISE 5.1.14.^{A} *Investigate the Gibbs phenomenon for the the* 2π*-periodic square wave function* $h = h(x)$, *defined for* $x \in (-\pi, \pi]$ *by*

$$h(x) = \begin{cases} 1, & \text{if } 0 \le x \le \pi; \\ -1, & \text{if } -\pi < x < 0. \end{cases}$$

5.1.3 *Computing the Fourier Series*

So far, we considered functions f defined on the interval $[-\pi, \pi]$, and that is sufficient from the theoretical point of view: if a function $f = f(y)$ is defined on the interval $[-L, L]$, then the function $F(x) = f(xL/\pi)$ is defined on the interval $[-\pi, \pi]$, and so $f(y) = F(y\pi/L)$, $S_f(x) = S_F(x\pi/L)$. In practice, it can be useful to have the explicit formulas for the Fourier coefficients and the Fourier series of f:

$$S_f(y) = \sum_{k=-\infty}^{\infty} c_k(f)e^{ik\pi y/L}, \text{ where}$$
$$c_k(f) = \frac{1}{2\pi}\int_{-\pi}^{\pi} f(xL/\pi)e^{-ikx}\,dx = \frac{1}{2L}\int_{-L}^{L} f(y)e^{-ik\pi y/L}\,dx. \tag{5.1.17}$$

Similarly, for real-valued functions f, we have $c_{-k}(f) = \overline{c_k(f)}$ and formulas (5.1.11), (5.1.12), and (5.1.13) become

$$S_f(y) = a_0 + \sum_{k=1}^{\infty}(a_k \cos(\pi k y/L) + b_k \sin(\pi k y/L), \text{ where}$$

$$a_0 = \frac{1}{2L}\int_{-L}^{L} f(y)\,dy,\; a_k = \frac{1}{L}\int_{-L}^{L} f(y)\cos(\pi k y/L)\,dy,\; k = 1, 2, \ldots;$$
$$b_k = \frac{1}{L}\int_{-L}^{L} f(y)\sin(\pi k y/L)\,dy,\; k = 1, 2, \ldots;$$
$$2a_0^2 + \sum_{k=1}^{\infty}(a_k^2 + b_k^2) = \frac{1}{L}\int_{-L}^{L} f^2(y)dy. \tag{5.1.18}$$

As in the case $L = \pi$, if f is piece-wise continuously differentiable then $S_f(y) = f(y)$ at the points y where the $2L$-periodic extension of f is continuous; the convergence is uniform if this extension of f is continuous everywhere on $\mathbb{R}$.

EXERCISE 5.1.15.[B] *Show that if the function f is extended with period $2L$ on $\mathbb{R}$, then all the integrals $\int_{-L}^{L}$ in (5.1.18) can be replaced with integrals $\int_{a-L}^{a+L}$ for an arbitrary $a \in \mathbb{R}$. Hint: if a function g satisfies $g(x) = g(x + 2L)$ for all $x \in \mathbb{R}$, then $\int_{a-L}^{a+L} g(x)dx = \int_{-L}^{L} g(x)dx + \int_{L}^{a+L} g(x)dx - \int_{-L}^{a-L} g(x)dx$; $\int_{L}^{a+L} g(x)dx = \int_{-L}^{a-L} g(y + 2L)dy = \int_{-L}^{a-L} g(y)dy$.*

If the function f is even ($f(x) = f(-x)$), then $b_k = 0$ for all k and the Fourier series contains only cosines; if the function is odd ($f(x) = -f(-x)$), then $a_k = 0$ for all k and the Fourier series contains only sines.

If the function is defined on the interval $[0, L]$, then there are three possibilities for constructing the Fourier series of f:

(1) Extend f on $\mathbb{R}$ with period L: $f(x+nL) = f(x)$, $x \in (0, L)$; the Fourier series will, in general, include $\cos(2\pi kx/L)$ and $\sin(2\pi kx/L)$.
(2) Extend f in an *odd* way onto $(-L, 0)$: $f(-x) = -f(x)$, $x \in (0, L)$; then consider the $2L$-periodic extension of the result; the Fourier series will include only $\sin(\pi kx/L)$ and is sometimes called the *odd half-range expansion*.
(3) Extend f in an *even* way onto $(-L, 0)$: $f(-x) = f(x)$, $x \in (0, L)$; then consider the $2L$-periodic extension of the result; the Fourier series will

include only $\cos(\pi kx/L)$ and is sometimes called the *even half-range expansion.*

If there is a possibility of choice, then the selection is dictated by Exercise 5.1.12: the more derivatives the periodic extension has, the better the convergence of the Fourier series. For a generic continuous function f, the even extension is guaranteed continuous and is therefore the best choice.

EXERCISE 5.1.16. *(a)^C Verify that the even extension of a continuous function is always continuous everywhere on $\mathbb{R}$. (b)^B Give an example of a function for which the odd extension is the best choice. (c)^C Is there an example of a function on $[0, L]$ for which the L-periodic extension is the best? Hint: drawing pictures helps in answering each of these questions.*

Let us briefly discuss the CONNECTION BETWEEN FOURIER SERIES AND SIGNAL PROCESSING. In signal processing, both the period $2L$ and the argument of f are measured in units of time, while the function $f = f(x)$ and each of the Fourier coefficients $c_k(f)$ are measured in volts. Then each $\omega_k = k\pi/L$ becomes a *frequency.* The collection of all ω_k corresponding to $c_k(f) \neq 0$ is called the (frequency) `spectrum` of the signal f, and it is essential to allow both positive and negative values of the frequencies. Since k is an integer, a periodic signal has a *discrete* spectrum. Once the spectrum of the signal is available, signal processing is carried out by removing or otherwise modifying the individual numbers $c_k(f)e^{i\omega_k t}$. Since $\int_{-L}^{L} |c_k(f)e^{i\omega_k t}|^2 dt = 2L|c_k(f)|^2$, the value $2L|c_k(f)|^2$ is proportional to the energy of the k-th frequency component over one period of the signal; the collection of numbers $(\omega_k, |c_k(f)|^2)$ is called the `power spectrum of` f. The physical interpretation of the Parseval identity, $\int_{-L}^{L} |f(x)|^2 dx = 2L\sum_{k=-\infty}^{\infty} |c_k(f)|^2$, is therefore that *the total energy of the signal over the period is equal to the sum of the energies of all the spectral components.* From this point of view, Parseval's identity makes perfect sense without any proofs: it just states the conservation of energy.

Sometimes, the Fourier series expansion of one function can be used to derive the expansions of several other functions. Along the way, we can also evaluate certain infinite sums using the Parseval identity and the results about the point-wise convergence of Fourier series. FOR EXAMPLE, consider the 2π-periodic function f defined by $f(x) = 1,\ 0 \le x < \pi$ and $f(x) = 0,\ -\pi \le x < 0$. This is an example of a `rectangular wave`; see Figure 5.1.1(a).

We have $c_0(f) = 1/2$ and, for $k \neq 0$, $c_k(f) = (2\pi)^{-1}\int_0^{\pi} e^{-ikx}dx =$

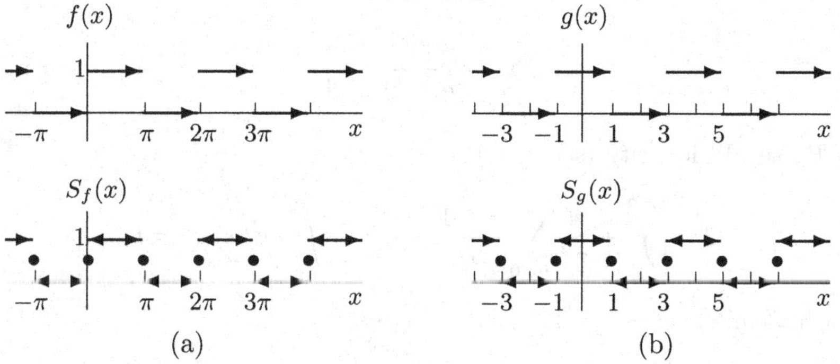

Fig. 5.1.1 Two Rectangular Waves

$(1-e^{-ik\pi})/(2\pi ki)$. Then, since $e^{ik\pi} = e^{-ik\pi}$ for all k,

$$S_f(x) = \frac{1}{2} + \sum_{k=1}^{\infty} \left(\frac{1-e^{ik\pi}}{\pi k}\right) \frac{e^{ikx}-e^{-ikx}}{2i} = \frac{1}{2} + \sum_{k=1}^{\infty} \frac{1-\cos(\pi k)}{\pi k} \sin(kx).$$

We note that $1-\cos(k\pi) = 0$ for even k and $1-\cos(k\pi) = 2$ for odd k, and so

$$S_f(x) = \frac{1}{2} + \frac{2}{\pi} \sum_{n=0}^{\infty} \frac{\sin(2n+1)x}{2n+1}. \tag{5.1.19}$$

By Theorem 5.1.5, we have $S_f(x) = f(x)$ for $x \neq \pi + 2\pi k$, and $S_f(\pi + 2\pi k) = 1/2$.

To continue our EXAMPLE, consider another rectangular wave $g = g(x)$ with period 4, so that $g(x) = 1$ for $-1 \le x < 1$, $g(x) = 0$ for $-2 \le x < -1$ and $1 \le x < 2$; see Figure 5.1.1(b). Then g is a result of horizontal shift and dilation of f: shift $x \to x + \pi/2$ makes the function f even, and then dilation $x \to \pi x/2$ makes the period equal to 4. In other words,

$$g(x) = f\left(\frac{\pi}{2}x + \frac{\pi}{2}\right)$$

(verify this!) and therefore

$$S_g(x) = S_f\left(\frac{\pi}{2}x + \frac{\pi}{2}\right) = \frac{1}{2} + \frac{2}{\pi} \sum_{n=0}^{\infty} \frac{(-1)^n}{2n+1} \cos\left(\frac{\pi}{2}(2n+1)x\right).$$

As a bonus, we can evaluate the following two infinite sums: $\sum_{n=0}^{\infty}(-1)^n/(2n+1)$ and $\sum_{n=0}^{\infty} 1/(2n+1)^2$. Indeed, by Theorem 5.1.5,

$S_g(0) = g(0)$ or $1 = 1/2 + (2/\pi)\sum_{n=0}^{\infty}(-1)^n/(2n+1)$, which means

$$\sum_{n=0}^{\infty} \frac{(-1)^n}{(2n+1)} = \frac{\pi}{4}. \tag{5.1.20}$$

By Parseval's identity (see (5.1.18)),

$$2\left(\frac{1}{2}\right)^2 + \frac{4}{\pi^2}\sum_{n=0}^{\infty}\frac{1}{(2n+1)^2} = \frac{1}{2}\int_{-2}^{2} g^2(x)dx = 1,$$

which means

$$\sum_{n=0}^{\infty}\frac{1}{(2n+1)^2} = \frac{\pi^2}{8}. \tag{5.1.21}$$

EXERCISE 5.1.17.^{C} *Use the Fourier series expansion (5.1.19) of the function f from Figure 5.1.1(a), to derive (5.1.20) and (5.1.21).*

To complete the EXAMPLE, let us consider the 2π-periodic function $h = h(x)$ so that $h(x) = |x|$ for $|x| \le \pi$; see Figure 5.1.2(b).

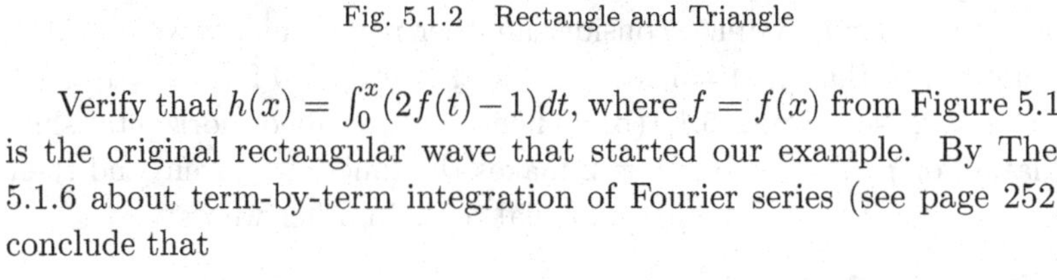

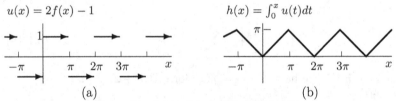

Fig. 5.1.2 Rectangle and Triangle

Verify that $h(x) = \int_0^x (2f(t)-1)dt$, where $f = f(x)$ from Figure 5.1.1(a) is the original rectangular wave that started our example. By Theorem 5.1.6 about term-by-term integration of Fourier series (see page 252), we conclude that

$$S_h(x) = \int_0^x (2S_f(t) - 1)dt = \frac{4}{\pi}\sum_{n=0}^{\infty}\frac{1}{(2n+1)^2} - \frac{4}{\pi}\sum_{n=0}^{\infty}\frac{\cos(2n+1)x}{(2n+1)^2}.$$

Even if we did not compute the value of the sum $\sum_{n=0}^{\infty} 1/(2n+1)^2$ before, we know that

$$\frac{4}{\pi}\sum_{n=0}^{\infty}\frac{1}{(2n+1)^2} = a_0 = \frac{1}{2\pi}\int_{-\pi}^{\pi} h(x)dx = \frac{1}{\pi}\int_0^{\pi} x dx = \frac{\pi}{2},$$

and so

$$S_h(x) = \frac{\pi}{2} - \frac{4}{\pi}\sum_{n=0}^{\infty}\frac{\cos(2n+1)x}{(2n+1)^2}.$$

It is true that using these computational tricks requires some experience. It is also true that using these tricks is much more efficient than computing the Fourier coefficients of the function h using formulas (5.1.12) on page 246.

EXERCISE 5.1.18.^{C} *Using Parseval's identity and the formula for S_h, verify that*

$$\sum_{n=0}^{\infty}\frac{1}{(2n+1)^4} = \frac{\pi^4}{96}.$$

EXERCISE 5.1.19.^{C} *For the function $f(x) = x$, $x \in [0,1]$, write the three possible Fourier series expansions (for the extension with period one, and for the even and odd extensions with period two.) Which expansion converges uniformly, and how can you figure it out* without computing *the corresponding Fourier coefficients? Suggestion: use the results of the above example as much as possible; in particular, try to limit your integration to computing $\int_{-1}^{1} x\sin(\pi n x)dx$ only.*

We finish this section by briefly discussing the APPLICATION OF FOURIER SERIES TO THE STUDY OF ORDINARY DIFFERENTIAL EQUATIONS. Consider the linear second-order equation $\ddot{y}(t) + \omega_0^2 y(t) = 0$, where $\omega > 0$ is a real number and $\ddot{y}(t)$ denotes the second derivative of y with respect to t. For example, equation (4.1.8) on page 188, describing the current in a series AC circuit with $R = 0$ and time-independent E has this form and $\omega_0 = 1/\sqrt{LC}$. Recall that the general solution of this equation is $y(t) = Ae^{i\omega_0 t} + Be^{-i\omega_0 t}$, and the numbers A, B are determined by the initial conditions; computations are often easier if we use complex exponentials rather than sines and cosines. This equation models a `harmonic oscillator`, that is, a system whose free motion (motion without outside forcing) is *undamped* (non-diminishing) oscillations; ω_0 represents the *proper* frequency of the system.

If an outside forcing $f = f(t)$ is applied to the system, then the corresponding equation becomes $\ddot{y}(t) + \omega_0^2 y(t) = f(t)$. For periodic function f, we can find the general solution of this equation by expanding f in a Fourier series and considering each term of the expansion separately. Indeed, if f is periodic with period $T = 2\pi/\omega$, then $f(t) = \sum_{k=-\infty}^{\infty} c_k(f)e^{2\pi i k t/T} =$

$\sum_{k=-\infty}^{\infty} c_k(f)e^{ik\omega t}$. If $k\omega \neq \omega_0$ for all $k > 0$, then, for every k, a particular solution of $\ddot{y}(t) + \omega_0 y(t) = e^{ik\omega t}$ has the form $y(t) = C_k e^{ik\omega t}$, which, after substitution, results in $C_k(-k^2\omega^2 + \omega_0^2) = 1$ or $C_k = 1/(\omega_0^2 - k^2\omega^2)$. By linearity of the equation, the general solution of $\ddot{y}(t) + \omega_0^2 y(t) = f(t)$ is therefore

$$y(t) = Ae^{i\omega_0 t} + Be^{-i\omega_0 t} + \sum_{k=-\infty}^{\infty} \frac{c_k(f)}{(\omega_0^2 - k^2\omega^2)} e^{ik\omega t}.$$

Note that the series on the right-hand side converges uniformly even if the Fourier series for f does not converge uniformly. Note also that $y(t)$ stays bounded for all t.

EXERCISE 5.1.20.[C] *Find the solution of* $\ddot{y}(t) + 4y(t) = h(t)$, *where* $h(t)$ *is the function shown on Figure 5.1.2, and* $y(0) = \dot{y}(0) = 0$.

If $\omega_0 = k\omega$ for some k, then we get a **resonance** and an unbounded solution: the particular solution of $\ddot{y}(t) + \omega_0^2 y(t) = e^{i\omega_0 t}$ has the form $y(t) = Cte^{i\omega_0 t}$. In practical terms, the system can break down if subjected to the external resonant forcing for sufficiently long time. As a result, knowledge of the Fourier series expansion of the external forcing is crucial to ensuring the stability of the system.

EXERCISE 5.1.21.[B] *Find the solution of* $\ddot{y}(t) + 9y(t) = h(t)$, *where* $h(t)$ *is the function shown on Figure 5.1.2(b), and* $y(0) = \dot{y}(0) = 0$.

Sometimes we know for sure that a given external periodic force will be acting on the system (example could be a train moving on a bridge and jumping on the rail joints). There are two ways to ensure that the system does not break down under this forcing:

- Adjust the proper frequency ω_0 of the system so that ω_0 is not a multiple of ω.
- Introduce *damping* (friction or other energy loss) into the system.

A damped system is described by the equation $\ddot{y}(t) + a^2\dot{y}(t) + \omega_0 y(t) = f(t)$, where $a > 0$; the larger the a, the stronger the **damping**. In a series AC circuit, the source of damping is the resistor R; see equation (4.1.8) on page 4.1.8. For small a, the free motion of the system, that is, the general solution of $\ddot{y}(t) + a^2\dot{y}(t) + \omega_0 y(t) = 0$, is *exponentially decaying oscillations*; for sufficiently large a, the free motion of the system has no oscillations at all.

EXERCISE 5.1.22.[A] *Verify that all solutions of* $\ddot{y}(t) + a^2\dot{y}(t) + 9y(t) = h(t)$, *where* $h(t)$ *is the function shown on Figure 5.1.2, remain bounded. How does the bound depend on* a*? Hint: the computations are easier with complex exponentials rather than with sines and cosines; see Section 4.1.3, page 187.*

This completes our discussion of the Fourier series. We will revisit the topic in the following chapter, where we use Fourier series to solve certain *partial* differential equations.

5.2 Fourier Transform

5.2.1 *From Sums to Integrals*

The Fourier series represents a function that is periodic. Is there a similar representation for functions that are not periodic? The following exercise provides a clue.

EXERCISE 5.2.1.[C] *For* $L > 1$, *consider a* $2L$*-periodic function* $f = f(x)$ *so that* $f(x) = 1$, $|x| \leq 1$ *and* $|f(x)| = 0$ *for* $1 < |x| \leq L$. *Verify that*

$$c_0(f) = \frac{1}{2L};\ c_k(f) = \frac{1}{2L}\,\frac{\sin(\pi k/L)}{\pi k/L},\ k = \pm 1, \pm 2, \ldots$$

are the Fourier coefficients of f. *For three different values of* L $(L = 5, 10, 50)$, *use a computer algebra system to plot the numbers* $2Lc_k(f)$ *versus* $\pi k/L$, $k = 0, \pm 1, \ldots \pm 5L$ *(try not to connect the points on your plots, and do not forget to multiply* $c_k(f)$ *by* $2L$*). Compare your plots with the graph of the function* $g(x) = \sin x/x$, $|x| \leq 5\pi$.

If you indeed do the above exercise (you do not even need a computer to get the main idea), you will notice that, as L increases, the suitably re-scaled graph of the Fourier coefficients approaches a continuous curve. In other words, as we increase the period of the function, the discreteness of the spectrum becomes less and less visible, and, in the limit, we get the curve $\sin x/x$, which could be called the *continuous spectrum* of the non-periodic function $f(x) = 1, |x| \leq 1$, $f(x) = 0, |x| > 1$. In what follows, we will try to formalize this observation.

Let $f = f(x)$ be a function defined on $(-\infty, \infty)$ and let $f_L(x)$ be the restriction of $f(x)$ to the finite interval $[-L, L]$. We assume that, for every $L > 0$, the function f_L satisfies the conditions of Theorem 5.1.5 on page 251, so that the Fourier series $S_{f_L}(x)$ of $f_L(x)$ converges. We rewrite formula

(5.1.17) on page 254 as follows:

$$f_L(x) = \sum_{k=-\infty}^{\infty} \left(\frac{1}{2\pi} \int_{-L}^{L} f_L(t) e^{-ik\pi t/L} dt \right) e^{ik\pi x/L} \cdot \frac{\pi}{L};$$

the equality holds at all points x where the $2L$-periodic extension of f_L is continuous. Let $\omega_k = k\pi/L$ and $\Delta\omega_k = \omega_{k+1} - \omega_k = \pi/L$. Then

$$f_L(x) = \sum_{k=-\infty}^{\infty} \left(\frac{1}{2\pi} \int_{-L}^{L} f_L(t) e^{-i\omega_k t} dt \right) e^{i\omega_k x} \cdot \Delta\omega_k.$$

Let us define the function $C_L = C_L(\omega)$, $\omega \in \mathbb{R}$, by

$$C_L(\omega) = \frac{1}{2\pi} \int_{-L}^{L} f(t) e^{-i\omega t} dt, \tag{5.2.1}$$

and also the function $F_L(\omega, x) = C_L(\omega) e^{i\omega x}$. Then

$$f_L(x) = \sum_{k=-\infty}^{\infty} F_L(\omega_k, x)\Delta\omega_k = \sum_{k=-N}^{N} F_L(\omega_k, x)\Delta\omega_k + \varepsilon_{N,L}(x), \tag{5.2.2}$$

where, for every x and L, $\varepsilon_{N,L}(x) \to 0$ as $N \to \infty$. Note that, for each x, the right-hand side of (5.2.2) is a Riemann sum, and it is natural to pass to the limit and turn this sum into an integral. This passage to the limit requires decreasing $\Delta\omega_k = \pi/L$, that is, increasing L. As $L \to \infty$, the value of $C_L(\omega)$ will converge to $C(\omega) = (1/2\pi) \int_{-\infty}^{\infty} f(t) e^{-i\omega t} dt$, *provided* $\int_{-\infty}^{\infty} |f(t)| dt$ exists. As a result, existence of this integral is the necessary additional assumption about the function f. With this assumption in place, we can now expect from (5.2.2) that, for each fixed N and x,

$$\begin{aligned} \lim_{L\to\infty} f_L(x) &= \lim_{L\to\infty} \sum_{k=-N}^{N} F_L(\omega_k, x)\Delta\omega_k + \lim_{L\to\infty} \varepsilon_{N,L}(x) \\ &= \int_{-N}^{N} F(\omega, x) d\omega + \varepsilon_N(x). \end{aligned}$$

So far, the only questionable step in our argument is our assumption that $\lim_{L\to\infty} \varepsilon_{N,L}(x)$ exists. In fact, we need to go even further and assume that, similar to $\varepsilon_{N,L}(x)$, we have $\lim_{N\to\infty} \varepsilon_N(x) = 0$ for all x. A more careful analysis shows that we can ensure these properties of ε, for example, with an additional condition $\int_{-\infty}^{\infty} |F(\omega, x)| d\omega < C$ with C independent of x.

Then, allowing $N \to \infty$, we conclude that $f(x) = \int_{-\infty}^{\infty} F(\omega, x) d\omega$ or

$$f(x) = \int_{-\infty}^{\infty} \left(\frac{1}{2\pi} \int_{-\infty}^{\infty} f(t) e^{-i\omega t} dt \right) e^{i\omega x} d\omega. \tag{5.2.3}$$

By analogy with Theorem 5.1.5 on page 251, we have the following result.

Theorem 5.2.1 *Assume that the function $f = f(t)$ is continuous and has a continuous bounded derivative everywhere on $\mathbb{R}$ except for a finite number of points. Also assume that $\int_{-\infty}^{\infty} |f(t)| dx < \infty$. Then the function*

$$I_f(t) = \int_{-\infty}^{\infty} \left(\frac{1}{2\pi} \int_{-\infty}^{\infty} f(s) e^{-i\omega s} ds \right) e^{i\omega t} d\omega. \tag{5.2.4}$$

is defined for all $t \in \mathbb{R}$ and

$$I_f(t) = \begin{cases} f(t), & \text{if } f \text{ is continuous at } t \\ \big(f(t^+) + f(t^-)\big)/2, & \text{if } f \text{ is not continuous at } t. \end{cases} \tag{5.2.5}$$

The proof is somewhat similar to the proof of Theorem 5.1.5, and we omit it. Notice that we do not have to assume that $\int_{-\infty}^{\infty} |F(\omega)| d\omega < \infty$.

The function I_f from (5.2.4) is called the `Fourier integral` of f. The `Fourier transform` of f, denoted either by $\widehat{f}$ or by $\mathcal{F}[f]$, is

$$\widehat{f}(\omega) = \mathcal{F}[f](\omega) = \frac{1}{\sqrt{2\pi}} \int_{-\infty}^{\infty} f(t) e^{-i\omega t} dt. \tag{5.2.6}$$

The `inverse Fourier transform` of f, denoted either by $\check{f}$ or by $\mathcal{F}^{-1}[f]$, is

$$\check{f}(\omega) = \mathcal{F}^{-1}[f](\omega) = \frac{1}{\sqrt{2\pi}} \int_{-\infty}^{\infty} f(t) e^{i\omega t} dt. \tag{5.2.7}$$

The condition $\int_{-\infty}^{\infty} |f(x)| dx < \infty$ is sufficient for the existence of both $\widehat{f}$ and $\check{f}$, because $|f(t) e^{-i\omega t}| = |f(t) e^{i\omega t}| = |f(t)|$ for all ω.

With the above definitions, $\mathcal{F}^{-1}[\mathcal{F}[f]](t) = I_f(t)$; under the conditions of Theorem 5.2.1, we have $\mathcal{F}^{-1}[\mathcal{F}[f]](t) = f(t)$, or

$$f(t) = \frac{1}{\sqrt{2\pi}} \int_{-\infty}^{\infty} \widehat{f}(\omega) e^{i\omega t} d\omega,$$

if f is continuous at t.

The reader should keep in mind that other scientific disciplines might use **other definitions of the Fourier transform**. For example, the following version of (5.2.4) is popular in engineering:

$$I_f(t) = \int_{-\infty}^{\infty} \left(\int_{-\infty}^{\infty} f(s) e^{-i\,2\pi\nu s} ds \right) e^{i\,2\pi\nu t} d\nu. \qquad (5.2.8)$$

The following exercise provides an explanation.

EXERCISE 5.2.2.^{C} *Given real numbers $A > 0$ and $B \neq 0$, define*

$$\widehat{f}_{A,B}(y) = \frac{1}{A} \int_{-\infty}^{\infty} e^{-iBxy} dx.$$

(a) Show that, under the assumptions of Theorem 5.2.20, we have

$$f(x) = \frac{A\,|B|}{2\pi} \int_{-\infty}^{\infty} \widehat{f}_{A,B}(y) e^{iBxy} dy$$

if f is continuous at x. (b) Verify that (5.2.8) corresponds to $A = 1$, $B = 2\pi$.

Another definition is related to the class of functions for which the Fourier transform is defined. We say that f is **absolutely integrable** on $\mathbb{R}$, and write $f \in L_1(\mathbb{R})$, if $\int_{-\infty}^{\infty} |f(x)| dx < \infty$. More generally, for a real number p satisfying $1 \leq p < \infty$, the set $L_p(\mathbb{R})$ is the collection of all functions f defined on $\mathbb{R}$ so that $\int_{-\infty}^{\infty} |f(x)|^p dx < \infty$. To summarize,

$$L_p(\mathbb{R}) = \left\{ f : \int_{-\infty}^{\infty} |f(x)|^p dx < \infty \right\}, \; 1 \leq p < \infty.$$

Strictly speaking, the integrals in these definitions are the Lebesgue integrals, as are all the integrals in this and the following two sections, but this should not stop the reader from proceeding, as the precise definition of either the Riemann or the Lebesgue integral is not necessary for understanding the presentation.

The results of the following exercise are useful to keep in mind, even if you do not do the exercise.

EXERCISE 5.2.3.^{A} *(a) Give an example of a function f from $L_p(\mathbb{R})$ so that $\lim_{|x|\to\infty} |f(x)| \neq 0$. Hint: let your function be zero everywhere except short intervals I_k around the points $x_k = k$, $k = 2, 3, 4, \ldots$, where the function is equal to 1 (draw a picture). For every p, the value of the integral $\int_{-\infty}^{\infty} |f(x)|^p dx$ is then $\sum_{k\geq 2} |I_k|$, the sum of the lengths of the intervals. Choose $|I_k|$ so that the sum is finite. (b) Give an example of a function from $L_1(\mathbb{R})$ that is not in $L_2(\mathbb{R})$.*

Hint: think $1/\sqrt{x}$ *for* $0 < x < 1$. *(c) Give an example of a function from* $L_2(\mathbb{R})$ *that is not in* $L_1(\mathbb{R})$. *Hint: think* $1/x$ *for* $x > 1$.

We now combine Theorem 5.2.1 with Theorem 5.1.5 on page 251 about the convergence of the Fourier series to prove the `Nyquist-Shannon sampling theorem`. The theorem shows that a `band-limited` signal f can be exactly recovered from its samples at equally spaces time moments; by definition, the signal f is called band-limited if there exists an $\Omega > 0$ such that the Fourier transform $\widehat{f}$ of f satisfies $\widehat{f}(\omega) = 0$ for $|\omega| > \Omega$.

Theorem 5.2.2 *If the Fourier transform* $\widehat{f} = \widehat{f}(\omega)$ *of a function* $f = f(t)$ *exists and is equal to zero for* $|\omega| > \Omega$, *and if both* f *and* $\widehat{f}$ *are continuous everywhere and have continuous derivatives everywhere except at finitely many points, then*

$$f(t) = \sum_{k=-\infty}^{\infty} \frac{\sin\left(\Omega(t - k\Delta t)\right)}{\Omega(t - k\Delta t)} f(k\Delta t), \text{ where } \Delta t = \frac{\pi}{\Omega}. \tag{5.2.9}$$

Proof. By Theorem 5.2.1 applied to f, with $\pi/\Omega = \Delta t$,

$$f(t) = \frac{1}{\sqrt{2\pi}} \int_{-\Omega}^{\Omega} \widehat{f}(\omega) e^{i\omega t}\, d\omega, \tag{5.2.10}$$

because, by assumption, $|\widehat{f}(\omega)| = 0$ for $|\omega| > \Omega$. By Theorem 5.1.5 applied to $\widehat{f}$,

$$\begin{aligned} \widehat{f}(\omega) &= \sum_{k=-\infty}^{\infty} \left(\frac{1}{2\Omega} \int_{-\Omega}^{\Omega} \widehat{f}(y) e^{-iky\Delta t} dy \right) e^{ik\omega\Delta t} \\ &= \frac{\sqrt{2\pi}}{2\Omega} \sum_{k=-\infty}^{\infty} f(-k\Delta t) e^{ik\omega\Delta t}, \end{aligned} \tag{5.2.11}$$

where the second equality follows from (5.2.10) with $t = -k\Delta t$. Substituting (5.2.11) in (5.2.10) and exchanging the order of summation and integration, we get

$$f(t) = \sum_{k=-\infty}^{\infty} \left(\frac{1}{2\Omega} \int_{-\Omega}^{\Omega} e^{ik\omega\Delta t + i\omega t} d\omega \right) f(-k\Delta t). \tag{5.2.12}$$

It remains to evaluate the integral in (5.2.12) and change the summation index from k to $-k$. □

EXERCISE 5.2.4.^{C} *Verify that one can take $\Delta t < \pi/\Omega$ and still have the result of the theorem. Hint: if $|\widehat{f}(\omega)| = 0$ for $|\omega| > \Omega$ and $\Omega_1 = (\pi/\Delta t) > \Omega$, then $|\widehat{f}(\omega)| = 0$ for $|\omega| > \Omega_1$.*

In 1927, the Swedish-born American scientist HARRY NYQUIST (1889–1976) discovered that, to recover a continuous-time signal from equally-spaced samples, the sampling frequency $2\pi/\Delta t$ should be equal to twice the bandwidth Ω of the signal (that is, $\Delta t = 2\pi/(2\Omega)$). In 1948, the American scientist CLAUDE ELWOOD SHANNON (1916–2001) put communication theory, including the result of Nyquist, on a firm mathematical basis.

Practical implementation of the sampling theorem is not straightforward, because no signal lasts infinitely long, while a signal that is zero outside of a bounded interval is not band-limited; see Exercise 5.2.12(a) on page 271. We discuss some of the related questions on page 277.

We conclude this section with a connection between the samples of a function and the samples of its Fourier transform. Take a continuously differentiable function $f \in L_1(\mathbb{R})$ and a real number $L > 0$, and assume that, for every x, the function $g(x) = \sum_{n=-\infty}^{\infty} f(x + 2Ln)$ is defined and is also continuously differentiable. This is true, for example, if there exists an $R > 0$ so that $f(x) = 0$ for $|x| > R$, because in this case the sum that defines g contains only finitely many non-zero terms. Note that g is periodic with period $2L$, and therefore, by (5.1.17) on page 254, $g(x) = \sum_{k=-\infty}^{\infty} c_k(g) e^{i\pi kx/L}$, where

$$
\begin{aligned}
c_k(g) &= \frac{1}{2L}\int_{-L}^{L} \sum_{n=-\infty}^{\infty} f(x+2Ln)e^{-i\pi kx/L}dx \\
&= \frac{1}{2L}\sum_{n=-\infty}^{\infty} \int_{2Ln}^{2L(n+1)} f(x)e^{-i\pi kx/L}dx \qquad (5.2.13)\\
&= \frac{1}{2L}\int_{-\infty}^{\infty} f(x)e^{-i\pi kx/L}dx = \frac{\sqrt{2\pi}}{2L}\widehat{f}(\pi k/L).
\end{aligned}
$$

Since $g(0) = \sum_{k=-\infty}^{\infty} c_k(g)$, we conclude that

$$
\sum_{n=-\infty}^{\infty} f(2Ln) = \frac{\sqrt{2\pi}}{2L} \sum_{k=-\infty}^{\infty} \widehat{f}(\pi k/L). \qquad (5.2.14)
$$

Equality (5.2.14) is called `Poisson's Summation Formula`, after S. D. Poisson.

EXERCISE 5.2.5. ^{A} *(a) Verify the computations in (5.2.13). (b) Taking* $f(x) = e^{-x^2}$, $t > 0$, *verify the* θ-**formula**:

$$\sum_{n=-\infty}^{\infty} e^{-4t\pi^2 n^2} = \frac{1}{\sqrt{2t}} \sum_{k=-\infty}^{\infty} e^{-k^2/(4t)}.$$

Then explain how to use such a formula for computing approximately the value of $\sum_{k=-\infty}^{\infty} e^{-ak^2}$ *for various* $a > 0$; *the problem of approximating this sum arises, for example, in statistics. Hint: how does the value* $a > 0$ *control the rate of convergence of the sequence* e^{-an^2}, $n \geq 1$, *to zero?*

5.2.2 Properties of the Fourier Transform

As in our discussion of Fourier series, we will study the general properties of the Fourier transform before doing any particular computations. To begin, let us mention those properties of the Fourier transform that have counterparts for the Fourier series; while the proofs are beyond the scope of our discussion, the results are believable given what we know about the Fourier series and the Fourier transform.

• The **Riemann-Lebesgue theorem**: If $f \in L_1(\mathbb{R})$, then $\widehat{f}$ is continuous on $\mathbb{R}$ and $\lim_{|\omega|\to\infty} |\widehat{f}(\omega)| = 0$.

• **Parseval's identity**: If f belongs to both $L_1(\mathbb{R})$ and $L_2(\mathbb{R})$, then $\widehat{f} \in L_2(\mathbb{R})$ and

$$\int_{-\infty}^{\infty} |f(t)|^2 dt = \int_{-\infty}^{\infty} |\widehat{f}(\omega)|^2 d\omega. \tag{5.2.15}$$

Equality (5.2.15) is also known as **Plancherel's Theorem**, after the Swiss mathematician MICHEL PLANCHEREL (1885–1967), who was the first to establish it.

EXERCISE 5.2.6.^{A} *Verify that (5.2.15) implies*

$$\int_{\infty}^{\infty} f(x)\,\overline{g(x)}\,dx = \int_{\infty}^{\infty} \mathcal{F}[f](\omega)\,\overline{\mathcal{F}[g](\omega)}\,d\omega \tag{5.2.16}$$

for all functions f, g *from* $L_2(\mathbb{R})$. *As usual,* $\overline{a}$ *is the complex conjugate of the number* a. *Hint: Apply (5.2.15) with* $f+g$ *instead of* f *to show that the real parts of the two sides of (5.2.16) are the same; then use* $f+ig$ *to establish the equality of the imaginary parts; the key relations are* $|a+b|^2 = |a|^2+|b|^2+2\Re(a\overline{b})$; $|a+ib|^2 = |a|^2+|b|^2-2i\,\Im(a\overline{b})$, *which you should verify too.*

If f represents a time signal, then $\widehat{f}$ is the **spectral density** of f, the continuous analog of the discrete spectrum we considered for periodic signals, see page 256. The values of $|\widehat{f}(\omega)|^2$ describe the distribution of energy among the different frequencies of the signal; Parseval's identity (5.2.15) represents conservation of energy.

The next property shows that the Fourier transform reduces differentiation to multiplication: if f is continuous and differentiable so that both f and f' are in $L_1(\mathbb{R})$ and $\lim_{|t|\to\infty}|f(t)| = 0$, then

$$\mathcal{F}[f'](\omega) = i\omega\mathcal{F}[f](\omega). \tag{5.2.17}$$

This follows after integration by parts:

$$\mathcal{F}[f'](\omega) = \frac{1}{\sqrt{2\pi}}\int_{-\infty}^{\infty} f'(t)e^{-i\omega t}dt = \frac{1}{\sqrt{2\pi}}f(t)e^{-i\omega t}\Big|_{-\infty}^{\infty}$$
$$-\frac{(-i\omega)}{\sqrt{2\pi}}\int_{-\infty}^{\infty} f(t)e^{-i\omega t}dt = i\omega\mathcal{F}[f](\omega).$$

EXERCISE 5.2.7.^{C} *Assume that f has two derivatives so that f, f', f'' all belong to $L_1(\mathbb{R})$, and also $\lim_{|t|\to\infty}|f(t)| = \lim_{|t|\to\infty}|f'(t)| = 0$. Show that*

$$\mathcal{F}[f''](\omega) = -\omega^2\mathcal{F}[f](\omega). \tag{5.2.18}$$

Hint: apply (5.2.17).

There is another operation, called **convolution**, that the Fourier transform reduces to multiplication. For two functions f, g from $L_1(\mathbb{R})$, we define their convolution $f * g$ so that

$$(f * g)(x) = \int_{-\infty}^{\infty} f(x-y)g(y)dy. \tag{5.2.19}$$

EXERCISE 5.2.8.^{C} *Verify that $(f * g)(x) = \int_{-\infty}^{\infty} g(x-y)f(y)dy$.*

It is easy to show that $f * g$ belongs to $L_1(\mathbb{R})$:

$$\int_{-\infty}^{\infty} |(f*g)(x)|\,dx \le \int_{-\infty}^{\infty}\int_{-\infty}^{\infty} |f(x-y)|\,|g(y)|\,dy\,dx$$

$$= \int_{-\infty}^{\infty}\left(\int_{-\infty}^{\infty} |f(x-y)|\,dx\right)|g(y)|\,dy = \left(\int_{-\infty}^{\infty}|f(x)|dx\right)\left(\int_{-\infty}^{\infty}|g(y)|dy\right).$$

Similar arguments show that if either f or g is bounded, then so is $f * g$. Using an advanced trick called *interpolation*, one can then prove that, for f from $L_p(\mathbb{R})$ and g from $L_q(\mathbb{R})$, the convolution $f*g$ is defined and belongs to $L_r(\mathbb{R})$, where $(1/p) + (1/q) = 1 + (1/r)$. The arguments, strictly speaking, work only with the Lebesgue integral; for an example how things can go bad with the Riemann integral, see Example C6 on p. 570 in the book *Fourier Analysis* by T. W. Körner, 1988.

The Fourier transform of the convolution is, up to a constant factor, the product of the Fourier transforms: if f, g belong to $L_1(\mathbb{R})$, then

$$\widehat{f * g} = \sqrt{2\pi}\, \widehat{f}\, \widehat{g}. \tag{5.2.20}$$

Indeed,

$$\begin{aligned}\widehat{f * g}(\omega) &= \frac{1}{\sqrt{2\pi}} \int_{-\infty}^{\infty} \int_{-\infty}^{\infty} f(s) g(x-s) e^{-i\omega x} ds dx \\ &= \frac{1}{\sqrt{2\pi}} \int_{-\infty}^{\infty} \int_{-\infty}^{\infty} f(s) g(y) e^{-i\omega (y+s)} dy ds\end{aligned}$$

$$= \left(\frac{1}{\sqrt{2\pi}} \int_{-\infty}^{\infty} f(s) e^{-i\omega s} ds \right) \left(\int_{-\infty}^{\infty} g(y) e^{-i\omega y} dy \right) = \sqrt{2\pi}\, \widehat{f}(\omega)\, \widehat{g}(\omega).$$

The next three properties of the Fourier transform are verified by direct computation.

LINEARITY OF THE FOURIER TRANSFORM. If f, g belong to $L_1(\mathbb{R})$ and a, b are real numbers, then

$$\mathcal{F}[af + bg] = a\,\mathcal{F}[f] + b\,\mathcal{F}[g]. \tag{5.2.21}$$

THE SHIFT FORMULA. If f belongs to $L_1(\mathbb{R})$, a is a real number, and $g(t) = f(t-a)$, then

$$\widehat{g}(\omega) = e^{ia\omega} \widehat{f}(\omega). \tag{5.2.22}$$

THE DILATION FORMULA. If f belongs to $L_1(\mathbb{R})$, $a > 0$ is a real number, and $h(t) = f(at)$, then

$$\widehat{h}(\omega) = \frac{1}{a} \widehat{f}(\omega/a). \tag{5.2.23}$$

EXERCISE 5.2.9. *(a)^C Verify the relations (5.2.21), (5.2.22), and (5.2.23). (b)^A Can we allow complex values of a in (5.2.22) and (5.2.23)?*

The next collection of equalities is as simple as it is useful:

$$\mathcal{F}^{-1}[f](\omega) = \mathcal{F}[f](-\omega) = \mathcal{F}[\tilde{f}\,](\omega), \ \ \mathcal{F}[\widehat{f}\,] = \tilde{f}, \tag{5.2.24}$$

where $\tilde{f}(x) = f(-x)$. All equalities in (5.2.24) follow directly from the definitions (5.2.6) and (5.2.7) after changing the sign of the variable of integration.

EXERCISE 5.2.10.[C] *Verify all equalities in (5.2.24).*

Together with the relation $\mathcal{F}^{-1}[\mathcal{F}[f]] = f$, equalities (5.2.24) have two practical benefits:

- *Two Fourier transforms for the price of one:*

$$\text{if } h(\omega) = \widehat{f}(\omega), \text{ then } \widehat{h}(\omega) = f(-\omega) \tag{5.2.25}$$

(verify this!) Once the definition of the Fourier transform is modified accordingly, the second equality will hold even if h is not absolutely integrable.

- *Two properties for the price of one:* every property of the Fourier transform has a dual property after the inverse Fourier transform is applied to both sides.

In particular, the dual property of (5.2.20) is

$$\widehat{f\,g} = \frac{1}{\sqrt{2\pi}}\,\widehat{f} * \widehat{g}, \tag{5.2.26}$$

assuming $fg \in L_1(\mathbb{R})$. Indeed, by taking the inverse Fourier transform on both sides of (5.2.20), we get $f * g = \sqrt{2\pi}\mathcal{F}^{-1}[\widehat{f}\,\widehat{g}]$. Now replace f, g with $\widehat{f}, \widehat{g}$, respectively, and use (5.2.24) to get

$$(\widehat{f} * \widehat{g})(x) = \sqrt{2\pi}\mathcal{F}^{-1}[\tilde{f}\,\tilde{g}](x) = \sqrt{2\pi}\mathcal{F}[\tilde{f}\,\tilde{g}](-x) = \sqrt{2\pi}\mathcal{F}[f\,g](x).$$

Similarly, the dual property of (5.2.17) is

$$\frac{d}{d\omega}\widehat{f}(\omega) = -i\mathcal{F}[g](\omega), \text{ where } g(x) = xf(x). \tag{5.2.27}$$

EXERCISE 5.2.11. [A] *Assuming that $\int_{-\infty}^{\infty}|xf(x)|dx < \infty$, verify (5.2.18) in two different ways: (i) By differentiating (5.2.6) and moving the derivative under the integral sign; (ii) By applying the inverse Fourier transform to both sides of (5.2.17).*

5.2.3 *Computing the Fourier Transform*

As the first EXAMPLE, let us compute the Fourier transform of the `rectangular pulse`

$$\Pi_L(t) = \begin{cases} 1, & |t| < L, \\ 0, & |t| > L, \end{cases} \tag{5.2.28}$$

where $L > 0$ is a real number; the definition of Π_L for $|t| = L$ is not important. We have

$$\widehat{\Pi}_L(\omega) = \frac{1}{\sqrt{2\pi}} \int_{-L}^{L} e^{-i\omega t} dt = \frac{1}{\sqrt{2\pi}} \frac{e^{iL\omega} - e^{-iL\omega}}{i\omega} = \sqrt{\frac{2}{\pi}} \frac{\sin \omega L}{\omega}.$$

By Theorem 5.2.1,

$$\Pi_L(t) = \frac{1}{\pi} \int_{-\infty}^{\infty} \frac{\sin \omega L}{\omega} e^{i\omega t} d\omega, \ |t| \neq L.$$

Notice that the integral on the right-hand side exists even though $\widehat{f}$ does not belong to $L_1(\mathbb{R})$, and, for $t = \pm L$, the value of the integral is $1/2$. Setting $t = 0$, we find $\int_{-\infty}^{\infty} \frac{\sin \omega L}{\omega} d\omega = \pi$ for all $L > 0$. Using Parseval's identity (5.2.15), we can evaluate another integral: $\int_{-\infty}^{\infty} \left(\frac{\sin \omega L}{\omega} \right)^2 d\omega = L\pi$.

EXERCISE 5.2.12. *(a)[B] Explain why a signal that is zero outside of a bounded time interval cannot be band-limited. Hint: if $f(t) = f(t)\Pi_L(t)$ for some $L > 0$, then $\widehat{f}(\omega)$ is connected to the convolution of $\widehat{f}(\omega)$ and $\frac{\sin \omega L}{\omega}$. Can this convolution be zero outside of a bounded interval in ω? (b)[C] Verify that if $f(x) = e^{-|x|}$, then $\widehat{f}(\omega) = \sqrt{2}/\big(\sqrt{\pi}\,(1 + \omega^2)\big)$. Hint: this is straightforward integration and complex number manipulation; do not try to find any connection with the rectangular pulse. (c)[C] Use (5.2.27) to conclude that the Fourier transform of $te^{-|t|}$ is $2i\sqrt{2/\pi}\,\omega/(1 + \omega^2)^2$.*

For the next EXAMPLE, consider the function $f(t) = e^{-t^2/2}$. Turns out, the Fourier transform does not change this function:

$$\widehat{f}(\omega) = e^{-\omega^2/2}. \tag{5.2.29}$$

The rigorous computation is somewhat long and uses complex integration; see Problem 5.7 on page 436. Here, we will give a more intuitive explanation. The key is the relation

$$\int_{-\infty}^{\infty} e^{-x^2/2} dx = \sqrt{2\pi}.$$

Indeed, writing $I = \int_{-\infty}^{\infty} e^{-x^2/2} dx$, we have $I^2 = \int_{-\infty}^{\infty} \int_{-\infty}^{\infty} e^{-(x^2+y^2)/2} dx dy$, and, after changing to polar coordinates, $I^2 = 2\pi \int_0^{\infty} e^{-r^2/2} r dr = 2\pi$. As a result,

$$\int_{-\infty}^{\infty} e^{-(x-a)^2/2} dx = \sqrt{2\pi} \tag{5.2.30}$$

for every *real* a. On the other hand,

$$\widehat{f}(\omega) = \frac{1}{\sqrt{2\pi}} \int_{-\infty}^{\infty} e^{-(t^2+2i\omega t)/2} dt,$$

and $t^2 + 2i\omega t = (t + i\omega)^2 + \omega^2$. Equality (5.2.29) follows immediately, if we assume that (5.2.30) holds for all *complex* numbers a (we should be more careful here because we are actually computing the integral in the complex plane; see Problem 5.7 for details). In any case, the result (5.2.29) is certainly worth remembering.

EXERCISE 5.2.13. ^{C} *Verify that if* $f(x) = e^{-ax^2}$, $a > 0$, *then* $\widehat{f}(\omega) = (2a)^{-1/2} e^{\omega^2/(4a)}$. *Hint: use (5.2.23).*

To conclude this section, let us discuss the *real*, as opposed to complex, form of the **Fourier integral**. On the right-hand side of (5.2.4), page 263, we rename one of the variables of integration from t to s and use the Euler formula for the complex exponential:

$$\begin{aligned} I_f(t) &= \frac{1}{2\pi} \int_{-\infty}^{\infty} \int_{-\infty}^{\infty} f(s) e^{-i\omega(s-t)} ds\, d\omega \\ &= \frac{1}{2\pi} \int_{-\infty}^{\infty} \int_{-\infty}^{\infty} f(s) \Big(\cos\big(\omega(s-t)\big) - i \sin\big(\omega(s-t)\big)\Big) ds\, d\omega. \end{aligned}$$

If $f(t)$ is real, then the imaginary part of the last expression must be zero. Therefore, $I_f(t) = (1/2\pi) \int_{-\infty}^{\infty} \int_{-\infty}^{\infty} f(s) \cos \omega(s-t)\, ds\, d\omega$. Since $\cos \omega(s-t)$ is an even function of ω,

$$I_f(t) = \frac{1}{\pi} \int_0^{\infty} \int_{-\infty}^{\infty} f(s) \cos \omega(s-t)\, ds\, d\omega.$$

Using the identity $\cos\big(\omega(s-t)\big) = \cos\,\omega t\,\cos\omega s + \sin\,\omega t\,\sin\,\omega s$, we get the *real form of the Fourier integral:*

$$\begin{aligned} f(t) = & \int_0^\infty \left(\frac{1}{\pi}\int_{-\infty}^\infty f(s)\cos\,\omega s\,ds\right)\cos\,\omega t\,d\omega \\ & + \int_0^\infty \left(\frac{1}{\pi}\int_{-\infty}^\infty f(s)\sin\,\omega s\,ds\right)\sin\,\omega t\,d\omega. \end{aligned} \tag{5.2.31}$$

For a function f on the half-line $(0,\infty)$, define the **Fourier cosine transform**

$$\mathcal{F}_c[f](\omega) = \hat{f}_c(\omega) = \sqrt{\frac{2}{\pi}}\int_0^\infty f(s)\cos\,\omega s\,ds, \tag{5.2.32}$$

and the **Fourier sine transform**

$$\mathcal{F}_s[f](\omega) = \hat{f}_s(\omega) = \sqrt{\frac{2}{\pi}}\int_0^\infty f(s)\sin\,\omega s\,ds. \tag{5.2.33}$$

If f is an even function, then $f(s)\sin\omega s$ is an odd function of s and the second term in (5.2.31) is zero. Hence, for even f,

$$I_f(t) = \int_0^\infty \sqrt{\frac{2}{\pi}}\hat{f}_c(\omega)\cos\omega t\,d\omega = \mathcal{F}_c[\hat{f}_c](t). \tag{5.2.34}$$

Thus, $\mathcal{F}_c^{-1} = \mathcal{F}_c$, and, for functions defined on $(0,\infty)$, the Fourier cosine transform represents the *even* extension of the function f to $\mathbb{R}$.

If $f(s)$ is an odd function, then $f(t) = \int_0^\infty \sqrt{2/\pi}\,\hat{f}_s(\omega)\sin\,\omega t\,d\omega = \mathcal{F}_s[\hat{f}_s](t)$. Thus, $\mathcal{F}_s^{-1} = \mathcal{F}_s$, and, for functions defined on $(0,\infty)$, the Fourier sine transform represents the *odd* extension of the function f to $\mathbb{R}$. Other conditions being equal, the choice between the even and odd extensions is determined by the smoothness properties of the result: the smoother the extension, the better.

Similar to (5.2.17), we have the following rules for transforms of derivatives; the reader is encouraged to verify these rules using integration by parts. Assume that f is continuous and $\displaystyle\int_0^\infty |f(x)|dx < \infty$. Also assume that $\lim_{x\to\infty}|f(x)| = 0$ and that, on every bounded interval, the function f has a bounded derivative everywhere except at finitely many points. Then

$$\mathcal{F}_c[f'] = \omega\mathcal{F}_s[f] - \sqrt{2/\pi}f(0),\ \ \mathcal{F}_s[f'] = -\omega\mathcal{F}_c[f].$$

If, in addition, the function f' has the same properties as f, then

$$\mathcal{F}_c[f''] = -\omega^2\mathcal{F}_c[f] - \sqrt{2/\pi}f'(0),\ \ \mathcal{F}_s[f''] = -\omega^2\mathcal{F}_s[f] + \sqrt{2/\pi}\,\omega f(0).$$

These results can be used to compute transforms without integration. FOR EXAMPLE, let us compute $\mathcal{F}_c[f]$ for $f(t) = e^{-at}$, where $a > 0$. Since $f'(t) = -ae^{-at}$ and $f'' = a^2 f$, we have $a^2\mathcal{F}_c[f] = -\omega^2\mathcal{F}_c[f] + a\sqrt{2/\pi}$ or

$$\mathcal{F}_c[f](\omega) = \sqrt{\frac{2}{\pi}}\,\frac{a}{a^2+\omega^2}.$$

EXERCISE 5.2.14. *(a)*^{B} *Use similar arguments to verify that, for* $f(t) = e^{-at}$, $\mathcal{F}_s[f](\omega) = \sqrt{2}\,\omega/\big(\sqrt{\pi}\,(a^2+\omega^2)\big)$. *(b)*^{A} *Compare the results with Exercise 5.2.12. Why can't we use the same trick to compute the Fourier transform of f using (5.2.18)? Hint: If you simply allow x to be negative, then $f \notin L_1(\mathbb{R})$; on the other hand neither the even nor the odd extension of f is differentiable twice.*

As a final comment, we mention that the Fourier transforms of rational functions can often be computed using residue integration, see page 229.

5.3 Discrete Fourier Transform

5.3.1 *Discrete Functions*

Looking back at our study of Fourier series and transforms, we realize that we have two parallel theories. One is for periodic functions represented by the Fourier series. The other is for integrable functions on $\mathbb{R}$ represented by the Fourier integral. Mathematics tries to avoid such separations as much as possible and aims at a unified theory. How, then, can we combine the Fourier series and Fourier integral? Clearly, a periodic function is characterized by a countable number of Fourier coefficients, and an $L_1(\mathbb{R})$ function, by its Fourier transform $\widehat{f}$, which is a function of a continuous variable ω. In general, there is little hope to represent a general function by a countably many values, and we therefore should try the other way and represent the discrete collection of the Fourier coefficients as a function of a continuous variable.

Given an integrable periodic function $f = f(t)$ with period $2L$, the collection $\{c_k(f),\ k = 0, \pm1, \pm2, \ldots\}$ of the Fourier coefficients is a *discrete* function defined only for $\omega = k\pi/L$. How to define this collection as a function of the *continuous* argument ω? The obvious definition, $\widehat{f}(\omega) = c_k(f)$ if $\omega = \pi k/L$, and $\widehat{f}(\omega) = 0$ otherwise, is not adequate, because any attempt to integrate the function $\widehat{f}$ will produce 0. An alternative definition, setting $\widehat{f}(\omega) = c_k(f)$ for $\pi(k-1/2)/L < \omega \le \pi(k+1/2)/L$, is nicely integrable, but looses the main spirit of the discrete signal: we expect $\widehat{f}$ to be zero everywhere except $\omega = \pi k/L$. To reconcile these seemingly irreconcilable objectives, take an integer $n > 1$ and define the function

$$\bar{\Pi}_n(x) = \begin{cases} n, & |x| < 1/(2n) \\ 0, & |x| > 1/(2n). \end{cases} \tag{5.3.1}$$

For large n, we have $\bar{\Pi}_n(x) = 0$ everywhere except in a small neighborhood of zero. On the other hand, $\int_{-\infty}^{\infty} \bar{\Pi}_n(x)dx = n(1/n) = 1$ for all n. It would be nice to pass to the limit as $n \to \infty$, but the limit, which should be equal to ∞ when $x = 0$ and zero otherwise, does not look like a function.

Let us try a different approach and consider the sequence of integrals $I_n = \int_{-\infty}^{\infty} f(x)\bar{\Pi}_n(x)dx$. If n is sufficiently large and f is continuous at zero, then, by the mean-value theorem for integrals (or a version of the rectangular rule), we get $I_n \approx f(0)$, and, in fact,

$$\lim_{n\to\infty} \int_{-\infty}^{\infty} f(x)\bar{\Pi}_n(x)dx = f(0). \tag{5.3.2}$$

EXERCISE 5.3.1.[C] *Verify (5.3.2) for every function that is integrable on* $[-1, 1]$ *and is continuous at zero. Hint:* $f(0) = \int_{-\infty}^{\infty} f(0)\bar{\Pi}_n(x)dx$*, and, by continuity, for every* $\varepsilon > 0$*, there exists an* N *so that, for all* $n > N$*,* $|f(x) - f(0)| < \varepsilon$ *if* $|x| < 1/(2n)$*.*

We can now use (5.3.2) to define $\lim_{n\to\infty} \bar{\Pi}_n$ as a "black box" (or a *rule*) that takes a continuous function as an input, and produce the value of that function at zero as the output. In mathematics, a rule that makes a number out of a function is called a `functional`. Thus, the limit of $\bar{\Pi}_n(x)$, as $n \to \infty$, is an functional. This functional is called `Dirac's delta function`, in honor of the British physicist PAUL ADRIEN MAURICE DIRAC (1902–1984), and is denoted by $\delta(x)$. Given what we know about the functions $\bar{\Pi}_n$, we have the following intuitive description of the delta function:

$$\delta(x) = 0, x \neq 0,\ \delta(0) = +\infty,\ \int_{-\infty}^{\infty} \delta(x)dx = 1.$$

The expression

$$\int_{-\infty}^{\infty} \delta(x)f(x)dx = f(0) \tag{5.3.3}$$

is both more precise and more convenient for calculations. Similarly, for a real number a, $\delta_a(x) = \delta(x-a)$ is a functional that takes in a continuous function and produces the value of the function at the point a:

$$\int_{-\infty}^{\infty} f(x)\delta(x-a)dx = f(a). \tag{5.3.4}$$

EXERCISE 5.3.2. *(a)*^{A} *Assume that the argument of the delta function is measured in some physical units, such as time or distance. Show that the delta function must then be measured in the reciprocal of the corresponding units. Hint: this follows directly from (5.3.3). (b)*^{C} *We can use (5.3.2) to compute the Fourier transform of the delta function. Verify that*

$$\mathcal{F}[\delta](\omega) = \frac{1}{\sqrt{2\pi}}. \tag{5.3.5}$$

(c)^B *Verify that*

$$\frac{2\pi}{L}\sum_{k=-\infty}^{\infty} \delta\left(y - \frac{2\pi k}{L}\right) = \sum_{k=-\infty}^{\infty} e^{iykL}. \tag{5.3.6}$$

Hint: treat both sides as functionals, multiply by a function $f = f(y)$ and integrate. Each term on the left will produce $f(2\pi k/L)$; each term on the right, $\sqrt{2\pi}\widehat{f}(-kL)$. Then use the Poisson summation formula (5.2.14) on page 266.

Coming back to the sequence of the Fourier coefficients $c_k(f)$ of a periodic function f, let

$$\widehat{f}(\omega) = \sqrt{2\pi}\sum_{k=-\infty}^{\infty} c_k(f)\delta(\omega - k\pi/L).$$

Then $f(\omega) \neq 0$ only when $\omega = k\pi/L$ for some k, and, by (5.3.4),

$$\frac{1}{\sqrt{2\pi}}\int_{-\infty}^{\infty} \widehat{f}(\omega)e^{i\omega t}d\omega = \sum_{k=-\infty}^{\infty} c_k(f)e^{itk\pi/L} = S_f(t),$$

the Fourier series of f (that was the reason for introducing the extra factor $\sqrt{2\pi}$ in the definition of $\widehat{f}(\omega)$; recall that if $\widehat{f}$ is the Fourier transform of f, then the right-hand side of the above equality is I_f, the Fourier integral of f). In other words, this function $\widehat{f}$ is the right choice.

To summarize, the natural representation for a discrete function of a continuous argument is a linear combination of Dirac's delta functions.

The resulting computational benefits by far outweigh the need to work with functionals.

Next, we will look at the spectrum of a discrete signal. Consider a continuous-time signal $f = f(t)$, sampled at equally spaced points $t_k = k\Delta t$, $k = 0, \pm 1, \pm 2, \ldots$. The corresponding discrete signal is

$$f_d(t) = \Delta t \sum_{k=-\infty}^{\infty} f(t_k)\delta(t - t_k), \tag{5.3.7}$$

where the factor Δt is introduced to preserve the dimension (units of measurement) of f; see Exercise 5.3.2.

EXERCISE 5.3.3.[B] *Verify that*

$$\widehat{f_d}(\omega) = \Delta t \sum_{k=-\infty}^{\infty} f(t_k)e^{-i\omega t_k} = \sum_{k=-\infty}^{\infty} \widehat{f}\left(\omega - \frac{2\pi k}{\Delta t}\right), \tag{5.3.8}$$

where $\widehat{f}$ is the Fourier transform of f. Hint: the first equality follows immediately from (5.3.4). Then write

$$\sum_{k=-\infty}^{\infty} f(t_k)e^{-i\omega t_k} = \sum_{k=-\infty}^{\infty} \int f(t)e^{-i\omega t}\delta(t - t_k)dt$$

and use (5.3.6) with $L = 2\pi/\Delta t$; feel free to exchange summation and integration.

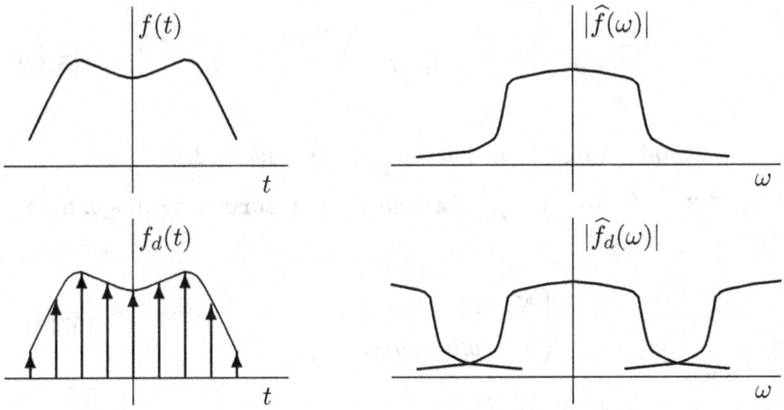

Fig. 5.3.1 Continuous and Discrete Signals

Equality (5.3.8) shows that the spectrum of the discrete signal is a periodic function with period $2\pi/\triangle t$, obtained by the periodic repetition of the original spectrum; see Figure 5.3.1. Recall that the **Nyquist-Shannon sampling theorem** on page 265 provides an example of this effect, where the samples of a signal with *bounded* spectrum represented the periodic extension of the spectrum; with the special selection of the sampling frequency, the extension had no overlaps. Exact recovery of the spectrum of f from the spectrum of f_d leads to exact recovery of f from f_d. If the function $\widehat{f} = \widehat{f}(\omega)$ does not vanish outside of a bounded interval, or if the sampling frequency is not high enough, then the periodic extension of $\widehat{f}$ contains overlaps, and the exact recovery of f from f_d by the Nyquist-Shannon sampling theorem is not possible. The error of recovery of f from f_d that is due to this spectral overlapping is called the **aliasing error**.

5.3.2 *Fast Fourier Transform (FFT)*

We derived the Fourier series expansion of a periodic function by computing the best mean-square approximation of the function by a trigonometric polynomial. Now let us formulate and solve a similar approximation problem for a discrete periodic function. Let $f_0, \ldots, f_{N-1}$, with $f_k = f(2Lk/N)$, be N samples of a $2L$-periodic function f. In discrete time, it is natural to replace the best mean-square approximation with the *best trigonometric interpolation:* find a trigonometric polynomial $P_N(t) = \sum_{k=0}^{N-1} c_k e^{i\pi kt/L}$ so that $P_N(2Ln/N) = f_n$, $n = 0, \ldots, N-1$. Our objective is therefore to find N numbers $c_0, \ldots, c_{N-1}$ from N relations

$$\sum_{k=0}^{N-1} c_k e^{i\,2\pi kn/N} = f_n,\ n = 0, \ldots, N-1. \tag{5.3.9}$$

To proceed, we take a hint from relation (5.1.2), page 242.

EXERCISE 5.3.4. ^{C} *(a) Verify the following* **discrete orthogonality relation**

$$\sum_{n=0}^{N-1} e^{-i\,2\pi n(m-k)/N} = \begin{cases} N, & m = k + \ell N,\ \ell = 0, \pm 1, \pm 2, \ldots \\ 0, & \textit{otherwise}. \end{cases} \tag{5.3.10}$$

Hint: $e^{-i\,2\pi n\ell N} = 1$ *for all* n, ℓ*; otherwise, you are summing* N *terms of a geometric series with the first term equal to one, and the ratio equal to* $e^{-i\,2\pi(m-k)/N}$*,* *so use the formula for the sum. (b) Conclude that* $c_m = F_m$*, where*

$$F_m = \frac{1}{N} \sum_{n=0}^{N-1} f_n e^{-i\,2\pi nm/N}, \tag{5.3.11}$$

and also

$$f_n = \sum_{k=0}^{N-1} F_k e^{i\,2\pi nk/N}, \quad n = 0, \ldots, N-1. \tag{5.3.12}$$

Hint: to derive (5.3.11), multiply both sides of (5.3.9) by $e^{-i\,2\pi mn/N}$*, sum over* n*, and use (5.3.10); (5.3.12) is the same as (5.3.9).*

The collection of numbers $F_n, n = 0, \ldots, N-1$, is called the `discrete Fourier transform` or `discrete Fourier coefficients` of the collection $f_n, n = 0, \ldots, N-1$.

Let us investigate the connection between the discrete Fourier transform and the Fourier series. Recall that the Fourier coefficients of f are $c_k(f) = \frac{1}{2L}\int_{-L}^{L} f(t)e^{-i\,\pi kt/L}dt$. If we assume that the function f is equal to its Fourier series for all t, then

$$f_n = f(2Ln/N) = \sum_{k=-\infty}^{\infty} c_k(f)e^{i\,2\pi kn/N}. \tag{5.3.13}$$

Substituting (5.3.13) in (5.3.11), we find

$$F_m = \frac{1}{N} \sum_{k=-\infty}^{\infty} c_k(f) \sum_{n=0}^{N-1} e^{-i\,2\pi(m-k)n/N} = \sum_{\ell=-\infty}^{\infty} c_{m+\ell N}(f), \tag{5.3.14}$$

where the second equality follows from (5.3.10). If N is even and f is `band-limited`, that is, $c_k(f) = 0$ for $|k| > N/2$, then (5.3.14) implies the following relation between F_m and $c_m(f)$:

$$F_m = \begin{cases} c_m(f), & m = 0, \ldots, \frac{N}{2} - 1, \\ c_{-(N-m)}(f), & m = \frac{N}{2}, \ldots, N-1. \end{cases} \tag{5.3.15}$$

If $c_k(f) \neq 0$ for infinitely many k, then the exact recovery of $c_k(f)$ from F_n is impossible due to the same `aliasing error` we discussed earlier in connection with the Nyquist-Shannon sampling theorem.

Let us now discuss the computational aspects of formula (5.3.11). Straightforward computation of all N numbers F_n according to (5.3.11) requires N multiplications and additions of complex numbers to compute each of the N coefficients. The total operation count is therefore N^2 complex flops (floating point operations), with one `complex flop` being one complex multiplication followed by one complex addition; in many applications this number of operations is inadmissibly large. The `Fast Fourier Transform`, or `FFT`, is a special algorithm that computes F_n in fewer than N^2 operations. There are many versions of FFT, but all of them rely on the same main idea; we will describe this idea next.

Introduce the following notation for the complex exponentials:

$$W_N^{kn} = e^{-i\,2\pi kn/N}. \tag{5.3.16}$$

Note that, on both sides of (5.3.16), kn denotes the product of k and n. Assume that N is a composite number so that $N = N_1N_2$ with integer $N_1, N_2 > 1$. Then we write

$$\begin{aligned} k &= N_1k_2 + k_1, \quad k_1 = 0, \ldots, N_1 - 1, \; k_2 = 0, \ldots, N_2 - 1, \\ n &= N_2n_1 + n_2, \quad n_1 = 0, \ldots, N_1 - 1, \; n_2 = 0, \ldots, N_2 - 1, \end{aligned} \tag{5.3.17}$$

and, using the properties of the exponential function,

$$\begin{aligned} W_N^{kn} &= W_N^{(N_1k_2+k_1)(N_2n_1+n_2)} \\ &= W_N^{Nk_2n_1} W_N^{N_1k_2n_2} W_N^{N_2k_1n_1} W_N^{k_1n_2} = W_{N_2}^{k_2n_2} W_{N_1}^{k_1n_1} W_N^{k_1n_2}. \end{aligned} \tag{5.3.18}$$

EXERCISE 5.3.5.^{C} *Verify all the equalities in (5.3.18). Hint:* $W_N^{N_1k_2n_2} = W_{N_2}^{k_2n_2}$, *etc.*

Using (5.3.18), we now find

$$F_n = \frac{1}{N}\sum_{k=0}^{N-1} f_k W_N^{kn} = \frac{1}{N}\sum_{k_1=0}^{N_1-1}\left(\sum_{k_2=0}^{N_2-1} f_{N_1k_2+k_1} W_{N_2}^{k_2n_2}\right) W_{N_1}^{k_1n_1} W_N^{k_1n_2}. \tag{5.3.19}$$

For each n, computing the inside sum in the last expression requires N_2 complex flops; the sum is then multiplied by $W_{N_1}^{k_1n_1}$, which results in less than N_2+1 complex flops (no addition involved). Finally, computation of the outside sum requires N_1 complex flops. With N coefficients to compute, the total number M of complex flops involved is $M < N(N_1+N_2+1) < N^2$. The reduction is noticeable even for moderate values of N. For example, if $N = 11\cdot 13$, then $M < 3575$, while $N^2 = 20449$, that is, $N^2/M > 5.72$. In other words, when $N = 11\cdot 13$, the above method results in a nearly six-fold reduction of the number of operations. If N_1 or N_2 is also composite, then a repeated application of this method results in even larger reduction of the number of operations. With a more careful analysis, one can show that if $N = 2^n$, then the above FFT requires $(N/2)\log_2 N$ complex multiplications and $N \log_2 N$ complex additions.

Formula (5.3.19) was known to Gauss, who used it in the early 1800s to simplify some of his astronomical computations. Somehow, the result was only published in Latin in 1865, and did not get much attention. A century later, as the advances in electronic computing prompted the development of digital signal processing, American mathematicians JAMES W. COOLEY (b. 1926) and JOHN W. TUKEY (1915–2000) re-discovered (5.3.19); see their paper *An algorithm for the machine calculation of complex Fourier series* in the journal *Mathematics of Computation*, Vol. 19, pages 297–301 (1965).

5.4 Laplace Transform

Recall that the Fourier transform is defined for absolutely integrable functions. For non-integrable functions, for example, those that are positive increasing, it is often impossible to define the Fourier transform, and the Laplace transform is used instead.

5.4.1 *Definition and Properties*

Let $f(x)$ be a real-valued function defined on $-\infty < x < \infty$ and suppose

$$f(x) = 0 \text{ for } x < 0. \tag{5.4.1}$$

Suppose further that, for some real value $s_f > 0$,

$$\int_0^{\infty} |e^{-s_f x} f(x)| dx < \infty. \tag{5.4.2}$$

Then $\int_0^{\infty} |e^{-sx} f(x)| dx < \infty$ for every *complex* number s with $\Re s > s_f$.

Definition 5.3 The `Laplace transform` of f is the function $F = F(s)$ of the complex variable s defined for $\Re s > s_f$ by the formula

$$F(s) = \int_0^{\infty} e^{-sx} f(x) dx. \tag{5.4.3}$$

In what follows, we will usually use a lower-case letter to denote a function, and the corresponding upper-case letter to denote the Laplace transform. Sometimes, we will also write $\mathcal{L}[f](s)$ for the Laplace transform of f.

EXERCISE 5.4.1. ^{C} *Verify that the Laplace transform is an analytic function in the half-plane* $\{s : \Re s > s_f\}$.

The reader might be familiar with the Laplace transform from a course in ordinary differential equations. Even though the transform bears the name of P.-S. Laplace, it was Oliver Heaviside, co-inventor of vector analysis, who developed the application of this transform to the study of differential equations; this application is known as the *operational*, or *Heaviside, calculus.*

One reason to mention Laplace transform in our discussion is the connection with Fourier transform. Indeed, if the function $f = f(x)$ is zero for $x < 0$ and $s_f \leq 0$, then the Fourier transform $\widehat{f}(\omega) = (2\pi)^{-1/2} \int_0^{\infty} f(x) e^{-i\omega x} dx$ of f is defined and

$$F(s) = \sqrt{2\pi}\widehat{f}(-is), \ \widehat{f}(\omega) = \frac{1}{\sqrt{2\pi}} F(i\omega). \tag{5.4.4}$$

Note that for every function f satisfying (5.4.1) and (5.4.2), and for every real number $s_0 > s_f$, the function $f_0(x) = f(x)e^{-s_0 x}$ has $s_{f_0} < 0$. As a result, for functions $f = f(x)$ that are equal to zero when $x < 0$, the Laplace transform is a generalization of the Fourier transform.

Below are the main properties of the Laplace transform.

$$\mathcal{L}[af_1 + bf_2] = aF_1 + bF_2, \ a, b \in \mathbb{C}; \tag{5.4.5}$$

$$\mathcal{L}[f'](s) = sF(s) - f(0+), \quad \mathcal{L}[xf](s) = -F'(s); \tag{5.4.6}$$

$$\mathcal{L}[f^{(n)}](s) = s^n F(s) - \left(\sum_{k=0}^{n-1} s^{n-k-1} f^{(k)}(0+) \right); \tag{5.4.7}$$

$$\mathcal{L}[x^n f](s) = (-1)^n F^{(n)}(s); \tag{5.4.8}$$

$$\text{If } g(x) = \int_0^x f(y)dy \text{ then } G(s) = \frac{F(s)}{s}; \tag{5.4.9}$$

$$\text{If } g(x) = f(ax), \ a > 0, \ \text{then } G(s) = \frac{1}{a} F(s/a); \tag{5.4.10}$$

$$\text{If } g(x) = f(x - x_0), \ x_0 > 0, \ \text{then } G(s) = e^{-x_0 s} F(s); \tag{5.4.11}$$

$$\text{If } h(x) = \int_0^x f(x-y)g(y)dy, \text{ then } H(s) = F(s)G(s) \tag{5.4.12}$$

If $f'(x_0)$ exists and $a > s_f$, then

$$f(x_0) = \frac{1}{2i\pi} \lim_{L\to\infty} \int_{a-iL}^{a+iL} F(s)e^{sx_0} ds. \tag{5.4.13}$$

In the above properties, all the functions involved must satisfy (5.4.1) and (5.4.2). Note that, to invert the Laplace transform according to (5.4.13), we can integrate along *any* vertical line in the half-plane $\{s : \Re s > s_f\}$. Two more properties of the Laplace transform are discussed in Problem 6.6, page 440.

EXERCISE 5.4.2. [B] *Verify (5.4.5)–(5.4.13). Hint: (5.4.5)–(5.4.12) are best verified directly by definition, (5.4.13), by using (5.4.4) and the inversion formula for the Fourier transform.*

EXERCISE 5.4.3.[C] *(a) Let $h = h(x)$ be* `Heaviside's function` *$h(x) = 1$, $x \geq 0$, $h(x) = 0$, $x < 0$. Show that $H(s) = 1/s$, $\Re s > 0$. (b) Let $F(x) = e^{-ax}$, $a \in \mathbb{R}$. Show that $F(s) = 1/(s+a)$, $\Re s > -a$.*

We review the operational calculus by an EXAMPLE. Let us solve the equation $y'' - 2y' + y = e^x$, with initial conditions $y(0) = 1$, $y'(0) = -1$. We have by (5.4.5) and (5.4.7): $s^2Y(s) - s + 1 - 2(sY(s) - 1) + Y(s) = (s-1)^{-1}$, $Y(s) = (s-1)^{-3} + (s-3)(s-1)^{-2}$. Consider the function $F(s) = (s-1)^{-1}$. We know that this is the Laplace transform of $f(x) = e^x$. We also notice that $F'(s) = -(s-1)^{-2}$ and $F''(s) = 2(s-1)^{-3}$. Then the property (5.4.8) implies that $(s-1)^{-3}$ is the Laplace transform of $(x^2/2)e^x$. Next,

$$\frac{s-3}{(s-1)^2} = \frac{s-1}{(s-1)^2} - \frac{2}{(s-1)^2} = \frac{1}{s-1} - \frac{2}{(s-1)^2},$$

and we conclude from the above calculations that $(s-3)(s-1)^{-2}$ is the Laplace transform of $e^x - 2xe^x$. Then the solution of the equation is

$$\boxed{y(x) = \left(\frac{x^2}{2} - 2x + 1\right) e^x}.$$

EXERCISE 5.4.4. [A] *Try to solve the following equation using the Laplace transform: $y'' + x^2y = e^{-x}$, $y(0) = y'(0) = 0$. Hint: do not be surprised if you do not succeed, but try to understand the reason.*

The **two-sided Laplace transform** $M_f(s) = \int_{-\infty}^{\infty} f(x)e^{-sx}dx$ can also be defined. In the theory of probability, this transform is known as the **moment generating function**, and the Fourier transform, as the **characteristic function**.

EXERCISE 5.4.5.[B] *Compute* $M_f(s)$ *if* $f(x) = e^{-x^2/2}$. *Hint: see page 272.*

Let us now look at the discrete version of the Laplace transform, known as the **Z-transform**.

Given the samples $f(k\triangle t)$, $k = 0, 1, \ldots,$ of a continuous signal $f = f(t)$, $t \geq 0$, we construct a discrete signal f_d similar to (5.3.4) on page 277:

$$f_d(t) = \triangle t \sum_{k=0}^{\infty} f(k\triangle t)\delta(t - k\triangle t), \tag{5.4.14}$$

where δ is the Dirac delta function (5.3.5); recall that in this section we always assume $f(t) = 0$ for $t < 0$. Taking the Laplace transform on both sides of (5.4.14) results in

$$F_d(s) = \triangle t \sum_{k=0}^{\infty} f(t_k)e^{-sk\triangle t}. \tag{5.4.15}$$

Similar to (5.3.8), we have

$$F_d(s) = \sum_{k=-\infty}^{\infty} F\left(s - i\frac{2\pi k}{\triangle t}\right). \tag{5.4.16}$$

That is, $F_d(s)$ is a periodic function with a *purely imaginary* period $i2\pi/\triangle t$.

EXERCISE 5.4.6.[B] *Verify (5.4.16). Hint: the same arguments as for (5.3.8); since* $f(t) < 0$ *for* $t < 0$, *integration and summation can start at* $-\infty$, *when necessary.*

Take a real number $a > s_f$ and write $s = a - i\omega$. For fixed a, the function $H(\omega) = F_d(a - i\omega)$ is periodic with period $2\pi/\triangle t$. Equality (5.4.15), which can be written as

$$H(\omega) = \triangle t \sum_{k=0}^{\infty} f(k\triangle t)e^{-ak\triangle t}e^{ik\omega\triangle t},$$

becomes a Fourier series expansion of H. Then

$$\triangle t f(k\triangle t)e^{-ak\triangle t} = \frac{\triangle t}{2\pi}\int_{-\pi/\triangle t}^{\pi/\triangle t} H(\omega)e^{-ik\omega\triangle t}d\omega,$$

and, after some algebra, we get

$$f(k\Delta t) = \frac{1}{2\pi i}\int_{a-i\pi/\Delta t}^{a+i\pi/\Delta t} F_d(s)e^{sk\Delta t}ds. \tag{5.4.17}$$

EXERCISE 5.4.7.^{B} *Verify (5.4.17).*

Let us introduce a new variable $z = e^{s\Delta t}$. Then the line segment $(a - i\pi/\Delta t, a + i\pi/\Delta t]$ becomes the circle C_a with center 0 and radius $e^{a\Delta t}$. Also $dz = \Delta t e^{s\Delta t}ds$. We now define the sequence $x = x(k), k > 0$, by $x(k) = f(k\Delta t)$ and the function $X = X(z)$ so that $X(z) = F_d(s)$ when $z = e^{s\Delta t}$. Then equalities (5.4.15) and (5.4.17) become

$$X(z) = \Delta t \sum_{k=0}^{\infty} x(k)z^{-k},\ x(k) = \frac{1}{2\pi i\Delta t}\oint_{C_a} X(z)z^{k-1}dz. \tag{5.4.18}$$

We call the function $X = X(z)$ the **Z-transform** of the sequence x. By construction, *the Z-transform is defined only for those sequences x that satisfy $x(k) = 0, k < 0$, and, in some definitions, Δt is taken to be* 1.

EXERCISE 5.4.8.^{B} *Verify that the circle C_a encloses all singularities of the function $X(z)z^{k-1}$ for every k.*

The following property of the Z-transform is useful in the study of finite-differece equations: if $y(k) = x(k - m)$ for some fixed number m, then $Y(z) = z^{-m}X(z)$. Indeed, keeping in mind that $x(k) = 0$ for $k < 0$, $Y(z) = \sum_{k=0}^{\infty} y(k)z^{-k} = \sum_{k=0}^{\infty} x(k - m)z^{-k} = \sum_{k=m}^{\infty} x(k - m)z^{-k} = \sum_{k=0}^{\infty} x(k)z^{-(k+m)} = z^{-m}X(z)$.

We conclude this section with the connection between the Z-transform and the Discrete Fourier Transform. Let $x = x(k),\ k = 0, \dots, N - 1$, be a finite sequence. Denote by $X = X(z)$ the Z-transform of this sequence, and by $X_k,\ k = 0, \dots, N - 1$, the coefficients (5.3.11) on page 279 of the Discrete Fourier Transform. The the definitions imply

$$X\left(e^{i2\pi k/N}\right) = NX_k. \tag{5.4.19}$$

EXERCISE 5.4.9.^{B} *Verify (5.4.19).*

5.4.2 *Applications to System Theory*

In this section, we will discuss the applications of the Laplace transform to the analysis of *linear autonomous dynamical systems.* A **linear** system produces a linear combination of outputs given the corresponding linear combination of inputs. A **autonomous** system does not explicitly involve the time variable.

A continuous-time autonomous dynamical system connects the input signal $x = x(t)$ with the output signal $y = y(t)$ via a relation

$$f\left(x(t), \dot{x}(t), \ddot{x}(t), \ldots, x^{(m)}(t), y(t), \dot{y}(t), \ddot{y}(t), \ldots, y^{(n)}(t)\right) = 0. \quad (5.4.20)$$

As usual, $\dot{x}$ denotes the first derivative with respect to t, and $x^{(k)}$, the derivative of order k. Given the input signal $x(t)$, $t \geq 0$, we can use (5.4.20) to find the output signal $y(t)$, $t > 0$, as long as the function f is reasonably good and we know the initial conditions $\overrightarrow{y}(0) = (y(0), \dot{y}(0), \ldots, y^{n-1}(0))$. A **physically implementable** system has $n > m$. In particular, neither the perfect amplification $y(t) = kx(t)$ nor differentiation $y(t) = \dot{x}(t)$ is implementable, even though both have theoretical interest, and can be implemented approximately. Roughly speaking, condition $n > m$ means that the output at time t depends only on the input up to time t. At the end of this section we will use the discrete-time analog of (5.4.20) to demonstrate the necessity of this condition.

The natural assumption about (5.4.20) is that zero input and zero initial conditions produce zero output, that is, $f(0, \ldots, 0) = 0$. Thus, the system has an equilibrium point at $(0, \ldots, 0)$. If we further assume that the function $f = f(x_0, \ldots, x_m, y_0, \ldots, y_n)$ has two continuous partial derivatives at zero, then we have a **linearization** of (5.4.20) as follows:

$$\begin{aligned} &a_0 x(t) + a_1 \dot{x}(t) + a_2 \ddot{x}(t) + \ldots + a_n x^{(m)}(t) \\ &\qquad = b_0 y(t) + b_1 \dot{y}(t) + b_2 \ddot{y}(t) + \ldots + b_n y^{(n)}(t), \end{aligned} \quad (5.4.21)$$

where $a_k = f_{x_k}(0, \ldots, 0)$, $b_k = -f_{y_k}(0, \ldots, 0)$ (recall that f_x is another notation for the partial derivative $\partial f/\partial x$). The system (5.4.21) is autonomous if and only if all the coefficients a_k, b_k do not depend on t. If $b_n \neq 0$, then y is uniquely determined from (5.4.21) given the input and the initial conditions.

EXERCISE 5.4.10. ^{B} *Using the Taylor series expansion, verify that (5.4.21) is an approximation of (5.4.20) when* $|x(t)|, |\dot{x}(t)|, \ldots, |x^{(m)}(t)|$, $|y(t)|, |\dot{y}(t)|, \ldots, |y^{(n)}(t)|$ *are all sufficiently small.*

In what follows, we consider the linear autonomous systems of the type (5.4.21), and assume that $b_n \neq 0$. Note that the non-homogenous n-order linear differential equation with constant coefficients corresponds to $a_0 = 1$, $a_1 = \ldots = a_m = 0$. Taking the Laplace transform of (5.4.21) and assuming that $x^{(k)}(0) = 0$ and $y^{(k)}(0) = 0$ for all k, we find $X(s)\sum_{k=0}^{m} a_k s^k = Y(s)\sum_{k=0}^{n} b_k s^k$, or

$$Y(s) = W(s)X(s), \text{ where } W(s) = \frac{\sum_{k=0}^{m} a_k s^k}{\sum_{k=0}^{n} b_k s^k}. \qquad (5.4.22)$$

The function $W = W(s)$ is called the `transfer function` of the linear system (5.4.21). In general, cancelling similar factors in the fraction (5.4.22) changes the transfer function and is therefore not allowed. Often, a linear system is identified with its transfer function, as in "consider the system W."

Definition 5.4 The `free motion` of system (5.4.21) is the solution $y = y(t)$ of the differential equation $\sum_{k=0}^{n} b_k y^{(k)}(t) = 0$ with some initial condition $(y(0), \dot{y}(0), \ldots, y^{n-1}(0))$. The system is called `stable` if every free motion eventually dies out: $\lim_{t\to\infty} |y(t)| = 0$ for every initial condition.

EXERCISE 5.4.11. *(a)^B Verify that (5.4.21) is stable if and only if, for every bounded function $x = x(t)$ and all initial conditions, the solution of $\sum_{k=0}^{n} b_k y^{(k)}(t) = x(t)$ stays bounded for all $t > 0$. (b)^C Verify that (5.4.21) is stable if and only if all the roots of the equation $\sum_{k=0}^{n} b_k s^k = 0$ have negative real parts. Hint (works for both parts): recall that the free motion of the system is a linear combination of the functions $t^\ell e^{s_k t}$, where s_k is a root of the above equation, $\ell = 0, \ldots, \ell_k$, and ℓ_k is the multiplicity of s_k.*

In other words, *the system is stable if and only if all the poles of the transfer function are in the (complex) left half-plane.* There are criteria of the stability of the system in terms of the coefficient $b_0, \ldots, b_n$, which fall outside the scope of our discussion. We only mention that the necessary condition for stability is either $b_0 > 0, \ldots, b_n > 0$ or $b_0 < 0, \ldots, b_n < 0$ (that is, all coefficients must be non-zero and have the same sign). Although not desirable, unstable systems can be used as long as the input signal and the initial conditions are periodically re-set to zero.

System (5.4.21) is the basic building block in the construction of more complicated linear systems. There are two main ways to connect two systems: `cascade` and `closed-loop`. In the CASCADE CONNECTION, the output $y_1(t)$ of one system of the type (5.4.21) is the input $x_2(t)$ of another, see the left side of Figure 5.4.1. In terms of the transfer functions, we have $Y_2(s) = W_2(s)X_2(s) = W_2(s)Y_1(s) = W_2(s)W_1(s)X_1(s)$, that is, the transfer function of the cascade system is the product of the individual transfer functions. One can certainly have a cascade of more than two systems.

EXERCISE 5.4.12.[C] *Verify that the cascade system is stable if and only if each of the individual systems is stable. Hint: use the result of the previous exercise.*

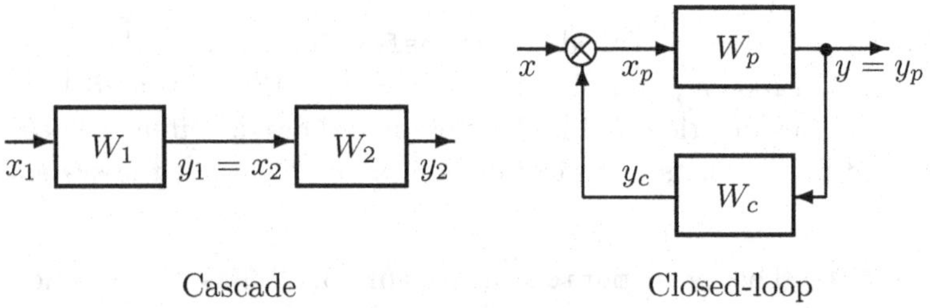

Fig. 5.4.1 Cascade and Closed-Loop Connections

In the `closed-loop` or `feed-back` connection, the two components are often called the `plant`, with transfer function $W_p(s)$, and the `controller`, with transfer function W_c, see the right side of Figure 5.4.1. The output $y = y_p$ of the plant is the input of the controller; the junction $\otimes$ combines the output y_c of the controller with the input x of the system to produce the input x_p of the plant.

If the signals x and y_c are added so that $x_p = x + y_c$, then $Y(s) = W_p(s)(X(s) + Y(s)W_c(s))$ and $W(s) = Y(s)/X(s) = W_p(s)/(1 - W_p(s)W_c(s))$ is the transfer function of the system. If the signals are subtracted so that $x_p = x - y_c$, then

$$W(s) = \frac{W_p(s)}{1 + W_p(s)W_c(s)} \tag{5.4.23}$$

(verify this). Notice that the signs of W_p and W_s do make a difference. For example, when the signals x and y_c are subtracted, changing the sign of W_c produces the transfer function $W_p(s)/(1 - W_p(s)W_c(s))$. Unlike the cascade system, individual stability of the components provides no information about the stability of the closed-loop system.

A closed-loop connection can approximate certain individual systems that are not physically implementable, that is, systems of the type (5.4.21) with $n < m$. For example, if in (5.4.23) we take $W_p(s) = K/(1 + s)$, and $W_c(s) = 1/(1 + s)$, then, as $K \to \infty$, we get $W(s) \approx 1 + s$.

EXERCISE 5.4.13.[C] *(a) Give an example of an unstable closed-loop system with stable plant and controller. (b) Verify that if $W_p(s) = 1/(1 + 2s)$ and*

$W_c = 1/(1+p)$, then the real parts of the poles of W are $-3/4$. In other words, the closed-loop system in this example is more stable than the plant itself. This is the idea behind negative feedback.

The following table summarizes the common types of controllers.

Type	Transfer Function $W_c(s)$
P	K_o
I	$K_i s^{-1}$
D	$K_d s$
PI	$K_o + K_i s^{-1}$
PID	$K_o + K_d s + K_i s^{-1}$

K_o, K_i, K_d are positive real numbers. Not surprisingly, the letters P,I,D, come from "proportional," "integral," and "derivative", respectively.

EXERCISE 5.4.14. [B] *Let $x = x(t)$ be the input of a PID controller and $y = y(t)$, the output. Assuming $x(0) = y(0) = 0$, find the relation between x and y.*

We conclude this section with a few words about discrete-time systems. The discrete analog of (5.4.21) is

$$y(k) = \sum_{m=0}^{M-1} a_m x(k-m) + \sum_{m=1}^{N} b_m y(k-m). \tag{5.4.24}$$

This system is always physically implementable: the input sequence $x(k)$, $k \geq 0$, and the initial conditions $y(0), \ldots, y(N-1)$ define the unique output sequence $y(k)$, $k \geq N$. An example of a system that is not physically implementable is at the end of this section.

Taking the z-transform on both sides of (5.4.24) and assuming that $x(k) = y(k) = 0$ for $k < 0$, we find $Y(z) = X(z)\sum_{m=0}^{M-1} a_m z^{-m} + Y(z)\sum_{m=1}^{N} b_m z^{-m}$, or

$$Y(z) = W(z)X(z), \text{ where } W(z) = \frac{\sum_{m=0}^{M-1} a_m z^{-m}}{1 - \sum_{m=1}^{N} b_m z^{-m}}. \tag{5.4.25}$$

We say that (5.4.24) is `stable` if, for every bounded input sequence x and for all initial conditions, the output sequence y stays bounded for all $k \geq 0$. Similar to continuous-time signals, one can show that (5.4.24) is *stable if and only if all poles of the transfer function W are in the unit disk* $\{z : |z| < 1\}$. An unstable system can still be used if the input and initial conditions are periodically reset to zero. As an EXAMPLE, consider the `digital integrator` $y(k) = x(k) + y(k-1)$. It is unstable (if $x(0) = y(0) = 0$ and $x(k) = 1$, $k \geq 1$, then $y(k) = k$), and its corresponding transfer function $W(z) = 1/(1-z^{-1})$ has a simple pole at $z = 1$. Still, a computer program implementing this system will work just fine as long as k is not allowed to get very large.

EXERCISE 5.4.15.[B] *Verify that if all* $b_m = 0$*, then (5.4.25) is stable.*

Given a function $W = W(z)$, the corresponding system that has W as the transfer function is *physically implementable* if the number of zeros of W is less than or equal to the number of poles (both poles and zeros are counted according to their multiplicities). FOR EXAMPLE, if $W(z) = z - 2$, then the corresponding difference equation (5.4.24) is $y(k) = x(k+1) - 2x(k)$. This system is not implementable, because at time k, only the values of $x(0), \ldots, x(k)$ are available. On the other hand, the transfer function $W(z) = 1 - 2z^{-1}$ corresponds to the difference equation $y(k) = x(k) - 2x(k-1)$ and is implementable.

Chapter 6

Partial Differential Equations of Mathematical Physics

A `partial differential equation` (PDE) describes a scalar or vector field using various operations of differentiation. Once a particular coordinate system is specified, the differentiation operations are expressed in term of the *partial derivatives* of the unknown field. We have already encountered several such equations: the equation of continuity (3.2.8) on page 154 for the scalar field ρ, the Poisson equation (3.2.15) on page 158 for the scalar field f, Maxwell's equations in vacuum (3.3.2)–(3.3.5) on page 164 for the vector fields $\boldsymbol{E}$, $\boldsymbol{B}$, and Maxwell's equations in material media (3.3.47)–(3.3.50) for the vector fields $\boldsymbol{E}, \boldsymbol{D}, \boldsymbol{B}, \boldsymbol{H}$. The objective of this chapter is to study these and other similar equations. We will only work with `classical solutions` of partial differential equations, that is, functions that have continuous derivatives required by the equation and satisfy the equation at every point. We intend our discussion to be the most basic introduction to partial differential equations; accordingly, for every equation, we will focus our attention on the following two questions: (a) where does the equation come from? (b) how to find a solution of the equation?

6.1 Basic Equations and Solution Methods

6.1.1 *Transport Equation*

In this section, we investigate one of the simplest partial differential equations and get familiar with two fundamental methods of solving partial differential equation: the method of characteristics and the method of variation of parameters.

Consider a continuum of point masses (particles) in $\mathbb{R}^n$, where $n = 1$, $n = 2$, or $n = 3$. Let $u = u(t, P)$ be the density of these particles at time

t and point $P \in \mathbb{R}^n$. If $\boldsymbol{v} = \boldsymbol{v}(t, P)$ is the velocity of the particle at point P and time t, and if there are no sources or sinks, then, according to the equation of continuity (3.2.8) on page 154, with ρ replaced by u, we have

$$u_t + \operatorname{div}(u\,\boldsymbol{v}) = 0, \tag{6.1.1}$$

where u_t is the partial derivative of u with respect to t.

EXERCISE 6.1.1.^{C} *Show that, if $\boldsymbol{v}$ is a constant vector (that is, $\boldsymbol{v}$ does not depend on t and P), then (6.1.1) becomes*

$$u_t + \boldsymbol{v} \cdot \operatorname{grad} u = 0. \tag{6.1.2}$$

Hint: use a suitable identity from the collection (3.1.30) on page 139.

Equation (6.1.2) is called the `transport equation`. Treating the variable t as time, it is natural to specify the density $f(P) = u(0, P)$ at time $t = 0$ and then use (6.1.2) to find u for all $t > 0$. This is called the `initial value problem`, or `Cauchy problem`, for (6.1.2).

Definition 6.1 Given a continuous scalar field f and a fixed vector $\boldsymbol{v}$, a `classical solution` of the initial value problem for (6.1.2) is a function $u = u(t, P)$, $t \geq 0$, $P \in \mathbb{R}^n$, with the following properties: (a) u is continuous for $t \geq 0$, $P \in \mathbb{R}^n$; (b) u is continuously differentiable for $t > 0$, $P \in \mathbb{R}^n$; (c) $u_t + \boldsymbol{v} \cdot \operatorname{grad} u = 0$ for $t > 0$, $P \in \mathbb{R}^n$; (d) $u(0, P) = f(P)$ for all $P \in \mathbb{R}^n$.

To solve the initial value problem, *we assume that the function f is continuously differentiable.*

Given the reference point O in $\mathbb{R}^n$, denote by $\boldsymbol{r} = \overrightarrow{OP}$ the corresponding position vector of the point P. A point mass that, at time $t = 0$, is at the point P, has the trajectory $\boldsymbol{R}(t) = \boldsymbol{r} + t\,\boldsymbol{v}$, $t \geq 0$. By (3.2.9) on page 154 (or by direct computation),

$$\frac{d}{dt} u(t, \boldsymbol{r} + t\,\boldsymbol{v}) = 0, \; t > 0. \tag{6.1.3}$$

EXERCISE 6.1.2.^{C} *Verify that (6.1.3) follows from (6.1.2).*

It follows from (6.1.3) that, given the initial position vector $\boldsymbol{r}$, the value of $u(t, \boldsymbol{r} + t\,\boldsymbol{v})$ does not depend on t; by continuity, we conclude that $\lim_{t \to 0+} u(t, \boldsymbol{r} + t\,\boldsymbol{v}) = u(0, \boldsymbol{r}) = f(\boldsymbol{r})$. Therefore, the function

$$u(t, \boldsymbol{r}) = f(\boldsymbol{r} - t\,\boldsymbol{v}) \tag{6.1.4}$$

is a solution of (6.1.2). For every line $\boldsymbol{R}(t) = \boldsymbol{r}_0 + t\boldsymbol{v}$, the value of the solution is the same at each point of the line and is equal to the value of the initial condition f at the point $\boldsymbol{r}_0$. This line is called a **characteristic** of equation (6.1.2). Note also that the solution function (6.1.4) is defined and is continuously differentiable for *all* $t \in \mathbb{R}$.

EXERCISE 6.1.3. *Verify that the function* $u(t,\boldsymbol{r}) = f(\boldsymbol{r} - t\boldsymbol{v})$ *satisfies (6.1.2) by computing* u_t *and* $\boldsymbol{v}\cdot \operatorname{grad} u$. *Hint: by the chain rule,* $\operatorname{grad} u = \operatorname{grad} f$, $u_t = -\boldsymbol{v} \cdot \operatorname{grad} f$.

Even though the computations so far implicitly assumed a three-dimensional space, the transport equation can be studied in one and two dimensions as well. In one dimension, $\boldsymbol{v} = b$, a real number, so that (6.1.2) becomes

$$u_t + bu_x = 0. \tag{6.1.5}$$

Given the initial density $f = f(x)$, the solution of (6.1.5) is $u(t,x) = f(x - bt)$, so that $u(t,x) = u(0, x - bt)$. If $b > 0$, then the initial density profile $f(x)$ moves to the right with speed b; see Figure 6.1.1. Physically, the fluid mass is transported to the right with speed $\|\boldsymbol{v}\| = b$.

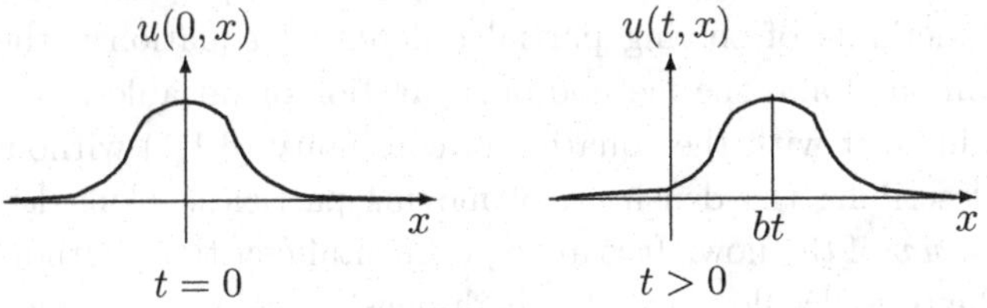

Fig. 6.1.1 Solution of the Transport Equation in One Dimension

EXERCISE 6.1.4. *Let* $f = f(x)$ *be a continuously differentiable function of* x, $b = b(t)$, *a continuous function of* t, *and* $B(t) = \int_0^t b(s)ds$. *Verify that the function* $u(t,x) = f(x - B(t))$ *solves the initial value problem* $u_t + b(t)u_x = 0$, $u(0,x) = f(x)$. *Generalize the result to the equation (6.1.2).*

Let us now consider the equation

$$u_t + \boldsymbol{v} \cdot \operatorname{grad} u = h, \quad u(0,P) = f(P), \tag{6.1.6}$$

where $f = f(P)$ and $h = h(t,P)$ are known functions; in terms of the original equation of continuity (3.2.8) on page 154, the function h represents the density of sources and sinks. We assume that both f and h are continuously differentiable. Similar to (6.1.3),

$$\frac{d}{dt}u(t,\boldsymbol{r}+t\,\boldsymbol{v}) = h(t,\boldsymbol{r}+t\,\boldsymbol{v}),\ t>0, \tag{6.1.7}$$

and therefore

$$u(t,\boldsymbol{r}) = f(\boldsymbol{r}-t\,\boldsymbol{v}) + \int_0^t h(s,\boldsymbol{r}+(s-t)\,\boldsymbol{v})\,ds \tag{6.1.8}$$

is a solution of (6.1.6). Representation (6.1.8) is an example of the `variation of parameters` formula.

EXERCISE 6.1.5. *(a)*^{C} *Verify that (6.1.8) follows from (6.1.7). (b)*^{B} *Verify that the function $u=u(t,x)$ defined by (6.1.8) is a solution of (6.1.6). (c)*^{A} *Let $\boldsymbol{v}=\boldsymbol{v}(t)$ and $a=a(t)$ be continuous functions. Find a solution of the initial value problem $u_t+\boldsymbol{v}(t)\cdot\operatorname{grad}u+a(t)u=h,\quad u(0,P)=f(P)$. Hint: solve the equation satisfied by the function $U(t,P)=u(t,P)\exp(\int_0^t a(s)ds)$.*

6.1.2 *Heat Equation*

In this section, we investigate the heat, or diffusion, equation, and introduce two other methods of solving partial differential equations: the Fourier transform method and the method of separation of variables.

We again start with the equation of continuity (6.1.1) without sources or sinks, describing the density u of moving particles. Consider the *flux density* $\boldsymbol{J}=u\,\boldsymbol{v}$ of the flow. It is an experimental fact that, without sources or sinks, the particles flow from the high-density regions to the low-density regions, that is, the direction of the flow is opposite to the direction of the density gradient. This type of flow is called `diffusion` and is described mathematically by the relation

$$\boldsymbol{J} = -a\operatorname{grad}u, \tag{6.1.9}$$

where $a>0$ is the `diffusion coefficient` of the medium. Relation (6.1.9) is known as `Fick's Law of Diffusion`, after the German scientist ADOLF EUGEN FICK (1829–1901), who introduced it in 1855 to describe the diffusion of gas through a fluid membrane. In 1905 A. Einstein derived (6.1.9) from a one-dimensional random walk model, with particles moving at random to the left or to the right. For general media, the value of a depends on the location and direction. In a homogeneous isotropic medium,

a is a constant number, so that, by combining (6.1.9) with the continuity equation $u_t + \operatorname{div} \boldsymbol{J} = 0$, we get the **diffusion equation**

$$u_t = a\nabla^2 u, \tag{6.1.10}$$

where $\nabla^2 u = \operatorname{div}(\operatorname{grad} u)$ is the Laplacian of u.

The diffusion equation (6.1.10) also describes HEAT FLOW. Let us consider a heated three-dimensional region. By the *Second Law of Thermodynamics*, heat flows in the direction of the decrease of temperature: if u denotes the temperature of the material, and $\boldsymbol{J}$, the heat flux density (measured in joules per unit area per unit time, $J\,m^{-2}\,s^{-1}$), then

$$\boldsymbol{J} = -c \operatorname{grad} u, \tag{6.1.11}$$

where c is the *thermal conductivity* of the material; for nonlinear materials, c can depend on temperature. *We will measure the temperature relative to some reference value that is* not *absolute zero, so that the values of u can be both positive and negative.* By Gauss's Theorem on page 152, for every region G of the material,

$$\iint_{\partial G} \boldsymbol{J} \cdot \widehat{\boldsymbol{n}}\, d\sigma = -\iiint_{G} \operatorname{div}(c \operatorname{grad} u) dV. \tag{6.1.12}$$

On the other hand, heat flow out of the region changes the temperature of the region, with positive flow outside reducing the temperature inside; without external sources of heat, the heat flux through the boundary of a small region of volume $\triangle V$ is equal to $-c_s \rho \triangle V\, u_t$, where c_s is the *specific heat* of the material, and ρ is the volume density (mass per unit volume). Thus, integration over G yields

$$\iint_{\partial G} \boldsymbol{J} \cdot \widehat{\boldsymbol{n}}\, d\sigma = -\iiint_{G} c_s \rho\, u_t\, dV. \tag{6.1.13}$$

For linear homogeneous isotropic materials, the values of c, c_s, ρ are constant, and, by equating the right-hand sides of (6.1.12) and (6.1.13), we conclude that the temperature u satisfies equation (6.1.10) with $a = c/(c_s\rho)$. As a result, equation (6.1.10) is often called the **heat equation**. Similar equations describe heat flow in one and two dimensions.

Another useful result related to heat transfer is **Newton's Law of Cooling**: *the rate of change of an object's temperature is proportional to the difference between the temperature of the object and the temperature of* *the surrounding medium.* Despite the name, the same law applies to both heating and cooling. Without going into the details, we mention that certain physical conditions must hold for this law to apply; in our discussions, we will always assume that these conditions do indeed hold.

EXERCISE 6.1.6.[A] *Verify that, for a homogeneous isotropic material, equations (6.1.12) and (6.1.13) imply (6.1.10). Hint: since G is an arbitrary part of the material, the equality of the integrals implies the equality of integrands; as always, we assume that all the functions involved are continuous. The argument is similar to the derivation of (3.2.8) on page 154; see Exercise 3.2.5.*

In the early 1780s, the French scientist ANTOINE-LAURENT DE LAVOISIER (1743–1794) introduced a model in which heat was a special liquid called `caloric` (from Latin *calor*, meaning "heat") that could be neither produced nor destroyed. In the early 1840s, the English physicist JAMES PRESCOTT JOULE (1818–1889) demonstrated experimentally the equivalence between heat and mechanical work and suggested the modern heat transfer model based on molecular interaction, thus refuting the *caloric* model. It is interesting that both models result in the same equation (6.1.10). In fact, the wrong idea of caloric led to many correct (and important) results in thermodynamics, such as the mathematical theory of heat engine. The importance of caloric and J. P. Joule in the development of thermodynamics has been recognized in the names of the two units to measure energy; one `calorie` is approximately 4.2 joules.

EXERCISE 6.1.7.[B] *The above derivation of the heat equation inside a region G of material assumed no heat loss due to contact with the outside, that is, the material was perfectly insulated. Now assume that the insulation is not perfect, and U_e is the temperature of the outside. Show that the heat equation (6.1.10) becomes*

$$u_t = a\nabla^2 u - C_e(u - U_e). \tag{6.1.14}$$

Hint: use Newton's Law of Cooling on page 295; note that $C_e = 0$ corresponds to perfect insulation, and, in general, C_e can be a function of t and x.

Let us now solve the INITIAL VALUE PROBLEM FOR THE HEAT EQUATION ON THE LINE:

$$u_t(t,x) = a\,u_{xx}(t,x),\ t>0,\ x\in\mathbb{R},\ u(0,x) = f(x), \tag{6.1.15}$$

where f is a bounded continuous function. Equation (6.1.15) mathematically models the temperature at time t and point x in a straight thin homogeneous rod or wire of infinite length and constant cross section, with no heat loss through the lateral surface; the function f defines the initial temperature at all points of the wire.

Definition 6.2 A `classical solution` of (6.1.15) is a function $u = u(t,x)$ with the following properties: (i) u is continuous for $t \in [0,+\infty)$, $x \in \mathbb{R}$, (ii) u has one continuous derivative in t and two continuous derivatives in x for $t \in (0,+\infty)$, $x \in \mathbb{R}$, (iii) $u_t = au_{xx}$ for all $t \in (0,+\infty)$, $x \in \mathbb{R}$, (iv) $\lim_{t\to 0^+} u(t,x) = f(x)$ for all $x \in \mathbb{R}$.

To solve (6.1.15), we carry out `a priori analysis` of the equation: we assume that a solution exists and has all the additional properties we might need, and then derive a representation formula for this solution in terms of the function f.

Suppose that, for every $t \geq 0$, the Fourier transforms of u, u_t, and u_{xx} exist in the variable x:

$$\widehat{u}(t,\omega) = \frac{1}{\sqrt{2\pi}} \int_{-\infty}^{\infty} u(t,x)e^{-i\omega x}dx.$$

Using property (5.2.18), page 268, of the Fourier transform of the derivatives, we apply the transform to (6.1.15) and get

$$\widehat{u}_t(t,\omega) = -a\omega^2\widehat{u}(t,\omega),\ t > 0,\ \widehat{u}(0,\omega) = \widehat{f}(\omega). \tag{6.1.16}$$

For each ω, (6.1.16) is a first-order linear ordinary differential equation for $\widehat{u}$ as a function of t; the solution of this equation is

$$\widehat{u}(t,\omega) = \widehat{f}(\omega)e^{-ta\omega^2}. \tag{6.1.17}$$

(Recall that the ordinary differential equation $y'(x) = Ay(x), y(0) = C$, where A, C are real numbers, has the solution $y(x) = Ce^{Ax}$). Using various properties of the Fourier transform and its inverse, we get the representation of u:

$$u(t,x) = \frac{1}{\sqrt{4\pi at}} \int_{-\infty}^{\infty} e^{-(x-y)^2/(4at)} f(y)dy. \tag{6.1.18}$$

EXERCISE 6.1.8. *(a)*^{C} *Verify (6.1.18). Hint: use (5.2.23) on page 269 and (5.2.29) on page 271 to conclude that the inverse Fourier transform of* $e^{-ta\omega^2}$ *is* $(2at)^{-1/2}e^{-x^2/(4at)}$. *Then recall that the product of Fourier transforms corresponds to the convolution of the original functions; see (5.2.20) on page 269.* *(b)*^{B} *Let* $h = h(t,x)$ *be a bounded continuous function. Using the Fourier transform, show that if*

$$u_t(t,x) = a\,u_{xx}(t,x) + h(t,x),\ t>0,\ x\in\mathbb{R},\ u(0,x)=f(x), \qquad (6.1.19)$$

then the solution is computed by the **variation of parameters** *formula*

$$\begin{aligned} u(t,x) = {} & \frac{1}{\sqrt{4\pi a t}}\int_{-\infty}^{\infty} e^{-(x-y)^2/(4at)} f(y)dy \\ & + \int_0^t \frac{1}{\sqrt{4\pi a(t-s)}}\int_{-\infty}^{\infty} e^{-(x-y)^2/(4a(t-s))} h(s,y)dy\,ds. \end{aligned} \qquad (6.1.20)$$

The function

$$K(t,x) = \frac{1}{\sqrt{4\pi a t}} e^{-x^2/(4at)},\ t>0,\ x\in\mathbb{R}, \qquad (6.1.21)$$

is called the (one-dimensional) **heat kernel**, or **fundamental solution** of the heat equation on the line. Note that this function is not defined for $t \le 0$.

EXERCISE 6.1.9.[A] *(a) Verify that, for every $t>0$,*

$$\int_{-\infty}^{\infty} K(t,x)dx = 1.$$

(b) Verify that, for every $t>0$, the function K has infinitely many continuous derivatives with respect to t and x, and $K_t = aK_{xx}$. (c) Verify that $\lim_{t\to 0+} K(t,x) = \delta(x)$, where δ is the delta-function; see page 275. Hint: the arguments are the same as in the proof of (5.3.2) on page 275. (d) Using the results of parts (a)–(c), verify that, for every continuous bounded function f, the function u defined in (6.1.18) is indeed a solution of (6.1.15), that is, $u_t = au_{xx}$ for $t>0$ and $x\in\mathbb{R}$, and $\lim_{t\to 0+} u(t,x) = f(x)$. Hint: it is OK to differentiate under the integral sign in (6.1.18).

Let us summarize the two main steps in the above arguments: *(a) We assumed that, for the given bounded continuous function f, a solution of (6.1.15) exists and has all the properties we need; then, using the Fourier transform, we showed that such a solution must have the form (6.1.18); (b) The reader who completed Exercise 6.1.9, verified that, conversely, for every bounded continuous function f, formula (6.1.18) defines a solution of (6.1.15); the reader who skipped the exercise, should at least pause and realize that this second step is indeed necessary.* We will follow these steps while analyzing many other equations, but will concentrate on the first step: we will derive a representation formula for the solution by assuming that the solution exists and has all the properties necessary for the derivation; we will call this step **a priori analysis** of the equation. As a rule, we will take for granted the converse, that is, given an equation, the function defined by the corresponding representation formula is indeed a solution of the equation. As both Exercise 6.1.9 and common sense suggest, the converse should be true, but a separate proof is required.

With Definition 6.1 and Definition 6.2 in mind, we will no longer provide similar definitions for other equations.

As a rule, the integral in (6.1.18) cannot be expressed using elementary functions, but there is a family of initial conditions for which integration in (6.1.18) is possible:

$$f(x) = e^{-bx^2}, \text{ for some } b > 0. \tag{6.1.22}$$

EXERCISE 6.1.10.^{C} *Verify that the solution of (6.1.15) with the initial condition (6.1.22) is*

$$u(t,x) = \frac{1}{\sqrt{4abt+1}} e^{-bx^2/(4abt+1)}, \quad t > 0.$$

Hint: use (6.1.17) together with the result of Exercise 5.2.13 on page 272.

Many important properties of the heat equation and its solution follow directly from formula (6.1.20). FOR EXAMPLE, the heat equation has the **infinite propagation speed** and the **instantaneous smoothing** property. To illustrate these properties, let $f(x) = \sin(x)$ for $|x| < \pi$ and $f(x) = 0$ for $|x| \geq \pi$. Thus, the initial temperature is non-zero only for $|x| < \pi$. Also, f is continuous for all x, but is not differentiable at the points $x = \pm\pi$. For the solution u we have

$$u(t,x) = \frac{1}{\sqrt{4\pi at}} \int_{-\pi}^{\pi} e^{-(x-y)^2/(4at)} \sin y \, dy. \tag{6.1.23}$$

In particular, $u(t,x) > 0$ for all $x \in \mathbb{R}$ and all $t > 0$, that is, the non-zero initial temperature immediately spreads to the whole line, although this non-zero value of u for $|x| > \pi$ is very small for small t. Also, $u(t,x)$ is infinitely differentiable, as a function of x, for all $t > 0$, that is, the temperature profile immediately becomes smooth. The reader who is not convinced that the function u is indeed smooth as a function of x, should change the variables in the integral (to get $\sin(x-y)$), and then use the appropriate trigonometric identity, together with the fundamental theorem of calculus.

EXERCISE 6.1.11.[B] *Assume that the initial condition f satisfies $0 < f(x) < 1$ and $\int_{-\infty}^{\infty} f(x)dx = 1$. Verify that the corresponding solution $u = u(t,x)$ of the heat equation satisfies $0 < u(t,x) < 1$ and $\int_{-\infty}^{\infty} u(t,x)dx = 1$ for all $t > 0$.*

Without going into the details, we mention that, with obvious modifications, formula (6.1.18) holds in all dimensions of the space variable x: if $u_t = a\nabla^2 u$, $t > 0$, and $u(0,x) = f(x)$, $x \in \mathbb{R}^n$, $n \geq 2$, where $f = f(x)$ is a bounded continuous function, then

$$u(t,x) = \frac{1}{(4\pi a t)^{n/2}} \int_{\mathbb{R}^n} e^{-|x-y|^2/(4at)} f(y)dy \tag{6.1.24}$$

is a **classical solution** of the equation.

Let us now consider a quite different problem of describing the temperature at time t and point x in a straight thin homogeneous wire of *finite length* L and constant cross section, with initial temperature given by the function $f = f(x)$, $x \in [0,L]$. We assume that the end points of the wire are kept at a constant temperature, equal to zero, and the wire is perfectly insolated, that is, there is no heat loss through the lateral surface of the wire. In other words, we consider the INITIAL-BOUNDARY VALUE PROBLEM FOR THE HEAT EQUATION ON THE INTERVAL $[0,L]$:

$$\begin{aligned} &u_t = a u_{xx}, \ t > 0, \ 0 < x < L; \\ &u(0,x) = f(x), \ 0 \leq x \leq L; \ \ u(t,0) = u(t,L) = 0, \ t \geq 0. \end{aligned} \tag{6.1.25}$$

We assume that the function f is continuously differentiable on $[0,L]$, and $f(0) = f(L) = 0$.

To solve the equation, we use a method called **separation of variables**, that is, we look for a solution in the form $u(t,x) = \sum_{k=1}^{\infty} G_k(t)H_k(x)$ for suitable functions G_k, H_k. Note that we cannot use Fourier transform on a bounded interval; Fourier series, on the other hand, could be an option, and the suggested form of the solution is indeed a series. Still, it is not completely clear why the solution should be of this particular form, and we will address the question later, when we discuss separation of variables in a general setting. For now, we start with the following **superposition principle**: if the functions $u_1 = u_1(t,x), \ldots, u_N = u_N(t,x)$ satisfy the heat equation $u_t = au_{xx}$ and the **boundary conditions** $u(t,0) = u(t,L) = 0$, then so does the sum $u_1 + \cdots + u_N$; boundary conditions that allow this superposition are called **homogeneous**. We also hope that the superposition principle holds for sums of infinitely many solutions, as long as the sum converges in the right sense. The *elementary solutions* $u_k(t,x) = G_k(t)H_k(x)$, with $H_k(0) = H_k(L) = 0$, are the easiest to study; once we have sufficiently many elementary solution, we will choose the functions G_k to satisfy the `initial condition` $u(0,x) = f(x)$.

Accordingly, we start by looking for a solution of the heat equation $u_t = au_{xx}$ in the form $u(t,x) = G(t)H(x)$, where

$$H(0) = H(L) = 0. \tag{6.1.26}$$

To satisfy the equation, we must have $G'(t)H(x) = aG(t)H''(x)$, or, assuming that neither G nor H is zero,

$$\frac{G'(t)}{aG(t)} = \frac{H''(x)}{H(x)}, \; t > 0, \;\; x \in (0, L). \tag{6.1.27}$$

Note that $G'(t)/(aG(t))$ depends only on t, while $H''(x)/H(x)$ depends only on x. Since t and x vary independently of each other, the only way to satisfy (6.1.27) is to have

$$\frac{G''(t)}{aG(t)} = r, \quad \frac{H''(x)}{H(x)} = r. \tag{6.1.28}$$

for some real constant r. The two ordinary differential equations in (6.1.28) are easily solvable, and we will start with the equation for H. Assume that $r > 0$, that is, $r = p^2$. The general solution of the equation $H''(x) - p^2H(x) = 0$ is $H(x) = Ae^{px} + Be^{-px}$; by (6.1.26), we must have

$$A + B = 0, \quad Ae^{pL} + Be^{-pL} = 0. \tag{6.1.29}$$

which implies $A = B = 0$. In other words, if $r > 0$, then the only solution of $H''(x) = r\,H(x)$ satisfying $H(0) = H(L) = 0$ is $H(x) = 0$ for all $x \in [0, L]$; this is not a solution we want.

EXERCISE 6.1.12.^{C} *Verify that (6.1.29) implies $A = B = 0$.*

Now assume that $r = 0$. Then $H''(x) = 0$ and $H(x) = Ax + B$. To find A and B, we again use (6.1.26): $H(0) = B = 0$, and $H(L) = AL + B = 0$, that is, $A = B = 0$. Once again, the only solution of $H''(x) = 0$ satisfying $H(0) = H(L) = 0$ is $H(x) = 0$ for all $x \in [0, L]$; this is not a solution we want.

Finally, assume that $r < 0$, that is $r = -\lambda^2$ for some $\lambda > 0$. The general solution of the equation $H''(x) + \lambda^2 H(x) = 0$ is $H(x) = A\cos\lambda x + B\sin\lambda x$;

by (6.1.26), we must have

$$A = 0, \quad B \sin \lambda L = 0. \tag{6.1.30}$$

To have H not equal to zero identically, we need $B \neq 0$; with no other restrictions on B, we take $B = 1$ for simplicity. Then

$$\lambda = \frac{\pi k}{L}, \ k = 1, 2, 3, \ldots, \tag{6.1.31}$$

are the possible values of λ, and

$$H_k(x) = \sin(\pi k x/L), \ k = 1, 2, 3, \ldots, \tag{6.1.32}$$

are solutions of $H''(x) = rH(x)$, $r < 0$, that are not identical zero and satisfy (6.1.26).

EXERCISE 6.1.13.[C] *Verify that (6.1.30) and $B \neq 0$ imply (6.1.31).*

We now return to (6.1.28), with $r = -(\pi k/L)^2$, and find the corresponding function G:

$$G_k(t) = G_k(0) e^{-a(\pi k/L)^2 t}, \ k = 1, 2, 3, \ldots; \tag{6.1.33}$$

recall that our plan is to satisfy the initial condition with a special choice of $G_k(t)$; apparently, it is the initial conditions $G_k(0)$ that will allow us to achieve that.

EXERCISE 6.1.14.[C] *Verify (6.1.33).*

By combining (6.1.32) and (6.1.33), we find the corresponding elementary solution

$$u_k(t, x) = G_k(0) e^{-a(\pi k/L)^2 t} \sin(\pi k x/L), \ k = 1, 2, 3, \ldots.$$

A possible solution of the original equation (6.1.25) is then

$$u(t, x) = \sum_{k=1}^{\infty} u_k(t, x) = \sum_{k=1}^{\infty} G_k(0) e^{-a(\pi k/L)^2 t} \sin(\pi k x/L).$$

Setting $t = 0$,

$$f(x) = \sum_{k=1}^{\infty} G_k(0) \sin(\pi k x/L),$$

which is the Fourier series expansion for the *odd* extension of f. Therefore, we define $f_o(x)$ by $f_o(x) = f(x)$, $x \in [0, L]$; $f_o(x) = -f(-x)$, $-L < x < 0$, and use formulas (5.1.18) on page 255 to conclude that

$$\begin{aligned} G_k(0) &= \frac{1}{L}\int_{-L}^{L} f_o(x)\sin(\pi kx/L)dx \\ &= \frac{2}{L}\int_0^L f(x)\,\sin(\pi kx/L)dx,\ k=1,2,\dots. \end{aligned} \tag{6.1.34}$$

The final formula for the solution of (6.1.25) does not look extremely attractive, but we present it for the sake of completeness:

$$u(t,x) = \sum_{k=1}^{\infty}\left(\frac{2}{L}\int_0^L f(y)\,\sin\left(\frac{\pi k}{L}y\right)dy\right)e^{-a\left(\frac{\pi k}{L}\right)^2 t}\,\sin\left(\frac{\pi k}{L}x\right). \tag{6.1.35}$$

EXERCISE 6.1.15. *(a)*^{C} *Find the appropriate formula in (5.1.18) and verify (6.1.34). (b)*^{B} *Verify that if f is continuously differentiable on $(0, L)$ and $f(0) = f(L) = 0$, then the series in (6.1.35) converges uniformly for $t \ge 0$ and $x \in [0, L]$. Hint: use the same arguments as in the proof of Theorem 5.1.4 on page 250 to conclude that $\sum_{k=1}^{\infty}|G_k(0)| < \infty$. Then note that $0 < e^{-(\pi k/L)^2 t} \le 1$ for all $t \ge 0$ and use the M-test of Weierstrass; see page 249. (c)*^{B} *Assume that the function f is bounded and Riemann integrable on $(0, L)$. Let $t_0 > 0$ be fixed. Verify that the series in (6.1.35) converges uniformly for $t \ge t_0 > 0$ and $x \in [0, L]$. Hint: if $|f| \le A$, then $|\int_0^L f(x)\sin(\pi kx/L)dx| \le AL$. (d)*^{A} *What conditions on f ensure that the function u defined by (6.1.35) is continuously differentiable in t and twice continuously differentiable in x for $t > 0$ and $x \in (0, L)$? Hint. Infinitely many derivatives exist even if f is not continuous; having $\sum_{k=1}^{\infty} k^2|G_k(0)| < \infty$ ensures that the first derivatives are uniformly bounded.*

Similar to the infinite line, the heat equation on the interval has the infinite propagation speed and the instantaneous smoothing property. FOR EXAMPLE, assume that $L = \pi$, $f(x) = 1$ for $\pi/4 < x < 3\pi/4$ and $f(x) = 0$ otherwise. Then

$$\frac{2}{L}\int_0^L f(y)\,\sin\left(\frac{\pi k}{L}y\right)dy = \frac{2}{\pi k}(\cos(\pi k/4) - \cos(3\pi k/4)).$$

The reader who is comfortable with uniform convergence will notice that the corresponding function u define by (6.1.35) has infinitely many derivatives in x for every $t > 0$.

While it is intuitively clear that the temperature will never drop below zero, because $f(x) \ge 0$, it is somewhat counterintuitive to expect the wire to heat up immediately at all points outside the interval $(\pi/4, 3\pi/4)$. Nonetheless, it is possible to show that $u(t,x) > 0$ for all $t > 0$ and $0 < x < \pi$, but the proof requires a lot of extra work; this proof does not rely on (6.1.35) and is beyond the scope of our discussion. We therefore conclude that *an explicit solution formula does not always provide all the answers, and is not always the ultimate prize in the study of partial differential equations.* As an application of (6.1.35), we encourage the reader to use a computer and plot the graphs of the corresponding partial sums

$$\sum_{k=1}^{N} \frac{2}{\pi k}(\cos(\pi k/4) - \cos(3\pi k/4))e^{-ak^2 t} \sin kx, \;\; x \in [0, \pi],$$

as functions of x for various values of a, t, and N.

EXERCISE 6.1.16.[A] *In the above example, we chose a discontinuous function f for the initial distribution of the temperature to simplify the computation of the Fourier coefficients of f. What is the value of $\lim_{t\to 0^+} u(t,x)$? Hint: $u(0^+, x) \neq f(x)$ at the points of discontinuity of f; see Theorem 5.1.5 on page 251. Similarly, one can consider initial conditions so that $f(0) \neq 0$ and/or $f(L) \neq 0$. The corresponding function u, as defined by (6.1.35), still satisfies $u_t = au_{xx}$ for all $t > 0$ and $x \in (0, L)$, but $u = u(t,x)$ is no longer continuous for $t \geq 0$, $x \in [0, L]$, and $u(0,x)$ might not be equal to $f(x)$ for all $x \in [0, L]$. Therefore, u is not a classical solution, and it is natural to call it a* **generalized solution** *of (6.1.25). One can extend the idea of the generalized solution even further and allow some flexibility not only in satisfying the initial condition, but in satisfying the equation itself. As with many other interesting topics, generalized solutions fall outside the scope of our discussion, even though we will encounter these solutions again in the section on the wave equation.*

Let us now summarize the main steps in the derivation of (6.1.35). The goal is to find a solution of (6.1.25) in the form $u(t,x) = \sum_{k=1}^{\infty} G_k(t)H_k(x)$.

Step 1. We want each function $u_k(t,x) = G_k(t)H_k(x)$ to satisfy the heat equation $u_t = au_{xx}$, and this leads to two ordinary differential equations (6.1.28).

Step 2. We want each function H_k to satisfy the boundary conditions $H_k(0) = H_k(L) = 0$, but not to vanish for all x, and this leads to the special values of the constant r in (6.1.28) — see (6.1.31), and the corresponding functions H_k — see (6.1.32).

Step 3. We go back to (6.1.28) to find the corresponding functions G_k —

see (6.1.33).
Step 4. We determine the values of $G_k(0)$ from the initial condition $u(0,x)$.
Step 5. We write the answer as (6.1.35).

We conclude this section with a discussion of other possible boundary conditions. Assume that the temperature of the end points is not equal to zero: $u(t,0) = U_0$, $u(t,L) = U_1$; for the initial temperature, we also assume that $f(0) = U_0$, $f(L) = U_1$. To solve the corresponding problem (6.1.25), we look for the temperature distribution $u = u(t,x)$ in the form $u(t,x) = v(t,x)+ax+b$, where $v_t = v_{xx}$, $v(t,0) = v(t,L) = 0$, $v(0,x) = f(x)-ax-b$. We only need to find the numbers a, b to satisfy the boundary conditions; note that we have two unknown numbers and two boundary conditions, while the function $g(t,x) = ax+b$ satisfies $g_t = g_{xx} = 0$ for all a, b. For $x = 0$, $b = U_0$; for $x = L$, $aL + b = U_1$ or $a = (U_1 - U_0)/L$.

EXERCISE 6.1.17. *(a)^C Verify that the above function u satisfies the initial and boundary conditions, and write an explicit formula for $u(t,x)$ similar to (6.1.35). (b)^B Verify that*

$$\lim_{t\to\infty} u(t,x) = U_0 + \frac{U_1 - U_0}{L}\, x,$$

that is, $g(t,x) = U_0 + \frac{U_1-U_0}{L}x$ is the **steady-state solution** *or steady-state temperature distribution. Note that $g(t,x) = U_0$ if $U_0 = U_1 = 0$, which is physically meaningful, because, with both ends kept at the same temperature, there should be no temperature gradient in equilibrium. (c)^C Explain why we cannot follow directly the same procedure as for equation (6.1.25) and look for the solution in the form $u(t,x) = \sum_{k=1}^{\infty} G_k(t)H_k(x)$ for suitable functions G_k, H_k. Hint. The boundary conditions are no loner homogeneous: if $u_1(t,0) = U_0$, $u_2(t,0) = U_0$, and $U_0 \neq 0$, then $u_1(t,0) + u_2(t,0) \neq U_0$. (d)^A Give a formal definition of a* **homogeneous boundary condition.**

Next, let us look more closely at the heat loss through the end points. Denote by U_e the outside temperature. By **Newton's Law of Cooling**, see page 295, applied at the point $x = 0$, $u_t(t,0) = c_1(u(t,0) - U_e)$, $c_1 \geq 0$, with $c_1 = 0$ corresponding to perfect insulation. Since the heat flux in one dimension is a scalar, we use equality (6.1.13) to rewrite Newton's Law of Cooling as $J(t,0) = -c_2(u(t,0) - U_e)$. Similarly, equality (6.1.11) at the point $x = 0$ becomes $J(t,0) = -c_3 u_x(t,0)$, $c_3 > 0$. Repeating the same arguments for the right end $x = L$, we conclude that

$$u_x(t,0) = C_0(u(t,0) - U_e), \quad u_x(t,L) = C_1(U_e - u(t,L)) \tag{6.1.36}$$

for some non-negative numbers C_0, C_1. **Perfectly insulated** ends correspond to $C_0 = C_1 = 0$ and the boundary conditions $u_x(t,0) = u_x(t,L) = 0,\ t \geq 0$.

EXERCISE 6.1.18.[B] *(a) Verify (6.1.36). (b) Verify that the boundary conditions (6.1.36) are homogeneous if either* $U_e = 0$ *or* $C_0 = C_1 = 0$. *(c) Assume that the temperature at the end points is always equal to* U_e, *the outside temperature. By (6.1.36), we will have* $u_x(t,0) = u_x(x,L) = 0$. *Explain why the correct choice of boundary conditions in this case is* $u(t,0) = u(t,L) = U_e$. *Hint: in this case,* $u_x(t,0) = u_x(x,L) = 0$ *is just a consequence of the boundary conditions* $u(t,0) = u(t,L) = U_e$.

EXERCISE 6.1.19.[B] *Consider the* **initial-boundary value problem** *describing the temperature in a fully insulated wire:*

$$\begin{aligned} &u_t = a u_{xx},\ t > 0,\ 0 < x < L; \\ &u(0,x) = f(x),\ 0 \leq x \leq L;\ \ u_x(t,0) = u_x(t,L) = 0,\ t \geq 0. \end{aligned} \tag{6.1.37}$$

(a) Derive a formula similar to (6.1.35) for the solution and show that $\lim_{t\to\infty} u(t,x) = (1/L)\int_0^L f(x)dx$. *Hint. Follow the same steps as in the derivation of (6.1.35); this time,* $H_k(x) = \cos(\pi k x/L)$, $k = 0,1,2,\ldots$, *and we take the even extension of* f. *(b) Let* $L = \pi$, $a = 1$, *and* $f(x) = 1$ *for* $\pi/4 < x < 3\pi/4$, $f(x) = 0$ *otherwise. Use a computer algebra system to plot the corresponding solution for* $t = 0.01, 0.1, 1$. *Can you predict this behavior of the solution theoretically by analyzing the Fourier series? (c) Define* $a_0 = (1/L)\int_0^L f(x)dx$, $a_k = (2/L)\int_0^L f(x)\cos(\pi k x/L)dx$, $k \geq 1$. *Verify that, for every* $t \geq 0$,

$$\frac{1}{2}\int_0^L |u(t,x)|^2 dx = 2L\, a_0^2 + L\sum_{k=1}^{\infty} a_k^2\, e^{-2a(\pi k/L)^2 t}. \tag{6.1.38}$$

Hint: the easiest way is to interpret the formula for the solution as a Fourier series and to use Parseval's identity. (d) Conclude that the heat equation is **dissipative**, *that is, the energy of the solution, represented by the left-hand side of (6.1.38), is decreasing in time even if there are no physical reasons for heat loss.*

EXERCISE 6.1.20.[B] *Assume that the wire is losing heat through the lateral surface, and the outside temperature is equal to* U_e. *By (6.1.14), page 296, the temperature* $u = u(t,x)$ *of the wire satisfies* $u_t = u_{xx} - C_e(u - U_e)$. *Find* u *if* $L = \pi$, $u(t,0) = u(t,\pi) = U_e$, *and* $u(0,x) = U_e + \sin \pi x$. *Hint: consider the function* $v(t,x) = (u - U_e)e^{ct}$ *for a suitable* c.

Other initial-boundary value problems for the heat equation on the interval are possible; see Problem 7.2 on page 440.

Let us now summarize the three main properties of the heat equation:
• Infinite propagation speed; see page 299.
• Instantaneous smoothing; see page 299.
• Dissipation of energy; see page 306.

6.1.3 *Wave Equation in One Dimension*

Both the transport equation and the heat equation are first-order in time, that is, contain only the first-order partial derivative of the unknown function with respect to the time variable. There are no immediate analogs of these equations for finite systems of points: by the Second Law of Newton, the ordinary differential equations we encounter in classical mechanics usually contain second-order time derivatives.

The *wave equation* arises when the Second Law of Newton is applied to a *continuum of points.* A rigid body, see Section 2.2.3, page 79, is a continuum of points, but, because of the *rigidity condition*, we are able to describe the motion of this continuum using a finite system of ordinary differential equations. In a non-rigid continuum of points, the distances between the points can change, and each point requires a separate second-order ordinary differential equation. Such an *infinite-dimensional dynamical system* is described by a partial differential equation; a natural conjecture is that, when applied to a continuum of points, the Second Law of Newton leads to a partial differential equation that is second-order in time.

As an example of a non-rigid continuum of points described by the wave equation, consider a VIBRATING STRING under the following assumptions:
• The string always stays in the same plane; accordingly, we place the string horizontally along the x-axis, and denote by $u(t,x)$ the displacement of the string in the vertical y-direction at the point x and time t.
• The vibrations are transverse: no parts of the string have a displacement along the x-axis.
• The motion of the string is smooth, that is, the function u is twice continuously differentiable in t and x.
• The vibrations are small, that is, both $|u(t,x)|$ and $|u_x(t,x)|$ are small for all t, x.
• The string is perfectly elastic, that is, there is no bending moment (resistance to bending) and so the tension vector at every point is tangent to

the string.

- The magnitude T of the tension vector is the same at all times and at all points of the string; this is a reasonable assumption when the vibrations are small.
- The string is homogeneous, that is, the mass ρ per unit length is constant.

Let us now consider the tension forces acting on a small piece of the string, see Figure 6.1.2.

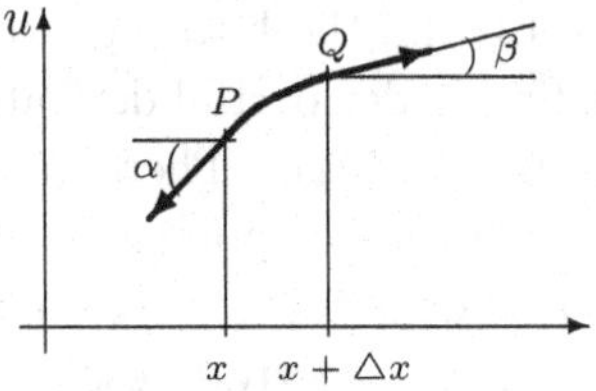

Fig. 6.1.2 Vibrating String

Since the vibrations are assumed transverse, we only need the forces in the vertical direction. Take two neighboring points on the string, P with coordinates $(x, u(t,x))$ and Q with coordinates $(x+\Delta x, u(t, x+\Delta x))$. Under the assumptions that the tension is tangent to the string and is constant in magnitude, we see from the picture that the vertical components of the tension at the points P and Q are $-T\sin\alpha$ and $T\sin\beta$, respectively. If $F = F(t,x)$ is the external vertical force, per unit length, acting on the string, and if Δx is sufficiently small, then the total vertical force acting on the PQ segment is $T(\sin\beta - \sin\alpha) + \int_x^{x+\Delta x} F(t,y)dy$. We assume that the function F is continuous; in general, F can depend on u. The linear momentum of the segment PQ is, by definition, $\int_x^{x+\Delta x} \rho u_t(t,y)dy$. Applying the Second Law of Newton and differentiating under the integral sign, we find

$$\int_x^{x+\Delta x} \rho u_{tt}(t,y)dy = T(\sin\beta - \sin\alpha) + \int_x^{x+\Delta x} F(t,y)dy,$$

and then, by the the mean-value theorem for the integrals,

$$\rho u_{tt}(t, x_1)\,\Delta x = T(\sin\beta - \sin\alpha) + F(t, x_2)\Delta x, \tag{6.1.39}$$

where x_1, x_2 are in the interval $[x, x+\Delta x]$.

EXERCISE 6.1.21.[B] *Verify (6.1.39).*

On the other hand, $\tan\alpha = u_x(t,x)$ and $\tan\beta = u_x(t,x+\Delta x)$ are the slopes of the string at the points P and Q. Since the vibrations are small, the angles α and β are small enough to allow the approximations $\tan\alpha \approx \sin\alpha$ and $\tan\beta \approx \sin\beta$. Replacing in (6.1.39) sines with tangents yields

$$\rho u_{tt}(t,x_1)\,\Delta x = T\big(u_x(t,x+\Delta x) - u_x(t,x)\big) + F(t,x_2)\Delta x.$$

Dividing by Δx and passing to the limit as $\Delta x \to 0$, we get

$$\rho u_{tt}(t,x) = T u_{xx}(t,x) + F(t,x).$$

Finally, we define $c^2 = T/\rho$ (so that c has the dimension of speed — verify this), and $h(t,x) = F(t,x)/\rho$, and conclude that the vibrations of the string are described by the `one-dimensional wave equation`

$$u_{tt}(t,x) = c^2 u_{xx}(t,x) + h(t,x). \tag{6.1.40}$$

While equation (6.1.40) is an interesting mathematical object to study in its own right, one would need experimental data to support the use of this equation as a modelling tool. Similar to other equations describing oscillations, (6.1.40) describes only small vibrations of the string, which leads to an approximation in the derivation. Fortunately, the experimental evidence is in favor of the equation as a good approximation to real vibrations.

Let us solve the WAVE EQUATION ON THE LINE, which corresponds to vibrations of an infinitely long string. By analogy with point-mass dynamics, the initial displacement $u(0,x)$ and the initial velocity $u_t(0,x)$ of the string should determine the position $u(t,x)$ for all $t > 0$. To begin, we assume that there are no external forces acting on the string: $h = 0$. Thus, we want to solve the `initial value problem`

$$\begin{aligned} &u_{tt}(t,x) = c^2 u_{xx}(t,x),\ t > 0,\ x \in \mathbb{R};\\ &u(0,x) = f(x),\ u_t(0,x) = g(x), x \in \mathbb{R}, \end{aligned} \tag{6.1.41}$$

for some functions f, g. Similar to the heat equation on the line, we will carry out an `a priori analysis` of (6.1.41) using the Fourier transform; recall that in this analysis we assume that all computations we want to perform can indeed be performed. To proceed, the reader should review the corresponding computations on page 297.

EXERCISE 6.1.22. [C] *By applying the Fourier transform to (6.1.41) with respect to x, show that the Fourier transform $\widehat{u} = \widehat{u}(t, \omega)$ of u satisfies*

$$\widehat{u}_{tt} = -c^2\omega^2\widehat{u},\ t > 0;\ \widehat{u}(0, \omega) = \widehat{f}(\omega),\ \widehat{u}_t(0, \omega) = \widehat{g}(\omega).$$

Conclude that

$$\widehat{u}(t, \omega) = \widehat{f}(\omega)\cos(c\omega t) + \frac{\widehat{g}(\omega)}{c\omega}\sin(c\omega t). \qquad (6.1.42)$$

Hint: if $y''(t) = -b^2 y(t)$, $b > 0$, then $y(t) = A\cos bt + B\sin bt$.

To recover the solution u from (6.1.42), we note that, by the Euler formula (more precisely, by (4.3.15) on page 214), $\cos(c\omega t) = (e^{ic\omega t} + e^{-ic\omega t})/2$. Then, by the shift property of the Fourier transform (5.2.22) on page 269, we conclude that the inverse Fourier transform of $\widehat{f}(\omega)\cos(c\omega t)$ is $(f(x - ct) + f(x + ct))/2$. Next, by (5.2.17) on page 268, the Fourier transform of the *antiderivative* $\int_0^t g(s)ds$ of g is $\widehat{g}(\omega)/(i\omega)$. Then, after replacing $\sin(c\omega t)$ by the Euler formula and using (5.2.22), we conclude that the inverse Fourier transform of $\big(\widehat{g}(\omega)/(c\omega)\big)\sin(c\omega t)$ is $(2c)^{-1}\int_{x-ct}^{x+ct} g(s)ds$. As a result,

$$u(t, x) = \frac{f(x - ct) + f(x + ct)}{2} + \frac{1}{2c}\int_{x-ct}^{x+ct} g(s)ds. \qquad (6.1.43)$$

Formula (6.1.43) is called **d'Alembert's solution** of the wave equation, after the French mathematician JEAN LE ROND D'ALEMBERT (1717–1783), who published the result in 1747.

EXERCISE 6.1.23.[A] *(a) Verify that (6.1.42) and the properties of the Fourier transform imply (6.1.43). (b) Verify that if f has two continuous derivatives, and g has one continuous derivative, then u is a classical solution of (6.1.41). For non-differentiable functions f, g, the corresponding u is a* **generalized solution** *of the wave equation; see also Exercise 6.1.16. (c) Let the functions $u = u(t, x)$ and $v = v(t, x)$ satisfy the system of equations $u_t - cu_x = v$, $v_t + cv_x = 0$ with initial conditions $u(0, x) = f(x)$, $v(0, x) = g(x) - cf'(x)$. Verify that u has the representation (6.1.43). Hint: use formulas (6.1.4) on page 292 and (6.1.8) on page 294. (d) According to part (c), we can write the wave equation as $D_+(D_-u) = 0$, where $D_+ = \partial/\partial t + c\,\partial/\partial x$, $D_- = \partial/\partial t - c\,\partial/\partial x$. What system corresponds to an equivalent representation $D_-(D_+u) = 0$?*

Unlike the heat equation, the wave equation has no smoothing property and has finite propagation speed. Indeed, assume that $g(x) = 0$. Then

(6.1.43) becomes $u(t,x) = (f(x-ct)+f(x+ct))/2$ (**travelling wave solution**) and u has continuous partial derivatives of order k if and only if $f = f(x)$ has continuous ordinary derivatives of order k. The finite propagation speed is illustrated by the following observation: if $f(x) = 0$ for $|x| > 1$, then $u(t,11) = 0$ for $0 < t < 10/c$. In general, the travelling wave solution consists of two waves, moving to the left $f(x+ct)$ and to the right $f(x-ct)$ with the same speed c. Let us emphasize that the individual points on the string do not move to the right or to the left, and move only up and down; it is the *disturbance* that travels along the string. This propagation of disturbance is characteristic of all wave motions.

EXERCISE 6.1.24.[C] *Consider the* **triangular pulse** *function*

$$f_0(x) = \begin{cases} 1-|x|, & |x| < 1 \\ 0, & |x| > 1, \end{cases}$$

(a) Assume that, in problem (6.1.41), $c = 1$, $f(x) = f_0(x)$, and $g(x) = 0$. Using your favorite programming language or a computer algebra system, write a program that plots the solution $u(t,x)$ of the wave equation, given by (6.1.43), as a function of x for a fixed $t \geq 0$. Plot the solution for $x \in [-5,5]$ and $t = 0, 0.5, 1, 2, 3$. (b) Repeat part (a) with $c = 1$, $f(x) = 0$, and $g(x) = f_0(x)$. Instead of a travelling wave, you will observe **wave diffusion**: *for every $x_0 \in \mathbb{R}$, there exists a $t_0 \geq 0$ so that $u(t,x_0) > 0$ for all $t \geq t_0$; in this particular example, you will have $u(t,x_0) = 1/2$ for all sufficiently large t.*

EXERCISE 6.1.25.[A] *Verify that the solution of the problem*

$$u_{tt} = c^2 u_{xx} + h(t,x),\ t > 0, x \in \mathbb{R},\ u(0,x) = u_t(0,x) = 0, \qquad (6.1.44)$$

is

$$u(t,x) = \frac{1}{2c}\int_0^t \left(\int_{x-c(t-s)}^{x+c(t-s)} h(s,y)dy \right) ds. \qquad (6.1.45)$$

Assume the necessary smoothness of the function h. This is yet another example of the **variation of parameters**.

We now consider a more realistic model, corresponding to a string of finite length L and fixed end points. In this case, we want to solve the

initial-boundary value problem

$$\begin{aligned} &u_{tt}(t,x) = c^2 u_{xx},\ t>0,\ x \in (0,L);\\ &u(0,x) = f(x),\ u_t(0,x) = g(x),\ x\in[0,L];\\ &u(t,0) = u(t,L) = 0,\ t \geq 0. \end{aligned} \tag{6.1.46}$$

We assume that the functions f, g are continuous on $[0, L]$ and $f(0) = f(L) = 0$.

EXERCISE 6.1.26. ^{C}. *Assume that $g(x) = 0$ for all x. it is physically reasonable to define the* **energy of the string** *as*

$$\mathcal{E}(t) = \frac{1}{2}\int_0^L (u_t^2(t,x) + c^2 u_x^2(t,x))dx; \tag{6.1.47}$$

the terms u_t^2 and $c^2u_x^2$ represent the kinetic and potential energy, respectively. Show that $d\mathcal{E}(t)/dt = 0$, that is, the energy is conserved, and, unlike the heat equation, the wave equation has no intrinsic dissipation of energy. Hint: differentiate under the integral sign to get $u_tu_{tt}+c^2u_xu_{tx}$, integrate by parts to transform the second term to $-c^2u_{xx}u_t$, and then use the equality $u_{tt} = c^2u_{xx}$.

Similar to the heat equation on the line, we will carry out an **a priori analysis** of (6.1.46) using **separation of variables** and following the steps on page 304. To proceed, the reader should review the corresponding computations on page 302.

We note that the boundary conditions in (6.1.46) are homogeneous, and therefore we should look for a solution in the form $u(t,x) = \sum_{k=1}^{\infty} G_k(t)H_k(x)$, with each function $u_k(t,x) = G_k(t)H_k(x)$ satisfying the wave equation and $H_k(0) = H_k(L) = 0$.

Step 1. We conclude that

$$G_k''(t) = rc^2G_k(t),\ H_k''(x) = rH_k(x), \tag{6.1.48}$$

for some real number r.

Step 2. We want each function H_k to satisfy the boundary conditions $H_k(0) = H_k(L) = 0$, but to be non-zero for some $x \in (0, L)$, and this leads to the special values of the number r:

$$r = -(\pi/L)^2k^2,\ k = 1, 2, \ldots, \tag{6.1.49}$$

and the corresponding functions H_k:

$$H_k(x) = \sin(\pi kx/L), k = 1, 2, \ldots. \tag{6.1.50}$$

Step 3. We go back to (6.1.48) to find the corresponding functions G_k:

$$G_k(t) = A_k \cos(c\pi kt/L) + B_k \sin(c\pi kt/L). \tag{6.1.51}$$

Step 4. We determine the values of A_k, B_k from the initial conditionx $u(0,x)$, $u_t(0,x)$:

$$\begin{aligned} A_k &= \frac{2}{L}\int_0^L f(x)\,\sin(\pi kx/L)dx, \\ B_k &= \frac{2}{ck\pi}\int_0^L g(x)\,\sin(\pi kx/L)dx, \ k = 1, 2, \dots. \end{aligned} \tag{6.1.52}$$

Step 5. We write the final answer:

$$\begin{aligned} u(t,x) = &\sum_{k=1}^{\infty}\left(\frac{2}{L}\int_0^L f(y)\,\sin\left(\frac{\pi k}{L}y\right)dy\right)\cos\left(\frac{c\pi k}{L}t\right)\,\sin\left(\frac{\pi k}{L}x\right) \\ &+\sum_{k=1}^{\infty}\left(\frac{2}{ck\pi}\int_0^L g(y)\,\sin\left(\frac{\pi k}{L}y\right)dy\right)\sin\left(\frac{c\pi k}{L}t\right)\,\sin\left(\frac{\pi k}{L}x\right) \end{aligned} \tag{6.1.53}$$

EXERCISE 6.1.27.^{C} *Verify the equalities (6.1.48)–(6.1.53).*

An elementary solution

$$u_k(t,x) = (A_k \cos(c\pi kt/L) + B_k \sin(c\pi kt/L)) \sin(\pi kx/L)$$

has three special names, depending on the area of applications: the `normal mode` of the string, the `standing wave`, and the `stationary wave`. The `fundamental mode` corresponds to $k = 1$; all other normal modes are called `overtones`. An overtone corresponding to $k = n$ has $n-1$ `nodes`, the points $x_\ell = (L/n)\ell$, $\ell = 1, \dots, n-1$, that are not the end points and where the string is not moving. The existence of the nodes is a characteristic feature of a stationary or standing wave; it is the fixed nodes that make a standing wave different from a travelling wave. The motion of the string in time produces sound vibrations, and $c/(2L)$ is the vibration frequency of the fundamental mode. Recalling that $c = \sqrt{T/\rho}$, where T is the tension and ρ is the mass per unit length, we see that, to increase the frequency of the sound produced, one has to increase T and to decrease ρ and L. Physical constraints prevent large variations in T: the string has to be tight enough for the wave equation to apply, while very large T will break the string. Accordingly, changing T corresponds to `tuning`, or small adjustments of the frequency, while the actual range of a string instrument is determined by the size of the string: going from double-bass to cello to viola to violin, we observe strings getting lighter and shorter.

Unlike the Fourier series solution (6.1.35), page 303, of the heat equation, formula (6.1.53) can be simplified and reduced to a form similar to (6.1.43). Indeed, denote by $f_o = f_o(x)$ the *odd* extension of the function $f(x)$ to the interval $(-L, 0)$, that is, $f_o(x) = -f(-x)$ for $x \in (-L, 0)$.

EXERCISE 6.1.28.[B] *Verify that*

$$\sum_{k=1}^{\infty} \left(\frac{2}{L} \int_0^L f(y) \sin\left(\frac{\pi k}{L} y\right) dy \right) \cos\left(\frac{c\pi k}{L} t\right) \sin\left(\frac{\pi k}{L} x\right) = \frac{S_{f_o}(x+ct) + S_{f_o}(x-ct)}{2}, \tag{6.1.54}$$

where S_{f_o} is the Fourier series of f_o. Hint: use the trigonometric identity $2\sin\alpha\cos\beta = \sin(\alpha+\beta) + \sin(\alpha-\beta)$ with $\alpha = \pi kx/L$ and $\beta = c\pi kt/L$. Then see page 255.

Next, define the function $G(x) = \int_0^x g(y)dy$, $x \in [0, L]$. Let G_e be the *even* extension of G to $(-L, 0)$, that is, $G_e(x) = G(-x)$ for $x \in (-L, 0)$.

EXERCISE 6.1.29.[A] *(a) Verify that*

$$\sum_{k=1}^{\infty} \left(\frac{2}{ck\pi} \int_0^L g(y) \sin\left(\frac{\pi k}{L} y\right) dy \right) \sin\left(\frac{c\pi k}{L} t\right) \sin\left(\frac{\pi k}{L} x\right) = \frac{S_{G_e}(x+ct) - S_{G_e}(x-ct)}{2c}, \tag{6.1.55}$$

where S_{G_e} is the Fourier series of G_e. Hint: integrate by parts to conclude that

$$\frac{2}{\pi k} \int_0^L g(y) \sin(\pi ky/L) dy = -\frac{2}{L} \int_0^L G(y) \cos(\pi ky/L) dy.$$

Then use the trigonometric identity $-2\sin\alpha\sin\beta = \cos(\alpha+\beta) - \cos(\alpha-\beta)$ with $\alpha = \pi kx/L$ and $\beta = c\pi kt/L$. (b) Verify that $G(x)$ can be replaced with $\int_{x_0}^x g(y)dy$ for an arbitrary $x_0 \in [0, L]$.

By combining (6.1.53), (6.1.54), and (6.1.55), we get an alternative representation formula for the solution of (6.1.46):

$$u(t,x) = \frac{S_{f_o}(x+ct) + S_{f_o}(x-ct)}{2} + \frac{S_{G_e}(x+ct) - S_{G_e}(x-ct)}{2c}, \tag{6.1.56}$$

$t \geq 0$, $x \in [0, L]$. Thus, the superposition of infinitely many standing waves in (6.1.53) can be represented as a sum of four travelling waves.

EXERCISE 6.1.30.[A] *Given a continuous function $h = h(t,x)$, consider the wave equation $u_{tt} = c^2 u_{xx} + h$, $t > 0$, $x \in (0, L)$, with zero initial and boundary conditions $u(0,x) = u_t(0,x) = 0$, $u(t,0) = u(t,L) = 0$. Derive the following analog of the* `variation of parameters` *formula (6.1.45) for the solution:*

$$u(t,x) = \frac{1}{2c}\int_0^t \Big(S_{H_e}(x + c(t-s)) - S_{H_e}(x - c(t-s))\Big)ds, \qquad (6.1.57)$$

where H_e is the even extension to $(-L, 0)$ of an antiderivative of h. Hint: Look for a solution in the form $u(t,x) = \sum_{k=1}^{\infty} G_k(t)\sin(\pi kx/L)$, substitute the sum in the equation, multiply by $\sin(\pi kx/L)$ for some fixed n and integrate from 0 to L. You get $G_n''(t) + (c\pi n/L)^2 G_n(t) = h_n(t)$ for suitable functions h_n. Note that the solution of the equation $y''(t) + a^2 y(t) = f(t)$ with $a > 0$ and initial conditions $y(0) = y'(0) = 0$ is $y(t) = (1/a)\int_0^t \sin(a(t-s))f(s)ds$.

Now assume that the function f is continuously differentiable on $[0, L]$ and $f(0) = f(L) = 0$. Denote by $\widetilde{f}_o$ the odd periodic extension of f (that is, the $2L$-periodic extension of the function f_o to the whole line). Similarly, assume that the function g is continuous on $[0, L]$ and denote by $\widetilde{G}_e$ the even periodic extension of an anti-derivative of g (that is, the $2L$-periodic extension of the function G_e to the whole line.) By Theorem 5.1.5, page 251, we can then re-write (6.1.56) in a simpler form similar to (6.1.43) on page 310:

$$u(t,x) = \frac{\widetilde{f}_o(x+ct) + \widetilde{f}_o(x-ct)}{2} + \frac{\widetilde{G}_e(x+ct) - \widetilde{G}_e(x-ct)}{2c}. \qquad (6.1.58)$$

EXERCISE 6.1.31.[B] *(a) Verify that $S_{f_o}(x) = \widetilde{f}_o$ if f is continuously differentiable on $[0, L]$ and $f(0) = f(L) = 0$. (b) Verify that $S_{G_e}(x) = \widetilde{G}_e$ if g is continuous on $[0, L]$. (c) Verify that if g is continuous on $[0, L]$ and $g(0) = g(L) = 0$, then $\widetilde{G}_e$ is the anti-derivative of the odd extension of g.*

EXERCISE 6.1.32.[C] *Assume that $c = 1$, $L = 2$, $f(x) = 1 - |x - 1|$, and $g(x) = 0$, $0 \le x \le 2$. (a) Write a computer program that plots the graph of the solution $u = u(t,x)$, defined by (6.1.58), as a function of x for different fixed values of t. (b) Note that, for fixed x_0, the function $u(t, x_0)$ is periodic in t. What is the period? Does the period depend on x_0?*

EXERCISE 6.1.33.[A] *Note that the right-hand side of (6.1.56) is defined for all functions f, g that are Riemann integrable on $[0, L]$. For such f, g, it is natural to call the right-hand side of (6.1.56) a* `generalized solution` *of*

(6.1.46), see also Exercise 6.1.16 on page 304. Find the conditions on the functions f, g ensuring that (6.1.56) is a `classical solution` *of (6.1.46). Hint: these conditions are much stronger than those leading to (6.1.58), so you can work with (6.1.58) from the beginning.*

Let us now summarize the three main properties of the wave equation:
- Finite propagation speed; see page 310.
- No smoothing; see page 310.
- Conservation of energy; see page 312.

6.2 Elements of the General Theory of PDEs

6.2.1 *Classification of Equations and Characteristics*

We start by classifying the partial differential equations (PDEs) according to the *order* of the derivatives in the equation. To simplify the presentation, we consider only two independent variables, which we denote by x and y, $x, y \in \mathbb{R}$. The unknown function in the equation is then $u = u(x, y)$. All the definitions in this section can be stated for an arbitrary number of independent variables; the names of the variables can also be different, for example, t, x instead of x, y.

To state the definitions, we assume that the function u has continuous partial derivatives of every order. A `partial derivative` of order $k \geq 1$ of the function $u = u(x, y)$ is an expression

$$D^k u = \frac{\partial^k u}{\partial x^{k_1} \partial y^{k_2}}, \; k_1 \geq 0, \; k_2 \geq 0, \; k_1 + k_2 = k;$$

by convention, $D^0 u = u$.

EXERCISE 6.2.1. ^{A} *Given $k \geq 1$, what is the largest possible number of distinct partial derivatives of order k for a smooth function $u = u(x, y)$? Hint: remember that $\partial^k u/(\partial x^{k_1} \partial y^{k_2}) = \partial^k u/(\partial y^{k_2} \partial x^{k_1})$.*

A `partial differential equation of order` N is an expression $F(x, y, u, Du, D^2 u, \dots, D^N u) = 0$, where F is a given function. In other words, a partial differential equation has order N if N is the highest order of the partial derivative of the unknown function appearing in the equation. The word *appearing* is essential and ensures that, for example, $0 \cdot u_{xxx} + u_x + u_t = 0$ is a first-order equation. Of course, not all possible partial derivatives of order N need to appear in the equation of order

N. FOR EXAMPLE, equations $u_x = u_y$ and $u_x + uu_y = 0$ are first-order, equations $u_x = u_{yy}$ and $u_{xx} + u_{yy} = 1$ are second-order, and equation $u_x + uu_y + u_{yyy} = 0$ is third-order.

Next, we classify the partial differential equations according to the *type of nonlinearity*. Again, for simplicity, we consider only two independent variables. A `linear` partial differential equation of order N has the form

$$\sum_{\substack{m,n\geq 0\\ m+n\leq N}} a_{mn}(x,y)\frac{\partial^{m+n}u(x,y)}{\partial x^m \partial y^n} = f(x,y), \tag{6.2.1}$$

where at least one function a_{mn} with $m+n = N$ is not equal to zero identically. The functions a_{mn} are called `coefficients` of the equation. This linear equation is called `homogeneous` if the function f is identically equal to zero; otherwise, the equation is `linear inhomogeneous`. A partial differential equation that does not have the form (6.2.1) is called `nonlinear`. FOR EXAMPLE, equations $u_x = u_y$ and $u_x = u_{yy}$ are linear homogeneous, $u_{xx} + u_{yy} = 1$ is linear inhomogeneous, and equations $u_x + uu_y = 0$, $u_x + uu_y + u_{yyy} = 1 + xy$ are nonlinear. Sometimes, a nonlinear equation becomes linear after a special change of the unknown function.

EXERCISE 6.2.2.^{A} *Let $u = u(t,x)$ satisfy the equation $u_t = u_{xx} - u_x^2$. Define $w(t,x) = e^{-u(t,x)}$. Verify that w satisfies the heat equation $w_t = w_{xx}$.*

Among the nonlinear equations, we distinguish between *semi-linear*, *quasi-linear*, and *fully nonlinear*. A `semi-linear equation` of order N has the form

$$\sum_{\substack{m,n\geq 0\\ m+n=N}} a_{mn}(x,y)\frac{\partial^{m+n}u(x,y)}{\partial x^m \partial y^n} + F_0(x,y,u,Du,D^2u,\ldots,D^{N-1}u) = 0; \tag{6.2.2}$$

in other words, all the partial derivatives of the highest order N appear linearly. A `quasi-linear equation` of order N has the form

$$\sum_{\substack{m,n\geq 0\\ m+n=N}} f_{mn}(x,y,u,Du,D^2u,\ldots,D^{N-1}u)\frac{\partial^{m+n}u(x,y)}{\partial x^m \partial y^n} + F_0(x,y,u,Du,D^2u,\ldots,D^{N-1}u) = 0; \tag{6.2.3}$$

in other words, all the partial derivatives of the highest order N can be multiplied by functions depending on the partial derivatives of lower order.

An equation is **fully nonlinear** if it cannot be represented in any of the forms (6.2.1), (6.2.2), (6.2.3). FOR EXAMPLE, $u_x + uu_y + u_{yyy} = 0$ is semi-linear, $u_x + uu_y = 0$ is quasi-linear, and $u_x + |u_y| = 0$ is fully nonlinear.

Let us now look more closely at LINEAR HOMOGENEOUS FIRST-ORDER EQUATIONS. Using a method called the **method of characteristics**, these equations are reduced to a system of ordinary differential equations. We illustrate the method for two independent variables x and y and equations of the type

$$a(x,y)u_x(x,y) + b(x,y)u_y(x,y) = 0, \tag{6.2.4}$$

where $a = a(x,y)$ and $b = b(x,y)$ are known continuous functions. Consider the system of ordinary differential equations for the unknown functions $x = x(t)$ and $y = y(t)$:

$$\dot{x}(t) = a(x(t), y(t)), \quad \dot{y}(t) = b(x(t), y(t)). \tag{6.2.5}$$

Recall that $\dot{x}$ denotes the derivative of x with respect to t. This system is called the **characteristic system** for equation (6.2.4). If the functions $x(t), y(t)$ satisfy (6.2.5), then the curve in $\mathbb{R}^2$ defined by the vector parametric equation $\boldsymbol{r}(t) = x(t)\,\hat{\imath} + y(t)\hat{\jmath}$ is called a **characteristic** or a **characteristic curve** of equation (6.2.4).

EXERCISE 6.2.3.^{C} *Define the function $g(t) = u(x(t), y(t))$, where $u = u(x,y)$ satisfies (6.2.4) and $(x(t), y(t))$ satisfy (6.2.5). (a) Show that $\dot{g}(t)$ is identically equal to zero.* Hint: use the chain rule. *(b) Conclude that $u(x(t), y(t))$ is constant, that is, the solution is constant on a characteristic curve..*

The detailed analysis of the method of characteristics is outside the scope of our discussion, and we simply state the result for equation (6.2.4): *If the characteristic curves can be written as the level sets $f(x,y) = c$ of some function f, then every solution of (6.2.4) has the form $u(x,y) = F(f(x,y))$, where $F = F(t)$ is a continuously differentiable function of one variable.* For EXAMPLE, consider the equation $yu_x + xu_y = 0$, with an extra condition $u(x,0) = x^2$. The corresponding characteristic system $\dot{x} = y$, $\dot{y} = x$ can be written as $d(x^2 - y^2)/dt = 0$ (verify this!) and therefore the characteristic curves are hyperbolas $x^2 - y^2 = c$. Therefore, the general solution of the equation is $u(x,y) = F(x^2 - y^2)$ for a continuously differentiable function F of one variable. Since $u(x,0) = F(x^2) = x^2$, we conclude that $u(x,y) = x^2 - y^2$. More generally, the function F can be identified from the values of the solution u on some curve in the (x, y)-plane, *as long as the curve is not of the form* $x^2 - y^2 = c$. In particular, there is no solution of the equation $yu_x + xu_y = 0$, satisfying $u(x, x) = x$. For more on the method of characteristics for the first-order partial differential equations, see Section 3.2 of the book *Partial Differential Equations* by L. C. Evans, 1998.

EXERCISE 6.2.4. *(a)*[C] *Solve the equation* $xu_x - yu_y = 0$, $u(1, y) = y^2$ *using the method of characteristics. Hint: the characteristic equation implies* $xy = c$, *so the solution is* $u(x, y) = x^2y^2$. *(b)*[C] *Solve the transport equation* $u_t + bu_x = 0$, *where* b *is a real number, using the method of characteristics, and show that the result coincides with formula (6.1.4) on page 292. (c)*[A] *Prove that* every solution *of* $yu_x + xu_y = 0$ *can be written as* $u(x, y) = F(x^2 - y^2)$ *for a continuously differentiable function* F. *Hint: it is obvious that* $u(x, y) = F(x^2 - y^2)$ *is a solution; the non-trivial part is showing that if* $u = u(t, x)$ *satisfies* $yu_x + xu_y = 0$, *then there exists a continuously differentiable function* F *such that* $u(x, y) = F(x^2 - y^2)$. *(d)*[A] *Verify that specifying the value of the solution on a characteristic curve makes it impossible to find a unique solution of equation (6.2.4). Hint: either you get more than one solution or no solutions at all.*

To conclude this section, we look more closely at SEMI-LINEAR SECOND-ORDER EQUATIONS IN TWO VARIABLES:

$$A(x, y)u_{xx} + 2B(x, y)u_{xy} + C(x, y)u_{yy} + F(x, y, u, u_x, u_y) = 0, \qquad (6.2.6)$$

where A, B, C, F are known functions. Consider the function

$$K(x, y) = \begin{vmatrix} A(x, y) & B(x, y) \\ B(x, y) & C(x, y) \end{vmatrix} = A(x, y)C(x, y) - B^2(x, y). \qquad (6.2.7)$$

The sign of the function K determines the *type* of the equation as follows. Equation (6.2.6) is called `elliptic` at the point (x_0, y_0) if $K(x_0, y_0) > 0$; equation (6.2.6) is called `hyperbolic` at the point (x_0, y_0) if $K(x_0, y_0) < 0$; equation (6.2.6) is called `parabolic` at the point (x_0, y_0) if $K(x_0, y_0) = 0$. We say that the equation is elliptic (hyperbolic, parabolic) in a set $G \subset \mathbb{R}^2$ if the equation is elliptic (hyperbolic, parabolic) at every point of the set.

EXERCISE 6.2.5.[C] *(a) Verify that the equations* $u_{xx} + u_{yy} = 0$, $u_{tt} = u_{xx}$, *and* $u_t = u_{xx}$ *are elliptic, hyperbolic, and parabolic, respectively, everywhere in* $\mathbb{R}^2$. *(b) Determine the type of the equation* $yu_{xx} + u_{yy} = 0$.

The ordinary differential equation for the unknown function $y = y(x)$,

$$A(x, y(x))(y'(x))^2 - 2B(x, y(x))y'(x) + C(x, y(x)) = 0, \qquad (6.2.8)$$

is called the **characteristic equation** for (6.2.6), and, for hyperbolic equations, can often be used to solve (6.2.6). FOR EXAMPLE, consider the equation $x^2u_{xx} - y^2u_{yy} = 0$. This equation is hyperbolic everywhere in $\mathbb{R}^2$, and the corresponding characteristic equation $x^2(y'(x))^2 - y^2 = 0$ becomes $y'(x) = \pm y/x$. The solution of $y'(x) = y/x$ is $y(x) = c_1x$ for some real c_1; the solution of $y'(x) = -y/x$ is $y(x) = c_2/x$ for some other real number c_2. Accordingly, we introduce new variables ξ, η by $\xi(x, y) = y/x$, $\eta(x, y) = xy$ and define a new function $v(\xi, \eta)$ so that $v(\xi(x, y), \eta(x, y)) = u(x, y)$. By the chain rule, $u_x = v_\xi\xi_x + v_\eta\eta_x = -yv_\xi/x^2 + yv_\eta$, $u_y = v_\xi/x + xv_\eta$, $u_{xx} = (y^2/x^4)v_{\xi\xi} - (2y^2/x^2)v_{\xi\eta} + y^2v_{\eta\eta}$, $u_{yy} = (1/x^2)v_{\xi\xi} + 2v_{\xi\eta} + x^2v_{\eta\eta}$ (verify the last three equalities!) The original equation $x^2u_{xx} - y^2u_{yy} = 0$ becomes $-4y^2v_{\xi\eta} = 0$, or $v_{\xi\eta} = 0$. Computing the anti-derivatives with respect to ξ and η, we conclude that $v(\xi, \eta) = F(\xi) + G(\eta)$, where F, G are twice continuously differentiable functions of one variable. Now we recall that $\xi(x, y) = y/x$, $\eta(x, y) = xy$, and so the general solution of the equation $x^2u_{xx} - y^2u_{yy} = 0$ is $\boxed{u(x, y) = F(y/x) + G(xy).}$ The functions F, G can be identified from suitable initial conditions; we leave it to the reader to work out the details.

We can now summarize the method of characteristics for second-order equations in two independent variables. If equation (6.2.6) is hyperbolic everywhere in $\mathbb{R}^2$, then the characteristic equation (6.2.8) has two families of solutions, which we write as $\xi(x, y) = c_1$, $\eta(x, y) = c_2$. As a rule, the change of variables $(x, y) \to (\xi, \eta)$ is such that we can solve the equations $q_1 = \xi(x, y), q_2 = \eta(x, y)$ for (x, y) in terms of (q_1, q_2). Then we introduce a new function $v(\xi, \eta)$ so that $v(\xi(x, y), \eta(x, y)) = u(x, y)$, and substitute this v into the original equation (6.2.6). After a fairly long computation, in which even a slight error can ruin everything, we transform (6.2.6) to $v_{\xi\eta} = \tilde{F}(\xi, \eta, v, v_\xi, v_\eta)$ for some function $\tilde{F}$. If the computations are carried out correctly, then the partial derivatives $v_{\xi\xi}$ and $v_{\eta\eta}$ cancel out, and $v_{\xi\eta}$ is the only second-order partial derivative that stays: the characteristic equation (6.2.8) ensures this cancellation of $v_{\xi\xi}$ and $v_{\eta\eta}$ for every hyperbolic equation! If the new equation is $v_{\xi\eta} = 0$, then $v(\xi, \eta) = F(\xi) + G(\eta)$ for two functions F, G, and so $u(x, y) = F(\xi(x, y)) + G(\eta(x, y))$ is the general solution of the original equation (6.2.6). Of course, $v_{\xi\eta} = 0$ is not the only equation that can be easily solved. For example, if the transformed equation is $v_{\xi\eta}+v_\xi=0$, then $v(\xi,\eta)=F(\xi)e^{-\eta}+G(\eta)$. Note the similarity between the two arbitrary functions F, G and the two arbitrary real numbers C_1, C_2 that appear in the general solution of a linear second-order ordinary differential equation.

EXERCISE 6.2.6. *(a)^C For the wave equation $u_{tt}=c^2u_{xx}$, verify that $\xi(t,x)=x-ct$, $\eta(t,x)=x+ct$, and so the general solution is $u(t,x)=F(x-ct)+G(x+ct)$. Determine the functions F, G from the initial conditions $u(0,x)=f(x)$, $u_t(0,x)=g(x)$ and verify that the result agrees with formula (6.1.43) on page 310. (b)^A Find the general solution of the equation $2x^2u_{xx}-3xyu_{xy}-2y^2u_{yy}+2xu_x-2yu_y$. Hint: it is $u(x,y)=F(y/\sqrt{x})+G(x^2y)$. (c)$^{A+}$ What initial conditions on the solution will make it possible to determine the functions F, G uniquely for the equation in part (b) Hint: similar to the first-order equations, consider a curve in the (x,y) plane, on which you prescribe the values of u and $\nabla u\cdot\widehat{\boldsymbol{n}}$ for a given continuous vector field $\widehat{\boldsymbol{n}}$ of unit vectors.*

If equation (6.2.6) is parabolic everywhere in $\mathbb{R}^2$, then equation (6.2.8) has only one family of solutions $\xi(x,y)=c$. The change of variables $(x,y)\to(x,\xi)$ and the new function $v(x,\xi)$ so that $v(x,\xi(x,y))=u(x,y)$ transform the equation to $v_{xx}=\bar{F}(x,\xi,v,v_x,v_\xi)$. Unlike the hyperbolic case, this equation is usually not easily solvable, except for the (not very interesting) trivial case $v_{xx}=0$. In particular, the method of characteristics does not lead to an alternative solution of the heat equation (verify this!)

If equation (6.2.6) is elliptic everywhere in $\mathbb{R}^2$, then equation (6.2.8) has no real-valued solutions, and so we cannot solve elliptic equations using the method of characteristics. The method of characteristics also does not work for equations that are of different types in different parts of $\mathbb{R}^2$, or for second-order equations in more than two independent variables.

6.2.2 *Variation of Parameters*

The objective of this section is to formulate the variation of parameters formula for general evolution equations. The level of abstraction in this section is somewhat higher than usual.

Formula (6.1.8) on page 294 for the solution of the inhomogeneous transport equation, formula (6.1.20) on page 298 for the solution of the inhomogeneous heat equation, and formula (6.1.45) on page 311 for the solution of the inhomogeneous wave equation suggest the existence of a general method of finding the solution of an inhomogeneous linear equation from the solution of the corresponding homogenous equation. To describe this general method we need several new notions that would allow us to create a unified framework for the analysis of different equations. For simplicity, we only consider equations in two independent variables, t and x, with $t \geq 0$ playing the role of time.

An **operator** $\mathcal{A}$ is a rule that transforms a function u into another function $\mathcal{A}[u]$. The operation of differentiation is an example of an operator. The Fourier transform and the Laplace transform are also operators.

Given a function $u = u(t, x)$, we define a *differential operator* $\mathcal{A}$ so that

$$\mathcal{A}[u](t, x) = \sum_{k=0}^{N} A_k(x) \frac{\partial^k u(t, x)}{\partial x^k}, \tag{6.2.9}$$

where $A_k = A_k(x)$ are bounded continuous functions. Note that if $f = f(x)$, then $\mathcal{A}[f]$ is a function of x only, and $\mathcal{A}[f](x) = \sum_{k=0}^{N} A_k(x) f^{(k)}(x)$, where $f^{(k)}$ denotes k-th derivative of f.

EXERCISE 6.2.7.^{B} *Verify that if $g = g(t)$ is a function depending only on t, then $\mathcal{A}[g](t, x) = A_0(x)$.*

A homogeneous linear **evolution equation** is a partial differential equation of the form

$$u_t(t, x) = \mathcal{A}[u](t, x). \tag{6.2.10}$$

Notice that both the transport equation $u_t + bu_x = 0$ and the heat equation $u_t = au_{xx}$ can be written in the form (6.2.10), with $\mathcal{A}[u](t, x) = -bu_x(t, x)$ and $\mathcal{A}[u](t, x) = au_{xx}(t, x)$, respectively.

Suppose that, for sufficiently many functions $f = f(x)$, equation (6.2.10) with initial condition $u(0, x) = f(x)$ has a *unique* classical solution $u = u(t, x)$. Then we define another operator, Φ, called the **solution operator** for equation (6.2.10). This operator takes the initial condition $f = f(x)$ and transforms it into the corresponding solution $u = u(t, x)$:

$$\Phi[f](t, x) = u(t, x). \tag{6.2.11}$$

In other words,

$$\frac{\partial \Phi[f](t, x)}{\partial x} = \mathcal{A}[\Phi[f]](t, x), \quad \Phi[f](0^+, x) = f(x). \tag{6.2.12}$$

For the transport equation $u_t + bu_x = 0$, the solution operator is $\Phi[f](t, x) = f(x - bt)$; for the heat equation on the line $u_t = au_{xx}$, $\Phi[f](t, x) =$

$\int_{-\infty}^{\infty} K(t, x-y)f(y)dy$, where K is the heat kernel (6.1.21), defined on page 298. As the example of the heat equation shows, it might not always be possible to simply set $t = 0$ in the formula for Φ. The uniqueness of the solution of (6.2.10), which we always assume but never prove, implies the `semi-group property` of the operator Φ:

$$\Phi[f](t+s, x) = \Phi[\Phi[f](s, \cdot)](t, x). \tag{6.2.13}$$

That is, for every $t, s \geq 0$, the solution at time $t+s$ with initial condition f is equal to the solution at time t with the initial condition equal to the solution at time s.

Now consider the inhomogeneous linear evolution equation

$$u_t(t, x) = \mathcal{A}[u](t, x) + h(t, x), \tag{6.2.14}$$

with zero initial condition, that is, $u(0, x) = 0$. Under certain assumptions on h, the function

$$u(t, x) = \int_0^t \Phi[h(s, \cdot)](t-s, x)ds \tag{6.2.15}$$

is a classical solution of (6.2.14). Formula (6.2.15) is the general `variation of parameters` formula.

Recall that, by definition, the operator Φ acts on functions that depend only on x. The dot in the notation $h(s, \cdot)$ indicates that we keep s fixed and consider h as a function of x only. In other words, for every fixed $s \geq 0$, the function $v(t, x) = \Phi[h(s, \cdot)](t, x)$, $t \geq s$, is the solution of the `initial value problem` $v_t(t, x) = \mathcal{A}[v](t, x)$, $t \geq s$, $v(s, x) = h(s, x)$. For example, if $\Phi[f](t, x) = f(x - bt)$, then $\Phi[h(s, \cdot)](t, x) = h(s, x - bt)$.

Here is an intuitive reason why (6.2.15) is true under some technical assumptions. Let us differentiate (6.2.15) with respect to t and apply the Fundamental Theorem of Calculus:

$$u_t(t, x) = \int_0^t \frac{\partial}{\partial t}\Phi[h(s, \cdot)](t-s, x)ds + \Phi[h(t, \cdot)](0^+, x).$$

On the other hand, (6.2.12), (6.2.15), and (6.2.9) imply

$$\begin{aligned}\int_0^t \frac{\partial}{\partial t}\Phi[h(s, \cdot)](t-s, x)ds &= \int_0^t \sum_{k=0}^{N} A_k(x)\frac{\partial^k}{\partial x^k}\Phi[h(s, \cdot)](t-s, x)ds \\ &= \sum_{k=0}^{N} A_k(x)\frac{\partial^k}{\partial x^k}\int_0^t \Phi[h(s, \cdot)](t-s, x)ds = \mathcal{A}[u](t, x);\end{aligned}$$

the technical conditions ensure that we can take the partial derivative with respect to x outside the time integral. Also, by (6.2.12), $\Phi[h(t,\cdot)](0^+,x) = h(t,x)$, and therefore $u_t = \mathcal{A}[u] + h$, as desired.

EXERCISE 6.2.8. [B] *Verify that formula (6.1.8) on page 294 and formula (6.1.20) on page 298 are particular cases of (6.2.15).*

EXERCISE 6.2.9.[A] *A more general operator $\mathcal{A}$ has the functions A_k depending on both t and x: $\mathcal{A}[u](t,x) = \sum_{k=0}^{N} A_k(t,x)\,\partial^k u(t,x)/\partial x^k$. Derive an analog of (6.2.15) for the corresponding evolution equation $u_t = \mathcal{A}[u]$. Hint: Again, consider the* **initial value problem** *$v_t = \mathcal{A}[v]$, $t > s$, $v(s,x) = h(s,x)$. The solution of this problem should now be written as $\Phi[h(s,\cdot)](t,s,x)$*

For equations with **homogeneous boundary conditions**, the boundary conditions are incorporated into the definition of the operator $\mathcal{A}$. For example, if we consider the evolution equation $u_t = \mathcal{A}[u]$ on the interval $(0,L)$ with boundary conditions $u(t,0) = u(t,L) = 0$, then we define the operator $\mathcal{A}$ as the rule (6.2.9) that applies *only* to the functions u that satisfy the boundary conditions $u(t,0) = u(t,L) = 0$. Thus, different boundary conditions result in different operators $\mathcal{A}$ and in different solution operators Φ, even if the rule (6.2.9) stays the same.

EXERCISE 6.2.10.[B] *Write (6.2.15) for the solution of the heat equation on $[0,L]$ with zero boundary conditions $u(t,0) = u(t,L) = 0$.*

The wave equation $u_{tt} = c^2 u_{xx}$, with initial conditions $u(0,x) = f(x)$, $u_t(0,x) = g(x)$, is second-order in time and can be reduced to (6.2.10) if we allow vector-valued unknown functions and matrix operators. Indeed, introduce the new unknown function $v = u_t$. Then the wave equation becomes

$$\begin{pmatrix} u_t(t,x) \\ v_t(t,x) \end{pmatrix} = \begin{pmatrix} 0 & 1 \\ \mathcal{A} & 0 \end{pmatrix} \begin{pmatrix} u(t,x) \\ v(t,x) \end{pmatrix}, \quad \begin{pmatrix} u(0,x) \\ v(0,x) \end{pmatrix} = \begin{pmatrix} f(x) \\ g(x) \end{pmatrix}, \tag{6.2.16}$$

where $\mathcal{A}[u](t,x) = c^2 u_{xx}$; for the sake of concreteness, we consider the equation on the whole line. The solution operator $\Phi[f,g]$ takes the initial conditions $u(0,x) = f(x)$, $v(0,x) = g(x)$, and makes the solution $u = u(t,x)$ of the equation according to formula (6.1.43) on page 310. By (6.2.15), the solution of the corresponding inhomogeneous equation

$$\begin{pmatrix} u_t \\ v_t \end{pmatrix} = \begin{pmatrix} 0 & 1 \\ \mathcal{A} & 0 \end{pmatrix} \begin{pmatrix} u \\ v \end{pmatrix} + \begin{pmatrix} 0 \\ h(t,x) \end{pmatrix},$$

with zero initial conditions, is

$$u(t,x) = \int_0^t \Phi[0, h(s,\cdot)](t-s,x)ds. \tag{6.2.17}$$

EXERCISE 6.2.11.[B] *(a) Verify that (6.2.17) coincides with formula (6.1.45) on page 311. (b) Verify that, for the wave equation on the interval with zero boundary conditions, (6.2.17) coincides with formula (6.1.57) on page 315.*

Formulas (6.2.15) and (6.2.17) apply to ordinary differential equations as well; the operator $\mathcal{A}$ in this case is multiplication by a real number or, for equations of second or higher order, by a matrix; the initial conditions are numbers. The reader can verify that, for the equation $y'(t) = ay(t) + h(t)$, $y(0) = 0$, formula (6.2.15) gives $y(t) = \int_0^t e^{a(t-s)}h(s)ds$; similarly, for the equation $y''(t) + a^2y(t) = h(t)$, $y(0) = y'(0) = 0$, formula (6.2.17) gives $y(t) = (1/a)\int_0^t \sin(a(t-s))h(s)ds$.

6.2.3 *Separation of Variables*

In the elementary interpretation, separation of variables is a method for finding *a solution* of some partial differential equations. FOR EXAMPLE, consider the equation $u_x + y(y-1)u_y = 0$ for the unknown function $u = u(x,y)$. To find a solution, we write $u(x,y) = F(x)G(y)$ and substitute in the equation to find $F'(x)G(y) + y(y-1)F(x)G'(y) = 0$ or $F'(x)/F(x) = -y(y-1)G'(y)/G(y) = c$ for some real number c. Then $F' = cF$ or $F(x) = F_0e^{cx}$ and $G'/G = -c(1/(y-1) - 1/y)$, so that $G(y) = G_0(y/(y-1))^c$. As a result, for every real numbers A, c, the function $u(x,y) = Ae^{cx}(y/(y-1))^c$ satisfies the equation $u_x + y(y-1)u_y = 0$.

EXERCISE 6.2.12.[B] *(a) What happens if we allow c to be complex, say, $c = i$? Hint: you get more solutions. (b) Many other solutions of this equation do not have the form $u(x,y) = F(x)G(y)$, for example, $u(x,y) = \ln(y-1) - \ln y - x$ or $u(x,y) = ye^x/(y-1) + y^2e^{2x}/(y-1)^2$. Can you see where these solutions come from?*

Notice that, at a certain step in the process, we get the equality of two expressions, each depending on only one variable. From this equality, we conclude that both expressions must be equal to the same number; we call this number the `separation constant`. In all our examples, the separation constant will be a real number. It can happen that the form of the solution depends on the separation constant. *For exam-*

ple, consider the equation $u_{xx} + u_{yy} = 0$. Writing $u(x, y) = F(x)G(y)$, we find $F''(x)/F(x) = -G''(y)/G(y) = c$. If $c = a^2 > 0$, then $F''(x) - a^2F(x) = 0$, $G''(y) + a^2G(y) = 0$, so that $F(x) = Ae^{ax} + Be^{-ax}$, $G(y) = C\sin(ax) + D\cos(ax)$ for some real numbers A, B, C, D, and $u(x, y) = (Ae^{ax} + Be^{-ax})(C\sin(ax) + D\cos(bx))$ is the corresponding solution. Verify that choosing the separation constant $c = -a^2$, $a > 0$, results in the solution $u(x, y) = (A\sin(ax) + B\cos(ax))(Ce^{ay} + De^{-ay})$. What solution do you get if $c = 0$?

With all the variety of choices, we certainly did not find *all* the solutions of $u_{xx} + u_{yy} = 0$: for example, $u(x, y) = x^2 - y^2$ satisfies the equation, but cannot be written in the form $F(x)G(y)$.

EXERCISE 6.2.13.^{C} *Find a solution of each of the following equations using separation of variables: (a)* $y^2u_x - x^2u_y = 0$, *(b)* $u_x + u_y = (x + y)u$, *(c)* $xu_{xy} + 2yu = 0$. *For each equation, match your answer to one of the expressions below:* $u(x, y) = Ae^{c(x-y)+(x^2+y^2)/2}$, $u(x, y) = Ae^{c(x^3+y^3)}$, $u(x, y) = Ax^ce^{-y^2/c}$; *in all cases, c is the separation constant.*

There are other ways to separate the variables. FOR EXAMPLE, we can look for the solution of the equation $u_{xx} + u_{yy} = 0$ in the form $u(x, y) = F(x) + G(y)$, which leads to $F''(x) + G''(y) = 0$ or $F''(x) = c$, $G''(y) = -c$, or $u(x, y) = A(x^2 - y^2)$ for every real number A.

The examples of the heat and wave equations on an interval (see pages 301 and 312) show that sometimes separation of variables can yield the *general solution* of the equation. This observation leads to a more sophisticated and a much more interesting interpretation of the method. To proceed with this interpretation, we need an abstract version of the Fourier series expansion, and for that we need some additional concepts. These concepts will also be used in our discussion of quantum mechanics and the Schrödinger equation in Section 6.4

A (complex) `vector space` $\mathbb{H}$ is a collection of objects, called *vectors*, with operations of addition and multiplication by a complex scalar so that the usual laws of arithmetic hold: for every $f, g, h \in \mathbb{H}$ and every $\alpha, \beta \in \mathbb{C}$ we have $\alpha f + \beta g = \beta g + \alpha f \in \mathbb{H}$, $\alpha(f + g) = \alpha f + \alpha g$, $(\alpha + \beta)f = \alpha f + \beta f$, $(\alpha\beta)f = \alpha(\beta f)$, $1f = f$, $f + (g + h) = (f + g) + h$. Also, there is a special element $\mathbf{0} \in \mathbb{H}$ so that $f + \mathbf{0} = f$ and $\alpha\mathbf{0} = \mathbf{0}$ (compare this with Definition 1.1 on page 6.) An alternative name for $\mathbb{H}$ is a `linear space over the field of complex numbers`. The complex numbers in this setting are called `scalars`.

A vector space $\mathbb{H}$ is called an **inner product space** if, for every $f, g \in \mathbb{H}$, there exists a unique complex number (f, g) with the following properties: (i) $(f, f) \geq 0$, with $(f, f) = 0$ if and only if $f = \mathbf{0}$, (ii) $(f, g) = \overline{(g, f)}$, (iii) $(af, g) = a(f, g)$, and (iv) $(f + g, h) = (f, h) + (g, h)$; as usual, $\overline{z}$ means the complex conjugate of the complex number z. A knowledgeable reader will recognize that this is almost a **Hilbert space**, sometimes called a pre-Hilbert space, but we do not want to complicate the presentation any further. The number (f, g) is called the **inner product** of f and g; the number (f, f), which is real by assumption, will be denoted by $\|f\|^2$. The elements f, g of $\mathbb{H}$ are called **orthogonal** if $(f, g) = 0$.

EXERCISE 6.2.14. ^{C} *Verify that each of the following is an inner product space: (a) the collection of complex-valued functions on an interval* (a, b) *so that* $\int_a^b |f(x)|^2 dx < \infty$, *with* $(f, g) = \int_a^b f(x)\overline{g(x)}dx$; *(b) the collection of complex-valued functions on a domain* $G \subset \mathbb{R}^2$ *so that* $\iint\limits_G |f(P)|^2 dA < \infty$, *with* $(f, g) = \iint\limits_G f(P)\overline{g(P)}dA$; *(c) the collection of complex-valued functions on a domain* $G \subset \mathbb{R}^3$ *so that* $\iiint\limits_G |f(P)|^2 dV < \infty$, *with* $(f, g) = \iiint\limits_G f(P)\overline{g(P)}dV$. *As usual,* $\overline{g}$ *is the complex conjugate of* g.

EXERCISE 6.2.15. ^{B} *Consider the collection* $\mathbb{H}$ *of sequences of complex numbers so that a sequence* $\boldsymbol{z} = \{z_1, z_2, \ldots\}$ *belongs to* $\mathbb{H}$ *if and only if* $\sum_{k=1}^{\infty} |z_k|^2 < \infty$. *By definition,* $\alpha\boldsymbol{z} = \{\alpha z_1, \alpha z_2, \ldots\}$, $\boldsymbol{z}+\boldsymbol{w} = (z_1+w_1, z_2+w_2, \ldots)$. *Verify that* $\mathbb{H}$ *is an inner product space, with* $(\boldsymbol{z}, \boldsymbol{w}) = \sum_{k=1}^{\infty} z_k\overline{w_k}$.

The notion of the inner product space allows us to forget about certain features of the problem, such as the number of spatial dimensions.

EXERCISE 6.2.16.^{A} *Prove the* **Cauchy-Schwartz inequality**:

$$|(f, g)| \leq \|f\| \, \|g\| \tag{6.2.18}$$

for every f, g *in an inner product space* $\mathbb{H}$. *Hint: use the same arguments as in the proof of (1.2.12) on page 15.*

A **linear operator** $\mathcal{A}$ on a complex vector space $\mathbb{H}$ is a rule that, to every element $f \in \mathbb{H}$, assigns a unique element $\mathcal{A}[f]$, also from $\mathbb{H}$, so that $\mathcal{A}[\alpha f + \beta g] = \alpha\mathcal{A}[f] + \beta\mathcal{A}[g]$ for all $f, g \in \mathbb{H}$ and for all complex numbers α, β. A linear operator $\mathcal{A}$ on an inner product space $\mathbb{H}$ is called

- **symmetric** (or **Hermitian**) if $(\mathcal{A}[f], g) = (f, \mathcal{A}[g])$;
- **positive** if $(\mathcal{A}[f], f) > 0$ for all $f \neq \mathbf{0}$.

EXERCISE 6.2.17. *(a)*[C] *Let* $\mathbb{H} = \mathbb{R}^n$ *with the usual inner product (see pages 7 and 14). Verify that an operator* $\mathcal{A}$ *on* $\mathbb{H}$ *is symmetric if and only if the representation of this operator in every orthonormal basis is a symmetric matrix.*
(b)[B] *Let* $\mathbb{H}$ *be an* n*-dimensional vector space over the field of complex numbers. (i) Define the inner product in* $\mathbb{H}$*. (ii) Verify that an operator* $\mathcal{A}$ *on* $\mathbb{H}$ *is Hermitian if and only if the representation of this operator in every orthonormal basis is a* **Hermitian matrix**,A*, that is, the entries of the transpose of the matrix are complex conjugates of the entries of the original matrix:* $A^T = \overline{A}$.

EXERCISE 6.2.18.[C] *Let* $\mathbb{H}$ *be the collection of twice continuously differentiable functions on* $(0,1)$ *such that* $f(0) = f(1) = 0$ *and* $(f,g) = \int_0^1 f(x)\overline{g(x)}dx$*. (a) Verify that the operator* $\mathcal{A}$ *defined by* $\mathcal{A}[f](x) = -f''(x)$ *is positive and symmetric on* $\mathbb{H}$*. (b) Verify that the operator* $\mathcal{A}$ *defined by* $\mathcal{A}[f] = f'$ *is linear, but is neither positive nor symmetric on* $\mathbb{H}$*. (c) Verify that the operator* $\mathcal{A}$ *defined by* $\mathcal{A}[f](x) = f(x) + \sin(\pi x)$ *is not a linear operator on* $\mathbb{H}$.

A collection $\Psi = \{\varphi_k,\ k \geq 1\}$ of elements of the inner product space $\mathbb{H}$ is called an **orthonormal system** if $\|\varphi_k\| = 1$ for all $k \geq 1$ and $(\varphi_k, \varphi_m) = 0$ for all $k \neq m$. An orthonormal system is called **complete** (or **total**) if, for every $f \in \mathbb{H}$,

$$\|f\|^2 = \sum_{k=1}^{\infty} |(f, \varphi_k)|^2. \tag{6.2.19}$$

EXERCISE 6.2.19.[C] *Consider the space* $\mathbb{H}$ *from Exercise 6.2.18. Verify that* $\Psi = \{\sqrt{2}\sin(\pi k x),\ k \geq 1\}$ *is a complete orthonormal system in* $\mathbb{H}$*. Hint: consider the odd extension of the functions from* $\mathbb{H}$ *and use Parseval's identity to verify (6.2.19).*

Let $\Psi = \{\varphi_k,\ k \geq 1\}$ be a complete orthonormal system in the inner product space $\mathbb{H}$. A **generalized Fourier series** of $f \in \mathbb{H}$ relative to Ψ is the series $S_f = \sum_{k=1}^{\infty}(f, \varphi_k)\,\varphi_k$.

EXERCISE 6.2.20.[B] *(a) For* $N \geq 1$*, define* $S_{N,f} = \sum_{k=1}^{N}(f, \varphi_k)\,\varphi_k$*. Show that* $\lim_{N\to\infty}\|f - S_{N,f}\|^2 = 0$*. Hint:* $\|f - S_{N,f}\|^2 = \|f\|^2 - 2(f, S_{N,f}) + \|S_{N,f}\|^2 = \|f\|^2 - 2\sum_{k=1}^{N}|(f,\varphi_k)|^2 + \sum_{k=1}^{N}|(f,\varphi_k)|^2$*. (b) Show that the generalized Fourier series of* f *is unique: if* a_k*,* $k \geq 1$*, is a collection of complex numbers such that* $\lim_{N\to\infty}\|f - \sum_{k=1}^{N} a_k\varphi_k\|^2 = 0$*, then* $a_k = (f, \varphi_k)$ *for* *all k. Hint: derive an analog of equality (5.1.5) on page 243.*

We now connect the notions of linear operator and orthonormal system. Let $\mathcal{A}$ be a linear operator on a vector space $\mathbb{H}$. An **eigenfunction**, or **eigenvector**, of $\mathcal{A}$ is an element $\varphi \in \mathbb{H}$ such that $\varphi \neq \mathbf{0}$ and there exists a complex number λ with the property $\mathcal{A}[\varphi] = \lambda\,\varphi$; this number λ is called an **eigenvalue** of $\mathcal{A}$ corresponding to φ. Conversely, we can say that φ is the eigenfunction of $\mathcal{A}$ corresponding to the eigenvalue λ. The collection of all eigenvalues of $\mathcal{A}$ is called the **point spectrum** of $\mathcal{A}$; in Latin, the word *spectrum* means "appearance," and, while we will only work with the point spectrum, there are also other types of spectrum. Notice that if φ is an eigenfunction corresponding to the eigenvalue λ, then so is $\alpha\,\varphi$ for every complex number α. Accordingly, we can always normalize φ to have $\|\varphi\| = 1$.

The following result connects linear operators with orthonormal systems.

Theorem 6.2.1 *Let $\mathcal{A}$ be a symmetric linear operator on an inner product space $\mathbb{H}$. Then all the eigenvalues of $\mathcal{A}$ are real and the eigenfunctions corresponding to different eigenvalues are orthogonal.*

Proof. Both statements follow by direct computation using the definition of the symmetric operator and the properties of the inner product. For the first statement,

$$\lambda\|\varphi\|^2 = (\mathcal{A}[\varphi], \varphi) = (\varphi, \mathcal{A}[\varphi]) = (\varphi, \lambda\varphi) = \overline{(\lambda\varphi, \varphi)} = \overline{\lambda}\,\|\varphi\|^2;$$

since $\|\varphi\| \neq 0$, we conclude that $\lambda = \overline{\lambda}$, that is, $\lambda \in \mathbb{R}$.

EXERCISE 6.2.21.[C] *Show that if $\lambda_1 \neq \lambda_2$, then $(\varphi_1, \varphi_2) = 0$. Hint: note that $\lambda_1(\varphi_1, \varphi_2) = (\mathcal{A}\varphi_1, \varphi_2) = (\varphi_1, \mathcal{A}\varphi_2) = \lambda_2(\varphi_1, \varphi_2)$.*

□

EXERCISE 6.2.22. [B] *Verify that all eigenvalues of a positive operator are positive.*

The eigenvalue λ_k is called **simple** if every two eigenfunctions corresponding to λ are linearly dependent, that is, scalar multiples of each other. If two or more linearly independent eigenfunctions correspond to the same eigenvalue, the eigenvalue is called **multiple**; the **multiplicity of an eigenvalue** is the number of linearly independent eigenfunctions corresponding to this eigenvalue. FOR EXAMPLE, consider the operator

$\mathcal{A}$ defined by $\mathcal{A}[f](x) = -f''(x)$ on the collection $\mathbb{H}$ of twice continuously differentiable functions f on $(-\pi, \pi)$ satisfying $f(\pi) = f(-\pi)$ and $f'(\pi) = f'(-\pi)$; $(f, g) = \int_{-\pi}^{\pi} f(x)\overline{g(x)}dx$ (verify that the operator is symmetric!) Then, for $k \neq 0$, both e^{ikx} and e^{-ikx} are eigenfunctions of $\mathcal{A}$ corresponding to the eigenvalue k^2. So the zero eigenvalue is simple and all others have multiplicity two. Notice that all eigenfunctions in this example are mutually orthogonal.

In general, let φ and ψ be two linearly independent eigenfunctions corresponding to the same eigenvalue λ (that is $\mathcal{A}[\varphi] = \lambda\varphi$, $\mathcal{A}[\psi] = \lambda\psi$, and $\varphi \neq \alpha\psi$ for every complex number α.) Then the functions φ and $\psi - \big((\psi, \varphi)/\|\varphi\|^2)\big)\varphi$ are orthogonal and are also eigenfunctions of $\mathcal{A}$ corresponding to the same eigenvalue λ (verify this!) A similar *orthogonalization procedure* applies if there are more than two linearly independent eigenfunctions corresponding to the same eigenvalue. Thus, *the eigenfunctions of a symmetric operator can be chosen to form an orthonormal system.*

A much more difficult question is whether the eigenfunctions of a symmetric operator form a *complete* orthonormal system. The following result shows that that the answer is positive for the `Dirichlet Laplacian`, that is, the Laplace operator $\mathcal{A} = \nabla^2$ on a bounded domain with zero boundary conditions, including the one-dimensional analog $\mathcal{A}[f](x) = f''(x)$ on a bounded interval. The proof is by far outside the scope of our discussion. In fact, there is hardly a single reference containing a complete proof.

Theorem 6.2.2 *Let G be either a bounded interval (a, b) in $\mathbb{R}$ or a bounded domain in $\mathbb{R}^n$, $n = 2, 3$, with a piece-wise smooth boundary ∂G. Let $\mathbb{H}$ be the collection of functions f that are twice continuously differentiable in G, continuously differentiable in the closure of G, and are equal to zero on the boundary ∂G of G; for $n = 1$ this last condition means $f(a) = f(b) = 0$. Define the inner product in $\mathbb{H}$ as in Exercise 6.2.14. Consider the operator $\mathcal{A} = \nabla^2$ on $\mathbb{H}$, or, for $n = 1$, $\mathcal{A}[f](x) = f''(x)$.*

Then

(1) Each eigenvalue λ_k of $\mathcal{A}$ has finite multiplicity. Moreover, the eigenvalues can be arranged so that $0 > \lambda_1 > \lambda_2 > \ldots$ and $\lim_{k\to\infty} |\lambda_k|k^{-2/n} = C$, where $C > 0$ depends only on the domain $G \subset \mathbb{R}^n$ and the dimension n of the space $\mathbb{R}^n$.

(2) The corresponding eigenfunctions φ_k, $k \geq 1$, are infinitely differentiable in G, can be chosen real-valued, and can be chosen to form a

complete orthonormal system in $\mathbb{H}$.

(3) For every function $f \in \mathbb{H}$, the series $\sum_{k=1}^{\infty}(f,\varphi_k)\varphi_k$ converges to f absolutely and uniformly in the closure of G.

EXERCISE 6.2.23. *(a)^C State the above theorem for $G = (0, L)$ and convince yourself that all the conclusions are indeed true. (b)^A Verify that, in the above theorem, the operator $-\mathcal{A}$ is positive and symmetric. Hint: For $n = 2, 3$ use Theorem 3.2.5 on page 159, to show that $-(\mathcal{A}f, f) = \|\operatorname{grad} f\|^2 = -(f, \mathcal{A}f)$.*

The conclusions of the above theorem hold for a more general operator $\mathcal{A}[f] = \operatorname{div}(p \operatorname{grad} f) + q\, f$ in a domain G with boundary conditions $a\, f + b\, df/dn = 0$ on the boundary of G, where a, b, p, q are sufficiently smooth functions and df/dn is the normal derivative of f on ∂G; for $n = 1$, the normal derivative on the boundary is replaced with the usual first-order derivative at the end points of the interval. The reader can verify that, for $p > 0$ and $q < 0$, the operator $-\mathcal{A}$ is symmetric and positive; see also Problem 7.7 on page 443.

We now describe the GENERAL METHOD OF SEPARATION OF VARIABLES for the abstract evolution equation

$$\frac{\partial u}{\partial t} = \mathcal{A}u + h(t),\ t > 0;\ u|_{t=0} = f, \tag{6.2.20}$$

where $\mathcal{A}$ is a linear operator on an inner product space $\mathbb{H}$, $h(t) \in \mathbb{H}$ for all $t \geq 0$, $f \in \mathbb{H}$, and we think of the unknown function $u = u(t)$ as having values in $\mathbb{H}$ for each $t \geq 0$. In many examples, the operator $-\mathcal{A}$ is symmetric and positive.

Step I. Find the eigenvalues λ_k and eigenfunctions φ_k of $\mathcal{A}$ and verify that $\Psi = \{\varphi_k,\ k \geq 1\}$ is a complete orthonormal system in $\mathbb{H}$ and $\lambda_k < 0$ for all sufficiently large k.

Step II. Expand the initial condition f and the right hand side h into the generalized Fourier series relative to the system Ψ: $f = \sum_{k=1}^{\infty} f_k\varphi_k$, $h(t) = \sum_{k=1}^{\infty} h_k(t)\varphi_k$. We also *suppose* that the solution u has the expansion $u = \sum_{k=1}^{\infty} u_k(t)\varphi_k$.

Step III. Substitute the series expansions into the equation to find

$$\sum_{k=1}^{\infty} u_k'(t)\varphi_k = \sum_{k=1}^{\infty} u_k(t)\mathcal{A}[\varphi_k] + \sum_{k=1}^{\infty} h_k(t)\varphi_k,$$

or, using $\mathcal{A}[\varphi_k] = \lambda_k \varphi_k$ and combining the terms,

$$\sum_{k=1}^{\infty} (u_k'(t) - \lambda_k u_k(t) - h_k(t))\varphi_k = 0;$$

during this analysis we assume that we can manipulate the sums as if the number of terms were finite. Uniqueness of the generalized Fourier series implies that *each* term in the last series must be zero:

$$u_k'(t) - \lambda_k u_k(t) - h_k(t) = 0, \; k = 1, 2, \ldots.$$

Step IV. Solve each ordinary differential equation with the corresponding initial condition $u_k(0) = f_k$ to conclude that

$$u_k(t) = f_k e^{\lambda_k t} + \int_0^t e^{\lambda_k (t-s)} h_k(s) ds. \tag{6.2.21}$$

The result $u = \sum_{k=1}^{\infty} u_k(t)\varphi_k$ is called a **generalized solution** of equation (6.2.20). The assumptions $f, h(t) \in \mathbb{H}$ and $\lambda_k < 0$ for all sufficiently large k imply that $\sum_{k=1}^{\infty} |u_k(t)|^2 < \infty$ for all $t \geq 0$ (verify this!) With additional information about the operator $\mathcal{A}$ and results such as Theorem 6.2.2, we can then investigate whether u is a classical solution.

EXERCISE 6.2.24. *(a)^C Convince yourself that the above procedure, when applied to the initial-boundary value problem (6.1.25) for the heat equation on page 300, results in (6.1.35). (b)^B Modify the above procedure to make it applicable to the initial-boundary value problem (6.1.46) for the wave equation on page 312, and verify that the result is (6.1.53). (c)^B Verify that for both the heat and wave equations in a bounded region G in $\mathbb{R}^n$ with zero boundary conditions the above procedure results in the eigenvalue problem $\nabla^2 u = \lambda u$ in G, $u = 0$ on ∂G. (d)^A Notice that equation (6.2.20) is inhomogeneous, and, by following the above procedure, we solve the equation directly. Find the solution operator Φ for the corresponding homogeneous equation and verify that the alternative method using the variation of parameters formula (6.2.15) produces the same solution of (6.2.20).*

As an example, in Section 6.3.2, we will compute the eigenvalues and eigenfunctions of the Dirichlet Laplacian in a rectangle and in a disk.

6.3 Some Classical Partial Differential Equations

We now turn to the study of the main partial differential equations of mathematical physics and engineering.

6.3.1 *Telegraph Equation*

The objective of this section is to derive the equation for a voltage signal propagating in a long homogenous wire, or cable, of uniform cross-section. The current in the cable can be time-dependent, and the cable can be placed in a conducting medium. Accordingly, beside the usual Ohm's Law for resistance, we need to take into account the self-inductance of the cable and current leakage due to imperfect insulation. We introduce the following notations: $V = V(t, x)$, the voltage in the cable at time t and point x; $I = I(t, x)$, the current along the cable; $j = j(t, x)$, the current per unit length leaking outside due to imperfect insulation; L, the self-inductance of the cable per unit length; r_c, the resistance of the cable, per unit length; r_o, the resistance of the insulation, per unit length; c_o, the capacitance of the insulation, per unit length. We assume that the cable is uniform and homogenous, and is surrounded by an isotropic homogeneous medium, so that the values of L, r_c, r_o, and c_o are all constants and do not depend on t and x; with obvious modifications, the following derivation also holds without this homogeneity assumption.

Consider a small segment $[x, x + \triangle x]$ of the cable. We approximate this segment by a lumped electric circuit, see Section 8.4, page 463, in Appendix: the resistance $r_c \triangle x$ and the inductance $L \triangle x$ of the segment, as well as the capacitance $c_o \triangle x$ and the resistance $r_o \triangle x$ of the corresponding segment of the insulation, are assumed to be concentrated at the point x. The resulting discrete collection of these lumped circuits approximates the behavior of the cable as $\triangle x \to 0$. The voltage drop across the segment $[x, x + \triangle x]$ is then

$$V(t, x) - V(t, x + \triangle x) = r_c\, I(t, x) \triangle x + L\, \frac{\partial I(t, x)}{\partial t} \triangle x, \tag{6.3.1}$$

where the first term on the right is Ohm's Law for voltage drop across a resistor, and the second is Faraday's Law for the voltage drop across an inductor (see Section 8.4 in Appendix). Dividing by $\triangle x$ and passing to the limit $\triangle x \to 0$, we conclude from (6.3.1) that

$$\frac{\partial V(t,x)}{\partial x} = -r_c I(t,x) - L\frac{\partial I(t,x)}{\partial t}. \tag{6.3.2}$$

The leakage current at x is the sum of the currents through the lumped parallel circuit of the resistance and capacitance of the insulation between x and $x + \triangle x$. Denoting the leakage current per unit length by j, we find

$$j(t,x)\triangle x = \frac{V(t,x)}{r_o}\triangle x + c_o\frac{\partial V(t,x)}{\partial t}\triangle x,$$

or, after dividing by $\triangle x$,

$$j(t,x) = \frac{V(t,x)}{r_o} + c_o\frac{\partial V(t,x)}{\partial t}. \tag{6.3.3}$$

To derive the relation between the currents I and j, we write the conservation of charge equality, also known as Kirchhoff's Current Law, for the currents entering and leaving the lumped circuit:

$$I(t, x + \triangle x) = I(t,x) - j(t,x)\triangle x.$$

After dividing by $\triangle x$ and passing to the limit $\triangle x \to 0$,

$$j(t,x) = -\frac{\partial I(t,x)}{\partial x}. \tag{6.3.4}$$

We now use (6.3.3) and (6.3.4) to eliminate I from (6.3.2). Differentiation of (6.3.4) with respect to t yields $j_t = -I_{xt}$. Differentiation of (6.3.2) with respect to x yields $V_{xx} = -r_c I_x - LI_{xt} = r_c j + L j_t$. Differentiation of (6.3.3) with respect to t yields $j_t = (V_t/r_o) + c_o V_{tt}$. Then we use (6.3.3) to express j in terms of V, and get the `telegraph equation` for the voltage $V = V(t,x)$:

$$V_{xx} = Lc_o V_{tt} + (r_c c_o + (L/r_o))V_t + (r_c/r_o)V. \tag{6.3.5}$$

EXERCISE 6.3.1.^{C} *(a) Verify that (6.3.2), (6.3.3), and (6.3.4) indeed imply (6.3.5). (b) Find the equation satisfied by I. Hint: it is the same as for V.*

To study the transmission of a voltage signal over a long cable, equation (6.3.5) is solved for $t > 0$, $x > 0$, with zero initial conditions $V(0,x) = V_t(0,x) = 0$ (meaning no signal at time $t = 0$) and the boundary condition $V(t,0) = f(t)$, where $f = f(t)$ is the signal at the transmitting end $x = 0$. Then $V(t,x)$ describes the signal being received at time t at point x. We will consider two special situations, the ideal cable and underwater cable, when equation (6.3.5) becomes the wave equation and the heat equation, respectively.

THE IDEAL (LOSSLESS) CABLE corresponds to $r_o = \infty$ (perfect insulation) and $r_c = 0$ (zero resistance in the cable). Define $c^2 = 1/(Lc_o)$. Then the voltage V satisfies the following `initial-boundary value problem`:

$$V_{tt} = c^2 V_{xx},\ t > 0,\ x > 0;\ V(0,x) = V_t(0,x) = 0,\ V(t,0) = f(t). \quad (6.3.6)$$

We solve this problem using the `method of reflection` by considering the *odd* extension of V to the whole line. Define the function $u = u(t,x)$, $t \geq 0$, $x \in \mathbb{R}$, so that $u(t,x) = V(t,x) - f(t)$ for $x \geq 0$ and $u(t,x) = -u(t,-x)$ for $x < 0$. We also assume that the function $f = f(t)$ is twice continuously differentiable and $f(0) = f'(0) = 0$. Then direct computations show that

$$u_{tt} = c^2 u_{xx} - f''(t)\,\mathrm{sign}(x),\ t > 0,\ x \in \mathbb{R};\ u(0,x) = u_t(0,x) = 0, \quad (6.3.7)$$

where the function sign(x) is defined by sign$(x) = 1, x > 0$, sign$(0) = 0$, sign$(x) = -1$, $x < 0$. Using formula (6.1.45) on page 311 and simplifying the corresponding integrals, we conclude that the solution of (6.3.6) is

$$V(t,x) = \begin{cases} f\left(t - \dfrac{x}{c}\right), & x < ct; \\ 0, & x \geq ct. \end{cases} \quad (6.3.8)$$

EXERCISE 6.3.2. *(a)*[C] *Verify the transformation of (6.3.6) to (6.3.7).* *(b)*[A] *Derive (6.3.8) from (6.1.45).*

Formula (6.3.8) shows that a signal in the ideal cable propagates with speed $c = 1/\sqrt{Lc_o}$ without changing shape: if at the point $x = 0$ a message is generated that lasts τ units of time, then the message will reach the recipient at the point $x = \ell > 0$ in ℓ/c units of time, and will take the same τ units of time to receive; see Figure 6.3.1.

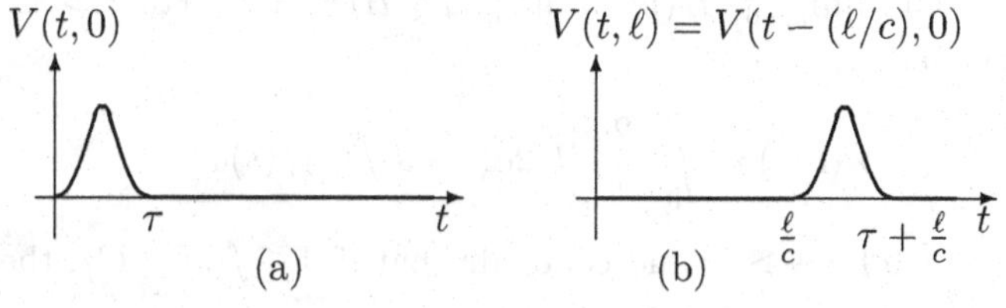

Fig. 6.3.1 Signal in Ideal Cable

THE UNDERWATER CABLE corresponds to $r_o = \infty$ (perfect insulation) and $Lc_oV_{tt} \ll r_cc_oV_t$ (effects of self-induction are small compared to the resistance), so that the terms Lc_oV_{tt} and $(r_c/r_o)V$ in (6.3.5) can be omitted. Define $a = 1/(r_cc_o)$. Then the voltage V solves the following `initial-boundary value problem`:

$$V_t = aV_{xx},\ t > 0,\ x > 0;\ V(0,x) = 0,\ V(t,0) = f(t). \tag{6.3.9}$$

Notice that this is the heat (diffusion) equation. This time the method of reflection transforms (6.3.9) to

$$u_t = au_{xx} - f'(t)\,\mathrm{sign}(x),\ t > 0,\ x \in \mathbb{R};\ u(0,x) = 0, \tag{6.3.10}$$

so that $V(t,x) = u(t,x) + f(t)$ for $x > 0$. By formula (6.1.20) on page 298,

$$V(t,x) = \int_0^t \frac{x}{\sqrt{4\pi a(t-s)^3}}\, e^{-x^2/(4a(t-s))} f(s)ds. \tag{6.3.11}$$

EXERCISE 6.3.3. *(a)*[C] *Verify that the method of reflection transforms (6.3.9) to (6.3.10). (b)*[A] *Derive (6.3.11) from (6.1.20). Hint: if $K(t,x)$ is the heat kernel (6.1.21), then one of the integrands in (6.3.11) is $-2aK_x(t-s,x) = -a\int_{-x}^{x} K_{xx}(t-s,y)dy = \int_{-x}^{x}(\partial K(t-s,y)/\partial s)dy$. Follow this chain of equalities backwards after transforming (6.1.20). (c)*[A+] *Derive (6.3.11) using the Laplace transform.*

Formula (6.3.11) shows that the propagation of a signal in an underwater cable is very different from (6.3.8). Indeed, assume that the initial signal $V(t,0) = f(t)$ is non-negative and lasts τ time units; see Figure 6.3.1(a). It follows from (6.3.11) that $V(t,x) > 0$ for all $t > 0, x > 0$. Moreover, we have $V(t,x) \leq C/x^2$ for some number C. Indeed, consider the function

$$H(y) = \frac{1}{2\pi y^{3/2}} e^{-1/(2y)},\ y > 0.$$

This function is positive and bounded: $\sup_{y>0} H(y) \leq C_H$ for some real number C_H, and $\lim_{y\to 0+} H(y) = \lim_{y\to+\infty} H(y) = 0$. For $t > \tau$, we can write (6.3.11) as

$$V(t,x) = \int_0^\tau \frac{2a}{x^2} H(2a(t-s)/x^2) f(s)ds,$$

because $f(s) = 0$ for $s > \tau$, and conclude that if $0 \leq f(t) \leq C_f$, then

$$0 \leq V(t,x) \leq C_f C_H \frac{2a\tau}{x^2}; \tag{6.3.12}$$

of course, this bound is interesting only for large x: it shows that the signal decays significantly at large x.

EXERCISE 6.3.4.[B] *(a) Verify (6.3.12). (b) With $V(t,0)$ as in Figure 6.3.1(a), denote by $\tau^*(x)$ the time during which the signal $V(t,x)$ stays above half its maximal value; the quantity $\tau^*(x)$ characterizes the time necessary to receive the original signal at x. Verify that, for x much larger than $\sqrt{a\tau}$, $\tau^*(x)$ is proportional to x^2/a and does not actually depend on τ.*

We conclude that, far away from the source, that is, for x much larger than $\sqrt{a\tau}$, the signal in the underwater cable is small and not localized in time at all, see Figure 6.3.2 where the original signal at $x = 0$ is a pulse of width τ.

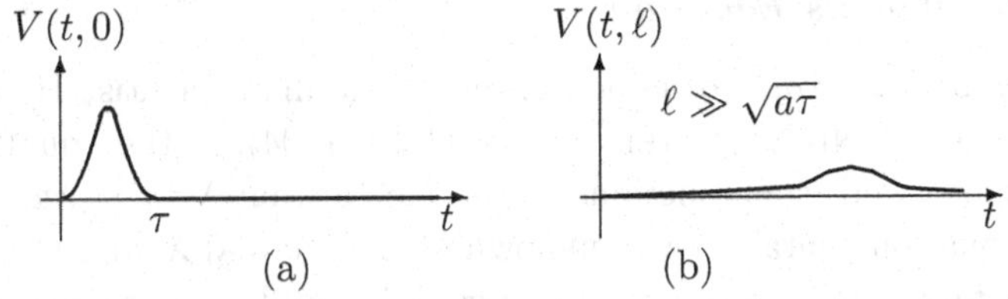

Fig. 6.3.2 Signal in Underwater Cable Far From The Source

As a result, reliable reception is possible only at distances that are small compared to $\sqrt{a\tau}$, where $a = 1/(r_c c_0)$ and τ is the characteristic time of the signal (the higher the frequency of the signal, the smaller the τ.) With small a, transmission over long distances will take a prohibitively long time. The first transatlantic cable between the UK and the US, completed in 1858, had fairly high resistance r_c. The resulting small value of a required more than 16 hours to transmit a 99-word greeting from Queen Victoria to the then-president James Buchanan. Needless to say, very few other messages were transmitted over that cable.

With little control over the insulation capacitance c_o, the only way to ensure high quality transmission at long distances was to increase a by decreasing r_c, the resistance of the cable material. This decrease was achieved by using better conducting material (copper of very high purity) and making a thicker cable. In the second transatlantic cable, completed in 1866, low r_c allowed transmission of about 10 words per minute; this cable earned £ 1000 during the first day of operation. Modern cables use amplifiers placed along the cable to overcome the signal decay.

Equation (6.3.9) for the underwater cable is sometimes called Kelvin's cable equation and was first derived by Sir WILLIAM THOMSON (LORD KELVIN), who also was a principal consultant for the second transatlantic cable project. You can read more about the history and mathematics of the transatlantic cable in Sections 62–66 of the book *Fourier Analysis* by T. W. Körner, 1988, where the interested reader can also find alternative derivations of formulas (6.3.8) and (6.3.11).

Beside the voltage in the underwater cable, equation (6.3.9) also models the conduction of nerve pulses (action potentials) along a passive nerve fiber. There are even molecular ion channels along a nerve fiber to overcome signal decay. For more on this topic, see the book *Theoretical Neurobiology, Volume 1: Cable Theory* by H. Tuckwell, 1988.

6.3.2 *Helmholtz's Equation*

`Helmholtz's equation`, so named after the German scientist HERMANN LUDWIG FERDINAND VON HELMHOLTZ (1821–1894), is the equation for the eigenvalues and eigenfunctions of the `Laplacian`: $\nabla^2 u = \lambda u$. Notice that this equation contains two unknown objects, u and λ, and the goal is to find the functions u that satisfy this equation and are not equal to zero *somewhere*, and such solutions can only exist for certain, a priori unknown, values of λ.

We will consider the `Dirichlet Laplacian`, that is, the Laplace operator with zero boundary conditions. The complete statement of the problem becomes as follows: given a bounded domain G with a piece-wise smooth boundary ∂G, find a function u and a number λ so that

$$\nabla^2 u = \lambda u \text{ in } G, \;\; u = 0 \text{ on } \partial G, u \neq 0 \text{ somewhere in } G. \tag{6.3.13}$$

Recall that this problem arises when we do separation of variables for the heat and wave equations; see Exercise 6.2.24 on page 332.

By Theorem 6.2.2 on page 330, we know, in particular, that (6.3.13) has infinitely many solutions (λ, u), that each λ is real and negative, and that u is infinitely differentiable in G and is real-valued. In what follows, we will find λ and u when G is either a rectangle or a disk.

PROBLEM (6.3.13) IN A RECTANGLE WITH SIDES a AND b; $a = b$ IS A POSSIBILITY. Introduce the cartesian coordinates (x, y) so that two opposite vertices of the rectangle are the points $(0,0)$ and (a,b) (draw a picture!) Then $u = u(x,y)$ and, by (3.1.31) on page 139, $\nabla^2 u(x,y) = u_{xx} + u_{yy}$. As a result, (6.3.13) becomes

$$\begin{aligned} &u_{xx}(x,y) + u_{yy}(x,y) = \lambda u(x,y), \;\; 0 < x < a, \;\; 0 < y < b; \\ &u(0,y) = u(x,0) = u(a,y) = u(x,b) = 0, \end{aligned} \tag{6.3.14}$$

and $u(x_0, y_0) \neq 0$ for some point $0 < x_0 < a,\ 0 < y_0 < b$.

We solve (6.3.14) by **separation of variables**. With $u(x,y) = v(x)w(y)$,

$$\frac{v''(x)}{v(x)} = \lambda - \frac{w''(y)}{w(y)},\ v(0) = v(a) = w(0) = w(b) = 0,\ v(x_0) \neq 0,\ w(y_0) \neq 0. \tag{6.3.15}$$

EXERCISE 6.3.5.^{C} *Verify (6.3.15).*

The left-hand side of the equation in (6.3.15) depends only on x, and the right-hand side, only on y. Therefore,

$$\frac{v''(x)}{v(x)} = \lambda - \frac{w''(y)}{w(y)} = q \tag{6.3.16}$$

for some real number q.

EXERCISE 6.3.6.^{C} *Verify that conditions $v''(x) = qv(x)$, $0 < x < a$, $v(0) = v(a) = 0$, $v(x_0) \neq 0$ for some $x_0 \in (0,a)$, imply*

$$q = -(\pi m/a)^2,\ v_m(x) = \sin(\pi m x/a),\ m = 1, 2, 3, \ldots. \tag{6.3.17}$$

Hint: use the same arguments as for the heat equation on page 301.

It follows from (6.3.15), (6.3.16), and (6.3.17) that $w''(y) = (\lambda - q)w(y)$, $w(0) = w(b) = 0$, $w(y_0) \neq 0$, so that

$$\lambda + (\pi m/a)^2 = -(\pi n/b)^2,\ n = 1, 2, 3, \ldots,\ w_n(y) = \sin(\pi n y/b). \tag{6.3.18}$$

Combining (6.3.17) and (6.3.18), we get the eigenvalues and eigenfunctions of the **Dirichlet Laplacian** in a rectangle:

$$\lambda_{m,n} = -\pi^2\left(\frac{m^2}{a^2} + \frac{n^2}{b^2}\right),\ u_{m,n}(x,y) = \sin\left(\frac{\pi m x}{a}\right)\sin\left(\frac{\pi n y}{b}\right), \tag{6.3.19}$$

$m, n = 1, 2, \ldots$. We can enumerate the ordered pairs (m,n) by assigning a unique positive integer $k = k(m,n)$ to every pair. For example, $k(1,1) = 1, k(1,2) = 2, k(2,1) = 3, k(2,2) = 4, k(3,1) = 4$, etc. Of course, there are many different enumerations. Similarly, we can enumerate the eigenvalues $\lambda_{m,n}$, but now, depending on a and b, different pairs of (m,n) can correspond to the same $\lambda_{m,n}$. FOR EXAMPLE, let $a = b = 1$. Then $\lambda_{1,2} = \lambda_{2,1} = -5\pi^2$.

EXERCISE 6.3.7. ^{A} *For a function $f = f(x,y)$, define its* **nodal line** *as the zero-level set: $\{(x,y) : f(x,y) = 0$. Let $a = b = 1$ and let u_λ be an eigenfunction of the Dirichlet Laplacian in the square, corresponding to the* *eigenvalue* λ. *Describe all the possible nodal lines of* u_λ *when (i)* $\lambda = -5\pi^2$; *(ii)* $\lambda = -65\pi^2$. *Hint:* $5 = 2^2 + 1^2$, $65 = 1^2 + 8^2 = 4^2 + 7^2$, *so that there are many possible functions* u_λ.

EXERCISE 6.3.8.$^{A+}$ *(a) Find a general rule for enumerating the ordered pairs* (m, n), $m, n = 1, 2, \ldots$. *You need two formulas: (i) for computing* $k(m, n)$ *and (ii) for computing* m, n *given* k. *(b) Assume* $a = b = 1$, *arrange* $|\lambda_{m,n}|$ *in increasing order:* $2\pi^2 < 5\pi^2 < 8\pi^2 < 10\pi^2 < 13\pi^2 < \ldots$, *and denote the resulting sequence by* $\mu_k, k \geq 1$. *Find* $\lim_{k\to+\infty}(\mu_k/k)$ *(by Theorem 6.2.2 on page 330, we know that the limit exists and is not equal to zero). Can you find this limit for the general rectangle with sides* a *and* b*? Hint: for the general rectangle, the limit is* $4\pi/(ab)$.

For given a, b, if $r = (m/a)^2 + (n/b)^2$, then the eigenvalue $\lambda_{m,n}$ is simple if and only if the equation $(x/a)^2 + (y/b)^2 = r$ has a unique solution (x, y) with positive integer x, y (verify this!) The question about multiplicity of a particular eigenvalue $\lambda_{m,n}$ therefore leads to a number-theoretic question about representing a positive integer as a sum of two squares. The following exercise is relatively simple.

EXERCISE 6.3.9.^{B} *Find the necessary and sufficient conditions on* a, b *so that all eigenvalues* $\lambda_{m,n}$ *are simple. Hint: first reduce the problem to* $b = 1$. *Then note that* $m^2 + a^2n^2 = m_1^2 + a^2n_1^2$ *implies* $a^2 = (m_1 - m)(m_1 + m)/((n - n_1)(n + n_1))$. *Then note that* $m_1 - m$ *and* $m_1 + m$ *are both odd or both even, and so are* $n - n_1$, $n + n_1$.

Analysis of multiple eigenvalues of the Dirichlet Laplacian in a rectangle leads us deeper into number theory. A theorem of C. G. J. Jacobi, published in 1829, describes the number of points n_r with `integer (positive, negative,or zero) coordinates on the circle` $x^2 + y^2 = r$ for a positive integer r: *if* N_1 *and* N_3 *are the number of divisors of* r *in the form* $4k + 1$ *and* $4k + 3$, *respectively,* $k = 0, 1, 2, \ldots$, *then* $n_r = 4(N_1 - N_3)$. FOR EXAMPLE, for $r = 45$, we find $N_1 = 4$ (the divisors of the form $4k + 1$ are $1, 5, 9, 45$), $N_3 = 2$ (the divisors of the form $4k + 3$ are $3, 15$), and so $n_{45} = 4 \cdot (4 - 2) = 8$; all eight points come from the representation $45 = 3^2 + 6^2$. To get a better understanding, verify the theorem for $r = 25$ and $r = 65$; for the proof, see the book *An Introduction to the Theory of Numbers* by G. H. Hardy and E. M. Wright, 1979. Note that, to use this theorem, you always include 1 in the list of the divisors of the form $4k + 1$; the number r should also be included in one of the lists if r is of the form $4k + 1$ or $4k + 3$.

EXERCISE 6.3.10.$^{A+}$ *Let a, b be such that the eigenvalue $\lambda_{m,n}$ can be multiple, and define $r = (m/a)^2 + (n/b)^2$. Denote by $M(r)$ the multiplicity of the eigenvalue $\lambda_{m,n} = -\pi^2 r$. Find a function $g = g(r)$ so that $0 < \limsup_{r\to\infty} M(r)/g(r) < \infty$. Can you choose the function g so that $\limsup_{r\to\infty} M(r)/g(r) = 1$? Hint: start with $a = b = 1$ and use the Jacobi theorem, keeping in mind that, of all the integer points (k, l) on the circle $x^2+y^2 = r$ you only need those satisfying $k, l > 0$.*

If you found the above discussion interesting, try to analyze the eigenvalues of the Dirichlet Laplacian in a rectangular box in three or more dimensions.

PROBLEM (6.3.13) IN A DISK OF RADIUS R. Introduce polar coordinates (r, θ) with the origin at the center of the disk: $x = r\cos\theta$, $y = r\sin\theta$. Then $u = u(r,\theta)$ and by (3.1.44) on page 147, $\nabla^2 u(r,\theta) = u_{rr}(r,\theta) + r^{-1}u_r(r,\theta) + r^{-2}u_{\theta\theta}(r,\theta)$. As a result, (6.3.13) becomes

$$\begin{aligned} &u_{rr}(r,\theta) + \frac{1}{r}\,u_r(r,\theta) + \frac{1}{r^2}\,u_{\theta\theta}(r,\theta) = \lambda u(r,\theta),\ 0 < r < R,\ 0 \le \theta < 2\pi, \\ &u(R,\theta) = 0,\ u(r_0,\theta_0) \neq 0 \text{ for some } 0 < r_0 < R,\ 0 \le \theta_0 < 2\pi. \end{aligned} \tag{6.3.20}$$

Before we proceed, note that $r = 0$, corresponding to the center of the disk, is a special point for the polar coordinate system, which means that some solutions of (6.3.20) might have a singularity for $r = 0$. On the other hand, the center of the disk is not a special point for the original problem (6.3.13), and we therefore add one more condition on the solutions of (6.3.20):

$$u = u(r,\theta) \text{ is infinitely differentiable for } 0 \le r < R. \tag{6.3.21}$$

We now solve (6.3.20) by `separation of variables`. With $u(r,\theta) = v(r)w(\theta)$,

$$\frac{r^2v''(r) + rv'(r)}{v(r)} - r^2\lambda = -\frac{w''(\theta)}{w(\theta)} \tag{6.3.22}$$

EXERCISE 6.3.11.^{C} *Verify (6.3.22).*

The left-hand side of (6.3.22) depends only on r, and the right-hand side, only on θ. Therefore, both sides must equal a constant q and

$$\begin{aligned} &r^2v''(r) + rv'(r) - (\lambda r^2 + q)v(r) = 0,\ v(R) = 0, \\ &w''(\theta) + q\,w(\theta) = 0. \end{aligned} \tag{6.3.23}$$

Exercise 6.3.12.[C] *(a) Verify (6.3.23). (b) Introduce a new variable* $s = \sqrt{|\lambda|}\, r$ *and the corresponding function* $V(s) = v(s/\sqrt{|\lambda|})$. *Keeping in mind that* λ *is negative, that is,* $\lambda = -(\sqrt{|\lambda|})^2$, *verify that*

$$s^2V''(s) + sV'(s) + (s^2 - q)V(s) = 0. \tag{6.3.24}$$

By (6.3.21), the function $V = V(s)$ must have all derivatives at $s = 0$, and, by Exercise 4.4.20(b) on page 236, such a solution of (6.3.24) exists if and only if

$$q = N^2, \quad N = 0, 1, 2, \ldots. \tag{6.3.25}$$

We now use power series to solve the resulting `Bessel's differential equation`

$$s^2V''(s) + sV'(s) + (s^2 - N^2)V(s) = 0. \tag{6.3.26}$$

Exercise 6.3.13.[C] *For* $N = 1, 2, \ldots,$ *and a complex variable* z *define the function*

$$J_N(z) = \sum_{k=0}^{\infty} \frac{(-1)^k}{k!(k+N)!} \left(\frac{z}{2}\right)^{2n+N}. \tag{6.3.27}$$

(a) Verify that this function is analytic everywhere. Hint: use the ratio test on page 207 to verify that the series on the right converges absolutely for all z. *(b) Verify that this function satisfies*

$$z^2J_N''(z) + zJ_N'(z) + (z^2 - N^2)J_N(z) = 0.$$

Hint: just plug the power series in the equation and collect the similar terms.

The function J_N is called `Bessel's function` of the first kind of order N; all in all, there are two kinds of Bessel's functions of a given order, and Bessel's function of the second kind of order N is one of the solutions of (6.3.26) with a singularity at zero.

It is known that, for each N, the equation $J_N(z) = 0$ has infinitely many positive solutions:

$$J_N(\alpha_n^{(N)}) = 0 \text{ with } 0 < \alpha_1^{(N)} < \alpha_2^{(N)} < \ldots. \tag{6.3.28}$$

For a proof of this and other properties of the Bessel functions, see Problem 7.5 on page 441.

Recall that $v(r) = V(r\sqrt{|\lambda|}) = J_N(r\sqrt{|\lambda|})$ and $v(R) = 0$. By (6.3.28), we conclude that $\lambda = -(\alpha_n^{(N)}/R)^2$. Also, by (6.3.23) and (6.3.28), $w(\theta) = \cos(N\theta)$ or $w(\theta) = \sin(N\theta)$. As a result, we get the eigenvalues and eigenfunctions of the Dirichlet Laplacian in a disk of radius R:

$$\lambda_{N,n} = -\left(\frac{\alpha_n^{(N)}}{R}\right)^2; \; u_{0,n}(r,\theta) = J_0\big(\alpha_n^{(0)} r/R\big),$$
$$u_{N,n}^{(1)}(r,\theta) = J_N\big(\alpha_n^{(N)} r/R\big)\cos(N\theta),\; u_{N,n}^{(2)}(r,\theta) = J_N\big(\alpha_n^{(N)} r/R\big)\sin(N\theta),$$
$$N = 0,1,2,\ldots,\; n = 1,2,\ldots. \tag{6.3.29}$$

EXERCISE 6.3.14. *(a)*[C] *What is the multiplicity of* $\lambda_{N,n}$? *Hint: it is either one or two.* *(b)*[B] *Verify the following* **orthogonality relation** *for the Bessel functions:*

$$\int_0^R J_N\big(\alpha_n^{(N)} r/R\big) J_N\big(\alpha_m^{(N)} r/R\big)\, r dr = 0$$

for $m \neq n$ *and* $N = 0,1,2\ldots$. *Hint: since* ∇^2 *is a symmetric operator, by Theorem 6.2.1 on page 329,* $(u_{N,n}^{(1)}, u_{N,m}^{(1)}) = 0$; *write the inner product in polar coordinates.* *(c)*[A] *Describe the possible* **nodal lines** *of the eigenfunctions, that is, the set of points where* $u = 0$. *Hint: keep in mind that, if* $N > 0$, *then, for every real* a, b, $au_{N,n}^{(1)} + bu_{N,n}^{(2)}$ *is an eigenfunction corresponding to* $\lambda_{N,n}$.

6.3.3 *Wave Equation in Two and Three Dimensions*

We begin with an abstract derivation of the wave equation. This derivation applies in $\mathbb{R}, \mathbb{R}^2$, and $\mathbb{R}^3$. Deformation of an elastic material is described by the stress and strain tensor fields. For a linear isotropic material, these two tensor fields are connected by Hooke's law $F = 2\mu D + \lambda D''$, see page 462 in Appendix. We make two additional assumptions:

- all points of the material are displaced in the direction of a fixed unit vector $\widehat{\boldsymbol{u}}$;
- $\lambda = 0$.

Denote by $u = u(t,P)$ the displacement of the point P at time t in the direction $\widehat{\boldsymbol{u}}$; for a two-dimensional membrane, such as a drumhead hit by a drumstick, this direction would be perpendicular to the surface of the membrane. For such displacements, the stress tensor field becomes a stress vector field $\boldsymbol{F}$, see Exercise 8.3.15(b) on page 462 in Appendix.

For the sake of concreteness, we carry out the derivation in a three-dimensional setting. Let Q be a point in the elastic material and consider a small region G_Q of the material containing the point P. The **stress** vector field $\boldsymbol{F}$, that is, force per unit area, stretches or compresses the material in G_Q. Thus, the force acting on G_Q is the flux of $\boldsymbol{F}$ through the boundary ∂G_Q of G_Q. If ρ is the density of the material, then the linear momentum of the region G_Q is $\iiint\limits_{G_Q} \rho(P) u_{tt}(t,P)dV$. Then the Second Law of Newton yields

$$\iiint\limits_{G_Q} \rho(P)u_{tt}(t,P)dV = -\iint\limits_{\partial G_Q} \boldsymbol{F}(t,P)\cdot\widehat{\boldsymbol{n}}\, d\sigma,$$

or, by Gauss's Theorem on page 152, $\iiint\limits_{G}(\rho\, u_{tt} + \operatorname{div}\boldsymbol{F})\, dV = 0$. Since the last equality holds around every point Q of the material, we conclude that

$$\rho(Q)\, u_{tt}(t,Q) + \operatorname{div}\boldsymbol{F}(t,Q) = 0. \tag{6.3.30}$$

In a linear homogeneous isotropic material, the density ρ is independent of x, and, by Hooke's Law (see Exercise 8.3.15(b)), $\boldsymbol{F} = -a \operatorname{grad} u$, where $a > 0$ is a constant. Then (6.3.30) becomes the wave equation $u_{tt} = c^2\nabla^2 u$, with $c^2 = a/\rho$.

EXERCISE 6.3.15. *(a)*^{B} *Carry out the above derivation for a homogeneous isotropic membrane and verify that $c^2 = T/\rho$, where T is the tension of the membrane and ρ is the mass per unit area; now T is measured in units of force per unit length. (b)*^{A} *Reconcile the above derivation with the derivation of the one-dimensional wave equation for the string.*

Consider the **initial value problem** for the wave equation in $\mathbb{R}^2$ or $\mathbb{R}^3$:

$$\begin{aligned} &u_{tt}(t,P) = c^2\nabla^2 u(t,P),\ t>0,\ P\in\mathbb{R}^n,\\ &u(0,P) = f(P),\ u_t(0,P) = g(P). \end{aligned} \tag{6.3.31}$$

If $n = 2$, then the solution of (6.3.31) is

$$u(t,P) = \frac{1}{2\pi ct}\iint\limits_{D(P,ct)} \frac{t\,g(Q) + f(Q) + \operatorname{grad} f(Q)\cdot\overrightarrow{PQ}}{\sqrt{c^2t^2 - |PQ|}}\, dA(Q), \tag{6.3.32}$$

where $D(P, ct)$ is the disk with center at the point P and radius ct:

$D(P, ct) = \{Q \in \mathbb{R}^2 : |PQ| \le ct\}$. If $n = 3$, then the solution of (6.3.31) is

$$u(t,P) = \frac{1}{4\pi c^2 t^2} \iint\limits_{S(P,ct)} \left(t\, g(Q) + f(Q) + \operatorname{grad} f(Q) \cdot \overrightarrow{PQ} \right) d\sigma(Q), \quad (6.3.33)$$

where $S(P, ct)$ is the surface of the sphere with center at the point P and radius ct: $S(P, ct) = \{Q \in \mathbb{R}^3 : |PQ| = ct\}$. The uniqueness of solution for the wave equation is known as **Huygens's principle**: *once the initial wave disturbance reaches a particular point, the point becomes the source of secondary waves.* Mathematically, Huygens principle means that the solution operator Φ for equation (6.3.30) satisfies the semi-group property (6.2.13) on page 323.

The interested reader can find the derivations of (6.3.32) and (6.3.33) in Section 2.4.1 of the book *Partial Differential Equations* by L. C. Evans, 1998. The derivation is rather ingenious: first, one shows that, if $n = 3$, then, for fixed P, the function $U(t, r; P) = \frac{1}{4\pi r} \iint\limits_{S(P,r)} u(t, Q) d\sigma(Q)$, as a function of t and r satisfies the one-dimensional wave equation on the half-line with zero boundary condition; the initial conditions for U are computed from the initial conditions for u. After solving this equation, the solution u of (6.3.31) is recovered from U by $u(t, P) = \lim_{r \to 0} U(t, r; P)/r$, resulting in formula (6.3.33). Then (6.3.32) is derived from (6.3.33) by artificially introducing the third dimension and then transforming the integrals over the sphere into integrals over the disk.

Formula (6.3.32) describes planar waves, such as ripples on the surface of the water. Formula (6.3.33) describes spatial waves, such as sound or radio. Figure 6.3.3 illustrates the radically different way the waves propagate in two and three dimensions. Suppose that the initial conditions $u(0, P) = f(P)$, $u_t(0, P) = g(P)$ are non-zero inside a bounded region G and are equal to zero outside of G. Let us look at the solution $u(t, P_0)$ at some fixed point P_0 not in G so that $u(t, P_0)$ is the disturbance produced by the initial displacement f and velocity g. The solution will be zero for $t < t_1 = |P_0P_1|/c$, that is, until the initial disturbance reaches the point. After that, in two dimensions the disk of integration will be covering the region for all $t > t_1$, and the disturbance at the point P_0 will be dying out for infinitely long time; in three dimensions, the surface of the sphere of integration will eventually stop intersecting the region and the disturbance at the point P_0 will stop completely after a finite time $t_2 = |P_0P_2|/c$; see Figure 6.3.3.

EXERCISE 6.3.16.[B] *How does the initial disturbance propagate when $n = 1$? Hint: use (6.1.43) on page 310; the pictures are different, depending on whether the initial velocity is zero or non-zero.*

Note that the circles on Figure 6.3.3 are *not* wave fronts; instead, they represent the regions of integration: disks in formula (6.3.32) and spherical surfaces in formula (6.3.33).

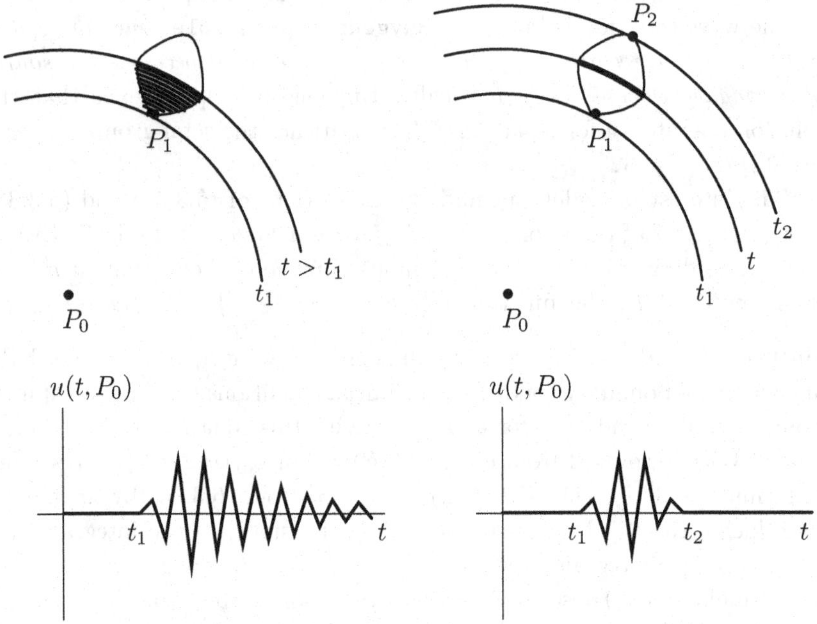

(a) Planar Wave (b) Spatial Wave

Fig. 6.3.3 Wave Propagation

We conclude this section with a brief discussion of the two-dimensional wave equation in a piece-wise smooth bounded region G in $\mathbb{R}^2$:

$$u_{tt}(t, P) = c^2 \nabla^2 u(t, P), \; t > 0, \; P \in G, \tag{6.3.34}$$

with initial conditions $u(0, P) = f(P)$, $u_t(0, P) = g(P)$ and zero boundary condition $u(t, P) = 0$ if $P \in \partial G$. This is a mathematical model of vibrations of a drum shaped as the region G.

Let φ_k, $k \geq 1$, be the eigenfunctions of the Dirichlet Laplacian in G with zero boundary conditions, and λ_k, $k \geq 1$, the corresponding eigenvalues: $\nabla^2 \varphi_k = \lambda_k \varphi_k$, $\varphi_k|_{\partial G} = 0$. Recall that $\lambda_k < 0$ and we can choose φ_k to satisfy $\iint\limits_G \varphi_k \varphi_m \, dA = 0$ for $m \neq k$. Following the steps on page 331, we conclude that

$$u(t, P) = \sum_{k=1}^{\infty} \left(A_k \cos(\sqrt{|\lambda_k|}\, ct) + B_k \sin(\sqrt{|\lambda_k|}\, ct) \right) \varphi_k(P), \qquad (6.3.35)$$

where

$$A_k = \frac{\iint\limits_G f(P) \varphi_k(P) \, dA}{\iint\limits_G |\varphi_k(P)|^2 \, dA}, \quad B_k = \frac{\iint\limits_G g(P) \varphi_k(P) \, dA}{\sqrt{|\lambda_k|}\, c \iint\limits_G |\varphi_k(P)|^2 \, dA}.$$

EXERCISE 6.3.17. *(a)^C Verify (6.3.35). (a)^B Write the expressions for A_k, B_k when G is (i) a rectangle with sides a, b, $a = b$ is a possibility; (ii) a disk of radius R (here, you can use the result of Problem 7.5(d) on page 441 to simplify some of the formulas). (c)^A Under what conditions on the functions f and g does (6.3.35) define a classical solution of (6.3.34)? (d)$^{A+}$ Can you tell the* `shape of the drum` *from its sound (assuming you have perfect pitch)? Hint: Since, for the given material and tension of the drum, the time frequencies are determined by $\sqrt{|\lambda_k|}$, you need to investigate whether there exist two different regions G_1, G_2, for which the corresponding collections of eigenvalues λ_k, $k \geq 1$, are the same.*

6.3.4 *Maxwell's Equations*

In this section, we will study `Maxwell's equations` in $\mathbb{R}^3$ by reducing them to a system of wave equations. By doing so, we will predict mathematically the existence of electromagnetic waves, just as Maxwell himself predicted in the early 1860s, nearly 20 years before such waves were discovered. This is a major example of the predictive power of mathematical models. In our derivations, we assume that the electric field $\boldsymbol{E}$ and the magnetic field $\boldsymbol{B}$ are functions of time and space, and have all the continuous derivatives we might need.

To begin, we consider the equations in vacuum, see page 164. With no currents or charges, these equations become

$$\operatorname{div} \boldsymbol{E} = 0, \ \operatorname{div} \boldsymbol{B} = 0, \ \operatorname{curl} \boldsymbol{E} = -\frac{\partial \boldsymbol{B}}{\partial t}, \ \operatorname{curl} \boldsymbol{B} = \mu_0 \varepsilon_0 \frac{\partial \boldsymbol{E}}{\partial t}. \qquad (6.3.36)$$

According to equality (3.1.32) on page 139, curl(curl $\boldsymbol{E}$) = grad(div $\boldsymbol{E}$) $-$ $\nabla^2\boldsymbol{E} = -\nabla^2\boldsymbol{E}$, because div $\boldsymbol{E} = 0$. Taking the curl on both sides of the third equation in (6.3.36) and then using the fourth equation, we get a system of three wave equations, one for each component of $\boldsymbol{E}$:

$$\frac{\partial^2 \boldsymbol{E}}{\partial t^2} = c^2\,\nabla^2 \boldsymbol{E}, \tag{6.3.37}$$

where $c = 1/\sqrt{\mu_0\varepsilon_0}$ is the propagation speed. Substitution of the numerical values for permittivity of vacuum $\varepsilon_0 = 8.85 \cdot 10^{-12}$ Coulombs/(Newton·meter^2) and permeability of vacuum $\mu_0 = 4\pi \cdot 10^{-7}$ Newtons/ampere^2 shows that c is the speed of light in vacuum: $c \approx 3 \cdot 10^8$ meters per second. This value was known *before* Maxwell derived (6.3.37) in 1864: from 1849 to 1862, J. B. L. Foucault, of the Foucault pendulum fame, together with another French physicist Armand Hippolyte Louis Fizeau (1819–1896), conducted numerous experiments on the propagation of light. In 1862, Foucault computed the value $c = 2.98 \cdot 10^8$ meters per second, less than 0.6% off the currently accepted $c = 299,792,458$ meters per second. Because of the appearance of c in equation (6.3.37), Maxwell suggested light to be an electromagnetic wave. While the wave theory of light goes back to a 1678 work "Traité de la lumiere" by C. Huygens, the notion of the *electromagnetic* wave was quite revolutionary, as it was only in the late 1880s that the German physicist Heinrich Rudolf Hertz (1857–1894) actually generated the first electro-magnetic wave, a wave in the radio frequency spectrum generated by accelerated electric charges; Hertz also demonstrated that these waves have many properties of light. Later experiments showed that every form of electromagnetic radiation — radio waves, infra-red radiation, visible light, ultra-violet radiation, X-rays, and gamma radiation — is produced by accelerated electric charges and travels in vacuum at the same speed c.

Incidentally, the particle theory of light has been around for a while as well, first proposed in 1669 by I. Newton, and in 1905 became a part of Einstein's theory of the photo-electric effect. Quantum mechanics and the theory of relativity provide a connection between particle and matter waves theories via the relation $h\nu = mc^2$, where ν is the frequency of the wave, m is the mass of the corresponding particle, and $h = 6.62 \cdot 10^{-34}$ joule-seconds is `Planck's constant`; see page 360 below.

A simple change of variables shows that the one-dimensional version of equation (6.3.37) is invariant under the `Lorentz transformation` but is not invariant under the Galilean transformation; see Problem 7.9 on page

446 for details. This invariance was crucial for the development of special relativity.

EXERCISE 6.3.18.[C] *Verify that the magnetic field* $\boldsymbol{B}$ *also satisfies (6.3.37). Hint: start by taking the curl of the last equation in (6.3.36).*

Thus, both $\boldsymbol{B}$ and $\boldsymbol{E}$ behave as waves, and the term "electromagnetic wave" is aptly chosen.

Let us now look for a solution of (6.3.37) in the form of a `planar, or plane, wave`

$$\boldsymbol{E}(t,P) = \boldsymbol{E}_0\, e^{i(\boldsymbol{k}\cdot\boldsymbol{r}-\omega t)}, \tag{6.3.38}$$

where $\boldsymbol{r} = \overrightarrow{OP}$ is the position vector of the point P relative to some fixed reference point O, $\boldsymbol{E}_0$ is a constant vector, i is the imaginary unit: $i = \sqrt{-1}$, $\boldsymbol{k}$ is a constant vector, called the `wave vector`, and ω is a real number, representing a frequency in time. Note that $\boldsymbol{E}$ is a complex-valued vector and, to recover the physical electric field, we can take the real part of $\boldsymbol{E}$. However, we can apply all operations and derivations to both the real and imaginary parts, and all equations remain valid.

EXERCISE 6.3.19. [C] *(a) Verify that* $\operatorname{grad}(\boldsymbol{k}\cdot\boldsymbol{r}) = \boldsymbol{k}$. *Hint: use cartesian coordinates. (b) Verify that, for* $\boldsymbol{E}$ *defined in (6.3.38), we have*

$$\operatorname{div}\boldsymbol{E} = i\,\boldsymbol{k}\cdot\boldsymbol{E},\ \ \frac{\partial^2\boldsymbol{E}}{\partial t^2} = -\omega^2\,\boldsymbol{E},\ \ \nabla^2\boldsymbol{E} = -\|\boldsymbol{k}\|^2\,\boldsymbol{E},\ \ \operatorname{curl}\boldsymbol{E} = i\,\boldsymbol{k}\times\boldsymbol{E}. \tag{6.3.39}$$

Hint: use suitable formulas on page 139.

From (6.3.39) it follows that

- $\boldsymbol{E}$ in (6.3.38) satisfies (6.3.37) if and only if $\|\boldsymbol{k}\| = \omega/c$.
- $\boldsymbol{E}$ in (6.3.38) satisfies (6.3.36) if and only if $\boldsymbol{k} \perp \boldsymbol{E}(t,P)$ for all t, P.
- With a suitable choice of the initial condition for $\boldsymbol{B}$ we have

$$\boldsymbol{B}(t,P) = \frac{1}{\omega}\,\boldsymbol{k}\times\boldsymbol{E}(t,P), \tag{6.3.40}$$

so that, for all t, P, the vector $\boldsymbol{E}(t,P)\times\boldsymbol{B}(t,P)$ has the same direction as $\boldsymbol{k}$ and $\boldsymbol{B}(t,P) \perp \boldsymbol{E}(t,P) \perp \boldsymbol{k}$ (draw a picture!)

EXERCISE 6.3.20. [C] *(a) Verify that if* $\boldsymbol{E}$ *is given by (6.3.38), then* $\boldsymbol{B} = \boldsymbol{k}\times\boldsymbol{E}/\omega$. *Hint:* $\partial\boldsymbol{B}/\partial t = -\operatorname{curl}\boldsymbol{E} = -i\,\boldsymbol{k}\times\boldsymbol{E}_0\, e^{i(\boldsymbol{k}\cdot\boldsymbol{r}-\omega t)}$, *then integrate in time. (b) Assuming that* $\boldsymbol{E}_0$ *is a real vector, find the real parts of* $\boldsymbol{E}(t,P)$ *and* $\boldsymbol{B}(t,P)$ *and note the term* $\cos(\boldsymbol{k}\cdot\boldsymbol{r}-\omega t)$ *in both cases.*

If we select the coordinate system so that $\boldsymbol{k} = k\,\hat{\boldsymbol{\kappa}}$ for some $k > 0$, then $\boldsymbol{E}$ becomes a travelling wave along the z-axis, $\boldsymbol{E} = \boldsymbol{E}_0 e^{ik(z-ct)}$ (recall that $kc = \omega$). The same is true for $\boldsymbol{B}$. Therefore, the planar electromagnetic wave (6.3.38), (6.3.40) propagates in the direction of the vector $\boldsymbol{k}$. For fixed $t = t_0$, a wavefront through a point $P_0 = (x_0, y_0, z_0)$, that is, the set of points $P = (x, y, z)$ at which the $\boldsymbol{E}(t_0, P) = \boldsymbol{E}(t_0, P_0)$ and $\boldsymbol{B}(t_0, P) = \boldsymbol{B}(t_0, P_0)$, satisfies $z = z_0$, which is a plane through P_0, perpendicular to $\boldsymbol{k}$. This observation explains the term "planar wave". As functions of time for fixed $\boldsymbol{r}$, the complex vectors $\boldsymbol{E}$ and $\boldsymbol{B}$ rotate around $\boldsymbol{k}$ clockwise, as seen from the tip of $\boldsymbol{k}$. This direction of rotation defines the `polarization` of the wave. If $\boldsymbol{E}(t, P) = \boldsymbol{E}_0 e^{-i(\boldsymbol{k}\cdot\boldsymbol{r}-\omega t)}$, the result will be a similar plane wave, but with opposite polarization. A directional dish antenna produces a nearly planar wave.

If $\nu = \omega/(2\pi)$ is the `linear frequency of the wave`, measured in Hertz (1Hz = 1/second), then $\|\boldsymbol{k}\| = 2\pi/\lambda$, where $\lambda = c/\nu$ is the wave length of the wave in meters. To generate and to receive a wave with wavelength λ, the corresponding emitting or receiving antenna must have the size of the order of λ. Note that we have $\nu\lambda = c \approx 3 \cdot 10^8$ m/s. The `spectrum of the electromagnetic radiation` is characterized by ν or λ of the wave. In particular,

RADIO WAVES have frequency ν from about $3 \cdot 10^3$ to $3 \cdot 10^{11}$ Hz, with the corresponding λ from about 10^5 to about 10^{-3} meters (100 km to 1 mm). For example, radio AM wave of frequency $\nu \approx 10^6$ Hz (such as broadcast of the AM1070 station) has wavelength $\lambda \approx 300$ m and a radio FM wave of frequency $\nu \approx 10^8$ Hz (such as broadcast of the FM105.1 station) has wavelength $\lambda \approx 3$ m;

INFRA-RED RADIATION has ν from about $3 \cdot 10^{11}$ to $4 \cdot 10^{14}$ Hz and the corresponding λ from about 10^{-3} to about $7.5 \cdot 10^{-7}$ meters;

VISIBLE LIGHT has ν from $3.75 \cdot 10^{14}$ Hz (red) to $7.5 \cdot 10^{14}$ Hz (violet) and the corresponding λ from about $8 \cdot 10^{-7}$ to about $4 \cdot 10^{-7}$ meters;

ULTRA-VIOLET RADIATION has ν from about $7.5 \cdot 10^{14}$ to about $7.5 \cdot 10^{16}$ Hz and the corresponding λ from about $4 \cdot 10^{-7}$ to about $4 \cdot 10^{-9}$ meters;

X-RAYS have ν from about $3 \cdot 10^{16}$ to about $3 \cdot 10^{19}$ Hz and the corresponding λ from about 10^{-8} to about 10^{-11} meters;

GAMMA-RADIATION has ν at least $3 \cdot 10^{18}$ Hz and the corresponding λ at most 10^{-10} meters, which is about the diameter of the hydrogen atom in the ground state.

Let us now consider Maxwell's equations with sources, either in vacuum or in material medium. We assume that the medium is homogeneous (has the same properties at every point), isotropic (has the same properties in every direction), and linear ($\boldsymbol{D}$ depends linearly on $\boldsymbol{E}$, $\boldsymbol{B}$ depends linearly on $\boldsymbol{H}$). We also disregard the possible dependence of the properties of the medium on the frequency of the electric and magnetic fields. Under these assumptions, we have $\boldsymbol{D} = \varepsilon\,\boldsymbol{E}$ and $\boldsymbol{H} = \boldsymbol{B}/\mu$ for some positive real numbers ε and μ, so that we can write equations (3.3.47)–(3.3.50) on page 178 using only the vectors $\boldsymbol{E}$ and $\boldsymbol{B}$ as the unknowns:

$$\operatorname{div}\boldsymbol{E} = \frac{\rho_f}{\varepsilon},\ \operatorname{div}\boldsymbol{B} = 0,\ \operatorname{curl}\boldsymbol{E} = -\frac{\partial \boldsymbol{B}}{\partial t},\ \operatorname{curl}\boldsymbol{B} = \mu\,\boldsymbol{J}_e + \mu\varepsilon\frac{\partial \boldsymbol{E}}{\partial t}. \quad (6.3.41)$$

The free charge density ρ_f and the external current $\boldsymbol{J}_e$ are assumed to be known; see pages 175 and 177 for more details. Notice that equation in vacuum are a particular case of (6.3.41) with $\varepsilon = \varepsilon_0$ and $\mu = \mu_0$.

To study (6.3.41), we introduce the `vector potential` $\boldsymbol{A} = \boldsymbol{A}(t, P)$ such that $\boldsymbol{B} = \operatorname{curl}\boldsymbol{A}$ and the second equation in (6.3.41) is automatically satisfied; see (3.1.30) on page 139. The third equation then yields

$$\operatorname{curl}(\boldsymbol{E} + \partial\boldsymbol{A}/\partial t) = \boldsymbol{0}. \quad (6.3.42)$$

Accordingly, we introduce the `scalar potential` $\varphi = \varphi(t, P)$ such that $\boldsymbol{E} + \partial\boldsymbol{A}/\partial t = -\operatorname{grad}\varphi$ and (6.3.42) is automatically satisfied; once again, see (3.1.30). By convention, this definition of φ has the minus sign in front of the gradient. As a result, we know $\boldsymbol{E}$ and $\boldsymbol{B}$ if we know $\boldsymbol{A}$ and φ.

EXERCISE 6.3.21.[B] *Verify that the values of $\boldsymbol{E}$ and $\boldsymbol{B}$ do not change if we take a sufficiently smooth scalar field ψ and change the values of $\boldsymbol{A}$, φ as follows:*

$$\boldsymbol{A} \to \boldsymbol{A} + \operatorname{grad}\psi,\ \varphi \to \varphi - \frac{\partial\psi}{\partial t}. \quad (6.3.43)$$

The function ψ is an example of a **`gauge`**. Choosing a particular ψ is called **`gauge fixing`**, and (6.3.43) is an example of a **`gauge transform`**. The study of gauge transforms for the **`Yang-Mills equations`** (the quantum-theoretical analogues of Maxwell's equations, see page 178) is an exciting area of research in both mathematics and physics.

It follows from the fourth equation in (6.3.41) that

$$\operatorname{curl}(\operatorname{curl}\boldsymbol{A}) = \operatorname{curl}\boldsymbol{B} = \mu\,\boldsymbol{J}_e + \mu\varepsilon\frac{\partial \boldsymbol{E}}{\partial t} = \mu\,\boldsymbol{J}_e + \mu\varepsilon\frac{\partial}{\partial t}\left(-\frac{\partial \boldsymbol{A}}{\partial t} - \operatorname{grad}\varphi\right).$$

On the other hand, according to equality (3.1.32) on page 139, $\operatorname{curl}(\operatorname{curl}\boldsymbol{A}) = \operatorname{grad}(\operatorname{div}\boldsymbol{A}) - \nabla^2\boldsymbol{A}$, so that

$$\operatorname{grad}\left(\operatorname{div}\boldsymbol{A} + \mu\varepsilon\frac{\partial\varphi}{\partial t}\right) + \mu\varepsilon\frac{\partial^2\boldsymbol{A}}{\partial t^2} = \nabla^2\boldsymbol{A} + \mu\,\boldsymbol{J}_e. \tag{6.3.44}$$

Given a particular choice of $\boldsymbol{A}$ and φ, we can always choose ψ in (6.3.43) so that

$$\operatorname{div}\boldsymbol{A} + \mu\varepsilon\frac{\partial\varphi}{\partial t} = 0. \tag{6.3.45}$$

This choice of ψ is called `Lorenz's gauge`, after the Danish physicist LUDWIG LORENZ (1829–1891), who has no connection to the Lorentz transform. Another popular choice of ψ ensures that $\operatorname{div}(\boldsymbol{A}) = 0$, and is called `Coulomb's gauge`; we used Coulomb's gauge before, see page 168.

EXERCISE 6.3.22.[C] *Verify that (6.3.45) is achieved if ψ satisfies*

$$\mu\varepsilon\frac{\partial^2\psi}{\partial t^2} = \nabla^2\psi + \operatorname{div}\boldsymbol{A} + \mu\varepsilon\frac{\partial\varphi}{\partial t}.$$

As a result, with no loss of generality we assume that (6.3.45) holds, and then (6.3.44) becomes an inhomogeneous wave equation for the vector potential $\boldsymbol{A}$:

$$\frac{\partial^2\boldsymbol{A}}{\partial t^2} = c_m^2\,\nabla^2\boldsymbol{A} + \frac{\boldsymbol{J}_e}{\varepsilon}, \tag{6.3.46}$$

where the propagation speed $c_m = 1/\sqrt{\varepsilon\mu}$. In a material medium, $\varepsilon > \varepsilon_0$ and $\mu > \mu_0$, so that $c_m < c$.

EXERCISE 6.3.23.[B] *Verify that, under assumption (6.3.45), the scalar potential φ satisfies*

$$\frac{\partial^2\varphi}{\partial t^2} = c_m^2\,\nabla^2\varphi + \frac{c_m^2\rho_f}{\varepsilon}, \tag{6.3.47}$$

Hint: from the definition of φ, $\nabla^2\varphi = -\operatorname{div}\boldsymbol{E} - \partial(\operatorname{div}\boldsymbol{A})/\partial t$. Then use (6.3.45) and the first equation in (6.3.41).

Since $\boldsymbol{B} = \operatorname{curl}\boldsymbol{A}$ and $\boldsymbol{E} = -\partial\boldsymbol{A}/\partial t - \operatorname{grad}\varphi$, it follows that $\boldsymbol{B}$ and $\boldsymbol{E}$ also satisfy inhomogeneous wave equations with the same propagation speed.

EXERCISE 6.3.24.[C] *Define a scalar field W and a vector field $\boldsymbol{S}$ as follows:*

$$W = \frac{1}{2}\left(\varepsilon\|\boldsymbol{E}\|^2 + \frac{\|\boldsymbol{B}\|^2}{\mu}\right), \quad \boldsymbol{S} = \frac{1}{\mu}\left(\boldsymbol{E}\times\boldsymbol{B}\right). \tag{6.3.48}$$

(a) Verify that W has the units of energy per unit volume, that is, joule/meter^3. Hint: see Section 8.5, page 465, in Appendix. (b) Verify that

$$\frac{\partial W}{\partial t} + \operatorname{div}\boldsymbol{S} = -\boldsymbol{E}\cdot\boldsymbol{J}_e. \tag{6.3.49}$$

Hint: use the last two equations in (6.3.41).

Equation (6.3.49) describes the balance of energy in an electromagnetic wave and is known as the `Poynting Theorem`, after the English physicist JOHN HENRY POYNTING (1852–1914), who published the result in 1884. The scalar field W is the energy of the electromagnetic field per unit volume and the vector $\boldsymbol{S}$, known as the `Poynting vector`, describes the flow of the energy in the electromagnetic wave. Notice a certain analogy between (6.3.49) and the equation of continuity (3.2.8) on page 154 or (3.3.12) on page 166.

6.3.5 *Equations of Fluid Mechanics*

The objective of this section is to introduce the main equations describing the motion of fluids; in what follows, "fluid" means any continuous medium, usually liquid or gas. There is also a good reason for placing this topic at the very end of the section on classical equations: with all the advances of modern science, and despite the long history of research on the equations, many types of fluid flows remain essentially a mystery, at least as far as the mathematical theory of the corresponding equations is concerned.

We already know one such equation, the equation of continuity; without sources or sinks (which we assume everywhere in this section), this equation is

$$\rho_t + \operatorname{div}(\rho\,\boldsymbol{u}) = 0, \tag{6.3.50}$$

where ρ is the density of the medium and $\boldsymbol{u}$ is the corresponding velocity field, see page 154. For three-dimensional flows, there are, in general, four unknowns in this equation: the density function ρ and the three components of the velocity vector $\boldsymbol{u}$. As a result, more equations are necessary to describe the flow. These equations come from the analysis of the dynamics of the fluid flow.

To derive some of these equations, we consider a particle in the medium with position vector $\boldsymbol{r} = \boldsymbol{r}(t)$. By assumption, $\boldsymbol{r}'(t) = \boldsymbol{u}(t, \boldsymbol{r}(t))$, and then, by (3.1.10) on page 126,

$$\boldsymbol{r}''(t) = \frac{\partial \boldsymbol{u}(t, \boldsymbol{r}(t))}{\partial t} + (\boldsymbol{r}'(t) \cdot \boldsymbol{\nabla})\boldsymbol{u}(t, \boldsymbol{r}(t)) = \frac{\partial \boldsymbol{u}(t, \boldsymbol{r}(t))}{\partial t} + (\boldsymbol{u} \cdot \boldsymbol{\nabla})\boldsymbol{u}(t, \boldsymbol{r}(t)).$$

The Second Law of Newton now yields

$$\rho(t, P)\left(\frac{\partial \boldsymbol{u}(t, P)}{\partial t} + (\boldsymbol{u} \cdot \boldsymbol{\nabla})\boldsymbol{u}(t, P)\right) = \boldsymbol{F}(t, P) \tag{6.3.51}$$

at every point P of the medium, where $\boldsymbol{F}$ is the total force density per unit volume. There are two main sources of the force $\boldsymbol{F}$, internal (for example, internal pressure and viscosity) and external (for example, outside pumping).

EXERCISE 6.3.25.^{A} *Show that the pressure $p = p(t, P)$ in the fluid results in the volume density of the force equal to $-\boldsymbol{\nabla} p$, the negative gradient of the pressure.*

For a three-dimensional flow, (6.3.51) and the equation of continuity (6.3.50) are four coupled partial differential equations. Along with the density ρ and the three components of the velocity $\boldsymbol{u}$, the pressure p is often an unknown quantity as well, and the thermodynamical equation of the state is then necessary to close the system. For ideal gases, the equation is $p = K\rho^{\gamma}$ for some known real numbers K and γ. Below, we discuss some particular cases of (6.3.51).

IDEAL, OR INVISCID, FLUID HAS NO VISCOSITY. The flow of an ideal fluid is described by the equation

$$\boldsymbol{u}_t + (\boldsymbol{u} \cdot \boldsymbol{\nabla})\boldsymbol{u} = -\frac{\boldsymbol{\nabla} p}{\rho}. \tag{6.3.52}$$

THE FLOW OF A VISCOUS FLUID dissipates energy at the molecular level; the larger the viscosity of the fluid, the higher the dissipation. Without going into the details, we postulate that, for many fluids, this dissipation is described mathematically by the force term $\nu \nabla^2 \boldsymbol{u}$, where ν is a positive number called the *coefficient of viscosity*. The resulting equation is then

$$\boldsymbol{u}_t + (\boldsymbol{u} \cdot \boldsymbol{\nabla})\boldsymbol{u} = \nu \nabla^2 \boldsymbol{u} - \frac{\boldsymbol{\nabla} p}{\rho}. \tag{6.3.53}$$

THE FLOW OF AN INCOMPRESSIBLE FLUID has density $\rho(t, P)$ constant for all time and all points in space; with no loss of generality, we assume that $\rho = 1$. From the equation of continuity we then conclude that $\boldsymbol{\nabla} \cdot \boldsymbol{u} = 0$, which is another equation for $\boldsymbol{u}$. While the density ρ is no longer unknown, the pressure $p = p(t, P)$, sustaining the flow and ensuring the `incompressibility condition`, is another unknown function in the resulting system of equations. For an ideal fluid, the system is

$$\boldsymbol{u}_t + (\boldsymbol{u} \cdot \boldsymbol{\nabla})\boldsymbol{u} = -\operatorname{grad} p, \ \operatorname{div} \boldsymbol{u} = 0. \tag{6.3.54}$$

For a viscous fluid, the system is

$$\boldsymbol{u}_t + (\boldsymbol{u} \cdot \boldsymbol{\nabla})\boldsymbol{u} = \nu \nabla^2 \boldsymbol{u} - \boldsymbol{\nabla} p, \ \boldsymbol{\nabla} \cdot \boldsymbol{u} = 0. \tag{6.3.55}$$

Equations (6.3.52) and (6.3.53), both with $\boldsymbol{\nabla} p = \boldsymbol{0}$, are often referred to as `Burgers equations`, after the Dutch physicist JOHANNES MARTINUS BURGERS (1895–1981), who, in the 1920s, introduced and studied the one-dimensional versions $u_t + uu_x = 0$ and $u_t + uu_x = \nu u_{xx}$. System (6.3.54) is `Euler's equations`; Leonhard Euler published major works on fluid mechanics in the 1750s. System (6.3.55) is one of the most famous in all of mathematics, and is known as the `Navier-Stokes equations`. The French civil engineer CLAUDE LOUIS MARIE HENRI NAVIER (1785–1836) proposed (6.3.55) in 1822 as a model of fluid flow; in the late 1840s and early 1850s, George Gabriel Stokes provided a rigorous derivation.

Note that if the velocity field $\boldsymbol{u}$ of an incompressible fluid has a potential V, that is, $\boldsymbol{u} = \boldsymbol{\nabla} V$, then $\boldsymbol{\nabla} \cdot \boldsymbol{u} = 0$ implies that V is a harmonic function: $\boldsymbol{\nabla}^2 V = 0$. By taking different harmonic functions V, letting $\boldsymbol{u} = \boldsymbol{\nabla} V$, and computing the resulting pressure from (6.3.54) or (6.3.55), one can model various flows. With the help of complex numbers and analytic functions, this approach is especially efficient in two dimensions, see Problem 5.3 on page 433. Problem 7.10 on page 447 presents some other basic techniques used in the study of the Euler and Navier-Stokes equations. For yet another interesting equation related to fluids, see Problem 7.11 on page 448.

EXERCISE 6.3.26.[A] *(a) Assume that* $\boldsymbol{u} = \boldsymbol{\nabla} V$. *Verify that* $2(\boldsymbol{u} \cdot \nabla)\boldsymbol{u} = \boldsymbol{\nabla}\big(\|\boldsymbol{u}\|^2\big)$. *Hint: use the results of Exercise 3.1.19, page 139. (b) Assume that* $\boldsymbol{u} = \boldsymbol{\nabla} V$ *and* $\boldsymbol{u}$ *satisfies (6.3.52) Show that* $V_t + (1/2)\|\boldsymbol{u}\|^2 = F$ *for some time-dependent scalar field* F *and find* F. *(c) Assume that* $\boldsymbol{u} = \boldsymbol{\nabla} V$ *and* $\nabla^2 V = 0$. *(i) Find the pressure* p *if* $\boldsymbol{u}$ *satisfies (6.3.54). (ii) Find the pressure* p *if* $\boldsymbol{u}$ *satisfies (6.3.55).*

The nonlinear term $(\boldsymbol{u} \cdot \boldsymbol{\nabla})\boldsymbol{u}$, present in all four equations (6.3.52)–(6.3.55), makes the mathematical analysis both interesting and hard. This analysis is especially hard in $\mathbb{R}^3$, and many basic problems related to fluid flows in three spatial dimensions remain open. In fact, existence of a classical solution for the Navier-Stokes equations (6.3.55) in $\mathbb{R}^3$ is literally a million-dollar question, being one of the seven `Millennium Problems`, announced by the Clay Mathematics Institute in 2000; for details, see the book *The Millennium Problems* by K. Devlin, 2002. Existence of a classical solution for the Navier-Stokes equations in $\mathbb{R}^2$ is known; Problem 7.10(c), page 447, shows why analysis in two dimensions is easier than in three dimensions.

EXERCISE 6.3.27. *(a)*[C] *Let* $\boldsymbol{u} = u_1(t,x,y,z)\,\hat{\boldsymbol{\imath}} + u_2(t,x,y,z)\,\hat{\boldsymbol{\jmath}} + u_3(t,x,y,z)\,\hat{\boldsymbol{\kappa}}$. *Write (6.3.55) as four scalar equations. Hint: the first equation is* $(u_1)_t + u_1(u_1)_x + u_2(u_1)_y + u_3(u_1)_z = \nu((u_1)_{xx} + (u_1)_{yy} + (u_1)_{zz}) - p_x$.
(b)[A] *Use the result of Exercise 6.2.2 on page 317 to solve the one-dimensional Burgers equation* $u_t + uu_x = u_{xx}$ *with initial condition* $u(0,x) = f(x)$. *Hint: if* $v_t + v_x^2 = v_{xx}$, *then* $u(t,x) = 2v_x(t,x)$.

6.4 Equations of Quantum Mechanics

6.4.1 *Schrodinger's Equation*

Schrödinger's equation is the fundamental equation in quantum mechanics. Unlike all the other partial differential equations we have encountered so far, Schrödinger's equation is not derived but rather postulated. Furthermore, unlike other fundamental equations, such as Newton's Second Law, it is much harder to argue that Schrödinger's equation is a generalization of the experimental facts. However, we will present a motivating example later.

We start by simply writing down one particular case of the equation and carrying out a basic mathematical analysis. Next, we briefly discuss the physical background, and finally, solve the equation for two particular physical models: the quantum oscillator and the hydrogen atom.

Mathematically, `Schrödinger's equation` is usually written as

$$iu_t + \nabla^2 u + Vu = 0, \tag{6.4.1}$$

where $i = \sqrt{-1}$ is the imaginary unit and V is a known function, usually called a *potential*. When $V = 0$, the equation can be solved using the Fourier transform similar to the solution of the heat equation; see page

297. Indeed, consider the **initial value problem**

$$iu_t(t,x) + u_{xx} = 0,\ t > 0,\ x \in \mathbb{R}; u(0,x) = f(x), \tag{6.4.2}$$

and assume that $\int_{-\infty}^{\infty} |f(x)|dx < \infty$. Then direct computations show that

$$u(t,x) = \frac{1}{\sqrt{4\pi i t}} \int_{-\infty}^{\infty} e^{i(x-y)^2/(4t)} f(y)dy. \tag{6.4.3}$$

EXERCISE 6.4.1. *(a)*[C] *Verify (6.4.3). Hint: repeat the computations leading to (6.1.18) on page 297. (b)*[C] *Let $U(t,x)$ be a solution of $U_t(t,x) = U_{xx}(t,x)$ and define $u(t,x) = U(it,x)$. Verify that u is a solution of $iu_t(t,x) + u_{xx}(t,x) = 0$. (c)*[B] *Assume that $\int_{-\infty}^{\infty} |f(x)|^p dx < \infty$ for $p = 1,2$. Show that*

$$\int_{-\infty}^{\infty} |u(t,x)|^2 dx = \int_{-\infty}^{\infty} |f(x)|^2 dx. \tag{6.4.4}$$

Hint: writing $(x-y)^2 = x^2 - 2xy + y^2$, interpret u in (6.4.3) as a Fourier transform of some function, and then use Parseval's identity (5.2.16) on page 267.

Similar to formula (6.1.24) on page 300, formula (6.4.3) can be extended to any number of spatial dimensions. The main difference from the heat equation is that Schrödinger's equation is **time-reversible** and the heat equation is not.

EXERCISE 6.4.2.[B] *Verify that formula (6.4.3) makes sense for both $t > 0$ and $t < 0$* with the same function f, *while formula (6.1.18) on page 297 usually makes sense only for $t > 0$. Hint: $|e^{ix^2/t}| = 1$ for all $t \neq 0$ and $e^{-x^2/t}$ grows very fast for $t < 0$.*

In physics, (6.4.1) is the fundamental equation of quantum mechanics, and is usually written as

$$i\hbar \frac{\partial \psi}{\partial t} = -\frac{\hbar^2}{2m} \nabla^2 \psi + V\psi, \tag{6.4.5}$$

where $\psi = \psi(t, \boldsymbol{r})$ is called a **wave function**, $\hbar = 1.05 \cdot 10^{-34}$ joule · seconds is the **reduced Planck's constant**, and m has the dimension of mass.

EXERCISE 6.4.3. [C] *Let the function $\psi = \psi(t,x)$ be a solution of $i\hbar\psi_t = -(\hbar^2/(2m))\psi_{xx}$. Find real numbers a, b so that the function $u(t,x) = \psi(at, bx)$ is a solution of $iu_t + u_{xx} = 0$.*

To explain the physical significance of all the components in (6.4.5), we review some relevant history and physics. Classical Newtonian mechanics, as embodied in Newton's Laws, especially his Second Law (see (2.1.1) on page 40) serves as an accurate mathematical model of the motions of physical systems in which particles or bodies move at speeds much smaller than the speed of light. For systems in which particles move at speeds near the speed of light, Newtonian mechanics must be replaced by Einstein's mechanics of special relativity, in which Newton's Second Law is replaced by equation (2.4.15) on page 102. We note that (2.4.15) approaches (2.1.1) in the limit as $c \to \infty$. Recall that special relativity holds only in a family of inertial frames moving with constant relative velocities, for example, with respect to the non-rotating frame of the fixed stars. For accelerating frames, for example those which rotate relative to the fixed stars, one must use a third model, the mechanics of general relativity (see (2.4.22) on page 107). This more general model is based on a non-Euclidean geometry. Still, the underlying mathematics of these three models is founded on the continuum of the real numbers. Quantities like energy are allowed to take on a continuum of real values. Furthermore, these three models of mechanics are usually applied to physical systems in which the motions take place on a macroscopic scale, say on the Earth's surface or in the solar system.

The mechanics of systems that operate on an atomic, or smaller, scale is not accurately described by any of these three models of mechanics. In various experiments on atomic structure involving electrons, neutrons and other particles moving on a sub-microscopic scale, it was discovered that physical quantities such as energy occur in discrete amounts, or quanta, rather than taking on the continuum of real number values. A new mathematical model of the mechanics of such sub-atomic particles is required. It is called `quantum mechanics`, in recognition of the discrete nature of key physical properties of systems of sub-atomic particles. However, this discrete nature is only one of the distinguishing characteristics of quantum mechanics.

Another, and more important, distinguishing characteristic is the abandonment of the deterministic mathematics inherent in the differential equations of the other three models and the adoption of a probability-based mathematics. This leads to a physical interpretation of mechanics in which quantities such as position and velocity can only be determined within a probability interval. This agrees with a phenomenon encountered in experiments and enunciated in the Heisenberg Uncertainty Principle formulated in 1927 by the German physicist WERNER KARL HEISENBERG (1901–1976): *It is not possible to determine simultaneously both position $\boldsymbol{x}$ and velocity $\boldsymbol{v}$, of a particle with arbitrarily small errors $\triangle \boldsymbol{x}$ and $\triangle \boldsymbol{v}$.* In terms of the momentum $\boldsymbol{p} = m\boldsymbol{v}$, the uncertainty principle is often stated as the inequality $\|\triangle\boldsymbol{x}\|\|\triangle\boldsymbol{p}\| \geq \hbar$, where $\hbar$ is the reduced Planck's constant. This uncertainty is not the result of the ordinary errors in measurement that occur with probability distributions in the deterministic models of mechanics. In those models, the physical quantities like position and velocity are, in principle, determined exactly, with zero error, by the equations of motion. In quantum mechanics, these quantities cannot ever be determined simultaneously with zero error, even theoretically. They can only be determined within computed probability bounds. The interesting mathematical feature is that these probabilities are specified by a partial differential equation — the Schrodinger equation. Elsewhere in this chapter, we saw how a partial differential equation, a continuum-based concept, is used to construct classical electromagnetic theory, fluid mechanics, and elasticity theory of continuous media. We will now show that a partial differential equation is also the basic mathematical tool for modelling the quantum mechanics of subatomic phenomena.

The discrete character of quantum mechanics originated in the famous 1901 paper by the German physicist MAX KARL ERNST LUDWIG PLANCK (1858–1947) on black-body radiation. In the paper he proposed that the energy radiation at a frequency ν in a heated black body occurs in discrete quantities, which are integer multiples of an energy `quantum` $h\nu$, where h is a physical constant, now called `Planck's constant`. Incidentally, the Latin word *quantus* means "how much." By experiments, the value $h = 6.626 \cdot 10^{-34}$ Joule-seconds has been determined; the reduced Planck's constant $\hbar = h/(2\pi)$. Einstein, in his 1905 paper on the photoelectric effect, proposed that the energy quantum can be viewed as a particle, now called a `photon`, which is emitted as the basic constituent of light. Einstein proposed further that light propagates as bundles of moving discrete photons rather than as continuous travelling waves, the electromagnetic waves implied by Maxwell's equations based on the real continuum; see page 348.

Furthermore, spectroscopic experiments on hydrogen gas that is made to glow in a discharge tube by applying an electric field detected spectral lines only in a discrete light spectrum. In 1913, the Danish physicist NIELS HENRIK DAVID BOHR (1885–1962) proposed a model of the hydrogen atom in which only a discrete, that is, countable, number of energy states are possible, since a transition between the different energy states results in an emission or absorption of a photon with energy $E = h\nu$; see the book *Introductory Quantum Mechanics* by R. Liboff, 2002, for more details.

In 1922, Einstein's photon theory of electromagnetic radiation received another experimental confirmation by the American physicist ARTHUR HOLLY COMPTON (1892–1962), who observed what is now known as the Compton effect: an increase in frequency of X-rays after their scattering by electrons. The Compton effect is consistent with the photon theory rather than the wave theory of X-rays. However, light passing through a double slit in a barrier and impinging on a screen exhibits patterns on the screen that are characteristic of the interference of two waves passing through the slits rather than of two bundles of photon particles. In other words, a number of experiments, when analyzed by methods of classical mechanics combined with the photon particle model, do not yield compatible results. To resolve the conflict between the photon and wave theories, it was suggested that photons, which behave as particles, may also be capable of some wave-like behavior. In fact, in 1923, the French physicist LOUIS-VICTOR PIERRE RAYMOND DE BROGLIE (1892–1987) conjectured that that even material particles in motion are capable of wave-like behavior, with the wave length depending on the particle's momentum $\boldsymbol{p} = m\,\boldsymbol{v}$.

To see how such matter waves might be possible consider the photon as a particle obeying special relativity mechanics. By (2.4.13) on page 102, a particle of relativistic mass m has total energy $\mathcal{E} = mc^2$. A photon has speed $\|\boldsymbol{v}\| = c$ and a finite energy $\mathcal{E} = h\nu$. The relativistic momentum of a particle is $\boldsymbol{p} = m\,\boldsymbol{v}$. Thus, for a photon, we get $\mathcal{E} = \|\boldsymbol{p}\|c = h\nu$ or $\nu = c\|\boldsymbol{p}\|/h$. For every wave, the wave length λ, the frequency ν, and the propagation speed v are connected by the basic relation $\lambda = v/\nu$. Applying this relation to the photon with $v = c$ and $\nu = c\|\boldsymbol{p}\|/h$, we find $\lambda = h/\|\boldsymbol{p}\|$. de Broglie made the bold conjecture that the relation $\lambda = h/\|\boldsymbol{p}\|$ holds more generally for any particle of mass m moving with velocity v and having momentum $\boldsymbol{p} = m\,\boldsymbol{v}$, that is, there is a particle wave length $\lambda_{db} = h/\|\boldsymbol{p}\|$, called the `de Broglie wave Length`, which suggests that there are matter waves of length λ_{db} associated with moving material particles. This equation has been verified by experiments on the diffraction of atomic particles by crystals. In 1925, the American physicists CLINTON JOSEPH DAVISSON (1881–1958) and LESTER HALBERT GERMER (1896–1971) demonstrated that beams of electrons are diffracted like classical light waves when they are scattered by the arrays of atoms in a crystal. Their results support the de Broglie wave length formula and, thereby, the implied existence of matter waves.

What is the nature of the matter waves predicted by de Broglie? Recall that in 1867 Maxwell predicted the existence of electromagnetic waves in which the electric and magnetic fields are the observable varying magnitudes in the waves. These waves are generated by accelerated electric charges, as was first done by Hertz (1880) and as is now routinely done by moving charges in an antenna. In sound waves, the air pressure varies in space and time. What observable physical quantity varies in particle matter waves? The rather unusual answer is that there is no such physical quantity. Instead, the theory of quantum mechanics postulates the existence of a rather abstract complex-valued functions $\psi(\boldsymbol{r}, t)$, called a wave function, which varies with the position $\boldsymbol{r}$ of a moving particle at time t and somehow embodies the state of both the matter and the matter wave of the particle. However, since position and velocity of atomic particles are subject to the uncertainty principle, ψ can only provide a probabilistic estimate of the state at time t. Furthermore, since ψ can have negative and complex values, it cannot serve as a probability function. In 1927, the German-born British physicist MAX BORN (1882–1970) suggested that $|\psi|^2 = \overline{\psi}\psi$ should be proportional to the probability of finding the particle at position $\boldsymbol{r}$ at time t. This turned out to be the correct interpretation of ψ. In quantum mechanics, the motion of a particle under the influence of external forces is governed by its wave function ψ. In the mathematical model of quantum mechanics, it is postulated that ψ is a single-valued and sufficiently smooth function. To qualify as a probability density function, $|\psi|^2$ must satisfy the usual integral relation for a probability density, $\int_G |\psi|^2(\boldsymbol{r}, t)dV = 1$, where G is the region in which the particle moves. It is further postulated that $\psi(\boldsymbol{r}, t)$ satisfies the partial differential equation (6.4.5), first proposed in 1924 by the Austrian physicist ERWIN RUDOLF JOSEF ALEXANDER SCHRÖDINGER (1887-1961). Like Newton's Laws, Schrödinger's equation has been substantiated by many experiments. To give some insight into the equation, we consider the one-dimensional motion of a photon of frequency ν and wave length λ. We assume that the wave function $\psi = \psi(t, x)$ of this photon is a plane wave $\psi(t, x) = Ae^{-2\pi i(\nu t - x/\lambda)}$ (compare this with equation (6.3.38) on page 349). Writing $\nu = E/h$, where E is the energy of the photon, and $\lambda = h/p$, where p is the momentum of the photon, we get $\psi(t, x) = Ae^{-2\pi i(Et-px)/h}$.

EXERCISE 6.4.4.^{C} *Verify that* $\psi_t = (-iE/\hbar)\,\psi$ *and* $\psi_{xx} = -(p^2/\hbar^2)\,\psi$, *where* $\hbar = h/(2\pi)$.

We now forget that the photon is a relativistic particle and write its energy E as the sum of the kinetic and potential energy, $E = (p^2/(2m)) + V$. Multiplying by ψ, we get $E\psi = (p^2/(2m))\psi + V\,\psi$. On the other hand, results of Exercise 6.4.4 suggest that $E\psi = i\hbar\psi_t$ and $p^2\psi = -\hbar^2\psi_{xx}$, that is, $i\psi_t = -(\hbar/(2m))\psi_{xx} + V\psi$, which is the one-dimensional version of equation (6.4.5) on page 357. Note that, even though we started with a function ψ describing a planar wave, the equation satisfied by ψ is not a wave equation.

Beside forgetting that photon is a relativistic particle, in the above computations we also ignored that $|\psi| = |A|$ is a constant, which means that $\int_{-\infty}^{\infty} |\psi(t,x)|^2 dx = \infty$ and $|\psi(t,x)|$ cannot be a probability density function on the real line. Nonetheless, the computations do provide a heuristic connection between waves and Schrödinger's equation.

Similar to classical mechanics, mathematical model of quantum mechanics is impossible without a set of *first principles*, known as axioms or postulates. These postulates of quantum mechanics were introduced in various forms in the late 1920s and early 1930s by the physicist P. A. M. DIRAC and two mathematicians, German HERMANN WEYL (1885–1955) and Hungarian-American JOHN VON NEUMANN (1903–1957). In this section, we consider *non-relativistic quantum mechanics*, that is, quantum systems in which the particles move at speeds much smaller than the speed of light.

To state the postulates, we consider an inner product space $\mathbb{H}$ over the complex numbers; see our previous discussions on page 327. In quantum mechanics, the elements of this space correspond to the wave functions ψ. These elements are denoted by $|\cdot\rangle$. If we think of $|\psi\rangle$ as a column vector with complex components, then $\langle\psi|$ corresponds to the row vector with complex conjugate components, and the inner product of two elements $|\psi_1\rangle$, $|\psi_2\rangle$ is natural to denote by $\langle\psi_1|\psi_2\rangle$. For example, we can have $\langle\psi_1|\psi_2\rangle = \int \overline{\psi}_1\, \psi_2\, dV$. In physics, $\langle\cdot|$ and $|\cdot\rangle$ are known as `Dirac's bra-ket notations`.

POSTULATES OF QUANTUM MECHANICS.

(1) A state of a quantum mechanical system is an element $|\psi\rangle$ of a complex inner product space $\mathbb{H}$, with $\langle\psi|\psi\rangle = 1$; such an element is called a `state vector`. The states satisfy the `superposition principle`: if $|\psi_1\rangle$ and $|\psi_2\rangle$ are two possible state vectors, then so is $|\psi_{12}\rangle = c_1|\psi_1\rangle + c_2|\psi_2\rangle$ for all complex numbers c_1, c_2 such that $\langle\psi_{12}|\psi_{12}\rangle = 1$.

(2) To every *observable* physical quantity characterizing a quantum mechanical system (position, momentum, energy, etc.) there corresponds a linear operator on $\mathbb{H}$ with a complete orthonormal system of eigenvectors in $\mathbb{H}$ and the symmetry property $\langle f|\mathcal{A}|g\rangle = \overline{\langle g|\mathcal{A}|f\rangle}$, where $\overline{}$ denotes complex conjugation and $|f\rangle$, $|g\rangle$ are arbitrary elements of $\mathbb{H}$.

(3) Let $|\psi\rangle$ be the state vector of a quantum mechanical system, and $\mathcal{A}$,

the operator corresponding to some observable physical quantity A. Then the observable values of A are the eigenvalues a_k of $\mathcal{A}$, and the probability of observing $A = a_k$ is equal to $|\langle\alpha_k|\psi\rangle|^2$, where α_k is the normalized eigenvector of $\mathcal{A}$ corresponding to a_k: $\mathcal{A}|\alpha_k\rangle = a_k|\alpha_k\rangle$, and $\langle\alpha_k|\alpha_k\rangle = 1$. In other words, the act of measurement randomly forces the system into a state having a state vector equal to one of the eigenvectors of $\mathcal{A}$; we call such a state a **detectable state**, an **observable state**, or an **eigenstate**.
(4) The time evolution of every state vector of a quantum mechanical system is described by **Schrödinger's equation**

$$i\hbar \frac{\partial}{\partial t}|\psi\rangle = \mathcal{H}|\psi\rangle, \tag{6.4.6}$$

where $i = \sqrt{-1}$ is the imaginary unit, $\hbar = h/(2\pi)$ is reduced Planck's constant, and $\mathcal{H}$ is a *symmetric* operator, called the **Hamiltonian** of the system; see page 328 for the definition of symmetric operator.

EXERCISE 6.4.5. ^{C} *Identify the space $\mathbb{H}$ and the operator $\mathcal{H}$ for equation (6.4.5) on page 357.*

Postulate 1 means that a state $|\psi\rangle$ of a quantum mechanical system is a unit vector in $\mathbb{H}$. This is consistent with the interpretation of $|\psi|^2 = \langle\psi|\psi\rangle$ as a probability density function. Without going into the details, we mention that Postulates (2) and (3) allow the possibility of *uncountably many* eigenfunctions of $\mathcal{A}$ and a *continuous spectrum.* In all our examples, we will have operators with a discrete (in fact, point) spectrum and countably many eigenvalues and eigenfunctions, all with finite multiplicity. While we can only observe the system in one of its *eigenstates*, that is, the states corresponding to the eigenfunctions of the observation operator, the superposition principle suggests that the system can also be in a combination of these eigenstates states. The inevitable philosophical difficulty is resolved by concluding that the *act of observing randomly forces the system into one of its eigenstates* Of course, the reality is much more complicated and superposition of states is a phenomenon without a clear physical interpretation; *quantum mechanics does not explain superposition but rather employs it as a useful mathematical concept.*

Superposition of states suggests very effective ways to represent and manipulate information: if a *classical* n-bit computer can only be in one of the 2^n states at a time, a corresponding *quantum* computer can be simultaneously in all of these states (you might have to think about it for a moment). As a result, a **quantum computer** can achieve an *exponen-*

tial increase in speed over a classical computer on certain computations. The problem with the implementation of such a computer is the difficulty in maintaining the quantum superposition of states, because the slightest disturbance from the outside, such as a stray photon, is interpreted as a measurement and sets the system into one of its eigenstates; this breakdown of quantum state structure due to outside influence is often referred to as **decoherence**. We take a closer look at quantum computing in Section 6.4.3 below.

If the Hamiltonian of the system does not depend on time, then we can look for a solution of (6.4.6) in the form

$$|\psi\rangle = e^{-iEt/\hbar}|\Psi\rangle, \tag{6.4.7}$$

where E is a real number and $|\Psi\rangle$ is a *stationary state*, that is, $\partial|\Psi\rangle/\partial t = 0$. The following exercise shows that each stationary state is an eigenvector of the operator $\mathcal{H}$, corresponding to the eigenvalue E; by Theorem 6.2.1 on page 329, we know that all eigenvalues of a symmetric operator are real. Accordingly, the first step in the study of a quantum system is to find the eigenvalues of the system Hamiltonian, as these eigenvalues correspond to the possible energy levels, or energy spectrum, of the system.

EXERCISE 6.4.6. ^{C} *(a) Verify that if $|\psi\rangle$ has the form (6.4.7) and satisfies (6.4.6), then $|\Psi\rangle$ satisfies the* **stationary** *or* **time-independent Schrödinger equation**

$$\mathcal{H}|\Psi\rangle = E|\Psi\rangle. \tag{6.4.8}$$

(b) Conversely, verify that (6.4.8) and (6.4.7) imply (6.4.6). The other postulates of quantum mechanics then suggest that equation (6.4.6) is reasonable to postulate as well.

In many quantum mechanical systems, the space $\mathbb{H}$ is a collection of suitable scalar functions on $\mathbb{R}^n$, $n = 1, 2, 3$ such that $\langle f|g\rangle = \int_{\mathbb{R}^n} f\,\overline{g}\,d\mathrm{m}$. From Theorem 2.3.1 on page 94 we know that the Hamiltonian of a point mass is the total (kinetic plus potential) energy of the point, $\mathcal{E}_K + U$. We also know that for a non-relativistic particle, the kinetic energy $\mathcal{E}_K = \|\boldsymbol{p}\|^2/(2m)$, where $\boldsymbol{p}$ is the momentum. Both momentum and kinetic energy are observable physical quantities. It is postulated in quantum mechanics that the operator corresponding to momentum is $-i\hbar\boldsymbol{\nabla}$, and $-(\hbar/(2m))\boldsymbol{\nabla}^2$ is the operator corresponding to the kinetic energy; our heuristic discussion of a photon on page 361 suggests that these postulates are reasonable. Then

$$\mathcal{H}|f\rangle = -\frac{\hbar}{2m}\nabla^2 f + Uf, \tag{6.4.9}$$

and (6.4.6) becomes (6.4.5) on page 357. Accordingly, much of the mathematical analysis of the Schrödinger equation is about finding conditions on the potential U so that the operator $\mathcal{A}[u] = -\nabla^2 u + U\,u$ has a discrete spectrum. Even in one space dimension, many open problems remain.

EXERCISE 6.4.7.[B] *Let $\mathbb{H}$ be the collection of complex valued functions defined on the interval $[0,1]$ so that every function from $\mathbb{H}$ is continuously differentiable on the interval and is equal to zero at the end points. Set $(f,g) = \int_0^1 f(x)\overline{g(x)}dx$. Verify that the operator $\mathcal{A}[f] = -if'(x)$ is symmetric on $\mathbb{H}$. Hint: integrate by parts and remember that we are working with complex numbers.*

We now consider two physical systems for which the Schrödinger equation can be solved explicitly: the linear quantum oscillator and the hydrogen atom. The explicit solutions are important for experimental validation of the quantum theory. In our analysis, we will encounter some of the equations from page 237. We leave it to the reader to fill in the details of most computations.

LINEAR QUANTUM OSCILLATOR. In classical mechanics, we encountered a **harmonic oscillator** in the study of the simple rigid pendulum (see (2.1.10) on page 42). Without external forces and no energy losses, such an oscillator is described by the second-order ordinary differential equation

$$\ddot{x}(t) + \omega^2 x(t) = 0, \tag{6.4.10}$$

where $x = x(t)$ represents the position of the particle. Another classical example is the oscillations of a point mass m attached to massless spring. If $x = x(t)$ is the displacement of m from the equilibrium position of the spring, then Newton's Second Law $F = m\ddot{x}(t)$ with `Hooke's Law` $F = -Kx$ yields (6.4.10) with $\omega = K/m$. Note also that $V = Kx^2/2 = m\omega^2x^2/2$ is the potential energy of the spring.

To motivate the quantum-mechanical analog of a harmonic oscillator, we note that (6.4.10) implies $d(\dot{x}^2(t) + \omega^2x^2(t))/dt = 0$. After integrating this equality in time and multiplying by $m/2$, we conclude that $H = (m\dot{x}^2/2) + (m\omega^2x^2)/2$ does not depend on time. With $p = m\dot{x}$, $(m\dot{x}^2/2) = p^2/(2m)$ is the kinetic energy, and $V = (m\omega^2x^2)/2$ is the potential energy. Therefore, H is the total energy of the oscillator.

For a quantum oscillator, the total energy H becomes a time-independent Hamiltonian $\mathcal{H}$, and, according to (6.4.9), the operator $\mathcal{H}$ should be defined by

$$\mathcal{H}|f\rangle = -\frac{\hbar^2}{2m}f''(x) + \frac{m\omega^2 x^2}{2}f(x),\ x \in \mathbb{R},$$

where, similar to the classical case, x has the dimension of space, m, mass, and ω, frequency (inverse time). The corresponding eigenvalue problem (6.4.8) becomes

$$-\frac{\hbar^2}{2m}\Psi''(x) + \frac{m\omega^2 x^2}{2}\Psi(x) = E\,\Psi(x). \tag{6.4.11}$$

Note that $a = \sqrt{\hbar/(m\omega)}$ has the dimension of the space and is the characteristic scale of the problem. Similarly, $\hbar\omega/2$ is the characteristic energy. These quantities appear naturally as we write the original equation using a dimensionless independent variable $z = x/a$. More precisely, a series of substitutions transforms (6.4.11) to

$$u''(z) - 2zu'(z) + (\lambda - 1)\,u(z) = 0, \tag{6.4.12}$$

where z is dimensionless (we can allow z to be complex, if we want), and

$$E = \frac{\hbar\omega}{2}\lambda,\ \Psi(x) = e^{-x^2/(2a^2)}\,u(x/a),\ a = \sqrt{\frac{\hbar}{m\omega}}. \tag{6.4.13}$$

EXERCISE 6.4.8.[C] *(a) Verify the transformation of (6.4.11) to (6.4.12). Hint: go step by step. First set* $\Psi(x) = v(x/a)$ *and verify that* $v''(z) - z^2v(z) + \lambda v(z) = 0$. *Then set* $v(z) = e^{-z^2/2}u(z)$. *(b) Verify that if* $u(z) = \sum_{k=0}^{\infty} u_k z^k$, *then*

$$u_{k+2} = \frac{2k - (\lambda - 1)}{(k+1)(k+2)}\,u_k. \tag{6.4.14}$$

Notice that (6.4.12) is a particular case of equation (4.4.45) on page 237. By Theorem 4.4.3 on page 233, we know that all solutions of (6.4.12) are analytic functions of z. Still, because of the connection between u and the wave function Ψ, we can only use those solutions of (6.4.12) that satisfy

$$\int_{-\infty}^{\infty} e^{-s^2/(2a^2)}|u(s/a)|^2 ds < \infty. \tag{6.4.15}$$

EXERCISE 6.4.9.[A] *(a) Verify that a solution* $u = u(z)$ *of (6.4.12) satisfies (6.4.15) if and only if* $u = u(z)$ *is a polynomial of degree* N, *which, in particular, implies that*

$$\lambda = 2N + 1, \quad N = 0, 1, 2, \ldots. \tag{6.4.16}$$

Hint: by (6.4.14), if $\lambda \neq 2N+1$ *so that* u *is not a polynomial, then, for large* k, $u_{k+2} \sim 2u_k/k$.

By combining (6.4.16) and (6.4.13), we conclude that the admissible energies of the linear quantum oscillator are $E_N = \hbar\omega(N + 1/2)$. In particular, the lowest energy level is $E_0 = \hbar\omega/2$. The corresponding *normalized* polynomial solutions u_N of (6.4.13) are

$$u_N(z) = \frac{(-1)^N}{\sqrt{2^N N! \sqrt{\pi}}}\, e^{z^2} \frac{d^N}{dz^N}\, e^{-z^2},$$

and are known as Hermite polynomials. The normalizing factor is chosen so that $u_N(z) = a_N z^N + \ldots$, with $a_N > 0$, and the corresponding wave function $\Psi_N(x)$ from (6.4.13) satisfies $\int_{-\infty}^{\infty} |\Psi_N(x)|^2 dx = 1$. The reader who likes detailed computations can verify these properties of Ψ_N.

According to `Bohr's correspondence principle`, a result obtained in quantum mechanics must converge to its classical counterpart in the limit as $\hbar \to 0$. For the quantum oscillator, this correspondence can be established rigorously; see the book *Introduction to Quantum Mechanics* by R. Liboff, 2002.

THE HYDROGEN ATOM. In this system we have one electron with charge $-e < 0$ moving around one proton with charge $e > 0$. Since the proton is more that 1800 times heavier than the electron, we will assume that the proton is not moving; for a more realistic and general setting, see Problem 7.8 on page 446. We also consider only the electrostatic interaction of the particles, so that the potential energy of the electron in the electric field produced by the proton is $U = -e^2/(4\pi\varepsilon_0 r)$, where r is the distance between the particles. Finally, we assume that the speed of the electron is much smaller than the speed of light so that non-relativistic mechanics can be applied; we will verify this assumption later; see (6.4.33), page 373.

Then the Hamiltonian operator is

$$\mathcal{H}|f\rangle = -\frac{\hbar^2}{2m}\nabla^2 f - \frac{e^2}{4\pi\varepsilon_0 r} f.$$

Note that the space now is three-dimensional. The corresponding eigenvalue problem (6.4.8) becomes

$$\frac{\hbar^2}{2m}\nabla^2\Psi + \frac{e^2}{4\pi\varepsilon_0 r}\Psi = -E\,\Psi. \tag{6.4.17}$$

Similar to the linear oscillator, equation (6.4.17) contains a characteristic scale a_0 and energy E_0 of the problem, which appear when we transform (6.4.17) into a mathematical form with dimensionless independent variables.

EXERCISE 6.4.10.[C] *Define the quantities*

$$a_0 = \frac{4\pi\varepsilon_0\hbar^2}{me^2},\quad E_0 = -\frac{\hbar^2}{2ma_0^2} = -\frac{me^4}{2\hbar^2(4\pi\varepsilon_0)^2}; \tag{6.4.18}$$

a_0 is known as the `Bohr radius` *of the hydrogen atom in the ground state, and E_0 is the energy of the electron in this ground state. The numerical values of the physical constants are as follows: $\hbar = 1.05 \cdot 10^{-34}$ J·s (joule · second); $m = 0.91 \cdot 10^{-30}$ kg, mass of the electron; $e = 1.6 \cdot 10^{-19}$ C (coulomb), the charge of the electron; $\varepsilon_0 = 8.85 \cdot 10^{-12}$ $C^2/(J{\cdot}m)$, electrical permittivity of free space. Using these values, verify that a_0 has the dimension of length, E_0 has the dimension of energy, and $a_0 = 0.53 \cdot 10^{-10}$ m, $E_0 = -13.6$ eV, where one* `electron-volt` *(eV) is $1.60 \cdot 10^{-19}$ joules. Hint: one joule is one kg · m^2/s^2.*

Given the spherical symmetry of the system, we introduce the spherical coordinates (r, θ, φ) so that $x = r\cos\theta\sin\varphi$, $y = r\sin\theta\sin\varphi$, $z = r\cos\varphi$. *Note that in the physics literature the role of the angles θ and φ is often switched.* Then $\Psi = \Psi(r, \theta, \varphi)$ and (6.4.17) becomes

$$\begin{aligned}\frac{\hbar^2}{2m}\left(\Psi_{rr} + \frac{2}{r}\Psi_r + \frac{1}{r^2\sin^2\varphi}\Psi_{\theta\theta} + \frac{\cos\varphi}{r^2\sin\varphi}\Psi_\varphi + \frac{1}{r^2}\Psi_{\varphi\varphi}\right)&\\ + \frac{e^2}{4\pi\varepsilon_0 r}\Psi + E\,\Psi = 0;&\end{aligned} \tag{6.4.19}$$

see formula (3.1.45) on page 147. We solve (6.4.19) by `separation of variables`.

EXERCISE 6.4.11.[C] *Verify that setting $\Psi(r, \theta, \varphi) = F(r)\Theta(\theta)\Phi(\varphi)$ and mul-*

tiplying through by $2mr^2/\hbar^2$ transforms (6.4.19) into

$$\frac{r^2F''(r)+2rF'(r)}{F}+\frac{\Theta''(\theta)}{\sin^2\varphi\,\Theta(\theta)}+\frac{\Phi''(\varphi)+\cot\varphi\,\Phi'(\varphi)}{\Phi(\varphi)} + \frac{2r}{a_0}-\frac{E}{E_0}\frac{r^2}{a_0^2}=0. \tag{6.4.20}$$

where a_0, E_0 are from (6.4.18).

By the usual separation argument, it follows from (6.4.20) that

$$\frac{r^2F''(r)+2rF'(r)}{F}+\frac{2r}{a_0}-\frac{E}{E_0}\frac{r^2}{a_0^2}=\beta, \tag{6.4.21}$$

$$\frac{\Theta''(\theta)}{\sin^2\varphi\,\Theta(\theta)}+\frac{\Phi''(\varphi)+\cot\varphi\,\Phi'(\varphi)}{\Phi(\varphi)}=-\beta, \tag{6.4.22}$$

where β is a constant. We start with equation (6.4.22).

EXERCISE 6.4.12.[C] *Argue that, for (6.4.22) to hold, it is necessary to have $\Theta''(\theta)/\Theta(\theta)=\alpha$ for some number α (real or complex).*

To proceed, we use some physical considerations. The spherical symmetry of the problem suggests that the wave function Ψ must be rotation-invariant, that is, $\Psi(r,\theta,\varphi)=\Psi(r,\theta+2\pi,\varphi+2\pi)$. Then $\Theta(\theta)=\Theta(\theta+2\pi)$, and therefore $\Theta''(\theta)=-n^2\Theta(\theta)$ for some integer n (positive, negative, or zero; if $n=0$, then we take $\Theta(\theta)$ equal to a constant). As a result,

$$\Theta(\theta)=c_1\cos n\theta+c_2\sin n\theta,\ c_1,c_2\in\mathbb{R},$$

and, with no loss of generality, we can take $c_1^2+c_2^2=1$. In fact, it is actually more convenient to write Θ as a complex exponential

$$\Theta(\theta)=e^{i\,n\theta}. \tag{6.4.23}$$

Coming back to (6.4.22), we find

$$\Phi''(\varphi)+\cot\varphi\,\Phi'(\varphi)=\left(\frac{n^2}{\sin^2\varphi}-\beta\right)\Phi(\varphi). \tag{6.4.24}$$

EXERCISE 6.4.13. *(a)[B] Let $v=v(z)$ be a function such that $v(\cos\varphi)=\Phi(\varphi)$. Verify that*

$$(1-z^2)v''(z)-2z\,v'(z)=\left(\frac{n^2}{1-z^2}-\beta\right)v(z). \tag{6.4.25}$$

(b)^B *Let* $w = w(z)$ *be a function such that* $v(z) = (1-z^2)^{n/2}w(z)$. *Verify that*

$$(1-z^2)w''(z) - 2(n+1)z\,w'(z) + (\beta - n(n+1))w(z) = 0. \qquad (6.4.26)$$

(c)^A *Note that (6.4.26) is a particular case of equation (4.4.43) on page 237. Using the power series method on page 233, verify that* **Legendre's differential equation** *(6.4.26) has a solution without a singularity at* $z_0 = \pm 1$ *if and only if*

$$\beta = \ell(\ell+1) \text{ for some non-negative integer } \ell \geq |n|,$$

and in that case $w = w(z)$ *is a polynomial, called a (generalized)* **Legendre polynomial** *and denoted by* $P_{\ell,n}(z)$. *For* $n = 0$, $P_{\ell,0}$ *is usually denoted by* P_ℓ *and is called the (usual) Legenedre polynomial. (d)*^{A} *Verify that, for* $n > 0$, $P_{\ell,n}(z) = P_{\ell,0}^{(n)}(z)$, *the n-th derivative of* $P_{\ell,0}$, *and we can take* $P_{\ell,0}(z) = d^\ell((1-z^2)^\ell)/dz^\ell$. *For more about the Legendre polynomials* $P_{\ell,0}$ *see Problem 4.4 on page 430.*

Once again, the physical content of the problem suggests that the wave function Ψ should not have any singularities away from $r = 0$, because otherwise the spherical symmetry of the problem would imply too many singularities. Therefore, the function $\Phi(\varphi)$ should not have any singularities at all. The reader who followed all three parts of the last exercise can now understand that, to avoid the singularities of Φ when $\varphi = 0, \pi$, we must take $\beta = \ell(\ell+1)$ in (6.4.22), and then

$$\Phi(\varphi) = \sin^n\varphi\, P_{\ell,n}(\cos\varphi). \qquad (6.4.27)$$

Finally, we consider equation (6.4.21). Once again, several substitutions are in order, and we can allow the dimensionless variable $z = r/a_0$ to be complex. We also define

$$\lambda = \sqrt{\frac{E}{E_0}},$$

where $E_0 < 0$ is from (6.4.18); since the electron is trapped near the proton and cannot fly away, we should expect the admissible energy values E to be negative.

EXERCISE 6.4.14.^{A} *(a) Let* $F(r) = (r/a_0)^\ell e^{-\lambda r/a_0} u(2\lambda r/a_0)$ *where* a_0, E_0 *are from (6.4.18) and* $\lambda = \sqrt{E/E_0}$. *Keeping in mind that* $\beta = \ell(\ell+1)$, *verify that the function $u = u(z)$ satisfies*

$$z\,u''(z) + (2\ell + 2 - z)u'(z) + ((1/\lambda) - (\ell + 1))u(z) = 0. \tag{6.4.28}$$

Hint: a multi-step procedure can simplify computations. For example, $F(r) = f(r/b)$, $f(z) = z^{\ell}V(z)$, $V(z) = e^{-\lambda z}W(z)$, $W(z) = u(2\lambda z)$. In particular, you should get $zV'' + 2(\ell+1)V' + (2 - \lambda^2 z)V = 0$ and $zW'' + (2(\ell+1) - 2\lambda z)W' + (2 - 2(\ell+1)\lambda)W = 0$.
(b) Equation (6.4.28) is a particular case of (4.4.47). Verify that (6.4.28) has a solution satisfying $\int_0^{\infty} |u(2\lambda s)|^2 e^{-2\lambda s} ds < \infty$ if and only if $1/\lambda = N$ for some positive integer number N. Verify that the corresponding solution u is a polynomial and can be taken to be $L_{N+\ell}^{(2\ell+1)}(z)$, see Exercise 4.4.23, page 238. Hint: if $u(z) = \sum_{k=0}^{\infty} u_k z_k$ and $1/\lambda$ is not a positive integer, then, with $A = (1/\lambda) - (\ell + 1)$,

$$u_{k+1} = \frac{k - A}{(k+1)(k + 2\ell + 2)} u_k \sim \frac{u_k}{k+1},\ k \to \infty,$$

and so $u(s) \sim e^s$. Note that $L_{N+\ell}^{(2\ell+1)}(z)$ is indeed the $(2\ell+1)$-st derivative of $L_{N+\ell}$.

Keeping in mind that the wave function must satisfy

$$\int_0^{\infty}\int_0^{\pi}\int_0^{2\pi} |\Psi(r,\theta,\varphi)|^2 r^2 dr\, d\theta\, \sin\varphi\, d\varphi < \infty,$$

we conclude that the function F must satisfy $\int_0^{\infty} r^2|F(r)|^2 dr < \infty$. The reader who followed both parts of the last exercise now understands that, to satisfy this integrability condition, we must take $\lambda = 1/N$ for some positive integer N, and with that, we actually complete the solution of the problem! Indeed, with $\lambda^2 = E/E_0$ and E_0 defined in (6.4.18), the admissible energy levels are

$$E_N = \frac{E_0}{N^2} = -\frac{me^4}{2\hbar^2(4\pi\varepsilon_0)^2}\frac{1}{N^2}. \tag{6.4.29}$$

In general, there are several `wave functions` $\Psi = \Psi(r,\theta,\varphi)$, corresponding to the energy level N:

$$\Psi(r,\theta,\varphi) = C_{N,\ell,n} e^{i\,n\theta} \sin^n\varphi\, P_{\ell,n}(\cos\varphi) \left(\frac{r}{b}\right)^{\ell} e^{-r/(a_0 N)} L_{N+\ell}^{(2\ell+1)}\left(\frac{2r}{a_0 N}\right), \tag{6.4.30}$$

where $i = \sqrt{-1}$, and, for fixed positive integer N, we have $\ell = 0, 1, \ldots, N-1$, $n = 0, \pm 1, \ldots, \pm\ell$; the numbers $C_{N,\ell,n}$ are chosen to normalize the wave function in a certain way, for example, so that

$$\int_0^\infty \int_0^{2\pi} \int_0^\pi |\Psi(t,\theta,\varphi)|^2 \sin\varphi \, d\varphi \, d\theta \, r^2 dr = 1. \tag{6.4.31}$$

The state of the hydrogen atom is therefore determined by three numbers: the **principal quantum number** N, the **azimuthal number** or **orbital number** ℓ, and the **magnetic number** n. In fact, these numbers are used to describe the state of all atoms, making quantum mechanics and Schrödinger's equation a fundamental part of atomic physics and physical chemistry. The azimuthal number determines the shape of the electron orbit and is often denoted by a letter, with s corresponding to 0, p, to 1, d, to 2, f to 3, etc. For a multi-electron atom, the notation $5s^2p^1$ means that, at the energy level $N = 5$, there are three electrons, two with $\ell = 0$ and one, with $\ell = 1$. The magnetic number determines the orientation of the orbit in space.

By the uncertainty principle, the location of the moving electron can never be known exactly, and, strictly speaking, there is no such thing as the radius of the atom. **Bohr's model** appeared in 1913, well before Schrödinger's equation and the uncertainty principle, and was much more simplified. According to Bohr's model, the electron stays at a fixed distance $a_N = N^2 a_0$ from the nucleus to achieve the energy level E_N; Bohr was able to deduce the formula (6.4.29) for the energy by analyzing the experimental data for the hydrogen spectrum. The value a_N, known as the N-th Bohr radius or the radius of the N-th **Bohr orbit**, provides a general measure of the distance between the electron and the proton at the energy level N. The corresponding wave function can be rather complicated. For more about the hydrogen and other atoms, see the book *Concepts of Modern Physics* by A. Beiser, 2002, for more details.

EXERCISE 6.4.15. *(a)^C By combining (6.4.23), (6.4.27), and the results of Exercise (6.4.14), verify the representation (6.4.30) of the wave function. (b)^A Given the value of $N \geq 1$, how many different states are there? (c)^B Verify that the* **ground state**, *corresponding to $n = \ell = 0$, $N = 1$, and satisfying (6.4.31), has the wave function*

$$\Psi_0(r,\theta,\varphi) = \frac{1}{\sqrt{\pi}} a_0^{-3/2} e^{-r/a_0}. \tag{6.4.32}$$

(d)^A Find the expected value of r in the ground state, that is, $\int_0^\infty \int_0^{2\pi} \int_0^\pi r |\Psi(t,\theta,\varphi)|^2 \sin\varphi \, d\varphi \, d\theta \, r^2 dr$. Hint: it is $3a_0/2$.

EXERCISE 6.4.16.[A] *(a) Energy E_0 in the ground state corresponds to the sum of the potential energy $-e^2/(4\pi\varepsilon_0 a_0)$ and the kinetic energy $mv_0^2/2$ of the electron. Verify that*

$$\frac{v_0^2}{c^2} = \frac{e^4}{(4\pi\varepsilon_0 c\hbar)^2} \approx \frac{1}{137^2}. \tag{6.4.33}$$

The dimensionless number $\alpha = e^2/(4\pi\varepsilon_0 c\hbar) \approx 1/137$ is called the `fine structure constant`. *(b) Verify that the speed v_{N-1} of the electron on the N-th Bohr orbit is $v_{N-1} = v_0/N$, $N = 1, 2, \ldots$ and conclude that the non-relativistic treatment of the problem is justified. (c) Relation (6.4.33), can be derived without solving the Schrödinger equation. Indeed, let a be the distance from the electron to the nucleus and p, the momentum of the electron. By the* `uncertainty principle`, *$p\,a \sim \hbar$. Then write $p = \hbar/a$ and $E(a) = (p^2/(2m)) - e^2/(4\pi\varepsilon a)$ to deduce that $\min_{a>0} E(a) = E_0 = E(a_0)$, from which (6.4.33) follows. Since the $\sim$ symbol in the uncertainty principle should be interpreted up to a constant factor (for example, taking $pa = 2\pi\hbar$ is just as reasonable), this approach leads only to an estimate of the type (6.4.33), and this* a priori *estimate is enough to conclude that non-relativistic theory can be applied.*

6.4.2 *Dirac's Equation of Relativistic Quantum Mechanics*

In this section we introduce Dirac's equation for an electron moving freely in space at speeds comparable with the speed of light. The presentation is intended as a very basic introduction to the subject, barely scratching the surface. To make the reading worthwhile, and to leave the reader with some sense of accomplishment, we compensate for this lack of breadth and depth by letting the reader carry out many of the computations.

We start by noting that Schrödinger's equation is first-order in time and second-order in space. As a result, it cannot be invariant under the Lorentz transformation and cannot account for relativistic effects.

EXERCISE 6.4.17.[B] *Verify that Schrödinger's equation (6.4.5) is not invariant under the Lorentz transformation. Hint: consider the one-dimensional case.*

Next, we outline the main idea of the derivation of Dirac's equation for a free relativistic quantum mechanics particle in three-dimensional Euclidean space. First of all, we re-name the usual cartesian coordinates (x, y, z) so that $x = x_1$, $y = x_2$, $z = x_3$. The goal is to obtain an equation that is first-order in time and looks like $i\hbar(\partial\psi/\partial t) = \mathcal{H}\psi$, where $\mathcal{H}$ is now a *relativistic Hamiltonian operator* of the particle. Recall (page 103) that the total energy $\mathcal{E}$ of a free relativistic particle with rest mass m_0 satisfies $\mathcal{E}^2 = m_0^2c^4 + c^2\|\boldsymbol{p}\|^2$, where $\boldsymbol{p}$ is the relativistic momentum. We still want $\mathcal{H}$ to represent the total energy and we want to continue using the operator $-i\hbar\partial/\partial x_k$ in connection with the component p_k of the momentum. This would lead to a somewhat strange expression for $\mathcal{H}$:

$$\mathcal{H} = \sqrt{m_0^2c^4 - \hbar^2c^2\sum_{k=1}^{3}\frac{\partial^2}{\partial x_k^2}}.$$

Even if we could make sense out of the square root, the result would still not provide equal treatment of the time and space variables.

Dirac's idea was to generalize the formula for the relativistic energy to the form $\mathcal{E} = m_0c^2\,\alpha_0 + c\boldsymbol{p}\cdot\boldsymbol{\alpha}$, where α_0 is a scalar and $\boldsymbol{\alpha}$ is a vector to be determined to obtain Lorentz invariance. This leads to the relativistic Hamiltonian operator in the form

$$\mathcal{H} = m_0c^2\alpha_0 + i\hbar c\sum_{k=1}^{3}\alpha_k\frac{\partial}{\partial x_k}$$

and a choice of α_0, α_k so that two applications of $\mathcal{H}$ result in

$$\mathcal{H}^2 = m_0^2c^4 - \hbar^2c^2\sum_{k=1}^{3}\frac{\partial^2}{\partial x_k^2}.$$

EXERCISE 6.4.18. *(a)*[C] *Verify that α_0, α_k must satisfy*

$$\alpha_0^2 = \alpha_k^2 = 1,\ \ \alpha_k\alpha_0 + \alpha_0\alpha_k = 0,\ \ k = 1, 2, 3;\ \ \alpha_k\alpha_n + \alpha_n\alpha_k = 0,\ \ n \neq k; \tag{6.4.34}$$

the reason for writing the conditions in this form will become clear soon. Hint: replace $\partial/\partial x_k$ with real numbers b_k. (b)[B] *Verify that there are no real or complex numbers α_0, α_k that satisfy (6.4.34). Hint: for numbers, multiplication is commutative: $\alpha_0\alpha_1 = \alpha_1\alpha_0$, etc.*

With no numbers α_0, α_k to produce the desired Hamiltonian, the only hope is to try higher dimensions and make α_0, α_k square matrices. Then conditions (6.4.34) become

$$\alpha_0^2 = \alpha_k^2 = I_{[N]},\ \ \alpha_k\alpha_0 + \alpha_0\alpha_k = 0_{[N]},\ \ \alpha_k\alpha_n + \alpha_n\alpha_k = 0_{[N]}, \tag{6.4.35}$$

for $n, k = 1, 2, 3$ and $n \neq k$, where $I_{[N]}$ is the $N \times N$ identity matrix and $0_{[N]}$ is the $N \times N$ zero matrix. The elements of the matrices can be complex numbers.

EXERCISE 6.4.19. *(a)*[C] *Verify that the matrices α_0, α_k must all be Hermitian ($\alpha_0^T = \overline{\alpha_0}$, etc.) Hint: the Hamiltonian must be a Hermitian operator.* *(b)*[A] *Verify that conditions (6.4.35) cannot hold if $N < 4$.*

If $N = 4$, then a possible choice of the matrices α_0, α_k, $k = 1, 2, 3$, is

$$\alpha_0 = \begin{pmatrix} 1 & 0 & 0 & 0 \\ 0 & 1 & 0 & 0 \\ 0 & 0 & -1 & 0 \\ 0 & 0 & 0 & -1 \end{pmatrix}, \ \alpha_1 = \begin{pmatrix} 0 & 0 & 0 & 1 \\ 0 & 0 & 1 & 0 \\ 0 & 1 & 0 & 0 \\ 1 & 0 & 0 & 0 \end{pmatrix}, \ \alpha_2 = \begin{pmatrix} 0 & 0 & 0 & -i \\ 0 & 0 & i & 0 \\ 0 & -i & 0 & 0 \\ i & 0 & 0 & 0 \end{pmatrix}, \ \alpha_3 = \begin{pmatrix} 0 & 0 & 1 & 0 \\ 0 & 0 & 0 & -1 \\ 1 & 0 & 0 & 0 \\ 0 & -1 & 0 & 0 \end{pmatrix}. \tag{6.4.36}$$

EXERCISE 6.4.20.[C] *(a) Verify that the matrices α_0 and α_k can be written in the block form:*

$$\alpha_0 = \begin{pmatrix} I_{[2]} & 0_{[2]} \\ 0_{[2]} & -I_{[2]} \end{pmatrix}, \ \alpha_k = \begin{pmatrix} 0_{[2]} & \sigma_k \\ \sigma_k & 0_{[2]} \end{pmatrix}, \ k = 1, 2, 3, \tag{6.4.37}$$

where $I_{[2]}$ is the 2×2 identity matrix, $0_{[2]}$ is the 2×2 zero matrix, and

$$\sigma_1 = \begin{pmatrix} 0 & 1 \\ 1 & 0 \end{pmatrix}, \ \sigma_2 = \begin{pmatrix} 0 & -i \\ i & 0 \end{pmatrix}, \ \sigma_3 = \begin{pmatrix} 1 & 0 \\ 0 & -1 \end{pmatrix}. \tag{6.4.38}$$

(b) Verify that the matrices σ_k have the following properties:

$$\begin{aligned} &\det(\sigma_k) = -1, \ \sigma_k^2 = I_{[2]}, \ k = 1, 2, 3; \ \sigma_k \sigma_n = -\sigma_n \sigma_k, \ k \neq n; \\ &\sigma_1 \sigma_2 = i\sigma_3, \ \sigma_3 \sigma_1 = i\sigma_2, \ \sigma_2 \sigma_3 = i\sigma_1. \end{aligned} \tag{6.4.39}$$

(c) Conclude that conditions (6.4.35) hold with $N = 4$.
(d) Verify that, for any real numbers b_0, b_1, b_2, b_3,

$$\left(b_0^2 + \sum_{k=1}^{3} b_k^2 \right) I_{[4]} = \left(b_0\, \alpha_0 + \sum_{k=1}^{3} b_k\, \alpha_k \right)^2. \tag{6.4.40}$$

The reader who followed the preceding discussion will now see that `Dirac's equation` is written as follows:

$$i\hbar \frac{\partial \psi(t, \boldsymbol{x})}{\partial t} = \alpha_0 m_0 c^2 \psi(t, \boldsymbol{x}) - i\hbar \sum_{k=1}^{3} \alpha_k \frac{\partial \psi(t, \boldsymbol{x})}{\partial x_k}, \tag{6.4.41}$$

where $\boldsymbol{x} = (x_1, x_2, x_3)$, with x_1, x_2, x_3 playing the role of x, y, z of the usual cartesian coordinates in $\mathbb{R}^3$; $\psi = \psi(t, \boldsymbol{x})$ is the column vector $(\psi_0(t, \boldsymbol{x}), \psi_1(t, \boldsymbol{x}), \psi_2(t, \boldsymbol{x}), \psi_3(t, \boldsymbol{x}))^T$, representing the unknown wave function; m_0 is the `rest mass` of the electron; $\hbar$ is the reduced Planck's constant; $i = \sqrt{-1}$ is the complex unit; c is the speed of light in vacuum; α_0 and α_k, $k = 1, 2, 3$, are the matrices from (6.4.36). Note that conditions (6.4.35) force the wave function to be four-dimensional, which is consistent with the theory of special relativity.

EXERCISE 6.4.21.$^{A+}$ *(a) Verify that equation (6.4.41) is invariant under the Lorentz transformation. (b) Show that, for $N = 4$, the choice (6.4.36) of the matrices is essentially unique (understanding the meaning of the word "essentially" is part of the question). (c) Can conditions (6.4.35) hold for $N > 4$?*

Matrices σ_k defined in (6.4.36) are called the `Pauli matrices`, so named after their creator, physicist WOLFGANG ERNST PAULI (1900–1958), who was born in Austria, became a US citizen in 1946, and later moved to Switzerland; his godfather was ERNST MACH, whence the middle name. These matrices first appeared in the study of the `spin` or `intrinsic magnetic moment` of the electron; the matrices $i\sigma_k$ also represent an infinitesimal rotation of three-dimensional space. The details, which lead to the theory of Lie groups and algebras, are beyond the scope of our discussion. Appearance of the Pauli matrices in a seemingly unrelated Dirac's equation further illustrates the power and beauty of abstract mathematical models.

The idea of electron spin originated from the efforts to explain the multiple splitting of spectral lines when a source of radiation is placed in a uniform and constant magnetic field. This splitting is known as the `Zeeman effect` and was first observed in 1896 by the Dutch physicist PIETER ZEEMAN (1865–1943). The Zeeman effect suggests that the electron has an intrinsic magnetic moment, in addition to its orbital moment, and the study of the splitting of the spectral lines makes it possible to estimate the value of this moment. A classical interpretation of spin considers the electron as a solid sphere rotating around one of its axes; since the sphere is charged, the rotation makes the electron a magnetic dipole with some magnetic moment. There are at least three problems with this interpretation: (a) by the uncertainty principle, we cannot think of the electron as a solid sphere; (b) if we indeed could imagine the electron as a rotating sphere, the speed of the rotation required to produce the observed magnetic moment would make some points on the surface of the sphere move with speeds much higher than the speed of light; (c) unlike classical angular momentum or magnetic moment, the spin is a purely discrete quantity: for the electron, the *observed* value of the projection of the spin on every line in $\mathbb{R}^3$ can take only two distinct values.

The discrete nature of spin was demonstrated in a 1922 experiment by two German physicists, OTTO STERN (1888–1969) and WALTER GERLACH (1889–1979). In this `Stern-Gerlach experiment`, a beam of heated silver atoms with their spinning outer electrons was made to pass through an appropriately oriented magnetic field $\boldsymbol{B}$. This field was strongly increasing in one direction, usually referred to as the z direction. The interaction between $\boldsymbol{B}$ and the magnetic dipoles of the spinning charges deflected the beam so that it hit a screen at different points in the z direction. While the classical theory predicts a continuous line of hits, the beam was essentially split into two, one up and one down, demonstrating that, once measured, the spin of the electron has only two possible values. The two values of the corresponding magnetic moment can be computed from the two observed deflections of the beam. The observable values of the electron spin s are taken to be $\pm 1/2$ (non-dimensional).

We will now show that Dirac's equation is consistent with the idea of the spin, and, in fact, puts the idea on a solid mathematical basis. We will also show how the equation leads to the prediction of the existence of anti-matter, that is, particles with the same mass as atomic particles but with opposite electric charge.

Let us forget the probabilistic interpretation of the wave function and simply look for a solution of equation (6.4.41) in the form of a plane wave

$$\psi(t, \boldsymbol{x}) = \boldsymbol{\Psi} e^{(ipx_3 - iEt)/\hbar}, \tag{6.4.42}$$

with both $\boldsymbol{\Psi}$ and p independent of t and $\boldsymbol{x}$. This solution should correspond to a free electron moving in $\mathbb{R}^3$ in the direction of the z axis and having constant momentum $p\,\hat{\boldsymbol{k}}$ and energy E.

EXERCISE 6.4.22.^{C} *(a) Verify that the function ψ from (6.4.42) satisfies (6.4.41) if and only if*

$$(m_0 c^2 \alpha_0 + pc\alpha_3)\boldsymbol{\Psi} = E\,\boldsymbol{\Psi}. \tag{6.4.43}$$

(b) Treating (6.4.43) as an eigenvalue problem, verify that there are two

eigenvalues,

$$E_{\pm} = \pm\sqrt{m_0^2c^4 + p^2c^2},$$

each of multiplicity two. With $K = \sqrt{m_0^2c^4 + p^2c^2} - m_0c^2$, *verify that the corresponding eigenvectors are* $(pc, 0, K, 0)^T$, $(0, pc, 0, -K)^T$ *for* E_+; $(-K, 0, pc, 0)^T$, $(0, K, 0, pc)$ *for* E_-. *In the non-relativistic limit as* $c \to \infty$, *we have* $K = p^2/(2m_0)$, *the kinetic energy of the electron.*

The above eigenvalues and eigenvectors have a deep physical meaning. In particular, the negative energy levels E_- suggest the existence of the `positron`, a particle with the same mass as the electron, but with the opposite electric charge. Positrons were detected in cosmic radiation by the American physicist Carl David Anderson (1905–1991) in 1932, soon after the prediction was made using Dirac's equation. The two eigenvectors at the given energy level correspond to the two possible observable values (eigenstates) of the spin of the electron (or positron).

For more details, including the discussion of the spins of other particles, see the books *Introductory Quantum Mechanics* by R. Liboff (2002) and *The Dirac Equation* by B. Thaller (1992).

Exercise 6.4.23.$^{A+}$ *(a) Write Dirac's equations when the space is one- and two-dimensional. (b) For each equation, find a particular solution in the form similar to (6.4.42).*

Recall that, in our derivation of (6.4.41), we wanted the equation to be first-order in time, and this requires us to work with the relativistic energy $E = \sqrt{m_0^2c^4 + \|\boldsymbol{p}\|^2c^2}$. Working with $\mathcal{E}^2$, that is, taking a formal square of operators on both sides of (6.4.41), leads to the operator $\frac{-\hbar^2}{2\pi}\frac{\partial^2}{\partial t^2}$ on the left-hand side and results in an equation

$$\psi_{tt} = c^2\nabla^2\psi - \frac{m_0^2c^4}{\hbar^2}\psi, \tag{6.4.44}$$

known as the `Klein-Gordon equation`, so named after the Swedish mathematician Oskar Klein (1894–1977) and the German physicist Walter Gordon (1893–1940). Note that (6.4.44) is second-order in time and space and the matrices α_0, α_k no longer appear. The unknown function ψ in (6.4.44) can be a scalar or a vector. Every solution of Dirac's equation (6.4.41) is necessarily a solution of (6.4.44), but not conversely. This fact is helpful in studying (6.4.41. Note that, for a photon or any other particle with zero rest mass, (6.4.44) is the usual wave equation.

EXERCISE 6.4.24.^{B} *Show that (6.4.44) is invariant under the Lorentz transformation. Hint: see Problem 7.9 on page 446.*

Now that we have discussed classical, relativistic, and quantum mechanics, we can take a broader view of the subject. Classical Newtonian mechanics fails at small distances (of the order of 10^{-10} meters) and/or at high speeds (approaching the speed of light in vacuum.) Quantum mechanics fixes many failures of classical mechanics at small distances. Similarly, the theory of special relativity fixes many failures of classical mechanics at high speeds. Still, there is no quantum theory of gravitation, no analog of general relativity for forces other than gravitation, and numerous unanswered questions at distances that are much smaller than 10^{-10} meters. Note in particular that, while analyzing the hydrogen atom, we used quantum *mechanics* but classical *electrodynamics* (Coulomb's Law, etc.) Quantum electrodynamics, or QED, takes over at much smaller distances, of the order of $e^2/(4\pi\varepsilon m_e c^2) \approx 2.8 \cdot 10^{-17} m$; see also a brief discussion of the subject on page 178. One of the objectives of the recently formulated `string theory` is to create a unified framework that would cover all natural phenomena at all distances and all speeds; possible references on the subject of string theory are the books [Johnson (2003)] and [Szabo (2004)].

6.4.3 *Introduction to Quantum Computing*

The time evolution of a non-relativistic quantum system is governed by the Schrödinger equation (6.4.5) on page 357. Then, setting up a a quantum system and measuring its state at a certain time should be equivalent to solving the corresponding Schrödinger's equation. Roughly speaking, this is the idea of the quantum computer, as suggested in 1981 by the American physicist RICHARD PHILLIPS FEYNMANN (1918–1988) in his talk at the First Conference on The Physics of Computation.

In this section, we discuss the basic mathematical questions related to quantum computing. We define the `quantum computer` with n quantum bits (or qubits) as a collection of n non-interacting, non-relativistic particles that evolve according to the laws of quantum mechanics. We also assume that each particle can be detected, or observed, in one of only two possible states. An example is n electrons, each having spin $1/2$ or $-1/2$. In what follows, we discuss the main properties of such a collection and how it may be possible to use these properties to perform computations.

We start with the easiest possible collection, consisting of a single particle. Denote the two detectable states of the particle by $|0\rangle$ and $|1\rangle$; we also call these states **observable states** or **eigenstates**. Let us emphasize the importance of these words in relation to the states: by the superposition principle, the number of *possible* states is much larger. We always assume that the eigenstates are orthonormal, that is, $\langle 1|1\rangle = \langle 0|0\rangle = 1$, $\langle 0|1\rangle = 0$.

EXERCISE 6.4.25.^{C} *(a) Verify that, by the superposition principle, the collection of all possible states of the particle is* $\alpha|0\rangle + \beta|1\rangle$, *where* α, β *are complex numbers so that* $|\alpha|^2 + |\beta|^2 = 1$. *(b) Verify that an equivalent description of the possible states of the particle is* $\cos\theta|0\rangle + e^{i\varphi}\sin\theta|1\rangle$, *that is, only two real numbers* θ *and* φ *are necessary for the complete description.*

Recall that a quantum mechanical system can modelled in an inner product space $\mathbb{H}$. The preceding exercise shows that, for a single particle with two observable states, this space is a two-dimensional vector space over the field of complex numbers. Recall that a two-dimensional space over the reals is denoted by $\mathbb{R}^2$. Similarly, the two-dimensional space over the complex numbers is denoted by $\mathbb{C}^2$. The basis vectors $|0\rangle$ and $|1\rangle$ are the eigenstates of the particle and represent the two detectable states. A **qubit** (short for **qu**antum **bit**) is a unit vector in $\mathbb{H}$ describing a possible state of the particle. It is important to keep in mind that a qubit is different from an eigenstate. The quantum computer operates on qubits. The following simple exercise shows how a single qubit computer can be more efficient than a classical computer.

EXERCISE 6.4.26.^{C} *Assume that the state* $|0\rangle$ *represents the number* 0 *and the state* $|1\rangle$ *represents the number* 1. *Let* f *be a linear function. How to use superposition to determine whether* $f(0)$ *is equal to* $f(1)$ *with only one function evaluation? Hint: compute* f *for the qubit* $(|1\rangle - |0\rangle)/\sqrt{2}$.

A classical computer, no matter how sophisticated, would need to compute separately the values of $f(0)$ and $f(1)$ to determine whether $f(0)$ is equal to $f(1)$. As a result, for this particular problem, the quantum computer reduces the number of operations by a factor of two. Later on, we will discuss practical implementations of the above idea, as well as its generalization, known as the **Deutsch-Jozsa algorithm**.

By the superposition principle and the product rule for probabilities, the corresponding inner product space $\mathbb{H}$ for a system of n non-interacting particles must be the collection of all linear combinations with complex coefficients of formal products of the vectors $|\psi_k\rangle \in \mathbb{H}_k$. In algebra, this space $\mathbb{H}$ is known as the tensor product, over the field of complex numbers, of the spaces $\mathbb{H}_k$:

$$\mathbb{H} = \mathbb{H}_n \otimes \mathbb{H}_{n-1} \otimes \cdots \otimes \mathbb{H}_1, \tag{6.4.45}$$

where $\mathbb{H}_k$ is the space for particle number k. For a product vector in $\mathbb{H}$, we use Dirac's notation and write $|\psi_n \ldots \psi_1\rangle = |\psi_n\rangle \otimes \cdots \otimes |\psi_1\rangle$. A natural orthonormal basis in this space is the set of all the product vectors of the form $|e_n e_{n-1} \ldots e_1\rangle = |e_n\rangle \otimes |e_{n-1}\rangle \otimes \cdots \otimes |e_1\rangle$, where $|e_k\rangle$ is one of the two eigenstates of particle number k. This basis has 2^n product vectors. Thus, $\mathbb{H}$ has dimension 2^n. By definition of the tensor product, any vector in $\mathbb{H}$ is a linear combination of these basis vectors.

Before we proceed, let us note that not all unit vectors from $\mathbb{H}$ can be written in the form $|\psi_n\rangle \otimes \cdots \otimes |\psi_1\rangle$ with $|\psi_k\rangle \in \mathbb{H}_k$. In quantum mechanics, the vectors that cannot be written as a tensor product are said to represent `entangled states`. Particles in an entangled state "feel" one another even without actual physical interaction — yet another counter-intuitive consequence of the superposition principle. Note also the order in which the tensor product of spaces is taken. For example, the space $\mathbb{H}_1 \otimes \mathbb{H}_2$ is different from $\mathbb{H}_2 \otimes \mathbb{H}_1$, even though there is a natural one-to-one correspondence between the elements of these spaces.

EXERCISE 6.4.27.^{C} *Let $n = 2$. Denote by $|0\rangle_k$, $|1\rangle_k$ an orthonormal basis in $\mathbb{H}_k$, $k = 1, 2$. (a) Verify that the product vectors*

$$|00\rangle = |0\rangle_2 \otimes |0\rangle_1,\ |10\rangle = |1\rangle_2 \otimes |0\rangle_1,\ |01\rangle = |0\rangle_2 \otimes |1\rangle_1,\ |11\rangle = |1\rangle_2 \otimes |1\rangle_1$$

form an orthonormal basis in $\mathbb{H} = \mathbb{H}_2 \otimes \mathbb{H}_1$. (b) Let $|\psi_k\rangle = \alpha_k|0\rangle_k + \beta_k|1\rangle_k$ be an element of $\mathbb{H}_k$, $k = 1, 2$. Show that the algebra of the tensor product gives

$$|\psi_2\rangle \otimes |\psi_1\rangle = \alpha_2\alpha_1|00\rangle + \alpha_2\beta_1|01\rangle + \beta_2\alpha_1|10\rangle + \beta_2\beta_1|11\rangle.$$

(c) Show that the vector $|00\rangle + |11\rangle$ cannot be written as $|\psi_2\rangle \otimes |\psi_1\rangle$ for any $|\psi_1\rangle \in \mathbb{H}_1$, $|\psi_2\rangle \in \mathbb{H}_2$. Hint: write $\psi_k = \alpha_k|0\rangle_k + \beta_k|1\rangle_k$, then conclude that $\alpha_1\beta_2 = \beta_1\alpha_2 = 0$, $\alpha_1\alpha_2 = \beta_1\beta_2 = 1$, which is impossible. Keep in mind that $|01\rangle \neq |10\rangle$.

Tensor product of spaces has some other important properties, which we investigate in the following exercise.

EXERCISE 6.4.28. *(a)*[A] *Show that* $\mathbb{C}^2 \otimes \mathbb{C}^2$ *is not the same as* $\mathbb{C}^4$. *Hint: look at how the vectors are added in each space. (b)*[B] *Similar to the single particle case, describe the vectors in* $\mathbb{H}$ *from (6.4.45) that correspond to states of the system. (c)*[C] *Define the inner product on the tensor product space. Hint: by definition, the vectors* $|e_n\rangle \otimes \cdots \otimes |e_1\rangle$, *with* $|e_k\rangle = |0\rangle$ *or* $|e_k\rangle = |1\rangle$, *form an orthonormal basis in* $\mathbb{H}$.

Next, we discuss the types of computations that are possible on a quantum computer. Recall the time evolution of every state $|\psi(t)\rangle$ of a quantum system is described by Schrödinger's equation

$$i\hbar \frac{\partial}{\partial t}|\psi(t)\rangle = \mathcal{H}|\psi(t)\rangle, \tag{6.4.46}$$

where $\mathcal{H}$ is a Hermitian operator on the corresponding inner product space $\mathbb{H}$; see page 363. For our n-qubit quantum computer, the space $\mathbb{H}$ is a finite-dimensional vector space; once the basis in the space is fixed, every symmetric operator becomes a `Hermitian matrix` (see Exercise 6.2.17(b) on page 328) and a partial differential equation for ψ becomes a system of 2^n ordinary differential equations for the components of ψ. We denote by H the matrix corresponding to the Hamiltonian $\mathcal{H}$. Recall that H is a Hermitian matrix if and only if $H^T = \overline{H}$, where $\overline{H}$ is the matrix whose entries are complex conjugates of the corresponding entries of H. The following exercise shows that a quantum computer can only perform a unitary transformation of the state vector.

EXERCISE 6.4.29.[C] *Let* H *be a Hermitian matrix. (a) Show that the matrix* $U = e^{-iH}$ *is* `unitary`, *that is,* $U^{-1} = \overline{U}^T$. *Hint: see (8.1.4) on page 454 in Appendix. (b) Verify that the solution of the finite-dimensional Schrödinger equation* $i\hbar \frac{d}{dt}|\psi(t)\rangle = H|\psi(t)\rangle$ *is* $|\psi(t)\rangle = e^{-iH/\hbar}|\psi(0)\rangle$.

To proceed, let us first review the main facts about the classical computer. In today's computers and digital communication devices, information is represented by strings (finite sequences) of bits. Mathematically, a `bit` is a binary digit, that is, a 0 or a 1. A string of n bits can be regarded as a vector with n components, each having value 0 or 1.

A bit string can encode numbers or other data. A byte is a string of eight bits and can encode alphanumeric characters (letters A-Z, punctuation and decimal numerals 0-9). For numerical computation, numbers can be represented as bit strings interpreted in binary (base 2) notation. For example, the bit string 1101 represents the decimal number

$1 \times 2^3 + 1 \times 2^2 + 0 \times 2^1 + 1 \times 2^0 = 13.$

A computation is a sequence of operations on bit strings. The operations can be arithmetic in base 2 or Boolean. An operation, often in hardware implementation, performed by a computer, classical or quantum, is often called a **gate**. A gate is called **reversible** if the corresponding operation is a **bijection**, that is, a one-to-one and onto correspondence between the sets of inputs and outputs. The mathematical fact that the solution operator of the finite-dimensional Schrödinger equation is unitary implies that all quantum gates are unitary and therefore are linear and reversible.

In classical computer hardware, a bit is physically implemented in one of several ways, for example, as

(1) the state of magnetization of a small area of a magnetic disc (hard drive);

(2) the state of electric charge of a capacitor in a cell on a random access memory chip (RAM);

(3) the state of a bistable transistor circuit known as a flip-flop on a static random access memory chip (SRAM);

(4) the state of connectivity of the oxide layers in a transistor on a read-only memory chip (ROM).

In a hard drive, bit strings are arranged as successive bits in tracks on the rotating disc. They remain in their state of magnetization unless changed by the magnetic read-write head scanning the disc tracks. In RAM, a bit string is arranged in a circuit connecting the cells on a memory chip and read out by a memory address system. Since charge leaks, the cells must be refreshed periodically. RAM memory is called volatile, and bits are not preserved when the power is turned off. In SRAM, the flip-flop state is maintained until changed by voltage inputs, but the stored bit is lost when power is turned off. In ROM, the transistor state is hard-wired, and the bits are permanent.

RAM, SRAM, and ROM are examples of memory chips. Besides the memory chips, there are other chips in a computer which perform the operations in a computation. A chip is fabricated as an integrated circuit of resistors, capacitors, and transistors formed on a wafer of layers of silicon semiconductors and metal oxides. The semiconductor circuit elements can be made to function either as resistors, capacitors, or transistors. Metallic contacts between circuit elements are made by photoetching a metallic layer using an appropriate mask. By improved miniaturization techniques in the manufacturing process it is now feasible to fabricate chips having many millions of circuit elements on a single chip.

The smaller a circuit element, the faster its switching time, that is, the time for executing Boolean operations on bits. With today's technology, the minimum size of a circuit element is limited by the wavelength of the light in the photo-lithography process used to form the wafer layers. Ultra-violet radiation can have a wavelength as small as $4 \cdot 10^{-9}$ meters and therefore can be used to produce elements smaller than one micron, or 10^{-6} meters, in size. Today's circuit elements are sub-micron in size, sometimes as small as 0.1 microns or 10^{-7} meters. So 10^{-7} meters can be taken as an approximate size of a bit on a chip.

In 1965, the American semi-conductor engineer GORDON EARL MOORE (born 1929), who later co-founded the Intel Corporation, made a prediction about the number of components on a silicon chip; this prediction became known as `Moore's Law`. One of the formulations of this law, which held remarkably accurate over the years, is that the number of transistors on a square inch of a silicon chip doubles every 1.5 years. If this trend continues, then, by the year 2040, the basic component of the chip will reach the size of 10^{-10} meters. At those scales, the laws of quantum mechanics must be taken into account even for a classical computer.

To conclude our review of classical computers, let us briefly review some classical logical gates, also known as **logical, or Boolean, operations**, defined on single bits or pairs of bits:

- `NOT`: $x \mapsto \neg x = 1 - x$, that is, $\neg x = 1$ if $x = 0$ and $\neg x = 0$ if $x = 1$.
- `AND`: $(x, y) \mapsto x \wedge y = \min(x, y)$, that is, $x \wedge y = 1$ if and only if $x = y = 1$.
- `OR`: $(x, y) \mapsto x \vee y = \max(x, y)$, that is, $x \vee y = 0$ if and only if $x = y = 0$.
- `XOR`: $(x, y) \mapsto x \oplus y = x + y \pmod 2$, that is, $x \oplus y = 0$ if and only if $x = y$.
- `NAND`: $(x, y) \mapsto \neg(x \wedge y)$.
- `NOR`: $(x, y) \mapsto \neg(x \vee y)$.

There is one other basic operation that is present in the classical computer: the operation `COPY`, or duplication of input. This operation takes one logical variable x as an input and produces two identical variables x as an output $x \mapsto (x, x)$.

We say that a gate is $m \times n$ if there are m input and n output variables. Of the above operations, only the NOT operation is reversible: for all others, the number of inputs is different from the number of outputs. The operations NOT, AND, OR, and COPY are enough to generate every Boolean function, that is, a function $f = f(x_1, \ldots, x_n)$ depending on n logical variables, for some arbitrary but fixed n, and taking only two possible values, 0 and 1. Accordingly, we call a gate **universal** if one or more of such gates can re-produce each of the operations NOT, AND, OR, COPY.

EXERCISE 6.4.30.[A] *(a) Show that the operations NOT, AND, OR, COPY are enough to generate every Boolean function. Hint: consider only the strings $x_n \cdots x_1$ on which f is equal to 1. (b) Explain where in the implementation the operation COPY is required.*

EXERCISE 6.4.31. *(a)*[C] *Verify that the 2×2 NAND/NOT gate $F_{N/N}$: $(x, y) \mapsto \big(\neg x, \neg(x \wedge y)\big)$ is universal but not reversible. Hint: for example, $F_{N/N}(F_{N/N}(x, 1)) = (x, x)$ provides the copy operation. (b)*[B] *Verify that the 2×2 XOR gate $(x, y) \mapsto (x, \oplus y)$ is reversible but not universal. (c)*[A] *Show that there is no 2×2 classical gate that is both reversible and universal. Hint: a universal 2×2 gate cannot be reversible.*

An example of a classical 3×3 universal and reversible gate is the **Toffoli gate**:

$$(x, y, z) \mapsto \big(x, y, z \oplus (x \wedge y)\big).$$

This gate was first suggested by TOMMASO TOFFOLI in a 1980 MIT Technical Report; as of 2005, T. Toffoli is at the Electrical and Computer Engineering Department of Boston University.

EXERCISE 6.4.32. [A] *Verify that the Toffoli gate is both universal and reversible.*

Let us now consider several basic quantum gates. Recall that a qubit is a two-dimensional vector, being an element of the space $\mathbb{C}^2$. Accordingly, we write $|1\rangle = (1, 0)^T$ and think of this state as a quantum analog of the logical value 1. Similarly, $|0\rangle = (0, 1)^T$ is the quantum analog of the logical value 0. A gate is then represented by a unitary matrix of appropriate size.

First, we look at the **Hadamard gate**, defined by the unitary matrix

$$U_H = \frac{1}{\sqrt{2}} \begin{pmatrix} 1 & 1 \\ 1 & -1 \end{pmatrix}.$$

EXERCISE 6.4.33. *(a)*[C] *Verify that U_H is a unitary (in fact, orthogonal) matrix: $U_H^{-1} = U_H^T$. (b)*[C] *Verify that $U_H|0\rangle = (|1\rangle - |0\rangle)/\sqrt{2}$ and $U_H|1\rangle = (|1\rangle + |0\rangle)/\sqrt{2}$. (c)*[B] *Explain how to use the results of part (b) to carry out the computation in Exercise 6.4.26 on page 380.*

The Hadamard gate is named after the French mathematician J.S. Hadamard, who certainly had nothing to do with quantum computers. The reason for the name is that U_H is, up to the factor of $1/\sqrt{2}$, a **Hadamard matrix**, that is, a square matrix whose entries are either 1 or -1 and whose rows are mutually orthogonal. The Hadamard matrices have a number of remarkable properties and are used in many areas of mathematics and engineering. In quantum computing, the Hadamard gate is used to create superposition of states; see Exercise 6.4.33.

EXERCISE 6.4.34.[A] *Verify that if H is a Hadamrd matrix of order n, then*

$$\begin{pmatrix} H & H \\ H & -H \end{pmatrix}$$

is a Hadamrd matrix of order $2n$.

Next, consider the Pauli matrix σ_1,

$$\sigma_1 = \begin{pmatrix} 0 & 1 \\ 1 & 0 \end{pmatrix},$$

and define the matrix

$$R_1(\theta) = e^{i\theta\sigma_1},$$

where θ is a real number.

EXERCISE 6.4.35. *(a)*[C] *Verify that, for a qubit $|x\rangle$, the operation*

$$|x\rangle \mapsto \sigma_1|x\rangle$$

corresponds to the classical NOT operation. (b)[B] *Verify that $R_1(\theta) = \cos\theta\, I_{[2]} + i\sin\theta\,\sigma_1$, where $I_{[2]}$ is the 2×2 identity matrix. Hint: $\sigma_1^2 = I_{[2]}$. (c)*[A] *Verify that if θ is not a rational multiple of π, that is, $\theta \neq \pi\, p/q$ for any integer p, q, then $\lim_{N\to\infty} R_1(N\theta/2) = R_1(\pi/2) = i\sigma_1$.*

The quantum analog of the Toffoli gate is the **Deutsch gate**:

$$(|x\rangle, |y\rangle, |z\rangle) \mapsto (|x\rangle, |y\rangle, |\tilde{z}\rangle),$$

where $|\tilde{z}\rangle = -iR_1(\theta/2)|z\rangle$ if $|x\rangle = |y\rangle = |1\rangle$ and $|\tilde{z}\rangle = |z\rangle$ otherwise; the real number θ is taken to be not a rational multiple of π. This gate was first introduced by the British physicist DAVID DEUTSCH (born in 1953), who, as of 2005, is at the Center for Quantum Computation, Clarendon Laboratory, Oxford University.

EXERCISE 6.4.36. *(a)*[C] *Explain how the Deutsch gate makes it possible to simulate a classical computer on a quantum computer. Hint: use the results* *of the previous exercise to get a Toffoli gate. (b)^A Explain why we do not take $\theta = \pi/2$ in the construction of the Deutsch gate. (c)^A In a suitable basis in $\mathbb{H} \otimes \mathbb{H} \otimes \mathbb{H}$, the Deutsch gate is represented by the unitary matrix, written in the block form as*

$$U_D = \begin{pmatrix} I_{[4]} & 0_{[6,2]} \\ 0_{[2,6]} & D \end{pmatrix}, \text{ where } D = \begin{pmatrix} i\cos(\pi\theta/2) & \sin(\pi\theta/2) \\ \sin(\pi\theta/2) & i\cos(\pi\theta/2) \end{pmatrix},$$

$I_{[4]}$ is a 6×6 identity matrix, and $0_{[n,k]}$ is an $n \times k$ zero matrix. Find the corresponding basis in $\mathbb{H} \otimes \mathbb{H} \otimes \mathbb{H}$ and verify that this matrix representation of the Deutsch gate is equivalent to its logical definition.

This completes our discussion of quantum gates, and we move on to QUANTUM ALGORITHMS. A quantum algorithm has several important differences from a classical algorithm. In particular, there are usually no IF/THEN commands, because it is impossible to determine the state of the quantum system without destroying the superposition of states: remember that a measurement randomly forces the system into one of the eigenstates. Also, the output of a quantum algorithm is typically random, although the correct answer has a higher probability than any other answer. As a result, a computation on a quantum computer can require several independent realizations.

Let us summarize three main quantum algorithms: the Deutsch-Jozsa algorithm, Grover's Search Algorithm, and Shor's Factorization Algorithm.

The `Deutsch-Jozsa algorithm` was proposed in 1992 by D. Deutsch and his fellow British-man, a computer scientist RICHARD JOZSA from the University of Bristol. This is a quantum algorithm with a deterministic output, that is, always produces the correct answer.

The algorithm solves the following problem. Consider a Boolean function $f = f(x_1, \ldots, x_n)$ of n binary inputs and two possible binary outputs. The function is known to be of one of two types: either it is *constant* (produces the same output for all inputs) or it is *balanced* (produces the two possible outputs, 0 and 1 on an equal number of inputs.) The problem is to determine the type of the function using as few evaluations of f as possible.

EXERCISE 6.4.37.^{A} *How many function evaluations would a classical computer require to solve this problem?*

The Deutsch-Jozsa algorithm generalizes the idea from Exercise 6.4.26 on page 380 and solves the problem with only one function evaluation. A curious reader can try and discover the algorithm independently or check pages 34–36 in the book Quantum Computation and Quantum Information by M. A. Nielsen and I. L. Chuang, 2000.

`Grover's Search Algorithm` was proposed in 1996 by Lov Kumar Grover, a physicist at Bell Labs. The algorithm solves the following problem. Consider a Boolean function $f = f(x_1, \ldots, x_n)$ of n binary inputs and two possible binary outputs. It is known that $f = 1$ only on one bit string, and the problem is to find that string; this problem is also known as searching an unsorted database. Grover's Search Algorithm solves this problem in about $\sqrt{n}$ steps and the probability to get a wrong answer after one realization of the algorithm is of order $1/n$; for one particular implementation of the algorithm, one can show that $\lim_{n\to\infty} N(n)/\sqrt{n} = \pi/4$, where $N(n)$ is the number of steps. For a detailed description of the algorithm, see the paper *From Schrodinger's Equation to Quantum Search Algorithm* by L. K. Grover in the American Journal of Physics, **69**(7): 769–777. Similar to the Deutsch-Jozsa algorithm, Grover's Search Algorithm is an example of a *provable* advantage of a quantum computer over the classical computer: one can rigorously prove that every algorithm for searching an unsorted database on a classical computer requires more steps than Grover's algorithm.

Exercise 6.4.38.^{C} *On average, how many steps are required to search an unsorted database on a classical computer? Hint: you need anywhere from 1 to n steps, depending on how lucky you are, and each number of steps is equally likely.*

`Shor's Factorization Algorithm` was proposed in 1994 by Peter Williston Shor, then a researcher at Bell Labs and now a Professor of Applied Mathematics at MIT. This algorithm allows fast factorization of integers, so fast, in fact, that, with its help, one can easily break most public encryption systems currently in use. As a result, the algorithm sparked the current interest in quantum computing outside the physics community. Originally published in 1994 in the Proceedings of the 35th Annual Symposium on Foundations of Computer Science, a more expanded description of the algorithm is in the paper *Polynomial-time algorithms for prime factorization and discrete logarithms on a quantum computer*, published in SIAM Journal on Computing, **26**(5):1484–1509, 1997.

Just as most quantum algorithms, Shor's algorithm is probabilistic, that is, provides a correct answer with high probability. This probability can be further increased by running the algorithm several times. The algorithm has a component that can be implemented on a classical computer. The number of steps required to factor an n-digit number using a modification of the original version of Shor's algorithm is of order $n^2 \ln n$; the number of steps for the best known fully deterministic algorithm on a classical computer is of order $e^{\sqrt{n \ln n}}$, although there is no proof that this bound cannot be improved. Note that, in estimating the number of steps in the factorization algorithm, the base in which the number is written does not matter. Also note that if K is a positive integer, then the number of digits of this number in any base is of order $\ln K$.

EXERCISE 6.4.39.^{C} *(a) How many digits in decimal representation does an n-bit binary number have? (b) Let $n = 256$, that is, consider the problem of factoring a 256-digit number. Evaluate $n^2 \ln n$ and $e^{\sqrt{n \ln n}}$.*

At this point, no problem is known that is solvable on a quantum computer but is not solvable on a classical computer; the only existing advantage of a quantum computer is (theoretically) higher efficiency for certain problems.

We conclude this section with a brief discussion of implementation problems for a quantum computer. HARDWARE IMPLEMENTATION of quantum computers is still in an experimental phase. There are several branches of quantum physics addressing this issue. One approach is based on Nuclear Magnetic Resonance (NMR). In a liquid-based NMR device, a qubit is implemented by the nuclear spin of an atom or a molecule of a liquid. The spin can be aligned or anti-aligned with an applied magnetic field $\boldsymbol{B}$. The orientation of spins can be changed by applying a short magnetic pulse, and the final result can be read out using the same method as in the Stern-Gerlach experiment (see page 377). Different shapes of the pulse result in different unitary transformations performed on the system of the spins. A quantum algorithm on an NMR device is thus a sequence of magnetic pulses.

So far, the liquid NMR technology was responsible for the main successes of practical quantum computing. In 1998, a 3-qubit computer implemented Grover's Search Algorithm; this was a collaborative effort of scientists at MIT, UC Berkeley, and Stanford University. In 2001, a 7-qubit computer implemented Shor's Factorization Algorithm to find the factors of the number 15; this was a collaborative effort of scientists at the IBM Almaden Research Center and Stanford University.

Other possible implementations of a quantum computer include solid-state NMR, superconductors, ions in electromagnetic traps, and optical devices.

Every implementation of a quantum computer must deal with the effects of **decoherence**, that is, loss of superposition due to the outside factors. These effects can limit the number of usable qubits. For example, a liquid NMR-based quantum computer at room temperatures can have about 10 usable qubits.

Mathematically, decoherence means that the system's Hamiltonian in the Schrödinger equation (6.4.46) is replaced with a different, and often unknown, operator describing not only the system but the interaction between the system and the environment; this uncontrolled change of the Hamiltonian inevitably leads to computation errors. The effects of decoherence can be minimized using quantum error-correcting codes and/or decoherence-free states. A quantum error-correcting code reduces the probability of error by introducing redundancy through additional qubits; see, for example, the paper *Good Quantum Error-Correcting Codes Exist* by A. R. Calderbank and P. W. Shor in the Physical Reviews A, **54**(2):1098–1106, 1996. The increase of the required number of qubits can be rather dramatic. For example, direct factorization of a 1000-bit number using Shor's algorithm would require slightly over 2000 qubits; with quantum error-correcting codes, the required number of qubits becomes somewhere between 10^{12} and 10^{18}. Alternatively, one could try to identify and use the decoherence-free subspaces, that is, states that are not affected by the interaction with the environment; see, for example, the paper *Decoherence-Free Subspaces and Subsystems* by D. A. Lidar and K. B. Whaley, in volume 622 of the Springer Lecture Notes in Physics, 2003.

The book Quantum Computation and Quantum Information by M. A. Nielsen and I. L. Chuang, 2000, is considered by many a standard general reference on all subjects related to quantum computing. Another good reference is the book *Classical and Quantum Computing* by Y. Hardy and W. H. Steeb, 2002.

6.5 Numerical Solution of Partial Differential Equations

Given the transport equation (6.1.2), we can easily compute the solution at every point in space and time from the initial condition using formula (6.1.4); see page 292. The solution formulas for the wave equation (6.1.43) on page 310 and (6.1.58) on page 315 require the antiderivative of the function g and are therefore less computation-friendly. The integrals appearing in the solution formulas for the heat equation (6.1.18) on page 297 or (6.1.35) on page 303 make the formulas even harder to use for computational purposes. Most other partial differential equation do not have any solution formula at all, making numerical approximation of the solution the only option. In what follows, we discuss the basic numerical methods of approximating the solution of a partial differential equation without using solution formulas.

6.5.1 *General Concepts in Numerical Methods*

In this section we introduce several basic notions of numerical analysis: the local and global approximation errors, the explicit and implicit methods, and numerical stability. We start with the approximation error and illustrate it with the `numerical quadrature` problem, that is, approximate numerical evaluation of the integral $I = \int_a^b f(t)dt$ for a continuous function f.

The basic methods of solving the numerical quadrature problem should be familiar from a one-variable calculus class. We will be interested in three particular methods: the left-point rule, the right-point rule, and the trapezoidal rule. If the interval $[a,b]$ is divided into M equal intervals of length $\tau = (b-a)/M$ and $t_m = \tau m$, $m = 0, \ldots, M$, then the three approximations of I are

$$I = \int_a^b f(t)dt \approx \begin{cases} I_r = \sum_{m=0}^{M-1} f(t_m)\,\tau & \text{(left − point rule)} \\ I_l = \sum_{m=1}^{M} f(t_m)\,\tau & \text{(right − point rule)} \\ I_p = \sum_{m=1}^{M} \big(f(t_m) + f(t_{m-1})\big)\tau/2 & \text{(trapezoidal rule)} \end{cases} \tag{6.5.1}$$

EXERCISE 6.5.1.[C] *(a) Draw three pictures illustrating each of the three rules. (b) Assume that the function f has two continuous derivatives on $[a,b]$. Define*

$$e_{r,m} = \left| \int_{t_{m-1}}^{t_m} f(t)dt - f(t_{m-1})\tau \right|,\ e_{l,m} = \left| \int_{t_{m-1}}^{t_m} f(t)dt - f(t_m)\tau \right|,$$
$$e_{p,m} = \left| \int_{t_{m-1}}^{t_m} f(t)dt - \big(f(t_{m-1}) + f(t_m)\big)\tau/2 \right|. \tag{6.5.2}$$

Show that there exists a number C independent of τ and n so that

$$e_{r,m} \le C\tau^2,\ e_{l,m} \le C\tau^2,\ e_{p,m} \le C\tau^3. \tag{6.5.3}$$

Hint: for $e_{r,n}$ and $e_{l,n}$ use the Taylor formula at t_{m-1}; for $e_{p,n}$, add the Taylor formulas at t_{m-1} and t_m. (c) In the notations of (6.5.1), show that there exists a number C_1 independent of τ and n so that

$$|I - I_r| \le C_1\tau,\ |I - I_l| \le C_1\tau,\ |I - I_p| \le C_1\tau^2. \tag{6.5.4}$$

Hint: $|I - I_r| \le Ne_{r,m} = (b-a)e_{r,m}/\tau$.

Formulas (6.5.2) give the `one-step`, or `local`, error of each method, while (6.5.4) gives the `multi-step`, or `global`, error. The number of steps is proportional to $1/\tau$, which leads to the loss of one power of τ in the global error bound, as compared to the local error. This observation does not rely on the particular nature of the numerical method and suggests the following general relation between the local error e and global error E for all multi-step numerical procedures with step τ: if $e \le C\tau^\gamma$ for some $\gamma > 1$, then $E \le C_1\tau^{\gamma-1}$; this relation does indeed hold for all the procedures we are going to study.

Next, we consider some numerical methods for *ordinary differential equation* to introduce the notions of explicit method, implicit method, and stability. Consider the equation $y'(t) = f(t, y(t)), t > 0$, $y(0) = y_0$, where the functions y and f can be vectors. We assume the function f satisfies the conditions that ensure the existence and uniqueness of the solution on every time interval, see Section 8.2 in Appendix. To find an approximation of this solution on an interval $[0, T]$, we select the `time step` $\tau = T/M$ for some positive integer M and `discretize` the interval $[0, T]$ by introducing the `grid points` $t_m = \tau m$, $m = 0, \ldots, M$. The *approximate solution* $\bar{y}_m$, $m = 0, \ldots, M$, is computed at the grid points t_m. Since the exact solution $y = y(t)$ satisfies $y(t_{m+1}) = y(t_m) + \int_{t_m}^{t_{m+1}} f(s, y(s))ds$, we see that the approximation $\bar{y}_m$ can be constructed using a suitable approximation of the definite integral. In particular, the left-point rule results in the `explicit Euler scheme` (or method)

$$\bar{y}_0^{(1)} = y_0,\ \bar{y}_{m+1}^{(1)} = \bar{y}_m^{(1)} + f(t_m, \bar{y}_m^{(1)})\tau, \tag{6.5.5}$$

the right-point rule, in the `implicit Euler scheme`

$$\bar{y}_0^{(2)} = y_0,\ \bar{y}_{m+1}^{(2)} = \bar{y}_m^{(2)} + f(t_{m+1}, \bar{y}_{m+1}^{(2)})\tau, \tag{6.5.6}$$

and the trapezoidal rule, in

$$\bar{y}_0^{(3)} = y_0, \; \bar{y}_{m+1}^{(3)} = \bar{y}_m^{(3)} + \frac{1}{2}\big(f(t_{m+1}, \bar{y}_{m+1}^{(3)}) + f(t_m, \bar{y}_m^{(3)})\big)\tau. \qquad (6.5.7)$$

Formulas (6.5.4) suggest the following global error bounds:

$$\begin{aligned} \max_{0\le m\le M} \|y(t_m) - \bar{y}^{(j)}(t_m)\| &\le C^{(j)}\tau, \; j = 1,2; \\ \max_{0\le m\le M} \|y(t_m) - \bar{y}^{(3)}(t_m)\| &\le C^{(3)}\tau^2; \end{aligned} \qquad (6.5.8)$$

recall that y can be a vector. The rigorous proof of (6.5.8) relies on the Taylor expansion and can be carried out by the interested reader. Note that both (6.5.6) and (6.5.7) are implicit methods, because $\bar{y}_{m+1}$ is not expressed explicitly in terms of $\bar{y}_m$ and must be computed by solving a certain equation. Even when $f(t, y) = Ay$ for a constant matrix A, this solution is usually computed numerically. The notion of *stability* explains the reason for using the implicit methods despite this added difficulty.

In general, stability of a numeral procedure characterizes the propagation, from one step to the next, of numerical errors that cannot be controlled, such as round-off errors. There are many different definitions of stability, so that the same method can be stable according to one definition and not stable according to another. The details are outside the scope of our discussion; see, for example, Chapter I, Section 5.1 of the the book *Finite Difference Methods for Partial Differential Equations* by G. Forsythe and W. Wasow, 2004. We will use a very special definition of stability under an additional assumption that the exact solution $y(t)$ stays bounded for all $t > 0$, that is, there exists a number $C_0 > 0$ independent of t so that $|y(t)| \le C_0$. In this situation, we say that a multi-step procedure with a fixed step τ is `stable` if there exits a number C_1 independent of m so that $|\bar{y}_m| \le C_1$ for all $m \ge 0$. In particular, our definition does not apply to equation $y'(t) = y(t)$. On the other hand, the following exercise shows that, for equation $y'(t) = -ay(t)$, $a > 0$, both implicit procedures (6.5.6) and (6.5.7) are always stable, while the explicit procedure (6.5.5) can be unstable.

EXERCISE 6.5.2.[C] *Consider the equation* $y'(t) = -ay(t)$, $y(0) = 1$, $a > 0$. *(a) Verify that, for* $m \ge 0$, $y_m^{(1)} = (1 - a\tau)^m$, $y_m^{(2)} = (1 + a\tau)^{-m}$, *and* $y_m^{(3)} = \big((2 - a\tau)/(2 + a\tau)\big)^m$. *(b) Verify that* $|y_m^{(1)}|$ *stays bounded for all* $m \ge 0$ *if and only if* $0 < \tau \le 2/a$, *while* $|y_m^{(2)}| \le 1$ *and* $|y_m^{(3)}| \le 1$ *for all* $m \ge 0$ *and all* $\tau > 0$.

To illustrate how stability affects computations, consider the equation $y'(t) = -1000y(t)$. One criterion for choosing the step τ is the accuracy of approximation, so let us assume that $\tau = 0.1$ achieves the required accuracy. The previous exercise shows that this time step is admissible for an implicit method, but the explicit method would require τ less than 0.002. In other words, computations with the explicit method would require at least 50 times as many steps as with the implicit method (for the sake of the argument, we ignore the fact that $y(0.1) = e^{-100}$ is equal to zero for all practical purposes). As a rule, all explicit methods have a `stability condition`, and satisfying this condition often requires a much smaller time step than would be needed for global error control. The resulting increase in computational complexity can be especially dramatic for partial differential equations, making explicit methods impractical. The implicit methods, on the other hand, are often *unconditionally stable*, and the extra computations related to solving the equation for $\bar{y}_{m+1}$ can be carried out without compromising the overall effectiveness of the method.

An alternative interpretation of (6.5.5)–(6.5.7) is the `finite difference approximation` of the first derivative:

$$y'(t_m) \approx \frac{y(t_{m+1}) - y(t_m)}{\tau} \approx \frac{\bar{y}_{m+1} - \bar{y}_m}{\tau} = F_m,$$

where F_m is a suitable approximation of $f(t_m, y(t_m))$. In our analysis of partial differential equations, we will also need the finite difference approximation of the second derivative:

$$y''(t) \approx \frac{y(t+\tau) - 2y(t) + y(t-\tau)}{\tau^2}. \tag{6.5.9}$$

EXERCISE 6.5.3.^{C} *Assume that the function $y = y(t)$ has four continuous derivatives on the interval $[0, T]$. Show that*

$$\sup_{\tau \le t \le T-\tau} \left| y''(t) - \frac{y(t+\tau) - 2y(t) + y(t-\tau)}{\tau^2} \right| \le C\tau^2, \tag{6.5.10}$$

for some number C independent of τ. Hint: add the Taylor expansions at the point t for $y(t+\tau)$ and $y(t-\tau)$.

There are many other methods of numerical quadrature beside (6.5.1), and there are many other numerical methods for solving ordinary differential equations beside (6.5.5)–(6.5.7), but the discussion of these methods is outside the scope of these notes.

6.5.2 *One-Dimensional Heat Equation*

In this section, we discuss the basic finite difference methods for the one-dimensional heat equation on an interval with zero boundary conditions:

$$\begin{aligned} &u_t(t,x) = a u_{xx}(t,x),\ 0 < t \le T,\ 0 < x < L, a > 0; \\ &u(0,x) = f(x),\ u(t,0) = u(t,L) = 0. \end{aligned} \tag{6.5.11}$$

We start by introducing the grid points in space: $x_n = nh$, $n = 0, \ldots, N$, where $N = L/h$. Denote by $U_n(t)$ an approximation of $u(t, x_n)$. By (6.5.9), we should expect

$$u_{xx}(t, x_n) \approx \frac{U_{n+1}(t) - 2U_n(t) + U_{n-1}(t)}{h^2}, \tag{6.5.12}$$

and so it is natural to define the $U_n(t)$ as the solutions of the system of ordinary differential equation

$$\frac{dU_n(t)}{dt} = \frac{a}{h^2}\left(U_{n+1}(t) - 2U_n(t) + U_{n-1}(t)\right), \tag{6.5.13}$$

$n = 1, \ldots, N-1$, $0 < t \le T$. The initial and boundary conditions imply

$$U_n(0) = f(x_n),\ n = 1, \ldots, N-1,\ U_0(t) = U_N(t) = 0. \tag{6.5.14}$$

EXERCISE 6.5.4.^{C} *(a) Let $U(t)$ be the column-vector $(U_0(t), \ldots, U_N)^T$; see page 12. For an integer number K and real numbers α, β, define the matrix $A_K^{[\alpha,\beta]}$ as a square $K \times K$ matrix with the number α along the main diagonal, the number β just above and below the diagonal, and zeros everywhere else:*

$$A_K^{[\alpha,\beta]} = \begin{pmatrix} \alpha & \beta & 0 & \cdots & 0 \\ \beta & \alpha & \beta & 0 \cdots & 0 \\ 0 & \ddots & \ddots & \ddots & \vdots \\ \vdots & & \ddots & \ddots & \beta \\ 0 & \cdots & \cdots & 0\ \beta & \alpha \end{pmatrix}. \tag{6.5.15}$$

Verify that equations (6.5.13) can be written in the matrix-vector form

$$\frac{dU(t)}{dt} = \frac{a}{h^2} A_{N+1}^{[-2,1]} U(t),\ 0 < t \le T. \tag{6.5.16}$$

(b) Let $U = (F(x_0), \ldots, F(x_N))^T$ be a column vector consisting of the samples of the function $F(x) = bx(x-L)$ for some real number b. Verify that $A_{[N+1]}^{[-2,1]} U = 2bh^2(1, \ldots, 1)^T$. Since $F''(x) = 2b$, we see that the matrix

$A_{N+1}^{[-2,1]}$ *results in an* `exact discretization` *for the quadratic polynomials satisfying the boundary conditions.*

A numerical solution of (6.5.16) will give us a numerical approximation of $u(t,x)$. Accordingly, let us discretize the time interval with step τ by setting $t_m = m\tau$, $m = 0, \ldots, M$, with $M = T/\tau$, and let

$$r = \frac{a\tau}{2h^2}. \tag{6.5.17}$$

By applying (6.5.5) to (6.5.16), we get the `explicit Euler scheme` for the heat equation:

$$\bar{u}^{(1)}(m+1) = \bar{u}^{(1)}(m) + 2rA_{N+1}^{[-2,1]}\bar{u}^{(1)}(m). \tag{6.5.18}$$

By applying (6.5.6) to (6.5.16), we get the `implicit Euler scheme` for the heat equation:

$$\bar{u}^{(2)}(m+1) = \bar{u}^{(2)}(m) + 2rA_{N+1}^{[-2,1]}\bar{u}^{(2)}(m+1). \tag{6.5.19}$$

By applying (6.5.7) to (6.5.16), we get the `Crank-Nicolson scheme` for the heat equation:

$$\bar{u}^{(3)}(m+1) = \bar{u}^{(3)}(m) + r\Big(A_{N+1}^{[-2,1]}\bar{u}^{(3)}(m) + A_{N+1}^{[-2,1]}\bar{u}^{(3)}(m+1)\Big). \tag{6.5.20}$$

This scheme was proposed in 1947 by the British physicists JOHN CRANK (b. 1916) and PHYLLIS NICOLSON (1917–1968). For each of the three schemes, the initial condition is $\bar{u}^{(i)}(0) = U(0)$.

EXERCISE 6.5.5.^{C} *(a) Note that each $\bar{u}^{(i)}(m)$ is a vector with components $\bar{u}_n^{(i)}(m)$, $n = 1, \ldots, N-1$. Write (6.5.18)–(6.5.19) in component forms. Hint: for example, $\bar{u}_n^{(1)}(m+1) = 2r\bar{u}_{n+1}^{(1)}(m) + (1-4r)\bar{u}_n^{(1)}(m) + 2r\bar{u}_{n-1}^{(1)}(m)$, with $\bar{u}_0^{(1)}(m) = \bar{u}_{N+1}^{(1)}(m) = 0$. (b) Verify that (6.5.20) can be written as*

$$A_{N+1}^{[1+2r,-r]}\bar{u}^{(3)}(m+1) = A_{N+1}^{[1-2r,r]}\bar{u}^{(3)}(m). \tag{6.5.21}$$

To study the accuracy of the approximations $\bar{u}^{(i)}$, assume that the odd $2L$-periodic extension of the initial condition $f = f(0)$ is twice continuously differentiable everywhere on the real line. Under this assumption, one can show that $u = u(t,x)$ is twice continuously differentiable with respect to x for all $0 \le t \le T$ and $\sup_{0\le t\le T,\ 0<x<L} |u_{xx}(t,x)| < \infty$; see Exercise 6.1.15 on page 303. Then inequality (6.5.10) suggests that

$$\max_{0\le t\le T, 1\le n\le N-1} |u(t,x_n) - U_n(t)| \le Ch^2 \tag{6.5.22}$$

for some positive number C independent of h. After that, inequalities (6.5.8) make the following error bounds believable:

$$\max_{0\leq m\leq M,\, 0\leq n\leq N} |u(t_m, x_n) - \bar{u}_n^{(i)}(m)| \leq C_1^{(i)}(\tau + h^2),\ i = 1, 2;$$
$$\max_{0\leq m\leq M,\ 0\leq n\leq N} |u(t_m, x_n) - \bar{u}_n^{(3)}(m)| \leq C_1^{(3)}(\tau^2 + h^2). \tag{6.5.23}$$

The real numbers $C_1^{(i)}$ do not depend on τ and h, but depend on T and L.

Being an explicit scheme, (6.5.18) has a **stability condition:** $r \leq 1/4$ or

$$\tau \leq \frac{h^2}{2a}; \tag{6.5.24}$$

see Problem 7.12 on page 448 for the proof. In particular, the larger the a, the smaller the time step. Also, to improve the resolution in space by a factor of $q > 1$ (that is, to reduce h by a factor of q), one has to increase the number of time steps by the factor of q^2.

On the other hand, both (6.5.19) and (6.5.20) are implicit and do not require any restrictions on r. Still, as (6.5.23) shows, if one reduces h by a factor of $q > 1$ in the implicit Euler scheme, then the time step must be reduced by the factor q^2 to achieve the q^2-fold reduction of the overall approximation error. For fixed T, the resulting increase of the number of steps in time is comparable to the explicit scheme.

What makes the **Crank-Nicolson scheme** special is a combination of three features: relative simplicity, unconditional stability, and quadratic accuracy in both time and space. For this scheme, a q-fold increase in the number of grid points in space and, with fixed T, a q-fold increase in the number of time steps, reduce the approximation error by the factor of q^{-2}.

While we have good reasons to believe that implicit methods (6.5.19), (6.5.20) should have no stability condition and that the error bounds (6.5.23) must be true, all these statements require proofs. Some of these proofs are outlined in Problem 7.12 on page 448.

EXERCISE 6.5.6. [B] *Write a computer program implementing the Crank-Nicolson scheme for the heat equation* $u_t = au_{xx} + H(t,x)$, $0 < t < T$, $0 < x < L$, *with initial condition* $u(0,x) = f(x)$ *and the boundary conditions* $b_1u(t,0) + b_2u_x(t,0) = h_1(t)$, $c_1u(t,L) + c_2u_x(t,L) = h_2(t)$, *where* b_1, b_2, c_1, c_2 *are real numbers and* h_1, h_2, H *are known functions. Test your program with* $a = 0.25$, $L = 1$, $T = 1$, $b_1 = c_1 = 1$, $b_2 = c_2 = 0$,

$h_1(t) = h_2(t) = H(t, x) = 0,$

$$f(x) = \begin{cases} 20x, & 0 \le x \le 1/2, \\ 20(1-x), & 1/2 \le x \le 1, \end{cases}$$

by comparing the approximate solution with the exact solution (6.1.35) on page 303. To use (6.1.35), compute the Fourier sine coefficients of f without using a computer, and take sufficiently many (say, 100) terms in the expansion. Hint: for general boundary conditions, values $\bar{u}_0(m)$ and $\bar{u}_N(m)$ are no longer zero. For the test problem, the exact solution tends to zero with time, and therefore you should expect the approximation error to decrease in time as well.

6.5.3 *One-Dimensional Wave Equation*

Our goal in this section is to study numerical approximation for solutions of the wave equation

$$\begin{aligned} &u_{tt}(t, x) = c^2 u_{xx}, \ 0 < t \le T, \ x \in (0, L); \\ &u(0, x) = f(x), \ u_t(0, x) = g(x), \ u(t, 0) = u(t, L) = 0. \end{aligned} \tag{6.5.25}$$

The explicit formula (6.1.56) on page 314 makes this study especially useful, as we can easily compare the exact and approximate solutions. The corresponding numerical schemes extend to more complicated hyperbolic equations which do not have explicit solutions.

There are three ways to solve (6.5.25) numerically using finite differences: (a) by discretizing the equation directly; (b) by writing the wave equation as an abstract evolution equation, see (6.2.16) on page 324, and then applying the same methods as for the heat equation; (c) writing the wave equation as a system of two transport equations

$$v_t + c v_x = 0, \ w_t - c w_x = 0 \tag{6.5.26}$$

and then discretizing the system.

EXERCISE 6.5.7.[C] *Verify that (6.5.26) is equivalent to $u_{tt} = c^2 u_{xx}$. What are the corresponding initial and boundary conditions for v, w, and how do you recover u from v, w? Hint: set $v = u_t - c u_x$ and $w = u_t + c u_x$.*

We restrict our discussion to the direct discretization of the wave equation. While not the easiest to generalize, this approach best serves our immediate goal of solving (6.5.26) numerically. The reader can then verify that many, but not all, of the resulting schemes can be obtained by discretizing the corresponding evolution system (6.2.16). For a discussion of hyperbolic systems in general and of (6.5.26) in particular, see Section 4.4 in the book *The Finite Difference Method in Partial Differential Equations* by A. R. Mitchell and D. F. Griffiths, 1980.

We start by discretizing the time and space intervals $[0, T]$ and $[0, L]$ with steps $\tau = T/M$ and $h = L/N$, respectively. Next, denote by $\bar{u}_n(m)$ the approximation of $u(m\tau, nh)$, $n = 0, \dots, N$, $m = 0, \dots, M$. Following (6.5.9) on page 394, we write

$$\begin{aligned} u_{tt}(m\tau, nh) &\approx \frac{\bar{u}_n(m+1) - 2\bar{u}_n(m) + \bar{u}_n(m-1)}{\tau^2}, \\ u_{xx}(m\tau, nh) &\approx \frac{\bar{u}_{n+1}(m) - 2\bar{u}_n(m) + \bar{u}_{n-1}(m)}{h^2}. \end{aligned} \tag{6.5.27}$$

Consider the matrix $A_{N+1}^{[-2,1]}$ as defined in (6.5.15) and introduce the column vector $\bar{u}(m) = (\bar{u}_0(m), \dots, \bar{u}_N(m))^T$; from the boundary conditions, $\bar{u}_0(m) = \bar{u}_N(m) = 0$. Since the equation is second-order in time, our approximation scheme can connect three vectors $\bar{u}(m-1)$, $\bar{u}(m)$, and $\bar{u}(m+1)$ in a symmetrical way using (6.5.27) as follows:

$$\begin{aligned} \bar{u}(m+1) - 2\bar{u}(m) + \bar{u}(m-1) = r^2 \big(\mu A_{N+1}^{[-2,1]} \bar{u}(m+1) \\ + (1-2\mu) A_{N+1}^{[-2,1]} \bar{u}(m) + \mu A_{N+1}^{[-2,1]} \bar{u}(m-1)\big), \end{aligned} \tag{6.5.28}$$

where $r = c\tau/h$, and $\mu \in [0, 1/2]$ is a parameter of the scheme. The three popular choices for μ are $0, 1/4$, and $1/2$. If $\mu = 0$, then (6.5.28) becomes

$$\bar{u}^{(1)}(m+1) = A_{N+1}^{[2(1-r^2), r^2]} \bar{u}^{(1)}(m) - \bar{u}^{(1)}(m-1), \tag{6.5.29}$$

which is an explicit method with the stability condition $r \leq 1$ or

$$\tau \leq h/c. \tag{6.5.30}$$

Unlike the corresponding condition for the heat equation (6.5.17), condition (6.5.30) requires τ to decrease linearly with h. Still, for large values of c, the required value of τ may result in a prohibitively large number M of time steps. Formula (6.5.30) is known as the `Courant-Friedrichs-Lewy condition` and was derived for a more general equation by three German mathematicians Richard Courant (1888–1972), Kurt Otto Friedrichs (1901–1982), and Hans Lewy (1904–1988); all three authors settled in the US by the end of the 1930s. The original paper with the result was in German and appeared in 1928. For a discussion of (6.5.30), see

Chapter 4 of the book *The Finite Difference Method in Partial Differential Equations* by A. R. Mitchell and D. F. Griffiths, 1980, or Sections 4.5 and 4.6 of the book *Finite Difference Methods for Partial Differential Equations* by G. E. Forsythe and W. R. Wasow, 2004.

EXERCISE 6.5.8. ^{B} *Assume that* $r = 1$ *so that* $h = c\tau$. *Verify that in this case (6.5.29) is an* **exact discretization** *of the wave equation: if* $u = u(t,x)$ *is the exact solution and* $\bar{u}_n^{(1)}(0) = u(0, nh)$, $\bar{u}_n^{(1)}(1) = u(\tau, nh)$, *then* $\bar{u}_n^{(1)}(m) = u(m\tau, nh)$ *for all* $m \geq 2$. *Explain why this observation does not necessarily make this particular scheme the best for solving the wave equation. Hint: notice that*

$$\bar{u}_n^{(1)}(m+1) = \bar{u}_{n-1}^{(1)}(m) + \bar{u}_{n+1}^{(1)}(m) - \bar{u}_n^{(1)}(m-1)$$

and use induction on m; *recall that the exact solution is* $u(t,x) = F(x+ct)+G(x-ct)$ *for suitable functions* F, G. *Can you achieve the equality* $\bar{u}_n^{(1)}(1) = u(\tau, nh)$ *in practice?*

If $\mu = 1/4$, then (6.5.28) becomes

$$A_{N+1}^{[2+r^2,-r^2/2]}\bar{u}^{(2)}(m+1) = A_{N+1}^{[2(2-r^2),r^2]}\bar{u}^{(2)}(m) - A_{N+1}^{[2+r^2,-r^2/2]}\bar{u}^{(2)}(m-1), \tag{6.5.31}$$

which is an implicit method, stable for all values of $r > 0$.

If $\mu = 1/2$, then (6.5.28) becomes

$$A_{N+1}^{[2(1+r^2),-r^2]}\bar{u}^{(3)}(m+1) = 4\bar{u}^{(3)}(m) - A_{N+1}^{[2(1+r^2),-r^2]}\bar{u}^{(3)}(m-1). \tag{6.5.32}$$

which is also an implicit method, stable for all values of $r > 0$; the matrix $A_K^{[\alpha,\beta]}$ is defined in (6.5.15). While computations in (6.5.31) and (6.5.32) require a solution of a linear system at each time step, the structure of the matrix $A_K^{\alpha,\beta}$ results in relatively easy computations. For more complicated equations, the extra computations required by the implicit methods are still more efficient than satisfying the stability condition of the explicit method.

The initial conditions $u_n^{(i)}(0)$ and $u_n^{(i)}(1)$, required in all these schemes, are computed as follows. For $m = 0$, we have $u_n^{(i)}(0) = u(0, nh) = f(nh)$, which is obvious. For $m = 1$, the ideal choice is $\bar{u}_n^{(i)}(1) = u(\tau, nh)$, but these values are unknown. On the other hand, we have $u(\tau, nh) \approx u(0, nh) + \tau u_t(0, nh) = f(nh) + \tau g(nh)$, and therefore we can take

$$\bar{u}_n^{(i)}(1) = f(nh) + \tau g(nh). \tag{6.5.33}$$

If the function f is sufficiently smooth, we can improve the quality of the approximation by observing that $u(\tau, nh) \approx u(0, nh) + \tau u_t(0, nh) +$

$(\tau^2/2)u_{tt}(0,nh)$. Using the wave equation $u_{tt}(0,nh) = c^2u_{xx}(0,nh) = c^2f''(nh)$, we have an alternative choice

$$\bar{u}_n^{(i)}(1) = f(nh)+\tau g(nh)+\frac{r^2}{2}\Big(f((n+1)h)-2f(nh)+f((n-1)h)\Big). \quad (6.5.34)$$

Inequality (6.5.10) on page 394 suggests that, for sufficiently smooth initial conditions f, g, the error bounds are

$$\max_{0\leq m\leq M,\ 0\leq n\leq N} |u(t_m, x_n) - \bar{u}_n^{(i)}(m)| \leq C^{(i)}(\tau^2 + h^2),\ i = 1,2,3; \quad (6.5.35)$$

the proof of these error bounds is beyond the scope of our discussion.

EXERCISE 6.5.9. *(a)^C Verify that (6.5.29), (6.5.31), and (6.5.32) are particular cases of (6.5.28). (b)^C Write (6.5.29), (6.5.31), and (6.5.32) in component form. Hint: for example, $\bar{u}_n^{(1)}(m+1) = r^2\bar{u}_{n-1}^{(1)}(m) + 2(1-r^2)\bar{u}_n^{(1)}(m) + r^2\bar{u}_{n+1}^{(1)}(m) - \bar{u}_n^{(1)}(m-1)$. (c)^A Write a computer program implementing (6.5.28) with $\mu \in [0,1/2]$ for the wave equation $u_{tt} = au_{xx}+H(t,x)$, $0<t<T$, $0<x<L$, with initial conditions $u(0,x) = f(x)$, $u_t(0,x) = g(x)$ and the boundary conditions $b_1u(t,0) + b_2u_x(t,0) = h_1(t)$, $c_1u(t,L) + c_2u_x(t,L) = h_2(t)$, where b_1, b_2, c_1, c_2 are real numbers and h_1, h_2, H are known functions. Test your program with $c = 2$, $L = 1$, $T = 2$, $b_1 = c_1 = 1$, $b_2 = c_2 = 0$, $h_1(t) = h_2(t) = H(t,x) = 0$, $g(x) = 0$, and with two initial displacements f_1, f_2, where $f_1(x) = x(1-x)$ and*

$$f_2(x) = \begin{cases} 20x, & 0 \leq x \leq 1/2, \\ 20(1-x), & 1/2 \leq x \leq 1. \end{cases}$$

Compare the approximate solution corresponding to $\mu = 1/2$ with the exact solution (6.1.56) on page 314. In each case try two different values of $\bar{u}(1)$, according to (6.5.33) and (6.5.34). Other conditions being equal, which selection results in smaller approximation error?

6.5.4 *The Poisson Equation in a Rectangle*

The objective of this section is to study numerical approximation for the solutions of the Dirichlet problem in the square:

$$u_{xx}(x,y) + u_{yy}(x,y) = 0,\ 0 < x,y < L, \quad (6.5.36)$$

with the boundary conditions $u(x,0) = f_1(x)$, $u(0,y) = g_1(y)$, $u(x,L) = f_2(x)$, $u(L,y) = g_2(y)$, where f_1, f_2, g_1, g_2 are known continuously differentiable functions on $[0, L]$, and

$$f_1(0) = g_1(0),\ f_1(L) = g_2(0),\ f_2(0) = g_1(L),\ f_2(L) = g_2(L). \tag{6.5.37}$$

EXERCISE 6.5.10.^{C} *What is the meaning of conditions (6.5.37)? Hint: draw a picture.*

We now construct an approximate solution of (6.5.36) using finite differences. For a positive integer N, define the mesh size $h = L/N$ and the mesh points $(x_m, y_n) = (mh, nh)$, $m, n = 0, \dots, N$. Denote by $\bar{u}_{m,n}$ the approximation of $u(mh, nh)$. Following (6.5.9) on page 394, we write $u_{xx}(mh, nh) \approx (\bar{u}_{m+1,n} - 2\bar{u}_{m,n} + \bar{u}_{m-1,n})/h^2$, $u_{yy}(mh, nh) \approx (\bar{u}_{m,n+1} - 2\bar{u}_{m,n} + \bar{u}_{m,n-1})/h^2$, and then, after multiplying by h^2, use (6.5.36) to get a system of linear equations

$$\bar{u}_{m+1,n} + \bar{u}_{m-1,n} + \bar{u}_{m,n+1} + \bar{u}_{m,n-1} - 4\bar{u}_{m,n} = 0,\ m, n = 1, \dots N - 1. \tag{6.5.38}$$

Note that this system is *not* homogeneous, because $4N$ of the values $\bar{u}_{m,n}$ are given by the boundary conditions:

$$\bar{u}_{m,0} = f_1(mh),\ \bar{u}_{m,N} = f_2(mh),\ \bar{u}_{0,n} = g_1(nh),\ \bar{u}_{N,n} = g_2(mh). \tag{6.5.39}$$

For example, the very first equation in the system, corresponding to $m = n = 1$, is

$$-4\bar{u}_{1,1} + \bar{u}_{2,1} + \bar{u}_{1,2} = -f(h) - g_1(h).$$

As a result, we can write (6.5.38) as an inhomogeneous system of linear $(N-1)^2$ equations in $(N-1)^2$ unknowns $U = (\bar{u}_{m,n},\ m, n = 1, \dots, N-1)$:

$$\mathbf{A}U = b, \quad \text{with the matrix } \mathbf{A} = \begin{pmatrix} B & I & 0 \cdots & 0 \\ I & B & I \cdots & 0 \\ 0 & I & B & I \ \vdots \\ \vdots & & \ddots & \ddots \ I \\ 0 & \cdots & & I \ B \end{pmatrix}, \tag{6.5.40}$$

where each matrix B is equal to $A_{N-1}^{[-4,1]}$, see (6.5.15) on page 395, and $I = A_{N-1}^{[1,0]}$ is the $(N-1) \times (N-1)$ identity matrix.

EXERCISE 6.5.11.^{C} *(a) Describe the components U_k, $k = 1, \dots, (N-1)^2$ of the vector U, that is, describe the rule that assigns $\bar{u}_{m,n}$ to the particular U_k. (b) Describe the components of the vector b.*

EXERCISE 6.5.12.[A] *(a) For what functions does the matrix* $\mathbf{A}$ *provide an* **exact discretization**, *similar to part (b) of Exercise 6.5.4? (b) Using the methods and results of Problem 7.12 on page 448, verify that the* $(N-1)^2$ *eigenvalues of the matrix* $\mathbf{A}$ *are* $\lambda_{m,n} = -4\big(\sin^2(\pi m/(2N)) + \sin^2(\pi n/(2N))\big)$, $m, n = 1, \ldots, N-1$, *and the components of the eigenvector* $\boldsymbol{x}_{m,n}$ *corresponding to* $\lambda_{m,n}$ *are* $\boldsymbol{x}_{m,n}(k,l) = \sin(k\pi m/N)\sin(l\pi n/N)$, $k, l = 1, \ldots, N-1$.

Unlike the heat and wave equations, there are no steps in time, as there is no time variable. As a result, there are no explicit or implicit methods and no stability issues that would have analogues in the previous two sections. Still, explicit and implicit methods and the questions of numerical stability appear in the process of solving numerically the system of equations (6.5.40). Solving (6.5.40) numerically, especially for large N, relies on advanced programming skills and numerical linear algebra; important as they are, these topics are beyond the scope of our discussion.

Inequality (6.5.10) on page 394 suggests that, if the exact solution $u = u(x, y)$ is sufficiently smooth, then

$$\sum_{0 \le m,n \le N} |\bar{u}_{m,n} - u(x_m, y_n)| \le Ch^2. \tag{6.5.41}$$

As with other error bounds, the proof is outside the scope of our discussion.

6.5.5 *The Finite Element Method*

For partial differential equations, there are two main types of numerical methods: finite difference and finite element. In the *finite difference* methods, we replace all the partial derivatives in the equation with the corresponding finite differences and get a *finite difference equation* for the approximation. The numerical schemes we have studied so far are all based on the finite difference method. In the *finite element* method, we approximate the solution by a finite linear combination of some known functions, and then use the equation to determine the *optimal*, in a certain sense, values of the coefficients in this linear combination.

In this section, we will briefly discuss the general ideas of the finite element method. There are two possible interpretations of the method, using either *weighted residuals* or a *variational formulation.* Both interpretations are best described for an abstract equation $\mathcal{A}\boldsymbol{x} = \boldsymbol{f}$, where $\mathcal{A}$ is a linear operator on an inner product space $\mathbb{H}$, $\boldsymbol{f} \in \mathbb{H}$ is a known element of $\mathbb{H}$, and

$\boldsymbol{x}$ is the unknown element; see pages 327–329 for the definitions.

According to the `method of weighted residuals`, we look for an *approximate solution* $\overline{\boldsymbol{x}}$ of the equation $\mathcal{A}\boldsymbol{x} = \boldsymbol{f}$ in the form $\overline{\boldsymbol{x}} = \sum_{k=1}^{N} a_k\varphi_k$, where a_k are real or complex numbers to be determined, and φ_k are known and fixed elements of $\mathbb{H}$. These elements are not necessarily a part of an orthonormal system in $\mathbb{H}$. We call $\overline{\boldsymbol{x}}$ a `finite element approximation` of the solution, based on the `finite elements` φ_k. In applications, the functions φ_k are often equal to zero outside of a small bounded region, whence the name "finite element."

By definition, the `residual` is $R(a) = \mathcal{A}\overline{\boldsymbol{x}} - \boldsymbol{f}$, and then the coefficients a_k are computed from the condition $(R(a), \psi_k) = 0$, $k = 1, \ldots, N$, where $\psi_k, k = 1, \ldots, N$, are known and fixed elements of $\mathbb{H}$, called `weights`, and $(\cdot, \cdot)$ is the inner product in $\mathbb{H}$. For linear operator $\mathcal{A}$, this leads to a system of N linear equations with n unknowns. The hope is that, as $N \to \infty$, the resulting approximation $\sum_{k=1}^{N} a_k\varphi_k$ will converge to the solution of the equation. Of course, the proof of this convergence is impossible without additional assumptions about the space $\mathbb{H}$, the operator $\mathcal{A}$, the finite elements φ_k and the weights ψ_k. Note also that, in general, the computation of a_k is not recursive: a different value of N will require new computations for *all* values of a_k, $k = 1, \ldots, N$.

`Galerkin's method`, suggested in 1915 by the Russian engineer BORIS GRIGORIEVICH GALERKIN (1871–1945), is a particular case of weighted residuals with a special selection of weights: $\psi_k = \varphi_k$ for all k. The idea of this method can be formulated alternatively as follows: if $x_0 = \sum_{m=1}^{\infty} a_m\varphi_m$ is the exact solution of $\mathcal{A}x = f$, then

$$(\mathcal{A}x_0, \varphi_k) = \left(\sum_{m=1}^{\infty} a_m \mathcal{A}\varphi_m, \varphi_k\right) = (f, \varphi_k)$$

for all $k \geq 1$. Then a natural approximation of x_0 is computed by truncating the above system.

EXERCISE 6.5.13.[B] *(a) Using the Galerkin method, write the system of equations for the coefficients a_k in the matrix-vector form. Hint: in components, $\sum_{m=1}^{N}(\mathcal{A}\varphi_m, \varphi_k)a_m = (f, \varphi_k)$. (b) How will the system simplify if φ_k form an orthonormal system and are eigenfunctions of the operator $\mathcal{A}$? Hint: it becomes diagonal.*

There is an alternative approach to solving the equation $\mathcal{A}\boldsymbol{x} = \boldsymbol{f}$, called the `variational approach`, when we interpret the solution $\boldsymbol{x}_0$ of the equation as the extremal point of some functional. A **functional** is a mapping from $\mathbb{H}$ to $\mathbb{R}$; an **extremal point** of a functional J is an element $x \in \mathbb{H}$ where J has a local minimum or maximum. Note that the element $\boldsymbol{x}_0$ is a solution of the equation $\mathcal{A}\boldsymbol{x} = \boldsymbol{f}$ if and only if $\|\mathcal{A}\boldsymbol{x}_0 - \boldsymbol{f}\| = 0$, where $\|\cdot\|$ is the norm in the space $\mathbb{H}$. In other words, the functional $J(\boldsymbol{x}) = \|\mathcal{A}\boldsymbol{x} - \boldsymbol{f}\|^2$ achieves its minimal value when $\boldsymbol{x} = \boldsymbol{x}_0$; for computational purposes, it is better to consider the square of the norm rather than the norm itself, and, since $J(\boldsymbol{x}) \geq 0$ for all $\boldsymbol{x}$, zero is indeed the minimal value of J. There can be other functionals whose extremal points correspond to the solution of the equation; for example, the *Dirichlet principle* on page 161, provides an alternative functional for the Poisson equation. For many equations, numerical computation of the extremal points of a corresponding functional is more efficient than solving the equation directly.

The variational approach leads to a different interpretation of the finite element method. Let the solution $\boldsymbol{x}_0$ of the equation $\mathcal{A}\boldsymbol{x} = f$ be the extremal point of a functional J. As before, we look for an approximate solution in the form $\overline{\boldsymbol{x}} = \sum_{k=1}^{N} a_k\varphi_k$, where $\varphi_1, \ldots, \varphi_N$ are the finite elements, that is, fixed elements of $\mathbb{H}$. Define the function $F(a_1, \ldots, a_N) = J(\overline{\boldsymbol{x}})$. Then F is a real-valued function of N variables, and a natural condition for choosing a_k is $\partial F/\partial a_k = 0$, $k = 1, \ldots, N$. In other words, the hope is that if $(a_1^*, \ldots, a_N^*)$ is a point of local minimum or maximum of F, then $\sum_{k=1}^{N} a_k^*\varphi_k$ is close to an extremal point of J. This method of solving the (infinite-dimensional) variational problems is sometimes called the **Rayleigh-Ritz method**, after the British physicist John William Strutt, Lord Rayleigh (1842–1919) and the Swiss physicist Walter Ritz (1878–1909), who published papers on the subject in 1870 and 1908, respectively. As with the weighted residuals, the computations of a_k^* are usually not recursive (any change in the number of finite elements requires a complete recalculation of all coefficients a_k), and any proof of convergence requires additional assumptions about the space $\mathbb{H}$, the operator $\mathcal{A}$, and the finite elements φ_k.

The next exercise shows the connection between the weighted residual and the variation interpretations of the finite element method.

Exercise 6.5.14. ^{B} *Consider the equation $\mathcal{A}\boldsymbol{x} = \boldsymbol{f}$ in an inner product space $\mathbb{H}$, and let $\overline{\boldsymbol{x}} = \sum_{k=1}^{N} a_k\varphi_k$ be an approximate solution. (a) Using the weighted residuals method with weights $\psi_k = \mathcal{A}\varphi_k$, write the system of equations for the coefficients a_k. (b) Consider the function $F(a_1, \ldots, a_N) = \|\mathcal{A}\bar{x} - f\|^2$. Verify that the system of equations $\partial F/\partial a_k = 0$, $k = 1, \ldots, N$,* *is the same as the one you obtained in part (a).*

EXERCISE 6.5.15.[A] *According to the* Dirichlet principle *on page 161, the solution of the Poisson equation* $\nabla^2 u = g$ *in* $G \subset \mathbb{R}^n$, $u = h$ *on* ∂G *minimizes the functional* $I(v) = (1/2)\int_G (\|\nabla v\|^2 + g\,v)\,d\mathfrak{m}$, *where* $d\mathfrak{m} = dA$ *or* $d\mathfrak{m} = dV$, *depending on whether* $n = 2$ *or* $n = 3$. *Write the system of linear equations for the coefficients in the corresponding finite element approximation of the solution. Hint: the number of spatial dimensions makes no difference, as long as you use the notation* $(u, v) = \int_G u\,v\,d\mathfrak{m}$. *Note that, for given* $\varphi_1, \ldots, \varphi_N$, *the function* $F(a_1, \ldots, a_N) = I\left(\sum_{k=1}^N a_k \varphi_k\right)$ *is quadratic in the* a_k.

The finite element methods applies also to evolution equations, such as $u(t) = u_0 + \int_0^t \mathcal{A}u(s)ds$. The weighted residual method is the typical interpretation in this setting, with the following modifications: (a) The coefficients a_k in the representation $\overline{u} = \sum_k^N a_k \varphi_k$ are functions of time; (b) The residual $R(t) = \sum_{k=1}^N a_k(t)\varphi_k - u_0 - \sum_{k=1}^N \int_0^t a_k(s)\mathcal{A}\varphi_k\,ds$ is also a function of time; (c) The condition is $(R(t), \psi_k) = 0$ for $k = 1, \ldots, N$ and $t \geq 0$. As a result, the functions $a_k = a_k(t)$ satisfy a system of ordinary differential equations. FOR EXAMPLE, consider Galerkin's method applied to the equation $u(t) = u_0 + \int_0^t \mathcal{A}u(s)ds$ under the assumption that φ_k, $k = 1, \ldots, N$, form an orthonormal system, that is $(\varphi_k, \varphi_m) = 0$, $k \neq m$, $(\varphi_k, \varphi_k) = 1$. The reader can verify that the corresponding ordinary differential equations for a_k are

$$a_k'(t) = \sum_{m=1}^N (\mathcal{A}\varphi_m, \varphi_k)a_m(t),\ t > 0,\ k = 1, \ldots, N,\ \ a_k(0) = (u_0, \varphi_k).$$

What happens if, in this setting, $\mathcal{A}\varphi_k = \lambda_k \varphi_k$ for all k?

A popular choice of finite elements is piece-wise linear functions: these are the easiest continuous functions. Since such functions have at most one derivative, some extra tricks must be used. We illustrate the idea on the easiest example, the Poisson equation on an interval

$$-u''(x) = f(x),\ x \in (0, 1),\ u(0) = u(1) = 0; \tag{6.5.42}$$

for more examples, see Problem 7.14 on page 449. Let $x_k = k/(N + 1) = kh$, $k = 0, \ldots, N + 1$, be a uniform partition of $(0, 1)$. Let $\varphi(x)$ be the "triangular" piecewise linear function

$$\varphi(x) = \begin{cases} 1 - |x|, & |x| < 1, \\ 0; & |x| \geq 1. \end{cases} \tag{6.5.43}$$

Stretching the definition of the derivative a little, we say that

$$\varphi'(x) = \begin{cases} -1, & 0 < x < 1; \\ 1, & -1 < x < 0; \\ 0, & \text{otherwise.} \end{cases}$$

Then define $\varphi_k(x) = \varphi((x - x_k)/h)$, $k = 1, \ldots, N$.

EXERCISE 6.5.16.^{C} *(a) Draw the graphs of* $\varphi_k(x)$ *when* $N+1 = 5$. *(b) Pretending that each* φ_k *has two derivatives, show that* $-\int_0^1 \varphi_k''(x)\varphi_m(x)dx = \int_0^1 \varphi_k'(x)\varphi_m'(x)dx$ *for all* $m, k = 1, \ldots, N$. *(c) With part (b) in mind, verify that the Galerkin approximation* $\sum_{k=1}^N u_k\varphi_k(x)$ *for (6.5.42) satisfies*

$$\sum_{k=1}^N u_k \int_0^1 \varphi_k'(x)\varphi_m'(x)dx = \int_0^1 f(x)\varphi_m'(x)dx, \quad m = 1, \ldots, N. \tag{6.5.44}$$

(d) Verify that (6.5.44) can be written as a system of linear equations $A_N^{[-2,1]}U = b$, *where* $U = (u_1, \ldots, u_N)^T$ *and the matrix* $A_N^{[-2,1]}$ *is defined in (6.5.15) on page 395. Identify the vector* b. *Note that the same matrix* A *appears in the finite difference approximation of (6.5.42). (e) Write a program implementing this finite element approximation and test it with* $N + 1 = 21$ *and* $f(x) = \pi^2 \sin(\pi x)$. *Compare the result with both the exact solution* $u(x) = \sin \pi x$ *and a finite difference approximation.*

The above example shows that implementation of a finite element methods is much more time consuming and labor-intensive than the implementation of a finite difference method. On the other hand, finite element methods are more flexible, and can be easily adjusted to various domain shapes and operators. A standard reference on the subject is the book *An Analysis of the Finite Element Method* by G. Strang and G. Fix, 1973.

Chapter 7

Further Developments and Special Topics

This chapter presents further development of certain subjects discussed in previous chapters. The material in this chapter is either of a more advanced nature or somewhat tangential to the main line of the earlier development, or else covers new material that could not be conveniently included earlier. We present this material in the form of problems with the aid provided here or in earlier chapters. We hope that some of the problems will have an element of fun.

7.1 Geometry and Vectors

This section collects problems related to Euclidean geometry, vector algebra, and spatial curves. Some problems are rather standard: Problem 1.2 is about using unit vectors to derive some trigonometric identities, Problem 1.6 is about skewed lines, and Problem 1.11 is about a practical way to compute torsion and curvature of a curve. Problems 1.10, 1.12, 1.13 address some other curve-related questions. Problems 1.3, 1.4, and 1.5 provide a deeper look into the cross product. Problem 1.7 is necessary given our discussion of linear dependence in various places throughout these notes. Problem 1.8 discusses alternative ways to measure distance in $\mathbb{R}^n$. Problem 1.9 might be helpful in understanding the mathematical meaning of a rigid motion. The geometric Problem 1.1 is just for fun.

PROBLEM 1.1. Because of the round shape of the Earth, we cannot see objects on the surface arbitrarily far away, even in the ocean and even with a spy-glass. Let L be the longest distance from which one can see an object of height H on the surface of the Earth from hight h above the Earth. Drive an approximate formula for L assuming that H and h are much smaller than the radius R of the Earth. How high above Honolulu must one get to see the California coast?

PROBLEM 1.2. Let $\widehat{\boldsymbol{u}} = \cos\alpha\,\hat{\boldsymbol{\imath}} + \sin\alpha\,\hat{\boldsymbol{\jmath}}$ and $\widehat{\boldsymbol{v}} = \cos\beta\,\hat{\boldsymbol{\imath}} + \sin\beta\,\hat{\boldsymbol{\jmath}}$, where $0 \le \alpha \le \beta \le 2\pi$.

(a) Verify that $\widehat{\boldsymbol{u}}$ and $\widehat{\boldsymbol{v}}$ are unit vectors.

(b) Compute $\widehat{\boldsymbol{u}} \cdot \widehat{\boldsymbol{v}}$.

(c) Plot the position vectors $\widehat{\boldsymbol{u}}$ and $\widehat{\boldsymbol{v}}$, interpreting α and β as the angles they make with the x axis. What is the angle between $\widehat{\boldsymbol{u}}$ and $\widehat{\boldsymbol{v}}$?

(d) Derive the trigonometric formula for $\cos(\beta-\alpha)$. Then take $\alpha \le 0 \le \beta$ and derive the formula for $\cos(\alpha+\beta)$.

PROBLEM 1.3. Complete the proof of Theorem 1.2.2 on page 20. *Hint: Equations (1.2.16) imply property (C2), that is,* $\boldsymbol{w} \cdot \boldsymbol{u} = 0$ *and* $\boldsymbol{w} \cdot \boldsymbol{v} = 0$. *Also, with* w_1, w_2, w_2 *as in (1.2.16), property (C4) follows. Finally, calculate* $\|\boldsymbol{w}\|^2 = w_1^2 + w_2^2 + w_3^2$ *and establish property (C1), that is,* $\|\boldsymbol{w}\|^2 = \|\boldsymbol{u}\|^2\|\boldsymbol{v}\|^2(1-\cos^2\theta)$, *where* $\cos\theta = (\boldsymbol{u}\cdot\boldsymbol{v})/(\|\boldsymbol{u}\|\|\boldsymbol{v}\|)$.

PROBLEM 1.4. (a) Prove `Lagrange's Identity`

$$(\boldsymbol{r}\times\boldsymbol{u})\cdot(\boldsymbol{v}\times\boldsymbol{w}) = (\boldsymbol{r}\cdot\boldsymbol{v})(\boldsymbol{u}\cdot\boldsymbol{w}) - (\boldsymbol{r}\cdot\boldsymbol{w})(\boldsymbol{u}\cdot\boldsymbol{v}).$$

Hint: By (1.2.30), $\boldsymbol{r}\cdot(\boldsymbol{u}\times\boldsymbol{v}) = \boldsymbol{v}\cdot(\boldsymbol{r}\times\boldsymbol{u})$. *Replace* $\boldsymbol{v}$ *by* $\boldsymbol{v}\times\boldsymbol{w}$ *in this equality to get* $\boldsymbol{r}\cdot\big(\boldsymbol{u}\times(\boldsymbol{v}\times\boldsymbol{w})\big) = (\boldsymbol{v}\times\boldsymbol{w})\cdot(\boldsymbol{r}\times\boldsymbol{u})$. *Now, apply* $\boldsymbol{r}\cdot$ *to (1.2.27).*

(b) Prove that $(\boldsymbol{r}\times\boldsymbol{u})\times(\boldsymbol{v}\times\boldsymbol{w}) = (\boldsymbol{r},\boldsymbol{u},\boldsymbol{w})\boldsymbol{v} - (\boldsymbol{r},\boldsymbol{u},\boldsymbol{v})\boldsymbol{w}$, where $(\boldsymbol{r},\boldsymbol{u},\boldsymbol{v})$ is the scalar triple product. *Hint: In (1.2.27), replace* $\boldsymbol{u}$ *by* $\boldsymbol{r}\times\boldsymbol{u}$.

PROBLEM 1.5. An alternative notation for the cross product $\boldsymbol{u}\times\boldsymbol{v}$ is $[\boldsymbol{u},\ \boldsymbol{v}]$. With this notation, equality (1.2.26) becomes

$$\big[\boldsymbol{u},\ [\boldsymbol{v},\ \boldsymbol{w}]\big] + \big[\boldsymbol{v},\ [\boldsymbol{w},\ \boldsymbol{u}]\big] + \big[\boldsymbol{w},\ [\boldsymbol{u},\ \boldsymbol{v}]\big] = \boldsymbol{0}. \tag{7.1.1}$$

There exist other operations $[\cdot,\ \cdot]$ with a similar property. For example, for two $n\times n$ matrices A, B, define $[A, B] = AB - BA$. Verify that, for every three such matrices A, B, C,

$$\big[A,\ [B,\ C]\big] + \big[B,\ [C,\ A]\big] + \big[C,\ [A,\ B]\big] = 0_{[n]}\,, \tag{7.1.2}$$

where $0_{[n]}$ is the zero matrix.

(a) Can you find another example of such an operation?

(b) Is it true that every operation with property (7.1.2) must satisfy $[x, y] = -[y, x]$?

PROBLEM 1.6. Two lines in $\mathbb{R}^3$ are called **skewed** if they are neither parallel nor intersecting. Let $\boldsymbol{r}_1(t) = \boldsymbol{r}_{10} + t\,\boldsymbol{d}_1$ and $\boldsymbol{r}_2(s) = \boldsymbol{r}_{20} + s\,\boldsymbol{d}_2$ be two lines in $\mathbb{R}^3$.

(a) Find the condition in terms of $\boldsymbol{r}_{01}$, $\boldsymbol{r}_{02}$, $\boldsymbol{d}_1$ and $\boldsymbol{d}_2$ for these lines to be skewed.

(b) Find the equation of the line that is perpendicular to both skewed lines and intersects them.

(c) Find the distance between two skewed lines.

(d) Given two skewed lines, there are two parallel planes, each containing one line. Find the equations of those planes.

PROBLEM 1.7. We say that the vectors $\boldsymbol{u}_1, \ldots, \boldsymbol{u}_k$ in $\mathbb{R}^n$ are *linearly dependent* if there exist real numbers $x_1, \ldots, x_k$ so that at least one of these numbers is not zero and $x_1\,\boldsymbol{u}_1 + \cdots + x_k\,\boldsymbol{u}_k = \boldsymbol{0}$.

(a) Verify that if at least one of the vectors $\boldsymbol{u}_1, \ldots, \boldsymbol{u}_k$ is the zero vector, then the vectors are linearly dependent.

(b) Verify that every four vectors in $\mathbb{R}^3$ are linearly dependent. *Possible hint: choose a cartesian coordinate system, and write the corresponding condition for linear dependence using the components of the vectors. How many equations are there? How many unknowns are there?*

(c) Verify that $n+1$ vectors in $\mathbb{R}^n$ are always linearly dependent.

(d) Verify that, if a vector space has n linearly independent vectors, but $n+1$ vectors are always linearly dependent, then the space is n-dimensional. *Hint: those n linearly independent vectors can be taken as a basis.*

PROBLEM 1.8. Define a **norm** in $\mathbb{R}^n$ as a function $\mathfrak{N}$ that, to every element $\boldsymbol{x}$ in $\mathbb{R}^n$, assigns a non-negative number $\mathfrak{N}(\boldsymbol{x})$ so that $\mathfrak{N}(\boldsymbol{x}) = 0$ if and only if $\boldsymbol{x} = \boldsymbol{0}$, and also $\mathfrak{N}(\boldsymbol{x}+\boldsymbol{y}) \le \mathfrak{N}(\boldsymbol{x}) + \mathfrak{N}(\boldsymbol{y})$ and $\mathfrak{N}(a\boldsymbol{x}) = |a|\mathfrak{N}(\boldsymbol{x})$ for every $\boldsymbol{x}, \boldsymbol{y} \in \mathbb{R}^n$ and every real number a.

(a) Fix a basis $\boldsymbol{u}_1, \ldots, \boldsymbol{u}_n$, and, for $\boldsymbol{x} = \sum_{k=1}^n x_k\,\boldsymbol{u}_k$ and $p \ge 1$, define $\|\boldsymbol{x}\|_p = \left(\sum_{k=1}^n |x_k|^p\right)^{1/p}$; $\|\boldsymbol{x}\|_\infty = \max_{1\le k\le n} |x_k|$. Verify that $\|\cdot\|_p$ and $\|\cdot\|_\infty$ are norms.

(b) Let $1 \le p, q \le \infty$ be numbers so that $(1/p)+(1/q) = 1$ (for example, if $p = 1$, then $q = \infty$). Show that, for all $\boldsymbol{x}, \boldsymbol{y}$ from $\mathbb{R}^n$, we have $|\boldsymbol{x}\cdot\boldsymbol{y}| \le \|\boldsymbol{x}\|_p\,\|\boldsymbol{y}\|_q$.

(c) For $1 \le p < r \le \infty$, find positive numbers c and C, possibly depending on p, r, and n, so that $c\|\boldsymbol{x}\|_r \le \|\boldsymbol{x}\|_p \le C\|\boldsymbol{x}\|_r$ for all $\boldsymbol{x} \in \mathbb{R}^n$. Is it possible to have c and/or C independent of n?

(d) Let $\mathfrak{N}_1$ and $\mathfrak{N}_2$ be two norms on $\mathbb{R}^n$. Show that there exist two positive numbers c, C, possibly depending on n, so that

$$c\,\mathfrak{N}_1(\boldsymbol{x}) \le \mathfrak{N}_2(\boldsymbol{x}) \le C\,\mathfrak{N}_1(\boldsymbol{x})$$

for all $\boldsymbol{x} \in \mathbb{R}^n$ *Hint: for every norm* $\mathfrak{N}$*, the set* $\{\boldsymbol{x} \in \mathbb{R}^n : \mathfrak{N}(\boldsymbol{x}) \le 1\}$ *is closed and bounded; every continuous function on such a set achieves its minimal and maximal values.*

(e) The norm $\|\cdot\|_2$ can be defined using an inner product; see formula (1.2.9) on page 14. Can other $\|\cdot\|_p$ norms be defined using an inner product?

PROBLEM 1.9. Show that every isometry is an orthogonal transformation followed by a parallel translation. Below is an outline of a possible proof.
Step 1. Denote the isometry transform by A and let $\boldsymbol{x}_0 = A(\boldsymbol{0})$. Define the transform S by $S(\boldsymbol{x}) = \boldsymbol{x} + \boldsymbol{x}_0$. Then S is the required parallel translation.
Step 2. Define the transform U by $U(\boldsymbol{x}) = A(\boldsymbol{x}) - \boldsymbol{x}_0$.
Step 3. By construction, U is an isometry and $U(\boldsymbol{0}) = \boldsymbol{0}$. Use this to show that U preserves the norm: $\|U\boldsymbol{x}\| = \|\boldsymbol{x}\|$, and then use the parallelogram law (1.2.10), page 15, to show that U preserves the dot product.
Step 4. Show that U is linear by verifying that $\|U(\boldsymbol{x}+\boldsymbol{y}) - U(\boldsymbol{x}) - U(\boldsymbol{y})\|^2 = 0$ and $\|U(\lambda\,\boldsymbol{x}) - \lambda\,U(\boldsymbol{x})\|^2 = 0$.

PROBLEM 1.10. Let $\mathcal{C} : \boldsymbol{r}(t),\ -1 < t < 1$, be a curve for which $\boldsymbol{r}'(t)$ exists for all t. Does this mean that $\mathcal{C}$ is rectifiable? Give a proof or provide a counterexample.

PROBLEM 1.11. Let $\mathcal{C}$ be a curve defined by the vector-valued function $\boldsymbol{r} = \boldsymbol{r}(t)$ that is at least three times continuously differentiable in t. Refer to page 31 and verify the following equalities:

$$\widehat{\boldsymbol{u}}(t) = \frac{\boldsymbol{r}'(t)}{\|\boldsymbol{r}(t)\|},\quad \widehat{\boldsymbol{p}}(t) = \frac{\widehat{\boldsymbol{u}}'(t)}{\|\widehat{\boldsymbol{u}}'(t)\|},\quad \widehat{\boldsymbol{b}}(t) = \widehat{\boldsymbol{u}}(t) \times \widehat{\boldsymbol{p}}(t),$$

$$\kappa(t) = \frac{\|\boldsymbol{r}'(t) \times \boldsymbol{r}''(t)\|}{\|\boldsymbol{r}'\|^3},\quad \tau(t) = \frac{(\boldsymbol{r}', \boldsymbol{r}'', \boldsymbol{r}''')}{\|\boldsymbol{r}'\|^2\,\|\boldsymbol{r}''\|^2 - (\boldsymbol{r}' \cdot \boldsymbol{r}'')^2}.$$

PROBLEM 1.12. Show that the torsion of a planar curve is identically zero. Is the converse true as well?

PROBLEM 1.13. Similar to the osculating plane (page 32), one can define the `osculating circle` as the circle that comes closest to the curve at a given point. In particular, the radius of this circle is the reciprocal of the curvature. Complete the definition of the osculating circle, and describe the set of points made by the centers of the osculating circles for the right-handed circular helix (1.3.13) on page 30.

7.2 Kinematics and Dynamics

The problems on kinematics and dynamics can easily fill several large volumes, even without solutions. In what follows, we present problems we believe are closely connected with the topics we discuss in the main text. In particular, Problem 2.1 suggests a rather comprehensive study of the motion in a central force field, with emphasis on the attracting inverse-square law. Problem 2.2 presents an extension of Problem 2.1, including the famous general relativity correction to the perihelion precession of Mercury. Problem 2.3 suggests a closer look at the nonlinear oscillator, leading to an elliptic integral representation of the solution. The next three problems are rather standard: Problem 2.4 is about the Parallel Axis Theorem, which provides an effective method to compute the moment of inertia; Problem 2.5 is about describing the orientation of a rigid body in space using the Euler angles; Problem 2.6 is about the kinetic energy of a rotating object. Problems 2.7 and 2.8 illustrate our discussion of a rigid body motion, for a rolling cylinder and a tumbling box, respectively; the tumbling box problem suggests a very easy and effective in-class demonstration. Finally, Problem 2.9, completes the study of a bead sliding on a helical wire; the study starts on page 90.

PROBLEM 2.1. MOTION IN A CENTRAL FORCE FIELD.

Consider a planar motion in polar coordinates under an attracting *central force field.* Gravity and electrostatic attraction are examples of central force fields. Choose polar coordinates in the plane of motion, and let the unit vector $\widehat{\boldsymbol{\kappa}}$ complete the couple $(\widehat{\boldsymbol{r}},\ \widehat{\boldsymbol{\theta}})$ to a right-handed triad. Then every central force acts radially toward the origin O at every point (r, θ) and therefore has the form $\boldsymbol{F} = -f(r)\widehat{\boldsymbol{r}}$ for some positive function $f = f(r)$. Assume that the frame $O(\widehat{\boldsymbol{r}},\ \widehat{\boldsymbol{\theta}},\ \widehat{\boldsymbol{\kappa}})$ is inertial.

(a) Using formulas (1.3.26) on page 36, obtain the equations of motion of a point mass m acted on by $\boldsymbol{F}$:

$$\ddot{r}(t) - r(t)\dot{\theta}^2(t) = -f/m \text{ (radial part);} \tag{7.2.1}$$

$$r(t)\ddot{\theta}(t) + 2\dot{r}(t)\dot{\theta}(t) = 0 \text{ (angular part.)} \tag{7.2.2}$$

(b) Apply formula (2.1.4) on page 40 and formula (1.3.24) on page 35, to compute the angular momentum $\boldsymbol{L}_O$ of the point mass about O. *Hint:*

$\boldsymbol{L}_O = mr\,\widehat{\boldsymbol{r}} \times r\omega\,\widehat{\boldsymbol{\theta}} = mr^2\omega\,\widehat{\boldsymbol{k}}$.

(c) Prove that the angular momentum is conserved, that is, $\boldsymbol{L}_O(t)$ is a constant vector, independent of time. *Hint: show that* $d\boldsymbol{L}_O(t)/dt = \boldsymbol{0}$.

(d) Multiply (7.2.2) by $r(t)$ and show that $r^2(t)\omega(t) = \alpha$, where $\omega(t) = \dot{\theta}(t)$ and α is a number independent of t. Then re-write (7.2.1) as

$$\ddot{r} - \frac{\alpha^2}{r^3} = -\frac{f}{m} = -f_1. \tag{7.2.3}$$

(e) In (7.2.3), set $r(t) = 1/u(\theta(t))$ and show that

$$\alpha^2 u^2(\theta)\big(u''(\theta) + u(\theta)\big) = f_1. \tag{7.2.4}$$

Hint: $\dot{r} = -\alpha du/d\theta$, $\ddot{r} = -\alpha^2 u^2 d^2u/d\theta^2$.

(f) Let e be a non-negative number, and ℓ, a positive number. Consider the following curve in polar coordinates:

$$r(1 - e\cos\theta) = \ell, \tag{7.2.5}$$

What curve is this and what are the geometric interpretations of e and ℓ? *Hint: The curve is a conic section,* $e = 0$ *corresponds to a circle,* $e \in (0,1)$, *to an ellipse,* $e = 1$, *to a parabola, and* $e > 1$, *to a hyperbola.*

(h) Now, suppose that $\boldsymbol{F}$ is an inverse square law field: $f(r) = C/r^2$, $C > 0$. Prove that the trajectory of m is a conic section and find the corresponding e, ℓ in terms of m, α, and C. *Hint. In (7.2.4), you get* $f_1 = c_1 u^2$. *Then solve the resulting equation to find* $r = r(\theta)$ *and use the results of part (f).*

(i) Conversely, suppose the central force field $\boldsymbol{F}$ is such that the corresponding trajectories are conic sections. Prove that $\boldsymbol{F}$ must be an inverse square law field. *Hint: Writing* $u = 1/r$, *use (7.2.5) to find* $d^2u/d\theta^2 + u$, *and then plug the result into (7.2.4).*

(j) Derive Kepler's Laws, stated on page 44. *Hint: For the second law, note that* $dA = (1/2)r^2 d\theta$ *is the area swept out by the radius vector* $r\widehat{\boldsymbol{r}}$ *moving through an angle* $d\theta$; $dA/dt = \dfrac{1}{2}r^2\omega$. *By part (b), this is* $\|\boldsymbol{L}_O\|/m$, *and therefore constant.*

(k) Describe the possible trajectories in the central field $\boldsymbol{F}(\boldsymbol{r}) = -r^{-\gamma}\widehat{\boldsymbol{r}}$ for all real $\gamma \neq 0$. What makes the value $\gamma = 2$ special? What trajectories do you get if the force is repelling rather than attractive?

PROBLEM 2.2. PERTURBED MOTION IN THE CENTRAL FIELD.

According to (7.2.4), the motion of a single planet around the Sun is described by equation

$$u''(\theta) + u(\theta) = A_0 \tag{7.2.6}$$

for some positive real number A_0, where $u(\theta(t)) = 1/r(t)$ and f_1 is an inverse-square force.

(a) For a planet in the Solar System, the gravitational attraction by the other planets is a small perturbation. As an example, consider the motion of planet Mercury perturbed by the Earth.

Assume that the Earth moves in a circular orbit with the Sun at the center, and consider only the radial (along the Mercury-Sun line) component of Earth's gravitational tug on Mercury. This changes the value of f_1 to $f_1 + \tilde{f}$ in equation (7.2.3) and in the subsequent equations. Find the value of $\tilde{f}$ and solve the resulting ordinary differential equation numerically using a suitable software package, for example, the MATLAB ode45 solver. The trajectory should change to a sequence of loops, each close to an ellipse, with the **`perihelion`**, the point of the closest approach to the Sun, changing location after each revolution. This is the classical **`perihelion precession`**, or **`shift`**. If you do not observe this picture right away, increase the value of $\tilde{f}$, which will now simulate the gravitational tugs from other planets. The observed value of this precession for Mercury is about $575''$ (575/3600 degrees) per century. The computed value of this precession using Newtonian mechanics, with the most careful account of all perturbations from all the planets is $532''$ per century. This discrepancy can only be explained by general relativity mechanics.

(b) General relativity leads to the following modification of equation (7.2.6):

$$u''(\theta) + u(\theta) = A_0 + \frac{3R_\circ}{2}u^2(\theta), \tag{7.2.7}$$

where $R_\circ$ is the Schwarzschild radius of the Sun (about 3km, see (2.4.36) on page 113.) For the derivation of (7.2.7) see page 116. If $\widetilde{u} = \widetilde{u}(\theta)$ is a solution of (7.2.6), then $(3R_\circ/2)\widetilde{u}^2(\theta)$ is very small compared to the number A_0. Thus, we can treat the term $(3R_\circ/2)u^2$ in (7.2.7) as a small perturbation of (7.2.6). Accordingly, we look for a solution of (7.2.7) in the form $u(\theta) = \widetilde{u}(\theta) + w(\theta)$, where $\widetilde{u}$ solves (7.2.6) and w is a small correction.
(i) Verify that

$$w''(\theta) + w(\theta) = (3R_\circ/2)(\widetilde{u}(\theta) + w(\theta))^2. \tag{7.2.8}$$

(ii) Neglecting the non-linear term $(3R_\circ/2)\,w^2$, verify that the linear approximation to (7.2.8) is $w''(\theta) + (1 - 3R_\circ\,\widetilde{u}(\theta))w(\theta) = (3R_\circ/2)\widetilde{u}^2(\theta)$, or, with $\widetilde{u}(\theta) = (1 + e\cos\theta)/\ell$ being the solution of (7.2.6),

$$w''(\theta) + (\mu\cos\theta + \lambda)w(\theta) = \frac{3R_\circ}{2\ell^2}(1 + e\cos\theta)^2. \tag{7.2.9}$$

This is an inhomogeneous, second-order linear ordinary differential equation with variable coefficients. Take the values of ℓ and e corresponding to Mercury's orbit(see below in part (iii)), and $R_\circ = 2.95$ km, solve this equation numerically using a suitable software package and observe the same changes of the trajectory as in part (a).
(iii) For $R_\circ/\ell$ much smaller than 1, we can neglect the terms of order $(R_\circ/\ell)^2$ and higher. Look for an approximate solution of equation (7.2.7) in the form $u(\theta) = \Big(1 + e\cos\big((1-\delta)\theta\big)\Big)/\ell$, with δ of order $R_\circ/\ell$, and verify that this leads to the following value of δ:

$$\delta \approx \frac{3R_\circ}{2\ell}. \tag{7.2.10}$$

To simplify computations, disregard the term of order e^2 compared with e; for Mercury, $e \approx 0.2056$. Then $2\pi\delta$ is the general relativity correction to the perihelion precession *in radians per revolution of the planet around the Sun.* The reader can plug in the numbers for Mercury ($T = 88$ Earth-days for the period of revolution around the Sun; $a = 0.579 \cdot 10^8$ km for the semi-major axis) and verify that the resulting number is about $43''$ *per century*; keep in mind that $\ell = a(1 - e^2)$ and use $R_\circ = 2.95$ km. This value of this number is exact to within the measurement error and was first computed in 1916 by A. Einstein. The computation provided the first validation of general relativity.

(c) *Discussion.* (i) The reason why Mercury is used in the analysis is that, for other planets in the Solar System, the perihelion precession is slower (the planets are further from the Sun, which decreases the general relativity perturbation) and harder to observe (the planets themselves move more slowly). For the Earth, general relativity predicts the perihelion precession of about $5''$ per century. For more on the subject, see the book *The Physical Foundations of General Relativity* by D. W. Sciama, 1969.
(ii) The homogenous version of (7.2.9), which is exactly (4.4.51) on page 237, is known as `Mathieu's differential equation`, after the French mathematician Claude Louis Mathieu (1783–1875), who used this equation in his astronomical studies well before the advent of general relativity.

An interested reader can follow Mathieu's reasoning by replacing the usual inverse square law with $f_1(r) = c_1 r^{-2} + c_2 r^{-4}$ and noticing that the corresponding equation of motion is of the form (7.2.7), but with some unknown number in place of $R_\circ$. This approach does not really *explain* anything, because there is no justification of the formula for f_1 and no way of computing the value of c_2; the best one can do is to fit the model to the data by estimating c_2 from the observations. It is the general relativity that justifies the model.

PROBLEM 2.3. THE SIMPLE RIGID PENDULUM.

Consider equation (2.1.9), page 42, of the simple rigid pendulum and assume there is no damping: $c = 0$. The equation becomes $\ddot{\theta} + (g/\ell)\sin\theta = 0$. We will see that this equation is *integrable by quadratures*.

(i) Set $\omega = \dot{\theta}$. Verify that $\omega \dfrac{d\omega}{d\theta} + \dfrac{g}{\ell}\sin\theta = 0$. *Hint: chain rule for* $\omega(\theta(t))$.

(ii) Verify that the time t as a function of θ is expressed using the **elliptic integral**:

$$t = \sqrt{\frac{\ell}{2}} \int_{\theta_0}^{\theta} \frac{1}{\sqrt{g\cos\theta + K_0 \ell}}\, d\theta,$$

where $K_0 = (\omega_0^2/2) - (g/\ell)\cos(\theta_0)$ and $\theta_0 = \theta(0)$. *Hint: from (i),* $\omega^2/2 = (g/\ell)\cos\theta + K_0 = (1/2)(d\theta(t)/dt)^2$.

(iii) Assuming that $\theta(0) = \theta_0$, $\dot{\theta}(0) = 0$, and $|\theta(t)|$ is small for all t, derive the familiar formula $\theta(t) = \theta_0 \cos(\omega_0 t)$, where $\omega_0 = \sqrt{g/\ell}$, from the elliptic integral representation of the solution.

PROBLEM 2.4. (a) Prove the **Parallel Axis Theorem** for the moment of inertia, I, of a mass M about an axis ℓ parallel to an axis through the center of mass and distance s between the two axes: $I = I_{CM} + s^2 M$, where I_{CM} is the moment of inertia about the center of mass. *Hint Choose an appropriate coordinate system* ($\hat{\imath}$, $\hat{\jmath}$, $\hat{k}$) *with* $\hat{k}$ *along the axis* ℓ *and such that* $x_{CM}^2 + y_{CM}^2 = s^2$. *Then* $I = \iiint_{\mathcal{R}} (x^2 + y^2)\rho\, dV$, *and* $I_{CM} = \iiint_{\mathcal{R}} [(x - x_{CM})^2 + (y - y_{CM})^2]\rho\, dV$, *where* $\rho = \rho(x, y, z)$ *is the density.*

(b) Use the result to give an alternative derivation of equation (2.2.45), page 83, for the distributed pendulum by considering rotation about the suspension point O rather than about the center of mass; see Figure 2.2.1, page 82. *Hint: work in the cartesian basis* ($\hat{\imath}$, $\hat{\jmath}$, $\hat{k}$) *at* O. *The moment of inertia* I_{zz} *about* $\hat{k}$ *is* $I_{zz}^* + M(\ell/2)^2$, *the angular momentum around* O *is* $I_{zz}\dot{\theta}$, *the torque around* O *is* $-(Mg\ell/2)\sin\theta\,\hat{k}$.

The Parallel Axis Theorem is also known as the **Huygens-Steiner theorem**, after C. Huygens and the Swiss mathematician JAKOB STEINER (1796–1863).

PROBLEM 2.5. THE EULER ANGLES.

Consider the trihedron system $S = (\widehat{\boldsymbol{u}}, \widehat{\boldsymbol{p}}, \widehat{\boldsymbol{b}})$ fixed on the airplane as described on page 32. Let $S_0 = (\hat{\boldsymbol{\imath}}, \hat{\boldsymbol{\jmath}}, \hat{\boldsymbol{\kappa}})$ be a cartesian coordinate system with the same origin O as S, but with orientation fixed in space. Assume that, at first, the corresponding basis vectors of S and S_0 are aligned. During the flight, the orientation of S relative to S_0 changes due to roll, yaw, and pitch of the airplane. One way to obtain this new orientation is to specify the **Euler angles** (ϕ, θ, ψ) by three rotations R_1, R_2, R_3 as follows. Starting from the aligned position, first R_1 rotates S about $\hat{\boldsymbol{\kappa}}$ through an angle ϕ. Then, R_2 rotates S about the new position of $\widehat{\boldsymbol{u}}$ through an angle θ. Finally, R_3 rotates S about the new position of $\widehat{\boldsymbol{b}}$ through an angle ψ.

(a) Verify the following matrix representation of each rotation:

$$R_1 = \begin{bmatrix} \cos\phi & \sin\phi & 0 \\ -\sin\phi & \cos\phi & 0 \\ 0 & 0 & 1 \end{bmatrix}, \; R_2 = \begin{bmatrix} \cos\theta & 0 & -\sin\theta \\ 0 & 1 & 0 \\ \sin\theta & 0 & \cos\theta \end{bmatrix}, \; R_3 = \begin{bmatrix} \cos\psi & \sin\psi & 0 \\ -\sin\psi & \cos\psi & 0 \\ 0 & 0 & 1 \end{bmatrix}.$$

(b) Prove that if a point P_0 has coordinates (x_0, y_0, z_0) in S_0 and coordinates (x, y, z) in S, then

$$\begin{pmatrix} x \\ y \\ z \end{pmatrix} = R_3 R_2 R_1 \begin{pmatrix} x_0 \\ y_0 \\ z_0 \end{pmatrix}. \tag{7.2.11}$$

(c) Let ω be the rotation vector of S relative to S_0. Prove that

$$\omega = \dot{\phi}\, \hat{\boldsymbol{\kappa}} + \dot{\theta}\, \widehat{\boldsymbol{p}}_0 + \dot{\psi}\, \widehat{\boldsymbol{b}} = \omega_1\, \widehat{\boldsymbol{u}} + \omega_2\, \widehat{\boldsymbol{p}} + \omega_3\, \widehat{\boldsymbol{b}}, \tag{7.2.12}$$

where $\widehat{\boldsymbol{p}}_0$ is the unit vector making angle ψ with the final position of $\widehat{\boldsymbol{p}}$. Show that

$$\widehat{\boldsymbol{p}}_0 = -\frac{1}{\sin\theta}\, \widehat{\boldsymbol{b}} \times \hat{\boldsymbol{\kappa}} = \cos\psi\, \widehat{\boldsymbol{p}} + \sin\psi\, \widehat{\boldsymbol{u}},$$
$$\widehat{\boldsymbol{k}} = -\cos\psi \sin\theta\, \widehat{\boldsymbol{u}} + \sin\psi \sin\theta\, \widehat{\boldsymbol{p}} + \cos\theta\, \widehat{\boldsymbol{b}}.$$

Substituting in (7.2.12) above, show that

$$\omega_1 = \dot{\omega} \sin\psi - \dot{\phi} \cos\psi \sin\theta, \; \omega_2 = \dot{\omega} \cos\psi + \dot{\phi} \sin\psi \sin\theta, \; \omega_3 = \dot{\omega} \cos\theta + \dot{\psi}.$$

(d) There are other choices for the Euler angles. For example, ϕ, θ, ψ can be the roll, pitch, and yaw angles. Find the corresponding rotation matrices R_1, R_2, R_3 and establish the analog of (7.2.11).

(e) How many different choices for the Euler angles are there?

PROBLEM 2.6. KINETIC ENERGY AND MOMENT OF INERTIA.

(a) A point mass m moving with speed v has kinetic energy $\mathcal{E}_K = (mv^2/2)$. Assume that m is moving in the $(\hat{\imath}, \hat{\jmath})$ plane around a circle of radius r with angular speed ω; see (1.3.28), page 36. (i) Show that $\mathcal{E}_K = (\omega^2/2)mr^2 = I\omega^2/2$. (ii) Show that the angular momentum of m is $\boldsymbol{L} = mr^2\omega\,\hat{\kappa} = I\omega\,\hat{\kappa}$. (iii) Conclude that $\mathcal{E}_K = \|\boldsymbol{L}\|\omega/2$.

(b) Now consider a system S of n masses m_i, $i = 1, \ldots, n$. If v_i is the speed of m_i, then the kinetic energy of the system is $\mathcal{E}_K^{(S)} = \sum_{i=1}^n (v_i^2/2)m_i$. (i) Verify that if each m_i is rotating about the same fixed axis with angular speed ω at a distance r_i from the axis, then $\mathcal{E}_K^{(S)} = (\omega^2/2)\sum_{i=1}^n m_i r_i^2 = I\omega^2/2$. (ii) Conclude that

$$\mathcal{E}_K^{(S)} = (1/2)\boldsymbol{\Omega}^T I_{CM}\boldsymbol{\Omega} = (1/2)\boldsymbol{\Omega}^T \mathcal{L}_{CM}, \tag{7.2.13}$$

where I_{CM} is the moment of inertia matrix in some basis $(\hat{\imath}, \hat{\jmath}, \hat{\kappa})$ (see (2.2.32), page 76), $\boldsymbol{\Omega}$ is the column-vector $(\omega_x, \omega_y, \omega_z)$ of components in $(\hat{\imath}, \hat{\jmath}, \hat{\kappa})$ of the corresponding rotation vector $\boldsymbol{\omega}$ ($\boldsymbol{\omega} = \omega_x\,\hat{\imath} + \omega_y\,\hat{\jmath} + \omega_z\,\hat{\kappa}$), and $\mathcal{L}_{CM}$ is the column-vector of components of $\boldsymbol{L}_{CM}$ in $(\hat{\imath}, \hat{\jmath}, \hat{\kappa})$. *Hint: choose* $(\hat{\imath}, \hat{\jmath}, \hat{\kappa})$ *so that* $\hat{\kappa} = \boldsymbol{\omega}/\omega$. *Then* $\omega_x = \omega_y = 0$ *and* $\boldsymbol{L}_{CMx} = -\omega I_{xz}$, $\boldsymbol{L}_{CMy} = -\omega I_{yz}$, $\boldsymbol{L}_{CMz} = \omega I_{zz}$. *Hence,* $\boldsymbol{\Omega}^T\mathcal{L}_{CM} = \boldsymbol{\omega}\cdot\boldsymbol{L}_{CM} = \omega^2 I_{zz} = \omega^2 I$. *On the other hand, the value of the dot product* $\boldsymbol{\omega}\cdot\boldsymbol{L}_{CM}$ *does not depend on the coordinate system.*

(c) Extend (7.2.13) to rigid bodies and to general rotations (that is, not necessarily around a fixed axis).

PROBLEM 2.7. ROLLING CYLINDER.

Suppose a homogeneous *solid*, circular cylinder of radius R and mass m is placed on an inclined plane. Let α be the angle of incline. Figure 7.2.1 shows a side view of the plane and the cylinder.

(a) Show that the moment of inertia of the cylinder about its axis is $I = mR^2/2$.

(b) Choose an inertial coordinate system $(\hat{\imath}, \hat{\jmath})$, fixed to the plane and non-inertial system $(\hat{\imath}_1, \hat{\jmath}_1)$ fixed to the cylinder and rotating with it. Let $\theta(t)$ be the angle between $\hat{\imath}$ and $\hat{\imath}_1$ at time t. The center of mass is at $x(t)\,\hat{\imath} + R\,\hat{\jmath}$, and the position of the cylinder at time t is specified by $(x(t), \theta(t))$. Assume that rolling occurs without slipping: $x(0) - x(t) = R\theta(t)$, $t \geq 0$.

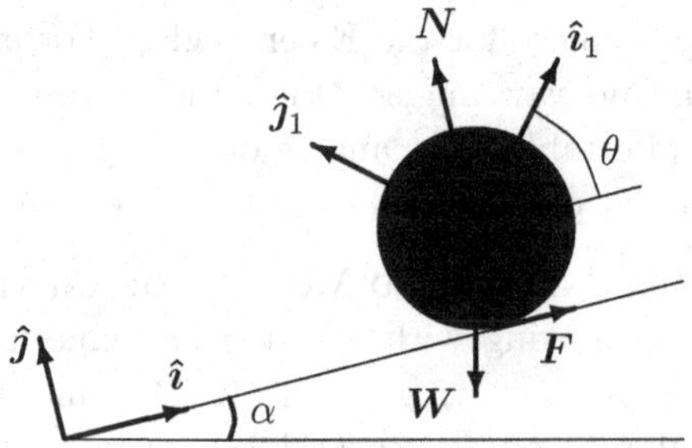

Fig. 7.2.1 Rolling Cylinder

Denote by $\boldsymbol{W}$ the weight of the cylinder, and by $\boldsymbol{F}$, $\boldsymbol{N}$, the tangential and the normal components of the reaction force of the plane in response to $\boldsymbol{W}$. Recall that $\|\boldsymbol{W}\| = mg$, where $g \approx 9.8$ meters/second^2 is the acceleration of free fall near the surface of the Earth. (i) Verify that the Second Law of Newton in $(\hat{\imath}, \hat{\jmath})$ yields $m\,\ddot{x}(t) = \|\boldsymbol{F}\| - mg\sin\alpha$, $\|\boldsymbol{N}\| = mg\cos\alpha$. (ii) Verify that the first two of the Euler equations in (2.2.35), page 79 result in $0 = 0$, and the third, in $(mR^2/2)\ddot{\theta} = R\,\|\boldsymbol{F}\|$. (iii) Verify that the no-slippage condition implies $\ddot{x} = -R\ddot{\theta}$. (iv) Conclude that $\|\boldsymbol{F}\| = (1/3)mg\sin\alpha$ and

$$\ddot{x} = (-2/3)g\sin\alpha, \quad \ddot{\theta} = (2/(3R))g\sin\alpha. \tag{7.2.14}$$

(v) By definition, the magnitude of the frication force in the direction of $\hat{\imath}$ is $\mu\|\boldsymbol{N}\|$, where μ is the `friction coefficient`. Conclude that $\|\boldsymbol{F}\| \le \mu\|\boldsymbol{N}\|$, or $\mu \ge (1/3)\tan\alpha$ means no slipping. (vi) Find t_0 so that $x(t_0) = 0$, that is, the cylinder reaches the bottom of the plane (remember that $x(0) = x_0$).

(c) Verify that the kinetic energy $\mathcal{E}_K$ and the potential energy V of the rolling cylinder are $\mathcal{E}_K = (mv^2/2) + (I\omega^2/2)$, $V = -mg(x_0 - x)\sin\alpha$, where $\omega = \dot{\theta}$, $v = \dot{x}$, and $x_0 - x = R\theta$. Ignoring heat loss due to friction, conservation of energy implies that the total energy $H = \mathcal{E}_K + V$ is constant in time. Differentiate H with respect to time to obtain an equation consistent with (7.2.14).

(d) Find the moment of inertia and derive the above equations of motion for a hollow cylinder with outer radius R and inner radius r, where $0 \le r \le R$. For what r will the cylinder roll the fastest?

Problem 2.8. The Tumbling Box

Consider the free motion of an object with three different values of the principal moments of inertia I^*_{xx}, I^*_{yy}, I^*_{zz}; see page 79. To simplify the notations, set $I^*_{xx} = a$, $I^*_{yy} = b$, $I^*_{zz} = c$.

(a) Verify that the system of the Euler equations (2.2.35), page 79, becomes

$$a\dot{\omega}^*_x = (b-c)\omega^*_y\omega^*_z,\ b\dot{\omega}^*_y = (c-a)\omega^*_z\omega^*_z,\ c\dot{\omega}^*_z = (a-b)\omega^*_x\omega^*_y, \qquad (7.2.15)$$

Hint: free motion means $\boldsymbol{T}_{CM} = \mathbf{0}$.

(b) For the sake of concreteness, let $a = 1$, $b = 2$, $c = 3$. (i) Verify that $(\omega^*_x(t))^2 + 4(\omega^*_y(t))^2 + 9(\omega^*_z(t))^2 = L$ for some number $L > 0$ independent of t. (ii) Taking $L = 1$, verify that the points $A_\pm = (\pm 1, 0, 0)$, $B_\pm = (0, \pm 1/2, 0)$, $C_\pm = (0, 0, \pm 1/3)$ are the only critical points of the system (7.2.15). *Hint: the only way for the right-hand side of (7.2.15) to vanish is to have two out of three omegas equal to zero; the remaining one must satisfy the equality from (i).* (iii) Linearizing the system around each critical point, verify that $A_\pm$ are centers, $B_\pm$ are saddles, and $C_\pm$ are centers. Conclude that, for the original nonlinear system, the points $A_\pm$ and $C_\pm$ are stable, while $B_\pm$ are unstable. *Hint: for A_+, the linearization is $\dot{\omega}^*_x = 0$, $\dot{\omega}^*_y = \dot{\omega}^*_z$, $\dot{\omega}^*_z = -\dot{\omega}^*_y/3$, which is a center in the (ω^*_y, ω^*_z) plane. Then notice from the last two equations in (7.2.15) that $(\omega^*_y(t))^2 + 3(\omega^*_z(t))^2$ does not depend on time, so the nonlinear system still has a center. A saddle always stays a saddle, as it is* **structurally stable**, *see, for example, Section 4.1 of the book* Differential Equation and Dynamical System *by L. Perko, 1991.* (iv) Visualize the conclusion of part (iii) by tossing up an object such as an empty cereal box, a tennis racquet, or your least favorite book (with the front and back covers taped together to prevent the book from opening). Observe the rotation wobbling when you spin the object around one of the axes. (v) What will happen if two of the sides of the box are the same? *Hint: nothing interesting.* (v) For a more detailed discussion and graphical illustrations, see the article *The Tumbling Box* by S. J. Colley in the *American Mathematical Monthly*, Vol. 94, pp. 62–68.

PROBLEM 2.9. Consider the example in Section 2.3.2 of a bead of mass m sliding down a frictionless helical wire given by the equation,

$$\boldsymbol{r}(q) = a\cos q\,\hat{\boldsymbol{\imath}} + b\sin q\,\hat{\boldsymbol{\jmath}} + cq\,\hat{\boldsymbol{\kappa}}, \qquad (7.2.16)$$

in an inertial frame. Let $\boldsymbol{N} = N_x\,\hat{\boldsymbol{\imath}} + N_y\,\hat{\boldsymbol{\jmath}} + N_3\,\hat{\boldsymbol{\kappa}}$ be the unknown reaction force exerted by the wire on m and let $\boldsymbol{F} = -mg\,\hat{\boldsymbol{\kappa}}$ be the force of gravity. Newton's equations are

$$m\ddot{x} = N_x,\quad m\ddot{y} = N_y,\quad m\ddot{z} = -mg + N_z, \qquad (7.2.17)$$

which with the constraint (7.2.16) give five equations for the six unknowns x, y, z, N_x, N_y, N_z. A sixth constraint equation is

$$\dot{\boldsymbol{r}} \cdot \boldsymbol{N} = 0, \qquad (7.2.18)$$

since there is no friction along the wire.

(a) Compute $\ddot{\boldsymbol{r}}(q)$ from (7.2.16) and express $\ddot{x}, \ddot{y}, \ddot{z}$ in terms of $q, \dot{q}, \ddot{q}$. Then use (7.2.17) and (7.2.18) to get

$$m\dot{x}\ddot{x} + m\dot{y}\ddot{y} + m\dot{z}(\ddot{z} + g) = 0. \qquad (7.2.19)$$

Finally, substitute the expressions for $\ddot{x}, \ddot{y}, \ddot{z}$ obtained in (7.2.19) to get the equation of motion.

(b) Show that equations (2.3.40), (2.3.41) reduce to the equation

$$\ddot{q}(a^2 \sin^2 q + b^2 \cos^2 q + c^2) + \dot{q}^2(a^2 - b^2) \cos q \sin q + cg = 0. \qquad (7.2.20)$$

which is the same as the equation of motion obtained in part (a).

(c) Equation (7.2.20) is a nonlinear ordinary differential equation (ODE). Use a numerical ODE solver to solve the equation with the initial conditions $q(0)$, $\dot{q}(0)$, corresponding to $\boldsymbol{r}(0) = b\,\hat{\boldsymbol{\jmath}} + (\pi/2)\,\hat{\boldsymbol{\kappa}}$ and $\dot{\boldsymbol{r}}(0) = \boldsymbol{0}$, taking $a = 1, b = 2$, $c = 1$.

(d) In the special case $a = b$ of a circular helix, equation (7.2.20) becomes

$$\ddot{q}(a^2 + c^2) + cg = 0. \qquad (7.2.21)$$

Solve this equation analytically with initial conditions $\boldsymbol{r}(0) = b\,\hat{\boldsymbol{\jmath}} + (\pi/2)\,\hat{\boldsymbol{\kappa}}$ and $\dot{\boldsymbol{r}}(0) = 0$, taking $a = c = 1$. Then solve (7.2.21) using an ODE solver and compare the results.

7.3 Special Relativity

The main part of this section, Problem 3.3, is the analysis of the Michelson-Morley experiment. Two shorter problems, Problem 3.1 and Problem 3.2, discuss the basic equations of relativistic kinematics and dynamics.

PROBLEM 3.1. RELATIVISTIC KINEMATICS.

Consider two frames O and O_1, with O_1 moving along the x-axis with constant velocity v relative to O; see Figure 2.4.1 on page 99. Let us write the Lorentz transformation as

$$x = a(x_1 + vt_1),\ y = y_1,\ z = z_1,$$
$$t = a(t_1 + bx_1),\ \text{where } a = \left(1 - (v^2/c^2)\right)^{-\frac{1}{2}},\quad b = v/c^2; \tag{7.3.1}$$

see page 100. Let $\dot{x} = dx/dt$ and $\dot{x}_1 = dx_1/dt_1$ be the velocities of a point mass m relative to O and O_1 respectively. Let $dy/dt = dz/dt = 0$. Thus, m moves parallel to the x and x_1 axes.

(a) In Newtonian kinematics the velocities would be related by the **Galilean relativity** relation $\dot{x} = \dot{x}_1 + v$. Show that in special relativity kinematics

$$\dot{x} = \frac{\dot{x}_1 + v}{1 + b\dot{x}_1}. \tag{7.3.2}$$

Hint: use (7.3.1) above to compute $dt/dt_1 = a(1 + b\dot{x}_1)$, $\dot{x} = a(\dot{x}_1 + v)(dt_1/dt) = \dfrac{(\dot{x}_1 + v)}{1 + b\dot{x}_1}$.

(b) Use (7.3.2) to prove that

$$1 - (\dot{x}/c)^2 = \frac{1}{a^2}\frac{(1 - (\dot{x}_1/c)^2)}{(1 + b\dot{x}_1)^2}. \tag{7.3.3}$$

(c) Prove that

$$\ddot{x} = \frac{\ddot{x}_1}{a^3(1 + b\dot{x}_1)^3}, \tag{7.3.4}$$

where $\ddot{x} = d^2x/dt^2$ and $\ddot{x}_1 = d^2x_1/dt_1^2$.

PROBLEM 3.2. RELATIVISTIC DYNAMICS.

If the point mass m moves along the $\hat{\imath}$-axis in the frame O, then, in the frame O, equation (2.4.15) becomes

$$\boldsymbol{F} = m_0 \frac{d}{dt}\frac{\dot{x}\,\hat{\imath}}{(1 - (\dot{x}/c)^2)^{\frac{1}{2}}}. \tag{7.3.5}$$

(a) Prove that (7.3.5) also holds in the frame O_1, that is,

$$\boldsymbol{F} = m_0 \frac{d}{dt_1}\frac{\dot{x}_1\,\hat{\imath}}{(1 - (\dot{x}_1/c)^2)^{\frac{1}{2}}}.$$

Hint: show that $dp/dt = dp_1/dt_1$, *where* $p_1 = (m_0\dot{x}_1/(1 - (\dot{x}_1/c)^2)^{\frac{1}{2}})$,

$$\frac{dp}{dt} = m_0\left(\frac{\ddot{x}}{(1 - (\dot{x}/c)^2)^{1/2}} + \frac{(\dot{x}/c)^2\ddot{x}}{\left(1 - (\dot{x}/c)^2\right)^{3/2}}\right)\hat{\imath} = \frac{m_0\ddot{x}\,\hat{\imath}}{\left(1 - (\dot{x}/c)^2\right)^{3/2}}.$$

(b) Use (7.3.3) and (7.3.4) to prove

$$\frac{dp}{dt} = \frac{m_0 \ddot{x}_1 i}{a^3(1+b\dot{x}_1)^3} \cdot \frac{a^3(1+b\dot{x}_1)^3}{(1-(\dot{x}_1/c)^2)^{3/2}} = \frac{dp_1}{dt_1}.$$

This proves that the relativistic form (7.3.5) of Newton's Second Law is invariant under the Lorentz transformation, that is, (7.3.5) holds in all relativistic inertial frames (moving with constant relative velocity).

(c) Verify that the corresponding non-relativistic equation (2.1.1) on page 40 is invariant under the Galilean transformation $x = x_1 + vt$, $t = t_1$. *Hint:* $\dot{x} = \dot{x}_1 + v$ *implies* $\ddot{x} = \ddot{x}_1$.

PROBLEM 3.3. THE MICHELSON-MORLEY EXPERIMENT.

(a) Recall (page 95) that the original intent of the Michelson-Morley experiment was to study the properties of *aether* (or *ether*), a hypothetical medium supposedly representing Newtonian absolute space in which the Earth moved with an "absolute" speed v and through which the electromagnetic waves were thought to be propagating. The idea going back to Maxwell is that this speed v can be determined experimentally using electromagnetic waves, such as light. To better understand this idea, let us first consider the propagation of sound waves.

Consider two frames, O and O_1. Frame O is filled with still air, the analog of the aether. Let O_1 move with speed v relative to O, say to the right along the x-axis of O. At a distance d from O_1 we fix a sound receiver, R, to the x_1-axis. So R also moves to the right, away from O with speed v (draw a picture!) At time zero, suppose O and O_1 coincide and a bell placed at O_1 is rung. Let v_s be the speed of sound in air.
(i) Calculate the time T it takes the sound wave front to reach R.
Hint: $v_s T = (d + vT)$.
(ii) Show that the relative propagation speed measured by an observer in frame O_1 is $v_{s_1} = d/T = v_s - v$.
(iii) Now, suppose that R moves toward O with speed v (i.e. in a frame O_2 moving to the left). Show that $T = (d - vT)/v_s$, and an observer in O_2 measures the relative propagation speed to be $v_{s_2} = d/T = v_s + v$. We can then determine v from v_{s_1} and v_{s_2}, because $v = (v_{s_2} - v_{s_1})/2$.
(iv) Let R be on the y_1-axis at distance d from O_1. Consider the same experiment as in part (i). Show that the relative propagation speed in O_1 is $v_{s_3} = (1 - v^2/v_s^2)^{1/2}\, v_s$, and express v in terms of v_{s3} and v_{s1}. The expression of v in terms of v_{s3} and v_{s1} is the main idea behind the Michelson-Morley experiment, and, as we mentioned earlier, goes back to Maxwell.

Hint: $v_{s_3} = d/T$, *and, by the Pythagorean Theorem,* $v_s^2 T^2 = d^2 + v^2 T^2$.

(b) Now let us go back to the Michelson-Morley experiment. In this part of the problem, we assume that the aether exists, so that, with light instead of sound, aether is the analog of air. The frame O_1 is the Earth moving with velocity v through the assumed aether, and the "absolute" frame O is assumed to be fixed. The x_1 axis is taken parallel to v. The propagation speed of light in O is $c = 3 \times 10^8$ m/sec. In the experiment, a light beam is sent in the assumed direction of v to a reflecting mirror placed on the x_1 axis at distance d_1 from O_1. Call this the "direct" beam. By another mirror at O_1 the direct beam is split into two, the second, or "transverse", beam being made to travel in a direction perpendicular to v to another reflecting mirror placed on the y_1 axis at distance d_2 from O_1. Thus, the two beams make round-trips from O_1 and back (draw a picture!)
(i) Show that the round-trip times T_1 for the direct beam and T_2 for the transverse beam are

$$T_1 = \frac{d_1}{c-v} + \frac{d_1}{c+v} = \frac{2cd_1}{c^2 - v^2}, \quad T_2 = 2d_2(c^2 - v^2)^{-1/2}.$$

(ii) Define $\Delta T_1 = T_{\text{direct}} - T_{\text{transverse}} = T_1 - T_2$. Verify that the predicted value of this time difference is

$$\Delta T_1 = \frac{2\big(d_1 - d_2(1 - v^2/c^2)^{1/2}\big)}{c(1 - v^2/c^2)}.$$

This time difference can be measured by observing the fringe pattern in an interferometer placed at O_1, since the reflected direct beam will arrive later at O_1, interfering with the earlier arriving reflected transverse beam.
(iii) Interchange the two beams by rotating the experimental apparatus through 90° and repeat the experiment. Show that the corresponding time difference is now

$$\Delta T_2 = \frac{2\big(d_1(1 - v^2/c^2)^{1/2} - d_2\big)}{c(1 - v^2/c^2)}.$$

(iv) Define the second difference $\Delta^2 T = \Delta T_1 - \Delta T_2$ and show that

$$\Delta^2 T = \frac{2(d_1 + d_2)\big(1 - (1 - v^2/c^2)^{1/2}\big)}{c(1 - v^2/c^2)}.$$

Verify that the values $d_1 = d_2 = 1$m used in the experiment and $v = 30$ km/s (Earth's orbital speed) predict a value $\Delta^2 T \approx 10^{-16}$ s. In the actual experiment, the observed value of $\Delta^2 T$, measured using an interferometer, was from about 1/10 to about 1/4 the predicted value.

(c) Verify that the theory of special relativity, including the non-existence of aether, is consistent with the null result $\Delta^2 T = 0$ of the Michelson-Morley experiment. *Hint: use the Lorentz transformation; you should get the* exact *equality* $\Delta^2 T = 0$.

7.4 Vector Calculus

Of all the numerous possible problems related to vector calculus, we suggest two standard problems, Problem 4.1 about the multi-dimensional Taylor formula, and Problem 4.2 about the rigorous proof of the representation formula for the solution of the Poisson equation; the solutions to these problems can be found in many books, although many of those books are graduate level. Problem 4.3 leads the reader through the derivation of the equation describing the lines of force of the electric dipole; the reader has probably seen the picture in a physics book and now gets a chance to understand the origin of that picture. Finally, Problem 4.4 shows how Legendre polynomials can appear in a somewhat unexpected setting.

PROBLEM 4.1. TAYLOR FORMULA FOR SCALAR FIELDS

Recall from one variable calculus that if $g = g(t)$ is a function having two continuous derivatives at $t = t_0$, then $g(t) \approx g(t_0) + g'(t_0)(t - t_0) + (1/2)g''(t_0)(t - t_0)^2$. If t_0 is a critical point of g, that is, $g'(t_0) = 0$, and $g''(t_0) \neq 0$, then the sign of the second derivative determines the type of the critical point: local minimum if $g''(t_0) > 0$, local maximum if $g''(t_0) < 0$. In what follows, we will briefly discuss how a similar analysis is carried out for scalar fields.

The role of the second derivative for a twice continuously differentiable scalar field f is played by a *linear transformation* A_P (see page 17), such that, for all points B sufficiently close to P,

$$\nabla f(B) \approx \nabla f(P) + A_P(\overrightarrow{PB}), \tag{7.4.1}$$

and

$$f(B) \approx f(P) + \nabla f(P) \cdot \overrightarrow{PB} + \frac{1}{2} A_P(\overrightarrow{PB}) \cdot \overrightarrow{PB}; \tag{7.4.2}$$

in other words, the linear transformation A_P is the derivative of ∇f at the point P. The point P_0 is called `critical` for the scalar function f if $\nabla f(P_0)$

is either equal to $\mathbf{0}$ or is not defined. For a twice continuously differentiable function f, approximation (7.4.2) near the critical point becomes

$$f(B) \approx f(P_0) + A_{P_0}(\overrightarrow{PB}) \cdot \overrightarrow{PB}.$$

We say that a linear operator A is *non-negative definite* if $A(\boldsymbol{v}) \cdot \boldsymbol{v} \geq 0$ for every non-zero vector $\boldsymbol{v}$. As a result, the critical point P_0 is a local minimum of f if A_P is non-negative definite.

(a) Verify that in $\mathbb{R}^2$ with cartesian coordinates, (7.4.2) becomes

$$f(x_0+x, y_0+y) = f(x_0, y_0)+\boldsymbol{\nabla} f(x_0, y_0)\cdot(x, y)^T + \frac{1}{2}(x, y)\, H(x_0, y_0)\,(x, y)^T, \tag{7.4.3}$$

where $(x, y)^T$ denotes the column vector (see page 12), and

$$H(x_0, y_0) = \begin{pmatrix} f_{xx}(x_0, y_0) & f_{xy}(x_0, y_0) \\ f_{yx}(x_0, y_0) & f_{yy}(x_0, y_0) \end{pmatrix}, \tag{7.4.4}$$

The matrix $H(x_0, y_0)$ is called the `Hessian matrix`, after the German mathematician LUDWIG OTTO HESSE (1811–1874). Use (7.4.3) to establish the `second partials test`: the point (x_0, y_0) is a local minimum of f if $\boldsymbol{\nabla} f(x_0, y_0) = \mathbf{0}$, $f_{xx}(x_0, y_0) > 0$, and $f_{xx}(x_0, y_0) f_{yy}(x_0, y_0) - f_{xy}^2(x_0, y_0) > 0$. *Hint: apply the Taylor expansion to the function* $g(t) = f(x_0 + tx, y_0 + ty)$ *at* $t = 0$.

(b) Using the Hessian matrix, define the directions of the fastest and slowest growth of the function at the point of local minimum. *Hint: if* $f_{xx} = 4$, $f_{yy} = 6$, *and* $f_{xy} = 0$, *then the fastest growth is along the* y *axis, and the slowest growth is along the* x *axis.*

(c) What does the Hessian matrix look like in $\mathbb{R}^3$? What are the analogous conditions for the local minimum?

(d) How to write the Taylor expansion of order three and higher?

PROBLEM 4.2. SOLVING THE POISSON EQUATION

Recall that our study of Maxwell's equations led us to a reasonable suspicion that the function U defined by (3.3.14) on page 167 solves the Poisson equation (3.3.15). In what follows, the interested reader will produce a rigorous proof of this statement and (once again) observe the difference between a physical suspicion and a mathematical proof.

In this problem, we fix a cartesian coordinate system in $\mathbb{R}^3$ and adopt somewhat more standard notations in advanced mathematics: $x = (x_1, x_2, x_3)$ or $y = (y_1, y_2, y_3)$ will denote a point in $\mathbb{R}^3$, $|x| =$

$\sqrt{(x_1)^2 + (x_2)^2 + (x_3)^2}$; a single integral $\int$ will denote the volume integral, and, depending on the variable of integration, dx or dy will denote the volume differential dV. For simplicity, we omit the constant ε_0.

Let $\rho = \rho(x)$ be a twice continuously differentiable function in $\mathbb{R}^3$ and assume that there exists a positive number R so that $\rho(x) = 0$ for all $|x| > R$. Define the function U by

$$U(x) = \frac{1}{4\pi} \int_{\mathbb{R}^3} \frac{\rho(y)}{|x-y|} dy. \tag{7.4.5}$$

(a) Verify that the function U is twice continuously differentiable everywhere in $\mathbb{R}^3$ and

$$\nabla^2 U(x) = \frac{1}{4\pi} \int_{\mathbb{R}^3} \frac{g(x-y)}{|y|} dy, \tag{7.4.6}$$

where $g(x) = \nabla^2 \rho(x)$. *Hint: after a change of variables,* $U(x) = \frac{1}{4\pi}\int_{\mathbb{R}^3} \frac{\rho(x-y)}{|y|} dy$. *Then verify directly that, in computing* $\lim_{\varepsilon\to 0}(U(x_1+\varepsilon, x_2, x_3) - U(x_1, x_2, x_3))/\varepsilon$ *and similar limits, it is possible to take the limit under the integral; remember, that by assumption on* ρ, *the integration is carried out over a bounded region.*

(b) Verify that $\boldsymbol{\nabla}^2 U(x) = -\rho(x)$ in $\mathbb{R}^3$ following these steps.

Step 1. Given an $a > 0$, we write $\mathbb{R}^3$ as the union of the two sets: $G_a = \{x \in \mathbb{R}^3 : |x| < a\}$ and $\overline{G}_a = \{x \in \mathbb{R}^3 : |x| \geq a\}$;

Step 2. Write the integral in (7.4.6) as a sum $A_a + B_a$, where A_a, B_a are the integrals over G_a and $\overline{G}_a$, respectively.

Step 3. By assumption, the function g is bounded. Use Exercise 3.1.24 on page 140 to conclude that $\lim_{a\to 0} A_a = 0$.

Step 4. Use Green's Formula (3.2.19) on page 159 with $f(x) = 1/(4\pi|x|)$ to write $B_a = I_a + J_a$, where I_a is the resulting *surface* integral over ∂G_a. Note that there is no other surface integral because $\rho(x) = 0$ for large $|x|$.

Step 5. Use Exercise 3.1.23, together with the assumed boundedness of $\|\boldsymbol{\nabla}\rho\|$ to conclude that $\lim_{a\to 0} I_a = 0$.

Step 6. Rewrite J_a again using (3.2.19). Since $\nabla^2(1/|x|) = 0$ in $\overline{G}_a$, only the surface integral over ∂G_a remains; using continuity of ρ, show that this surface integral tends to $\rho(x)$ as $a \to 0$.

Step 7. Combine the above steps to conclude the proof.

(c) State and prove the analogs of (a) and (b) in $\mathbb{R}^2$. *Hint: instead of* $1/(4\pi|x|)$, *use* $(-1/(2\pi)) \ln |x|$.

(d) Can you relax the regularity assumptions about ρ? For example, can you prove the statements in (a) and (b) assuming that ρ is only once continuously differentiable? *Hint: it is possible to prove (a) and (b) under even weaker assumptions on ρ.*

PROBLEM 4.3. LINES OF FORCE OF THE ELECTRIC DIPOLE.

Recall (page 167) that the lines of force of an electric field $\boldsymbol{E}$ are the collection of the solutions of the differential equation $\dot{\boldsymbol{r}}(t) = \boldsymbol{E}(\boldsymbol{r}(t))$ for all possible initial conditions. An equivalent form of the equation is $\dot{\boldsymbol{r}}(t) = -\boldsymbol{\nabla} U(\boldsymbol{r}(t))$, where U is the potential of the field. For the electric dipole, the potential is given by (3.3.24), page 170. We introduce a cartesian coordinate system $(\hat{\imath}, \hat{\jmath}, \hat{k})$ so that the coordinates of the positive and negative charges are $(a, 0, 0)$ and $(-a, 0, 0)$ respectively.

(a) Convince yourself that the electric field of the dipole has cylindrical symmetry, that is, does not change when rotated by any angle about the x-axis, on which the charges lie. As a result, it is enough to find the lines of force in the $(\hat{\imath}, \hat{\jmath})$ plane. *Hint:*

$$\boldsymbol{E}(\boldsymbol{r}) = \frac{q}{4\pi\varepsilon_0}\left(\frac{\boldsymbol{r}-\boldsymbol{r}_1}{\|\boldsymbol{r}-\boldsymbol{r}_1\|^3} - \frac{\boldsymbol{r}-\boldsymbol{r}_2}{\|\boldsymbol{r}-\boldsymbol{r}_2\|^3}\right).$$

(b) Following part (a), we set $\boldsymbol{r} = x\,\boldsymbol{i} + y\,\boldsymbol{j}$, $\boldsymbol{r}_1 = a\,\boldsymbol{i}$, $\boldsymbol{r}_2 = -a\,\boldsymbol{i}$. (i) Verify that the corresponding equations for the lines of force in the $(\hat{\imath}, \hat{\jmath})$ plane is

$$\frac{dx(t)}{dt} = \frac{q}{4\pi\varepsilon_0}\left(\frac{x(t)-a}{\big((x(t)-a)^2+y^2(t)\big)^{3/2}} - \frac{x(t)+a}{\big((x(t)+a)^2+y^2(t)\big)^{3/2}}\right),$$

$$\frac{dy(t)}{dt} = \frac{qy(t)}{4\pi\varepsilon_0}\left(\frac{1}{\big((x(t)-a)^2+y^2(t)\big)^{3/2}} - \frac{x(t)+a}{\big((x(t)+a)^2+y^2(t)\big)^{3/2}}\right).$$

(ii) Set $u = (x+a)/y$, $v = (x-a)/y$ to get $(du/dv) = \big((1+u^2)/(1+v^2)\big)^{3/2}$. *Hint: compare the expressions for dy/dx obtained from the above equations and by direct differentiation of $x = a(u+v)/(u-v)$, $y = 2a(u-v)$.* (iii) Conclude that the force lines are defined as the level curves by the following equation:

$$\begin{aligned}(x+a)\sqrt{(x-a)^2+y^2} &- (x-a)\sqrt{(x+a)^2+y^2}\\ &= C\sqrt{\big((x-a)^2+y^2\big)\big((x+a)^2+y^2\big)}\end{aligned}$$

where C is real number. *Hint:* $\int ds/(1+s^2)^{3/2} = s/\sqrt{1+s^2}$. (iv) Plot several lines of force using a computer algebra system. Note that the points $(-a, 0)$ and $(a, 0)$ are always on a line of force, as follows from the equation, but not all computer systems will recognize it. Also, the selected range of values for x, y can make a big difference in the quality of the picture.

PROBLEM 4.4. In (3.3.26) on page 171, we approximated $1/\|\boldsymbol{u}-\boldsymbol{v}\|$ when $\|\boldsymbol{v}\|/\|\boldsymbol{u}\|$ is small. In what follows, we get the exact expansion for $1/\|\boldsymbol{u}-\boldsymbol{v}\|$ in powers of $s=\|\boldsymbol{v}\|/\|\boldsymbol{u}\|$, which holds for all $s<1$.

(a) For $|t|<1$ and $n\geq 1$, define the functions $P_n=P_n(t)$ so that

$$\frac{1}{\sqrt{1-2tx+x^2}}=\sum_{n=0}^{\infty} P_n(t)\,x^n,\ |x|\leq 1 \tag{7.4.7}$$

(i) Convince yourself that each $P_n(t)$ is a polynomial in t.
(ii) Verify that $(n+1)P_{n+1}(t)=(2n+1)tP_n(t)-nP_{n-1}(t)$.
(iii) Verify that $(1-t^2)P_n''(t)-2tP_n'(t)+n(n+1)P_n(t)=0$.

(b) Verify that

$$\frac{1}{\|\boldsymbol{u}-\boldsymbol{v}\|}=\frac{1}{\|u\|}\sum_{n=1}^{\infty} P_n(\cos\theta)\,s^n,$$

where $s=\|\boldsymbol{v}\|/\|\boldsymbol{u}\|<1$ and $\cos\theta=\boldsymbol{u}\cdot\boldsymbol{v}/\big(\|\boldsymbol{u}\|\,\|\boldsymbol{v}\|\big)$. *Hint:* $\|\boldsymbol{u}-\boldsymbol{v}\|^2=\|\boldsymbol{u}\|^2(1-2s\cos\theta+s^2)$.

(c) *Discussion.* The polynomials P_n, $n\geq 0$, are called `Legendre's polynomials`. These polynomials appear in the analysis of the Laplace operator in spherical coordinates, see page 370. Also, see Exercise 4.4.24, page 238.

7.5 Complex Analysis

For many readers, the presentation in this book could be the first systematic treatment of complex numbers. Accordingly, in Problem 5.1 we discuss some algebraic and number-theoretic ideas used in the construction of complex numbers; an interested reader is encouraged to check with an abstract algebra textbook for basic definitions. Problem 5.2 discusses quaternions, which were the only graduate-level mathematical topic studied in American Universities in the 1880s, but now are all but forgotten. Problem 5.3 discusses a connection between vector fields in the plane and analytic functions; this connection is the foundation of numerous applications of complex numbers in the study of two-dimensional electrostatic fields and two-dimensional fluid flows. Problem 5.4 is an addition to the topic of conformal mappings. Problem 5.5 discusses one useful method of computing a power series expansion. Problem 5.6 introduces the subject of zeros of an analytic function. Problem 5.7 leads the reader through a rigorous computation of the Fourier transform of the normal probability density; a non-rigorous computation is on page 272. Finally, Problem 5.8 introduces a standard but somewhat advanced tool in residue integration, known as Jordan's Lemma.

PROBLEM 5.1. SOME ALGEBRA AND NUMBER THEORY.

(a) A rational number can be represented using an eventually periodic decimal fraction. For example, $1/3 = 0.333\ldots$, which we will denote by $0.(3)$. Indeed, $0.(3) = 0.3\sum_{k=0}^{\infty} 10^{-k} = 0.3/(1-0.1) = 0.3/0.9 = 1/3$, where we used the formula for the sum of the geometric series. Note that the same argument yields $0.(9) = 1$, therefore, by convention, we avoid infinite tails of nines, and use tails of zeros instead.

Find and verify the decimal expansions for $1/6$ and $1/7$.

(b) Use the division algorithm to prove that every rational number has a decimal expansion that is eventually periodic.

(c) Use the formula for the sum of the geometric series to show that every eventually periodic decimal fraction represents a rational number.

(d) Prove that $\sqrt{2}$ is not rational. *Hint: go by contradiction. If $\sqrt{2} = m/n$ in the lowest terms, then $m^2 = 2n^2$ and both m and n must be multiples of 2.*

(e) Define the set $\mathbb{Q}[\sqrt{2}\,]$ as the collection of real numbers of the form $a + \sqrt{2}b$, where a, b are rational numbers. Verify that this set is a field (that is, every element has an additive inverse, and every nonzero element has a multiplicative inverse; consult an algebra textbook for the precise definition of the field). *Hint: at some point, you have to use the result of part (d) to make sure everything works out as it should.*

(f) Fix a non-zero integer d, positive or negative, so that d is not divisible by a square, that is, n^2 does not divide d for every integer n. Consider the numbers $a + \sqrt{d}\,b$ with integer a, b; the collection of all these numbers is denoted by $\mathbb{Z}[\sqrt{d}\,]$. Show that $\mathbb{Z}[\sqrt{d}\,]$ is a *ring*, that is, the numbers $a+\sqrt{d}\,b$ can be added, subtracted, and multiplied just as the ordinary integers; check an algebra book for the precise definition.

(g) Consider the ring $\mathbb{Z}[\sqrt{d}\,]$ from part (f). An element u in $\mathbb{Z}[\sqrt{d}\,]$ is called a `unit` or an `invertible element` if $uv = 1$ for some $v \in \mathbb{Z}[\sqrt{d}\,]$. Verify that $\sqrt{2} \pm 1, \pm 1$ are units in $\mathbb{Z}[\sqrt{2}]$. Are there any other units in $\mathbb{Z}[\sqrt{2}\,]$?

(h) Once again consider the ring $\mathbb{Z}[\sqrt{d}\,]$. Note that all ordinary integers are in $\mathbb{Z}[\sqrt{d}\,]$ and some of them can be factored in unusual ways. For example, in $\mathbb{Z}[\sqrt{2}\,]$ we have $2 = (2-\sqrt{2})(2+\sqrt{2})$. Still, since $2 \pm \sqrt{2} = \sqrt{2}(\sqrt{2} \pm 1)$, and, as we saw in part (g), $\sqrt{2} \pm 1$ are units in $\mathbb{Z}[\sqrt{2}\,]$, the number 2 cannot actually be factored in a way different from $2 = 2 \cdot 1$.

(Even in $\mathbb{Z}$, you could write $2 = 2 \cdot (-1) \cdot (-1)$, but you would not consider it a different factorization of the number 2.) On the other hand, $7 = (3 - \sqrt{2})(3 + \sqrt{2})$ is a non-trivial factorization that is not available in $\mathbb{Z}$. Show that if we do not count the extra factorizations involving units, then every element of $\mathbb{Z}[\sqrt{2}\,]$ has a `unique factorization`, up to a permutation of factors.

(i) For some d, certain elements of $\mathbb{Z}[\sqrt{d}\,]$ have more than one distinct factorization. For example, $9 = 3 \cdot 3 = (2 + i\sqrt{5})(2 - i\sqrt{5})$ in $\mathbb{Z}[i\sqrt{5}\,] = \mathbb{Z}[\sqrt{-5}\,]$ (the only units in $\mathbb{Z}[i\sqrt{5}]$ are ± 1. Find another example of non-unique factorization with a different d. Can you find a condition on d to ensure unique factorization in $\mathbb{Z}[\sqrt{d}\,]$? *Hint: the last question is hard, and the complete answer is not known. For the background, see the books* Algebra *by M. Artin, 1991, and* Algebraic Number Theory *by E. Weiss, 1998*

PROBLEM 5.2. MULTIPLICATION AND DIVISION IN HIGHER DIMENSIONS.

Multiplication and division are well defined for the real numbers, and the complex numbers extend these operations to the Euclidean space $\mathbb{R}^2$: $(x_1, y_1)(x_2, y_2) = (x_1x_2 - y_1y_2, x_1y_2 + y_1x_2)$, $(x_1, y_1)/(x_2, y_2) = (x_1, y_1)(x_2, -y_2)/(x_2^2 + y_2^2)$. This multiplication is both associative and commutative.

Can one define multiplication and division in $\mathbb{R}^n$ for $n > 2$? The answer is "no" for all n *except* $n = 4$ and $n = 8$.

For $n = 4$, the multiplication is defined as follows:

$$(x_1, x_2, x_3, x_4)(y_1, y_2, y_3, y_4) = (z_1, z_2, z_3, z_4),$$
$$z_1 = x_1y_1 - x_2y_2 - x_3y_3 - x_4y_4,\; z_2 = x_1y_2 + x_2y_1 + x_3y_4 - x_4y_3, \quad (7.5.1)$$
$$z_3 = x_1y_3 - x_2y_4 + x_3y_1 + x_4y_2,\; z_4 = x_1y_4 + x_2y_3 - x_3y_2 + x_4y_1.$$

By analogy with complex numbers, we write $X = x_1 + ix_2 + jx_3 + kx_4$ for suitable i, j, k; the corresponding object X is called a `quaternion`. Verify the following statements:

(a) The multiplication defined by (7.5.1) is consistent with the product XY of quaternions, computed using the usual distributive laws and the following multiplication table:

	i	j	k
i	-1	k	$-j$
j	$-k$	-1	i
k	j	$-i$	-1

The first term in the product comes from the left-most column, the second, from the top row. For example, $ij = k$, $ji = -k$, $kj = -i$. In particular, *multiplication of quaternions is non-commutative.*

(b) Multiplication is associative: $(XY)Z = X(YZ)$.

(c) If we define $|X| = \sqrt{x_1^2 + x_2^2 + x_3^2 + x_4^2}$, then $|XY| = |X|\,|Y|$.

(d) If we define $\overline{X} = x_1 - ix_2 - jx_3 - kx_4$, then $X\overline{X} = \overline{X}X = |X|^2$. As a result, the definition of division for quaternions is $X/Y = X\overline{Y}/|Y|^2$.

Discussion. In $\mathbb{R}^8$, the corresponding objects are called *octonions;* multiplication of octonions is both non commutative and non-associative. For more details, see the book *On Quaternions and Octonions* by J. H. Conway and D. A. Smith, 2003. Quaternions were discovered by W. R. Hamilton on October 16, 1843, after some 15 years of intensive work. The formulas $i^2 = j^2 = k^2 = ijk = -1$ he carved that day in the stone of Broome (or Brougham) Bridge across the Royal Canal in Dublin might still be there. Octanions were discovered shortly after, on December 26, 1843, by Hamilton's friend JOHN T. GRAVES (1806–1870). Still, the British mathematician ARTHUR CAYLEY (1821–1895) re-discovered octanions on his own and published the first paper on the subject in 1845. As a result, octanions are sometimes called the Cayley numbers or Cayley algebra.

PROBLEM 5.3. VECTOR FIELDS IN THE PLANE.

Let $\boldsymbol{F} = P(x,y)\,\boldsymbol{i} + Q(x,y)\,\boldsymbol{j}$ be a continuously differentiable vector field in $\mathbb{R}^2$, defined in cartesian coordinates. In applications, $\boldsymbol{F}$ usually represents the velocity $\boldsymbol{v}$ of moving fluid or the intensity $\boldsymbol{E}$ of electrostatic field. We introduce the following objects:

- the divergence of $\boldsymbol{F}$, as a scalar field $\operatorname{div}\boldsymbol{F} = P_x + Q_y$;
- the `vorticity` of $\boldsymbol{F}$, as a scalar field $\omega_{\boldsymbol{F}} = Q_x - P_y$;
- a `streamline` of $\boldsymbol{F}$, as a curve $x = x(t)$, $y = y(t)$ satisfying the system of ordinary differential equations $x'(t) = P(x(t), y(t))$, $y'(t) = Q(x(t), y(t))$;
- a critical, or stationary, point of $\boldsymbol{F}$, as a point (x_0, y_0) such that $\boldsymbol{F}(x_0, y_0) = \boldsymbol{0}$.

A vector field $\boldsymbol{F}$ is called

- `solenoidal`, if $\operatorname{div}\boldsymbol{F} = 0$;
- `irrotational`, if $\omega_{\boldsymbol{F}} = 0$;
- `harmonic`, if it is both solenoidal and irrotational.

Note that, if we consider a vector field $\widetilde{\boldsymbol{F}}$ in $\mathbb{R}^3$ defined by $\widetilde{\boldsymbol{F}}(x,y,z) = \boldsymbol{F}(x,y) + 0\,\hat{\boldsymbol{\kappa}}$, then $\operatorname{curl}\widetilde{\boldsymbol{F}} = (\omega_{\boldsymbol{F}})\,\hat{\boldsymbol{\kappa}}$. Note also that a streamline is a trajectory of a point mass m moving in the plane so that the velocity of m at point (x,y) is $\boldsymbol{F}(x,y)$.

(a) Let G be a simply connected domain. Verify that $\operatorname{div} \boldsymbol{F} = 0$ in G if and only if there exists a scalar field U, called the `stream function` of $\boldsymbol{F}$, so that every streamline of $\boldsymbol{F}$ in G is a level set of U.

(b) Let G be a simply connected domain. Verify that $\omega_{\boldsymbol{F}} = 0$ in G if and only if there exists a scalar field V, called the `velocity potential` of $\boldsymbol{F}$ so that $\boldsymbol{F} = \operatorname{grad} V$ in G. A level set of the velocity potential is called an `equipotential line` of the field $\boldsymbol{F}$.

(c) Assume that $\boldsymbol{F}$ is a harmonic vector field in a simply connected domain G. (i) Show that $\boldsymbol{F}$ has both a stream function and a velocity potential. (ii) Let U and V be a stream function and a velocity potential of $\boldsymbol{F}$, respectively. Show that U and V are conjugate harmonic functions. The analytic function $f(x+iy) = V(x,y) + iU(x,y)$ is called the `complex potential` of the harmonic vector field $\boldsymbol{F}$. (iii) Verify that if $f = f(z)$ is a complex potential of a harmonic vector field $\boldsymbol{F} = P\,\hat{\imath} + Q\,\hat{\jmath}$, then $\boldsymbol{F} = \overline{f'}$, in the sense that $P = \Re f'$ and $Q = -\Im f'$.

(d) Assume that (x_0, y_0) is a critical point of a harmonic vector field $\boldsymbol{F}$. Show that there exist a positive integer number n, a positive real number δ, and a function $g = g(z)$ so that g is analytic at 0, $g(0) = z_0 = x_0 + iy_0$, $g'(0) \neq 0$, and the complex potential f of $\boldsymbol{F}$ satisfies $f(g(z)) = f(z_0) + z^n$ for $|z| < \delta$.

(e) According to part (c), every harmonic vector field is completely characterized by its vector potential, and conversely, every analytic function is a vector potential of a harmonic vector field. For each of the following analytic functions, find the corresponding vector field and sketch a few of its streamlines and equipotential lines. (i) $f(z) = Az$, where A is a complex number. (ii) $f(z) = z^2$; explain why the corresponding vector field can model fluid flow around a corner. (iii) $f(z) = c \ln z$, where c is a real number; explain why the corresponding vector field can model a source if $c > 0$ and a sink if $c < 0$. (iv) $f(z) = ic \ln z$, where c is a real number; explain why the corresponding vector field can model a vortex. (v) $f(z) = A \ln z$, where A is a complex number. What can this vector field model? (vi) $f(z) = c/z$, where c is a real number; explain why the corresponding vector field can model the electric field of a point dipole. What complex potential can model the electric field of a regular (non-point) dipole? (vii) $f(z) = z + z^{-1}$; explain why the corresponding vector field can model fluid flow around a cylinder of radius 1. How would you model flow around a cylinder of radius R?

PROBLEM 5.4. MORE ON CONFORMAL MAPPINGS.

(a) The function $w(z) = (az+b)/(cz+d)$ is called **linear-fractional**. Verify that if $ad - bc > 0$ and $c \neq 0$, then (i) the mapping defined by w is conformal everywhere except $z_0 = -d/c$; (ii) the image of a line or a circle is a line or a circle (line can become a circle and a circle, a line).

(b) Consider the function $w(z) = \dfrac{1}{2}\left(z + \dfrac{1}{z}\right)$. (i) Verify that it defines a conformal mapping for all $z \neq \pm 1$. (ii) Show that this function maps a circle $|z| = a$, $a \neq 1$, to an ellipse with foci at ± 1 and an opposite orientation; (iii) Show that this function maps a line $\arg(z) = \alpha$ to a hyperbola with foci at ± 1. (iv) Let $\mathcal{C}_1$ be circle with center at $z_1 = i/8$ and passing through the points ± 1. Let $\mathcal{C}_2$ be a circle of radius $\sqrt{1 + (1/32)}$ so that $\mathcal{C}_1$ touches $\mathcal{C}_2$ from the inside at the point -1 (draw a picture). Find the images of $\mathcal{C}_1$ and $\mathcal{C}_2$ under w. *Hint: the image of $\mathcal{C}_1$ has empty interior.*

(c) *Discussion.* The function w in (a) defines the **Möbius transformation**, after A. F. MÖBIUS. The collection of these transforms is a *group*, and is widely used in algebraic geometry and related areas of mathematics. The function w in (b) is called **Zhukovsky's function**, after the Russian mathematician NIKOLAI YEGOROVICH ZHUKOVSKY (1847–1921), who used it to calculate the lift of an airplane wing. The image of the circle $\mathcal{C}_2$ in part (iv) does resemble the profile of a wing and is sometimes called **Zhukovsky's airfoil**.

PROBLEM 5.5. Consider a function $g(z) = \sum_{k \geq 0} a_k (z - z_0)^k$ and assume that the power series converges for $|z - z_0| < R$ for some $R > 0$. Let $\mathcal{C}(z_0, z_1)$ be a piece-wise smooth curve starting at z_0, ending at z_1, and always staying inside the disk $|z - z_0| < R$. Using the same arguments as in Step 2 of the proof of Theorem 4.3.4, page 211, show that

$$\int_{\mathcal{C}(z_0, z_1)} g(z)dz = \sum_{k \geq 0} \frac{a_k}{k+1}(z_1 - z_0)^{k+1}.$$

Since the function g is analytic, the integral does not depend on the particular choice of the curve $\mathcal{C}(z_0, z_1)$.

Use the result to find the power series expansion of

$$f(z) = \int_{\mathcal{C}_{0,z}} \frac{\cos \zeta - 1}{\zeta}\, d\zeta$$

at $z_0 = 0$.

PROBLEM 5.6. (a) Let $z_*, z_1, z_2, \ldots$ be complex numbers in a domain G of the complex plane so that $\lim_{n \to \infty} z_n = z_*$ and let f be a function analytic in G so that $f(z_n) = 0$ for all $n \geq 1$. Prove that $f(z) = 0$ in some neighborhood of the point z_*. *Hint: by continuity,* $f(z^*) = 0$. *Then consider the Taylor series of* f *at* z_*. (b) Consider the function $f(z) = \sin(1/z)$ which is analytic for all $z \neq 0$ and is equal to zero for $z_n = 1/(\pi n)$. Is this a contradiction of the statement of part (a)?

PROBLEM 5.7. Verify that $I = \int_{-\infty}^{\infty} \exp\left(-\frac{x^2}{2} - ix\omega\right) dx = \sqrt{2\pi} e^{-\omega^2/2}$, when ω is a real number. Here is a possible argument to follow.

Step 0. $|e^{-x^2/2 - ix\omega}| = e^{-x^2/2}$, and $\int_{-\infty}^{\infty} e^{-x^2/2} = \sqrt{\pi}/2$ (known from calculus, or see page 272). This means I exists.

Step 1. $x^2 + i2x\omega = (x + i\omega)^2 + \omega^2$, so $I = e^{-\omega^2/2} \lim_{A\to\infty} \int_{\mathcal{C}_A} e^{-z^2/2} dz$, where $\mathcal{C}_A$ is the line segment from the point $z_1 = -A + i\omega$ to $z_2 = A + i\omega$, parallel to the real axis.

Step 2. Consider the boundary $\mathcal{C}_{A,1}$ of the rectangle with vertices at the points $\pm A$ and $\pm A + i\omega$ so that the orientation of $\mathcal{C}_{A,1}$ is consistent with the orientation of $\mathcal{C}_A$ (draw a picture). Then $\oint_{\mathcal{C}_{A,1}} e^{-z^2/2} dz = 0$ for every A.

Step 3. Show that the integrals of $e^{-z^2/2} dz$ along the vertical sides of $\mathcal{C}_{A,1}$ tend to zero as $A \to \infty$.

Step 4. Conclude that $\lim_{A\to\infty} \int_{\mathcal{C}_A} e^{-z^2/2} dz = \int_{-\infty}^{\infty} e^{-x^2/2} = \sqrt{\pi}/2$.

PROBLEM 5.8. JORDAN'S LEMMA.

(a) The following result, known as `Jordan's Lemma`, allows evaluation of integrals of the type (4.4.20), page 229, using residue integration when the degree of Q is only one unit bigger than the degree of P.

Consider a function $g = g(z)$ and the semi-circle $\mathcal{C}_R = \{z : |z| = R, \Im z > 0\}$. Assume that g is continuous for $\Im z \geq 0$ and $\lim_{R\to\infty} \max_{z\in\mathcal{C}_R} |g(z)| = 0$. Show that $\lim_{R\to\infty} \int_{\mathcal{C}_R} g(z) e^{i\alpha z} dz = 0$ for every $\alpha > 0$.

Hint: $|e^{i\alpha z}| = e^{-\alpha R \sin\varphi}$ *on* $\mathcal{C}_R$, *and* $\sin\varphi \geq 2\varphi/\pi$ *for* $0 \leq \varphi \leq \pi/2$.

(b) Use the result to show that $\int_{-\infty}^{\infty} x \sin x/(1 + x^2)\, dx = \pi/e$. Note that the integral does not converge without the $\sin x$ factor.

7.6 Fourier Analysis

The main part of this section is Problem 6.2, which outlines the proof of the theorem about point-wise convergence of the Fourier series; the details can be found in several sections of the book [Körner (1989)]. An almost immediate consequence of the theorem is the possibility to approximate a continuous function with polynomials, which the reader can establish by solving Problem 6.3. Problem 6.1 presents a standard construction illustrating no connection between mean-square and point-wise convergence. Problem 6.4 discusses some fine points related to the uniform convergence of Fourier series. Problem 6.5 is about a useful characterization of Dirac's delta function. Problem 6.6 presents two interesting properties of the Laplace transform.

PROBLEM 6.1. For $n \geq k \geq 1$ define function $f_{n,k} = f_{n,k}(x)$ so that

$$f_{n,k}(x) = \begin{cases} 1, & \dfrac{k \quad 1}{n} \leq x \leq \dfrac{k}{n}, \\ 0, & \text{otherwise.} \end{cases}$$

For $N \geq 1$, define $g_N(x) = f_{n,k}(x)$ if $N = (n(n-1)/2) + k$.
(a) Verify that the sequence $\{g_N, N \geq 1\}$ coincides with the sequence

$$\{f_{1,1}, f_{2,1}, f_{2,2}, f_{3,1}, f_{3,2}, f_{3,3}, \ldots\}.$$

(b) Draw the graph of g_N for $1 \leq N \leq 6$. (c) Show that $\lim_{N\to\infty} \int_0^1 g_N^2(x)dx = 0$. *Hint: the value of the integral is* $1/n$. (d) Show that if $0 \leq x_0 \leq 1$, then $\lim_{N\to\infty} g_N(x_0)$ does not exist. *Hint: for every* $n \geq 1$, *there exists at least one* k *so that* $1 \leq k \leq n$ *and* $f_{n,k}(x_0) = 1$. (e) Consider the sequence $h_n(x) = \sqrt{n}$, $0 < x < 1/n$; $h_n(x) = 0$ for $1/n \leq x < 1$. Show that $\lim_{n\to\infty} h_n(x) = 0$ for all $x \in (0,1)$, but $\lim_{x\to\infty} \int_0^1 |h_n(x)|^2 dx = 1$.

PROBLEM 6.2. Prove Theorem 5.1.5 on page 251 by following the steps below.

Step 1. Verify that $f^+ = f(x^+)$ and $f^- = f(x^-)$ indeed exist for all x. *Hint: you only need to worry about points where* f *is not continuous, If* x_0 *is such a point, then there exist numbers* a, b *so that* $a < x_0 < b$ *and* f *has a continuous bounded derivative on* (a, x_0) *and on* (x_0, b). *Use the mean-value theorem and the Cauchy criterion for the existence of the limit to show that, for every sequence of positive number* $\varepsilon_n \to 0$ *as* $n \to \infty$, *the limits of* $f(x_0 + \varepsilon_n)$ *and* $f(x_0 - \varepsilon_n)$ *exist.*

Step 2. Show that $|c_k| < A/(|k|+1)$ for all k. *Hint: we did it in the proof of Theorem 5.1.6.*

Step 3. Introduce the following notations:

$$S_{f,N}(x) = \sum_{k=-N}^{N} c_k e^{ikx}, \quad \sigma_{f,N} = \frac{1}{N+1} \sum_{k=0}^{N} S_{f,k}(x).$$

Show that if f is continuous at x, then $\lim_{N\to\infty} \sigma_{f,N}(x) = f(x)$, and the convergence is uniform if f is continuous everywhere (this is the key step in the whole proof). *Hint: show that*

$$\sigma_{f,N}(x) = \frac{1}{2\pi}\int_{-\pi}^{\pi} f(y)K_N(x-y)dy,$$

where

$$K_N(x) = \sum_{k=-N}^{N} \frac{N+1-|k|}{N+1} e^{ikx} = \frac{1}{N+1}\left(\frac{\sin\left((N+1)x/2\right)}{\sin(x/2)}\right)^2.$$

Then show that $(1/2\pi)\int_{-\pi}^{\pi} K_N(x)dx = 1$ *for all* N, *while* $\lim_{N\to\infty} K_N(x) = 0$ *uniformly on the set* $\delta < |x| < \pi$ *for every* $\delta > 0$. *As a result, if* f *is continuous at* x, *then, for large* N,

$$\frac{1}{2\pi}\int_{-\pi}^{\pi} f(y)K_N(x-y)dy - f(x) = \frac{1}{2\pi}\int_{-\pi}^{\pi} (f(y)-f(x))K_N(x-y)dy \approx 0.$$

The function K_N is called the **Fejér kernel**, after the Hungarian mathematician LIPÓT FEJÉR (1880–1959), who, in 1900, showed that a continuous function can always be reconstructed from its Fourier coefficients using $\sigma_{f,N}$.

Step 4. For integer $M > N \geq 0$, define

$$\sigma_{f,N,M}(x) = \frac{1}{M-N}\left((M+1)\sigma_{M,f}(x) - (N+1)\sigma_{N,f}(x)\right).$$

Show that if f is continuous at x, then, for every fixed positive integer k, we have

$$\lim_{N\to\infty} \sigma_{f,kN,(k+1)N}(x) = f(x),$$

and the convergence is uniform if f is continuous everywhere. *Hint: this is easy; just use the result of the previous step.*

Step 5. Show that there exists a positive number B such that, for all x and for all positive integer k, M, N satisfying $kN < M < (k+1)N$, we have

$$|\sigma_{f,kN,(k+1)N}(x) - S_{f,M}| \leq B/k.$$

Hint: first show that $|\sigma_{f,kN,(k+1)N}(x) - S_{f,M}| \leq \sum_{kN\leq l\leq (k+1)N} |c_l|$, *and then use* $|c_l| \leq A/(|l|+1)$.

Step 6. By combining the results of Steps 4 and 5, show that $S_f(x) = f(x)$ if f is continuous at x, and the Fourier series converges to f uniformly if f is continuous everywhere. This takes care of convergence at the points of continuity of f.

The idea now is to verify the theorem for a special discontinuous function, and then use the result to handle the general case.

Step 7. Consider the 2π-periodic function g so that $g(x) = x$ for $|x| < \pi$; it does not matter how you define g at the points $\pm\pi$, for example, put g equal to zero. Then g is continuous everywhere on $\mathbb{R}$ except the points $x = \pi + 2\pi n$, $n = 0, \pm 1, \pm 2, \ldots$ (draw a picture). The result of Step 6 implies $S_g(x) = g(x)$ for all $x \neq \pi + 2\pi n$, and direct computations show that $S_g(\pi) = 0 = \big(g(\pi^+)+g(\pi^-)\big)/2$. In other words, the complete theorem holds for the function g.

Step 8. Assume that f is not continuous at $x = \pi$. Define the function F so that $F(\pi) = \big(f(\pi^+) + f(\pi^-)\big)/2$ and $F(x) = f(x) + \big(f(\pi^+) - f(\pi^-)\big)g(x)/(2\pi)$, where g is from the previous step. Show that F is continuous at π, that is, $F(\pi^+) = F(\pi^-) = F(\pi)$ (keep in mind that, by periodicity, $F(\pi^+) = F(-\pi^+)$). Then use the results of Steps 6 and 7 to conclude that $S_f(\pi) = \big(f(\pi^+) + f(\pi^-)\big)/2$.

Step 9. Finally, modify the arguments in Step 8 to consider a point other than π.

PROBLEM 6.3. APPROXIMATION BY POLYNOMIALS.

(a) Show that, for every continuously differentiable function $f = f(x)$ on a bounded interval $[a, b]$ and for every $\varepsilon > 0$, there exists a polynomial $P = P(x)$ such that $\max_{a \le x \le b} |f(x) - P(x)| < \varepsilon$. *Hint: Consider the Fourier series for the function* $g(t) = f(a^* \cos t + b^*)$, $-\pi < t \le \pi$, *where* $a^* = (b-a)/2$, $b^* = (a+b)/2$; *note that* $g(-\pi) = g(\pi) = f(a)$. *Then prove by induction that* $\cos(nt) = T_n(\cos t)$ *for some polynomial* T_n *of degree* n.

(b) Next, use Step 4 from the previous problem to show that, for every CONTINUOUS function $f = f(x)$ on a bounded interval $[a, b]$ and for every $\varepsilon > 0$, there exists a polynomial $P = P(x)$ such that $\max_{a \le x \le b} |f(x) - P(x)| < \varepsilon$. That is, extend the result of part (a) to functions that are only continuous on $[a, b]$.

(c) *Discussion.* The polynomials T_n, $n \ge 1$ from part (a) satisfy $T_{n+1}(x) = 2xT_n(x) - T_{n-1}(x)$. They are the **Chebyshev polynomials** of the first kind, and, for each $n \ge 0$, T_n is the polynomial solution of the differential equation $(1 - z^2)w''(z) - zw'(z) + n^2 w(z) = 0$, that is, equation (4.4.43), page 237, with $\mu = 0$, $\nu = -1$, $\lambda = n^2$. The result of part (b) is known as **Weierstrass's polynomial approximation theorem.**

PROBLEM 6.4. Consider the 2π-periodic function $g = g(x)$ so that $g(x) = x$, $|x| < \pi$ and $g(\pm\pi) = 0$. The Fourier series for g does not converge uniformly to g on $[-\pi, \pi]$, because otherwise S_g would be a continuous function. Does the Fourier series for g converge uniformly on $(-\pi, \pi)$? *Hint: no, for, if it did, the convergence at the points $\pm\pi$ would imply uniform convergence on $[-\pi, \pi]$. Try to understand why.*

PROBLEM 6.5. Let $\varphi = \varphi(x)$ be a function with the following properties: (a) $\int_{-\infty}^{\infty} \varphi(x)dx = 1$; (b) $\lim_{|x|\to\infty} \varphi(x) = 0$. Show that $\lim_{n\to\infty} n\varphi(nx) = \delta(x)$, the Dirac delta function; see page 275. In other words, show that, for every continuous function $f = f(x)$ that is equal to zero for all sufficiently large $|x|$, $\lim_{n\to\infty} n \int_{-\infty}^{\infty} f(x)\varphi(nx)dx = f(0)$.

PROBLEM 6.6. Assume that $f = f(t)$ is a bounded continuous function on $[0, +\infty)$ and $\lim_{x\to\infty} f(t) = A$. Denote by $F = F(s)$ the Laplace transform of f. Show that $\lim\limits_{s\to 0, s>0} sF(s) = A$, $\lim\limits_{s\to\infty} sF(s) = f(0)$.

7.7 Partial Differential Equations

Three problems in this section are of a fundamental nature: Problem 7.5 discusses Bessel's functions in some detail; Problem 7.7 gives a fairly complete account of the Sturm-Liouville problem; Problem 7.14 introduces finite elements in two dimensions. There are two more problems related to numerical methods that are rather comprehensive: Problem 7.12 outlines the stability analysis of finite difference schemes, and Problem 7.13 connects the finite difference schemes with the Fast Fourier Transform. Problem 7.9 establishes the invariance of Maxwell's equations under the Lorentz transformation and also introduces the subject of electromagnetic radiation. Problems 7.1, 7.2, 7.3, 7.4, 7.6, and 7.8 are extensions of the examples discussed in the main text, although these extensions are not always easy. Problem 7.10 discusses some basic methods used in the analysis of the Euler and Navier-Stokes equations. Problem 7.11, although related to a deep and complicated subject, is mostly for fun.

PROBLEM 7.1. Derive the equation describing the temperature $u = u(t, x)$ in a homogenous thin wire of constant cross-section if the lateral surface is not insulated. *Hint: if $U_o = U_o(t, x)$ is the outside temperature, then $u_t = au_{xx} + C(u - U_o)$.*

PROBLEM 7.2. Use separation of variables to find a solution of the following initial-boundary value problem:

$$\begin{aligned} &u_t = u_{xx},\ t > 0,\ 0 < x < L; u(0,x) = f(x),\ 0 \le x \le L; \\ &a\,u(t,0) + b\,u_x(t,0) = A(t),\ c\,u(t,L) + d\,u_x(t,L) = B(t),\ t \ge 0, \end{aligned} \tag{7.7.1}$$

where a, b, c, d are real numbers so that $a^2 + b^2 > 0$, $c^2 + d^2 > 0$, and $A = A(t)$, $B = B(t)$ are known continuous functions.

PROBLEM 7.3. Find a solution of the initial-boundary value problem

$$\begin{aligned} &u_{tt} = c^2 u_{xx},\ t > 0, \quad 0 < x < L; \\ &u(0,x) = u_t(0,x) = 0,\ 0 < x \le L; \\ &u(t,0) = F(t),\ u(t,L) = 0,\ t \ge 0, \end{aligned} \tag{7.7.2}$$

Is a standing wave possible?

Note that (7.7.2) models a string with one end fixed and the other moving up and down according to $F(t)$.

PROBLEM 7.4. Consider the telegraph equation

$$U_{xx}(t,x) = aU_{tt}(t,x) + bU_t(t,x) + cU(t,x), \quad t > 0,$$

with positive real numbers a, b, c, and initial conditions $U(0,x) = f(x)$, $U_t(0,x) = g(x)$ (see page 334. Find the formula for U when
(a) $x \in \mathbb{R}$;
(b) $x \in (0, +\infty)$ with boundary condition $U(t,0) = F(t)$;
(c) $x \in (0, L)$ with boundary conditions $U(t,0) = F_1(t)$, $U(t,L) = F_2(t)$.

PROBLEM 7.5. BESSEL'S FUNCTIONS.

(a) Let $V = V(s)$ be a solution of (6.3.26) on page 342, with $q = N^2$. Verify that the function $F(s) = \sqrt{s}\,V(s)$ satisfies

$$F''(s) + \left(1 - \frac{q - (1/4)}{s^2}\right) F(s) = 0. \tag{7.7.3}$$

Use this result to conclude that the Bessel function J_N has infinitely many positive zeroes and that $\lim_{x\to\infty} |J_N(x)| = 0$. *Hint:* $J_N(s) = F(s)/\sqrt{s}$.

(b) Use the power series representation of J_N (see (6.3.27) on page 342) to verify the following equalities:

$$\begin{aligned}
&J_N(-z) = (-1)^N J_N(z),\\
&(z^n J_N(z))' = z^N J_{N-1}(z), \quad (z^{-N} J_N(z))' = -z^{-N} J_{N+1}(z),\\
&zJ_N'(z) = NJ_N(z) - zJ_{N+1}(z) = -NJ_N(z) + zJ_{N-1}(z), \qquad (7.7.4)\\
&\int_{\mathcal{C}(0,z)} \zeta^N J_{N-1}(\zeta)d\zeta = z^N J_N(z),\ \mathcal{C}(0,z) \text{ is a path from 0 to } z.
\end{aligned}$$

(c) Verify that

(i) all zeroes of J_N are real;

(ii) $J_N(x) = 0$ if and only if $J_N(-x) = 0$;

(iii) For $N > 0$, J_N has a zero of multiplicity N at $x = 0$;

(iv) If $J_N(x) = 0$ and $x \neq 0$, then $J_N'(x) \neq 0$ and $J_M(x) \neq 0$ for all $M \neq N$.

(d) Let $\alpha_k^{(N)}$ be a positive zero of J_N. Verify that

$$\int_0^R \Big(J_N(\alpha_k^{(N)} r/R)\Big)^2 r dr = \frac{R^2}{2} J_{N+1}^2(\alpha_k^{(N)}). \qquad (7.7.5)$$

(e) Show that

$$J_N(x) = \frac{1}{\pi}\int_0^{\pi} \cos(N\theta - x\sin\theta)d\theta \qquad (7.7.6)$$

and use the result to show that $\lim_{|x|\to\infty} |J_N(x)| = 0$. *Hint: for (7.7.6) with $N = 0$, consider $\int_{-\pi}^{\pi} e^{ix\sin\theta} d\theta$ and note that $\int_{-\pi}^{\pi} \sin^{2k}\theta d\theta = \big(2\pi(2k)!\big)/(2^k k!)^2$.*

(f) All other solutions of equation (6.3.26) on page 342 have a singularity at zero. What is the type of the singularity for different N? *Hint: start with (4.4.39) on page 236.*

PROBLEM 7.6. (a) Find the eigenvalues and eigenfunctions of the Dirichlet Laplacian in a ball of radius R; see (6.3.13) on page 338. Find the multiplicity of each eigenvalue.

(b) What sound is produced by a hollow sphere? In other words, solve $u_{tt} = \nabla^2 u$ with $u = u(t,\theta,\varphi)$, $\theta \in [0, 2\pi], \varphi \in [0,\pi]$ in spherical coordinates. How does your answer depend on the radius of the sphere?

(c) What sound is produced by a solid ball of radius R? In other words, solve $u_{tt} = \nabla^2 u$ with $u = u(t,r,\theta,\varphi)$ in spherical coordinates $0 \le r \le R$, $\theta \in [0, 2\pi], \varphi \in [0,\pi]$. How does your answer depend on the radius of the ball?

Hint: for each problem, review the computations related to the hydrogen atom starting on page 368. See also Section 17 of the book Boundary Value Problems and Orthogonal Expansions *by C. R. MacCluer, 1994.*

PROBLEM 7.7. THE STURM-LIOUVILLE PROBLEM IN ONE DIMENSION.

Each of the following three eigenvalue problems is known as the **Sturm-Liouville problem**, after the French mathematicians JACQUES CHARLES FRANÇOIS STURM (1803–1855) and J. Liouville, who pioneered the work on the subject in the 1830s:

$$\frac{1}{w(x)}\left(p(x)u'(x)\right)' + q(x)u(x) = \lambda\, u(x), \tag{7.7.7}$$

$$P(x)u''(x) + Q(x)u'(x) + R(x)u(x) = \lambda\, u(x), \tag{7.7.8}$$

$$-u''(x) + V(x)u(x) = \lambda\, u(x). \tag{7.7.9}$$

In each case, we assume that x is in a bounded interval (a, b), the real-valued functions w, p, p', q, P, Q, R, V are continuous on (a, b), the functions w, p, P satisfy $w(x) > 0$, $0 < p(x) \le C$, $0 < P(x) \le C$ for all $x \in (a, b)$ and some $C > 0$, and the real-valued function u satisfies the boundary conditions

$$c_1\, u(a) + c_2\, u'(a) = 0, \quad c_3\, u(b) + c_4\, u'(b) = 0; \tag{7.7.10}$$

we will see that sometimes, instead of one or both boundary conditions, it is enough to require only the boundedness of the solution u near the corresponding end point. The real numbers c_1, c_2, c_3, c_4 satisfy $c_1^2 + c_2^2 > 0$, $c_3^2 + c_4^2 > 0$, and are sometimes chosen as $c_1 = \sin\alpha, c_2 = \cos\alpha$, $c_3 = \sin\beta, c_4 = \cos\beta$.

(a) *Equivalence of the three equations (7.7.7)–(7.7.9).*
(i) Verify that (7.7.7) is a particular case of (7.7.8). (ii) Let u be a solution of (7.7.8) satisfying both boundary conditions (7.7.10). Assume that the functions P, Q are continuously differentiable on (a, b) and define $A(x) = \int_{x_0}^{x}(Q(t)/P(t))dt$, where $x_0 \in [a, b]$. Show that $v(x) = e^{A(x)/2}u(x)$ satisfies (7.7.7) and find the corresponding functions w, p, q. What are the boundary conditions satisfied by v? Can it happen that one or both of those conditions disappear? *Hint: start by dividing by P in (7.7.8), and then do the substitution* $u = ve^{-A/2}$. (iii) Let $u''(x) + (\ell(x) + \lambda r(x))u(x) = 0$ and assume that the function $r = r(x)$ is twice continuously differentiable and positive on $[a, b]$. Define $\tau(x) = \int_a^x \sqrt{r(t)}\, dt$, and let the function $v = v(x)$ be such that $u(x) = v(\tau(x))r^{-1/4}(x)$. Verify that v satisfies (7.7.9), that is, under certain conditions, we have the equivalence between (7.7.8) and (7.7.9). What are the boundary conditions for v if u satisfies both boundary conditions in (7.7.10)?

(b) *Properties of eigenvalues and eigenfunctions.*
(i) Let $\kappa = \kappa(x)$ be a **weight function** on (a, b), that is, κ is continuous

on (a, b) and satisfies $\kappa(x) > 0$, $x \in (a, b)$. Let $\mathbb{H}_\kappa$ be the collection of real-valued functions f that are twice continuously differentiable on $[a, b]$ and satisfy $\int_a^b |f(x)|^2\kappa(x)dx < \infty$ together with the boundary conditions (7.7.10). Verify that if, for $f, g \in \mathbb{H}_\kappa$, we define

$$(f, g)_\kappa = \int_a^b f(x)g(x)\kappa(x)dx,$$

then $\mathbb{H}_\kappa$ becomes an inner product space. (ii) Show that if we take the weight function $\kappa(x) = w(x)$, then the operator $\mathcal{A}_1$ defined by (7.7.7) is symmetric on $\mathbb{H}_\kappa$. More precisely, denote by $\mathcal{A}_1[u](x)$ the left-hand side of (7.7.7) and verify that $(v, \mathcal{A}_1[u])_\kappa = (\mathcal{A}_1[v], u)_\kappa$ (that is, $\int_a^b w(x)v(x)\mathcal{A}_1[u](x)dx = \int_a^b w(x)u(x)\mathcal{A}_1[v](x)dx$) for every $u, v \in \mathbb{H}_\kappa$. Note that no boundary condition is necessary at a and/or b if p vanishes at a and/or b. *Hint: as you integrate by parts with a non-vanishing p, note that (c_1, c_2) is a nontrivial solution of the system $c_1u(a) + c_2u'(a) = 0, c_1v(a) + c_2v'(a) = 0$, which means $u(a)v'(a) = u'(a)v(a)$.* (iii) Denote by $\mathcal{A}_2[u](x)$ the left-hand side of (7.7.8) and define

$$W(x) = \frac{1}{P(x)} \exp\left(\int_{x_0}^x \frac{Q(t)}{P(t)}dt\right), \quad x_0 \in [a, b]. \tag{7.7.11}$$

Verify that if we take the weight function $\kappa(x) = W(x)$, then $\mathcal{A}_2$ is a symmetric operator on $\mathbb{H}_\kappa$. Give three examples when one or both boundary conditions (7.7.10) can be avoided (in each example, you must provide three specific functions P, Q, R). (iv) Verify that if we take the weight function $\kappa(x) = 1$ for all $x \in (a, b)$, then the corresponding operator in (7.7.9) is symmetric on $\mathbb{H}_\kappa$. (v) Conclude that, for each of the three problems, the eigenvalues are real and the eigenfunctions corresponding to different eigenvalues are orthogonal in the appropriate space $\mathbb{H}_\kappa$. *Hint: see Theorem 6.2.1 on page 329.* (vi) Show that if at least one of the boundary conditions (7.7.10) is necessary for the symmetry of the corresponding operator $\mathcal{A}$, then each eigenvalue is simple, that is, any two eigenfunctions corresponding the same eigenvalue are multiples of each other. *Hint: if two linearly independent eigenfunctions correspond to the one eigenvalue, then all solution of the corresponding second-order equation must satisfy this condition. Still, there are should be solutions that do not satisfy this condition.* (vii) For each of the three equations, find sufficient conditions to have all eigenvalues negative. *Hint: go over the integration by parts; for (7.7.7), you need $w(x)q(x) \le -c < 0$.* (viii) For each of the three equations, determine when the `periodic boundary conditions` $u(a) = \pm u(b)$, $u'(a) = \pm u'(b)$ (you take either $+$ or $-$ in both equalities) result in a symmetric operator. *Hint: go over the integration by parts; for (7.7.7), you need $p(a) = p(b)$.* Show that, with periodic boundary conditions, eigenvalues can be multiple, but the multiplicity is at most two.

(c) *Particular examples.* Consider equations (4.4.40) and (4.4.43)–(4.4.47); see page 236. Write each equation in the form (7.7.7); you can assume that everything is real-valued.

(d) *Discussion.* A Sturm-Liouville problem is called **regular** if
• in (7.7.7), the functions w, p, q are bounded on $[a, b]$ and $p(x) \geq c > 0$, $w(x) \geq c > 0$;
• in (7.7.8), the functions R, W are bounded on $[a, b]$ and $W(x) \geq c > 0$, where W is from (7.7.11);
• in (7.7.9), $\int_a^b |V(x)|dx < \infty$.
In most other cases, including the unbounded interval (a, b) or vanishing of the functions p and/or w in (7.7.7), the problem is called **singular**. For the regular problem, the conclusions of Theorem 6.2.2 on page 330 hold. Thus, every twice continuously differentiable function satisfying either the boundary conditions (7.7.10) or the periodic boundary conditions, is represented by a generalized Fourier series of the eigenfunctions u_k of the regular Sturm-Liuoville problem, and the series converges uniformly on $[a, b]$. For example, for (7.7.7), the series expansion is

$$f(x) = \sum_{k=1}^{\infty} f_k u_k(x), \quad f_k = \frac{\int_a^b w(x)f(x)u_k(x)dx}{\int_a^b w(x)u_k^2(x)dx}.$$

This property of the regular Sturm-Liuoville problem is known as **Steklov's Theorem**, after the Russian mathematician VLADIMIR ANDREEVICH STEKLOV (1864–1926), who made major contributions to the subject in the early 1900s. The proof of Steklov's Theorem can be found in many advanced text books that cover ordinary and/or partial differential equations; see, for example, [Levitan and Sargsjan (1991)] or [Shu (1987)]. More generally, if the function f is piece-vice continuously differentiable on $[a, b]$, then the sum of the Sturm-Liouville eigenfunction expansion of f coincides on (a, b) with the sum of the Fourier series for f. While an analog of Steklov's Theorem holds for many singular problems, the statement and the proof must usually be carried out on a case-by-case basis. The book *Sturm-Liouville and Dirac Operators* by B. M. Levitan and I. S. Sargsjan, 1991, provides more information about the Sturm-Liouville problem (7.7.9), both regular and singular. Numerous versions of the Sturm-Liouville problem on $(0, +\infty)$ and $(-\infty, +\infty)$ are presented as problems in the book *Linear Op-*

erator, Part II: Spectral Theory, by N. Dunford and J. Schwartz, see pages 1551–1576 in Wiley Classic Library Publication of 1988.

PROBLEM 7.8. A GENERALIZATION OF THE HYDROGEN ATOM MODEL.

Consider a charged particle A moving around a charged particle B. Assume that the charge of the particle A is $-e < 0$, and of particle B, Ze for some $Z > 0$. Also assume that the mass of the particles is m_A and m_B, respectively. Considering only electrostatic force between the particles, and no relativistic effects, find the admissible energy levels of the particle A and the corresponding wave functions. *Hint: computations are almost identical to the analysis of the hydrogen atom; see page 368. There are two changes: (a) m is replaced with $m^* = m_A m_B/(m_A + m_B)$; (b) the potential $U = -Ze^2/(4\pi\varepsilon_0 r)$. As a result, $E_N = -Z^2 m^* e^4/(2(4\pi\varepsilon_0)^2\hbar^2 N^2)$.*

PROBLEM 7.9. MAXWELL'S EQUATIONS AND THE LORENTZ TRANSFORMATION.

(a) We saw that Maxwell's equations in vacuum can be reduced to a system of wave equations with the propagation speed c equal to the speed of light in vacuum; see, for example, (6.3.37) on page 348. In what follows, we will see that the one-dimensional wave equation $u_{tt} = c^2 u_{xx}$, which is a special case of (6.3.37), is invariant under the Lorentz transformation, but is not invariant under the Galilean transformation. (i) Consider the wave equation $u_{tt} = c^2 u_{xx}$, $t > 0$, $x \in \mathbb{R}$. Define new variables t_1, x_1 so that $t = \alpha(t_1 + bx_1)$, $x = \alpha(x_1 + vt_1)$, where $\alpha = (1 - v^2/c^2)^{-1/2}$, $b = v/c^2$, and v is a positive real number; this is the Lorentz transformation, see page 100. Let $\tilde{u}(t_1, x_1) = u(t(t_1, x_1), x(t_1, x_1))$. Show that $\tilde{u}_{t_1 t_1} = c^2 \tilde{u}_{x_1 x_1}$. *Hint: use the chain rule, starting with $\tilde{u}_{t_1} = u_t(\partial t/\partial t_1) + u_x(\partial x/\partial t_1) = \alpha\, u_t + \alpha v\, u_x$, etc.* (ii) Define new variables t_1, x_1 so that $x = x_1 + vt$, $t = t_1$; this is the Galilean transformation, which is the Lorentz transformation with $c = \infty$. Let $\tilde{u}(t_1, x_1) = u(t(t_1, x_1), x(t_1, x_1))$. Find the equation satisfied by $\tilde{u}$.

(b) Consider an inertial frame O and a frame O_1 moving relative to O along the $\hat{\imath}$-direction with speed v; see Figure 2.4.1 on page 99. Let $\boldsymbol{E} = E^{(x)}\,\hat{\imath} + E^{(y)}\,\hat{\jmath} + E^{(z)}\,\hat{\kappa}$ and $\boldsymbol{B} = B^{(x)}\,\hat{\imath} + B^{(y)}\,\hat{\jmath} + B^{(z)}\,\hat{\kappa}$ be the electric and magnetic fields as measured by an observer in frame O. Similarly, let $\boldsymbol{E}_1$ and $\boldsymbol{B}_1$ be the same fields as measured by an observer in frame O_1. (i) Find the relation between $\boldsymbol{E}, \boldsymbol{B}$ and $\boldsymbol{E}_1, \boldsymbol{B}_1$. *Hint: apply Theorem 8.3.2 on page 460 to the tensor $F^{ik} = \mathfrak{g}^{ij}\mathfrak{g}^{k\ell}F_{j\ell}$; see page 461. For example, you get $E_1^{(x)} = E^{(x)}$, $E_1^{(y)} = \alpha(E^{(y)} - vB^{(z)})$, etc., where $\alpha = (1 - (v/c)^2)^{-1/2}$.* (ii) Verify that the vectors $\boldsymbol{E}_1$ and $\boldsymbol{B}_1$ satisfy Maxwell's equations in frame

O_1. *Hint: you also need to apply the Lorentz transformation to the space and time coordinates and to the quantities* ρ, $\boldsymbol{J}$. *To transform* ρ *and* $\boldsymbol{J}$, *apply Theorem 8.3.2 to the type* $(1,0)$ *tensor* $J = (\boldsymbol{J}, \rho)$.

(c) Once again, consider two frames, O and O_1, from Figure 2.4.1. Assume that a point charge q is fixed at the origin of frame O_1. (i) Verify that the electric field $\boldsymbol{E}(t,x,y,z)$ produced by this moving charge, as measured by an observer in O at point (x,y,z) and time t, is

$$\boldsymbol{E} = \frac{q\alpha}{4\pi\varepsilon_0} \frac{(x-vt)\,\hat{\imath} + y\,\hat{\jmath} + z\,\hat{\kappa}}{\left(\alpha^2(x-vt)^2 + y^2 + z^2\right)^{3/2}},$$

where $\alpha = (1-(v/c)^2)^{-1/2}$. *Hint: use the results of part (b). Note that* $\boldsymbol{E}_1 = (q/(4\pi\varepsilon_0))(x_1\hat{\imath}_1 + y_1\hat{\jmath}_1 + z_1\hat{\kappa}_1)/(x_1^2+y_1^2+z_1^2)^{3/2}$, $x_1 = \alpha(x-vt)$ *and* $\boldsymbol{B}_1 = 0$. (ii) Find the magnetic field $\boldsymbol{B}$ produced by this moving charge, as measured by an observer in O at point (x,y,z) and time t. (iii) For a fixed point (x^*,y^*,z^*) and various values of v, plot $\|\boldsymbol{E}(t,x^*,y^*,z^*)\|$ and $\|\boldsymbol{B}(t,x^*,y^*,z^*)\|$ as functions of t. You can use a computer algebra system. What do you observe as v approaches c?

(d) *Discussion.* Results of part (c) show that an electric charge moving with constant velocity generates electromagnetic radiation, but not in the form of a planar wave as discussed on page 349. A planar wave, or a good approximation of it, is produced by *accelerating* charges. For more details on the subject of electromagnetic radiation see the book *Electromagnetism* by G. Pollack and D. Strump, 2002.

PROBLEM 7.10. ANALYZING THE EULER AND NAVIER-STOKES EQUATIONS.

In this problem, we discuss some methods of studying equations (6.3.54) and (6.3.55) on page 355. We always assume the incompressibility condition $\boldsymbol{\nabla}\cdot\boldsymbol{u} = 0$.

(a) Verify that if $\boldsymbol{u}$ satisfies either (6.3.54) and (6.3.55), then

$$\nabla^2 p = -\operatorname{div}\left((\boldsymbol{u}\cdot\boldsymbol{\nabla})\,\boldsymbol{u}\right).$$

Explain how this result can be used to eliminate pressure from either (6.3.54) or (6.3.55).

(b) Define $\boldsymbol{\omega} = \operatorname{curl}\boldsymbol{u}$. (i) Verify that if $\boldsymbol{u}$ satisfies (6.3.54), then $\boldsymbol{\omega}$ satisfies

$$\boldsymbol{\omega}_t + (\boldsymbol{u}\cdot\boldsymbol{\nabla})\,\boldsymbol{\omega} = (\boldsymbol{\omega}\cdot\boldsymbol{\nabla})\,\boldsymbol{u}.$$

(ii) Find the corresponding relation for $\boldsymbol{\omega}$ if $\boldsymbol{u}$ satisfies (6.3.55).

(c) In this part of the problem we consider equations (6.3.54) and (6.3.55) in $\mathbb{R}^2$. If $\boldsymbol{u}$ is a vector field in a plane, then we write $\boldsymbol{\omega} = \omega\, \boldsymbol{k}$, where ω is the **vorticity** of $\boldsymbol{u}$ and $\boldsymbol{k}$ is a unit vector perpendicular to the plane. This leads to a significant simplification of relations from part (b).
(i) Verify that if $\boldsymbol{u}$ satisfies (6.3.54), then ω satisfies

$$\omega_t + (\boldsymbol{u}\cdot\boldsymbol{\nabla})\,\omega = 0.$$

(ii) Find the corresponding relation for ω if $\boldsymbol{u}$ satisfies (6.3.55).

PROBLEM 7.11. KDV EQUATION AND SOLITONS.

The equation

$$u_t(t,x) + u_{xxx}(t,x) + 6u(t,x)u_x(t,x) = 0 \tag{7.7.12}$$

is known as the **Korteweg-de Vries**, or KdV, equation, after the Dutch mathematicians DIEDERIK JOHANNES KORTEWEG (1848–1941) and GUSTAV DE VRIES (1866–1934), who proposed the equation in 1895 as a model of propagation of long surface waves in a narrow and shallow channel. This equation was part of de Vries's doctoral dissertation, supervised by Korteweg. Note that (7.7.12) contains the uu_x term, characteristic of many other equations describing fluids.

(a) Verify that if $v = v(t,x)$ satisfies $v_t(t,x) + v_{xxx}(t,x) + av(t,x)v_x(t,x) = 0$ for some real number $a \neq 0$, then $u = (a/6)v$ satisfies (7.7.12).

(b) Verify that (7.7.12) has a solution

$$u_{(c)}(t,x) = \frac{c}{2}\operatorname{sech}^2\big(\sqrt{c}(x-ct)/2\big),$$

where $c > 0$ is a real number and $\operatorname{sech}(s) = 1/\cosh(s) = 2/(e^s + e^{-s})$. *Hint: if $u(t,x) = f(x-ct)$ is a solution of (7.7.12), then $3f^2 + f'' - cf = b$ for some real number b.*

The function $u_{(c)}$ is an example of a **soliton**, a single travelling wave with many interesting and unusual properties. For example, the propagation speed c of $u_{(c)}$ is proportional to the amplitude.

PROBLEM 7.12. STABILITY OF FINITE DIFFERENCE APPROXIMATIONS.

(a) Consider the matrix $A_K^{[-2,1]}$, that is, a tri-diagonal $K \times K$ matrix with -2 along the main diagonal and 1 above and below the diagonal; see (6.5.15) on page 395. Verify that the eigenvalues of the matrix are

$$\lambda_k^{(K)} = -2(1-\cos\theta_k) = -4\sin^2(\theta_k/2),\ \theta_k = \frac{k\pi}{K+1},\quad k = 1,\ldots,K.$$

Hint: if $(x_1, \ldots, x_K)^T$ *is an eigenvector, then we need a nontrivial solution of* $x_{m+1} + (\lambda - 2)x_m + x_{m-1} = 0$. *Try* $x_m = e^{i\theta m}$, *with* $i = \sqrt{-1}$, *to conclude that* $\lambda = 2(1 - \cos\theta)$, $\sin(K+1)\theta = 0$, *and* $x_m = \sin(m\theta)$.

(b) Prove the stability condition (6.5.24) on page 397 for the explicit Euler scheme (6.5.18). *Hint: (6.5.18) means* $u^{(1)}(m+1) = A_{N+1}^{[1-4r,2r]}u^{(1)}(m)$, *and the eigenvalues of the matrix* $A_{N+1}^{[1-4r,2r]}$ *are* $1+2r\lambda_k^{(N-1)}$; *(6.5.24) is equivalent to* $|1 + 2r\lambda_k^{(N+1)}| \le 1$ *for all* k.

(c) Prove that the implicit Euler scheme (6.5.19) and the Crank-Nicolson scheme (6.5.20) on page 396 are stable for all $r > 0$. *Hint: look at the eigenvalues of the matrices* $\left(A_{N+1}^{[1+4r,-2r]}\right)^{-1}$ *and* $\left(A_{N+1}^{[1+2r,-r]}\right)^{-1} A_{N+1}^{[1-2r,r]}$.

(d) Prove that the numerical scheme (6.5.28) on page 399 for the wave equation is stable if and only if $\mu \ge 1/4$. *Hint: reduce the scheme to the second-order finite difference equation* $ax_{n+1} + bx_n + cx_{n-1}$, *where* a, b, c *are eigenvalues of the suitable matrices; see parts (a) and (b). The solution of this equation stays bounded if and only if both roots of the equation* $ax^2 + bx + c = 0$ *satisfy* $|x| \le 1$; *complex roots are allowed.*

PROBLEM 7.13. NUMERICAL METHODS AND FAST FOURIER TRANSFORM.

(a) Consider the explicit Euler scheme for the heat equation; see (6.5.18) on page 396. Verify that the solution vector $\bar{u}^{(1)}(m)$ at step m can be written as

$$\bar{u}_n^{(1)}(m) = \sum_{k=0}^{N} c_k(1 + 2r\lambda_n)^m w_{k,n}.$$

Once you find the numbers $w_{k,n}$, which are connected to the eigenvectors of the matrix $A_N^{[-2,1]}$, you will see how $\bar{u}^{(1)}$ can be computed using the FFT. *Hint: start by looking for the solution in the form* $\bar{u}_n^{(1)}(m) = \sum_{k=1}^{N} v_{k,m}w_{k,n}$.

(b) Extend this approach to other numerical methods for the heat, wave, and Poisson equations.

PROBLEM 7.14. MORE ABOUT FINITE ELEMENTS.

(a) Use the results of Exercise 6.5.16, page 407, to solve the heat equation from Exercise 6.5.6, page 397, using the Galerkin method with piecewise linear functions φ_k.

(b) In two dimensions, an analog of the function φ from (6.5.43) on page 407 is constructed as follows. Let G be the *hexagon* in $\mathbb{R}^2$ with vertices $(0, \pm 1), (\pm 1, 0), (1, -1), (-1, 1)$ (draw a picture), and let $\varphi = \varphi(x, y)$ be the function such that $z = \varphi(x, y)$ is the graph of the pyramid with base

G_0, one vertex at $(0,0,1)$, and six faces. In particular, $\varphi(x,y) = 0$ if the point (x,y) is not in G. Find the formula for $\varphi = \varphi(x,y)$ when $(x,y) \in G$. *Hint: you need six formulas, one for each face of the pyramid, and, for each face,* $\varphi(x,y) = 1 + ax + by$ *with suitable* a, b. *For example, for the face with vertices at* $(0,0,1), (1,0,0), (0,1,0)$, *we have* $\varphi(x,y) = 1 - x - y$.

(c) Consider the Poisson equation $-\nabla^2 u(x,y) = f(x,y)$ in a square $(0,1) \times (0,1)$ with zero boundary conditions. Let $x_m = mh = m/N$, $y_n = nh = n/N$, $m,n = 0,\ldots,N$, be a grid and, using the hexagon function φ from part (b), define the functions

$$\varphi_{m,n}(x,y) = \varphi\big((x - x_m)/h, (y - y_n)/h\big), \quad m,n = 1,\ldots,N-1.$$

Let $G_{m,n}$ be the hexagon on which $\varphi_{m,n} \neq 0$. Note that $G_{m,n}$ and $G_{k,l}$ overlap if and only if $m = k, n = l \pm 1$ or $m = k \pm 1, n = l$, and the overlap region in each case is a parallelogram of area h^2. (i) For $N = 6$, draw the picture of the square divided into the hexagons $G_{m,n}$ and observe how different hexagons overlap. (ii) Verify that the Galerkin approximation using the functions $\varphi_{k,m}$ leads to a system of linear equations $AU = b$ with the same matric A as for the finite difference approximation (6.5.40) on page 402. Identify the corresponding vector b (it is different from the one in (6.5.40)).

Chapter 8

Appendix

8.1 Linear Algebra and Matrices

While the reader is expected to know the basic definitions from linear algebra, we review the main points below for the sake of completeness. For a more detailed account of the material in this section, see a linear algebra text book, for example, [Mirsky (1990)].

An $m \times n$ **matrix** is a rectangular array of numbers with m rows and n columns. In the notations of matrices, $A = (a_{ij})$ means that a_{ij} is the element in row number i and column number j; A^T is the **transpose of** A, that is, rows of A are columns of A^T: $A^T = (a_{ji})$. A **square matrix** has the same number of rows and columns. A matrix A is called **symmetric** if $A = A^T$; a symmetric matrix is necessarily square. A **diagonal matrix** $A = (a_{ij})$ is a square matrix with zeroes everywhere except on the main diagonal, that is, $a_{ij} = 0$ for $i \neq j$. A **row vector** is an $1 \times m$ matrix $(x_1, \ldots, x_m)$; a **column vector** is an $n \times 1$ matrix $(x_1, \ldots, x_n)^T$. The elements of $\mathbb{R}^n$, $n \geq 2$, are usually considered *column vectors*. The **identity matrix** I is a square matrix that has ones on the main diagonal and zeros everywhere else. For 3×3 square matrices,

$$A = \begin{pmatrix} a_{11} & a_{12} & a_{13} \\ a_{21} & a_{22} & a_{23} \\ a_{31} & a_{32} & a_{33} \end{pmatrix}, \quad A^T = \begin{pmatrix} a_{11} & a_{21} & a_{31} \\ a_{12} & a_{22} & a_{32} \\ a_{13} & a_{23} & a_{33} \end{pmatrix}, \quad I = \begin{pmatrix} 1 & 0 & 0 \\ 0 & 1 & 0 \\ 0 & 0 & 1 \end{pmatrix}. \qquad (8.1.1)$$

The sum of two matrices of the same size is defined componentwise: if $C = A + B$, then $c_{ij} = a_{ij} + b_{ij}$. The product of two matrices $C = AB$ is defined by $c_{ij} = \sum_{k=1}^{m} a_{ik}b_{kj}$; in particular, the number of columns in A must be the same as the number of rows in B. For example, if both A and B are 3×3, then the element of AB in the second row and first column is

$a_{21}b_{11} + a_{22}b_{21} + a_{23}b_{31}$. If $\boldsymbol{x}$, $\boldsymbol{y}$ are column vectors of the same size n, then $\boldsymbol{x} \cdot \boldsymbol{y} = \boldsymbol{x}^T \boldsymbol{y}$ and, for every $n \times n$ matrix, $(A\boldsymbol{x}) \cdot \boldsymbol{y} = \boldsymbol{x} \cdot (A^T \boldsymbol{y})$. In general, $AB \neq BA$ even for square matrices. All other rules of multiplication hold, as long as the product is defined. In particular, for every three $n \times n$ matrices A, B, C, we have $(AB)C = A(BC)$ and $(A+B)C = AC + BC$. If the elements of the matrix A are differentiable functions of t, then we define the matrix dA/dt by differentiating each element of the matrix A.

The **determinant** $|A|$ of a 2×2 matrix A is defined by follows:

$$\begin{vmatrix} a_{11} & a_{12} \\ a_{21} & a_{22} \end{vmatrix} = a_{11}a_{22} - a_{12}a_{21}.$$

For example,

$$\begin{vmatrix} 1 & 2 \\ 3 & 4 \end{vmatrix} = 1 \cdot 4 - 2 \cdot 3 = 4 - 6 = -2.$$

For a 3×3 matrix, the determinant is

$$\begin{vmatrix} a_{11} & a_{12} & a_{13} \\ a_{21} & a_{22} & a_{23} \\ a_{31} & a_{32} & a_{33} \end{vmatrix} = a_{11} \begin{vmatrix} a_{22} & a_{23} \\ a_{32} & a_{33} \end{vmatrix} - a_{12} \begin{vmatrix} a_{21} & a_{23} \\ a_{31} & a_{33} \end{vmatrix} + a_{13} \begin{vmatrix} a_{21} & a_{22} \\ a_{31} & a_{32} \end{vmatrix}.$$

EXERCISE 8.1.1. *Verify that*

$$\begin{vmatrix} 1 & 2 & 3 \\ 4 & 5 & 6 \\ 7 & 8 & 9 \end{vmatrix} = 0. \tag{8.1.2}$$

By induction, the determinant can be defined for every square $n \times n$ matrix. Both $|A|$ and $\det A$ are used to denote the determinant of the matrix A. The main facts about the determinant are (a) the determinant of the product of two square matrices of the same size is the product of determinants: $\det(AB) = (\det A)(\det B)$; (b) $\det A^T = \det A$; (c) $\det A = 0$ if and only if the columns of the matrix A are linearly dependent as column vectors. Note that in (8.1.2) the middle column is half the sum of the other two.

The **inverse** of a square matrix A is the matrix A^{-1} satisfying $AA^{-1} = A^{-1}A = I$; such a matrix exists, and is necessarily unique, if and only if the determinant of A is not equal to zero. A square matrix A is called **nonsingular** if A^{-1} exists.

EXERCISE 8.1.2.^{C} *(a) Verify that* $(A^T)^T = A$ *and* $(AB)^T = B^T A^T$*. (b) Verify that* A *is a non-singular matric if and only if* $A\boldsymbol{x} \neq \mathbf{0}$ *for every vector* $\boldsymbol{x} \neq \mathbf{0}$*. (c) Assume that the entries of the matrices* A*,* B *are differentiable functions of* t*. Verify that* $d(A^T)/dt = (dA/dt)^T$ *and* $d(AB)/dt = (dA/dt)B + A(dB/dt)$*.*

Definition 8.1 A square matrix A is called `orthogonal` if

$$A^T A = AA^T = I. \tag{8.1.3}$$

If A is an orthogonal matrix, then equality (8.1.3) implies the following properties of A:

- $A^{-1} = A^T$.
- The columns vectors of A are an orthonormal set and so are the row vectors.
- The matrix A preserves the inner product and the norm: $(A\boldsymbol{u}) \cdot (A\boldsymbol{v}) = \boldsymbol{u} \cdot (A^T A\boldsymbol{v}) = \boldsymbol{u} \cdot \boldsymbol{v}$, $\|A\boldsymbol{r}\| = \|\boldsymbol{r}\|$.
- The determinant of A is either 1 or -1.

In particular, an *orthogonal matrix represents an orthogonal transformation* of $\mathbb{R}^3$.

EXERCISE 8.1.3. *Verify the above properties of the orthogonal matrix.*

Let $\mathfrak{U} = (\boldsymbol{u}_1, \ldots, \boldsymbol{u}_n)$ be a basis in $\mathbb{R}^n$. A column vector $\boldsymbol{x} = (x_1, \ldots, x_n)^T$ is called a `representation` of $\boldsymbol{x} \in \mathbb{R}^n$ in the basis $\mathfrak{U}$ if $\boldsymbol{x} = \sum_{k=1}^n x_k \boldsymbol{u}_k$. Note that the same element $\boldsymbol{x}$ has, in general, different representations in different bases; by the inevitable abuse of notation, we use the same symbol for the vector and its representation in a basis. A matrix A is a `representation` of a linear transformation $\mathcal{A}$ of $\mathbb{R}^n$ in the basis $\mathfrak{U}$ if $\mathcal{A}(\boldsymbol{x}) = A\,(x_1, \ldots, x_n)^T$ for every $\boldsymbol{x} \in \mathbb{R}^n$.

EXERCISE 8.1.4. ^{C} *(a) Show that the square matrix* $A = (a_{ij})$ *is a representation of a linear transformation* $\mathcal{A}$ *in a basis* $\mathfrak{U}$ *if and only if* $\mathcal{A}(\boldsymbol{u}_k) = \sum_{i=1}^n a_{ik} \boldsymbol{u}_i$*,* $i = 1, \ldots, n$*. In particular, every* $n \times n$ *matrix is a representation of a linear transformation of* $\mathbb{R}^n$ *in some basis, and if the basis* $\mathfrak{U}$ *is* `orthonormal`*, that is,* $\boldsymbol{u}_i \cdot \boldsymbol{u}_j = 1$ *for* $i = j$ *and* $\boldsymbol{u}_i \cdot \boldsymbol{u}_j = 0$ *for* $i \neq j$*, then* $a_{jk} = (\mathcal{A}(\boldsymbol{u}_k)) \cdot \boldsymbol{u}_j$*. (b) Let* A *and* $\widetilde{A}$ *be the representations of the same linear transformation in the bases* $\mathfrak{U}$ *and* $\widetilde{\mathfrak{U}}$ *with the same origin. (i) Show that there exists an invertible matrix* B *so that* $\widetilde{A} = B^{-1}AB$*. (ii) Describe the matrix* B*. (iii) Show that* B *is orthogonal if both* $\mathfrak{U}$ *and* $\widetilde{\mathfrak{U}}$ *are orthonormal. Hint:* $\mathcal{A}(\boldsymbol{u}_k) = \sum_m a_{mk} \boldsymbol{u}_m$*,* $\mathcal{A}(\widetilde{\boldsymbol{u}}_k) = \sum_m \widetilde{a}_{mk}\, \widetilde{\boldsymbol{u}}_m$*; if* $\widetilde{\boldsymbol{u}}_k = \sum_m b_{mk} \boldsymbol{u}_m$*,* *then* $\sum_k b_{ik}\tilde{a}_{kj} = \sum_k a_{ik}b_{kj}$ *for all* i, j.

An **eigenvalue** of a square matrix A is a *complex* number λ so that the equation $A\boldsymbol{x} = \lambda\boldsymbol{x}$ has a solution $\boldsymbol{x}$ that is not a zero vector; this solution is called an **eigenvector** corresponding to the eigenvalue λ. All n eigenvalues of an $n \times n$ matrix are roots of the degree n equation $\det(A - \lambda I) = 0$, where *det* is the determinant of the matrix. As a result, A and A^T have the same eigenvalues.

EXERCISE 8.1.5.[C] *(a) Verify that if* $A = A^T$*, then all eigenvalues of* A *are real and eigenvectors corresponding to different eigenvalues are orthogonal. (b) Assume that* A^{-1} *exists. Verify that* λ *is an eigenvalue of* A *if and only if* $1/\lambda$ *is an eigenvalue of* A^{-1}*. (c) Verify that all eigenvalues of an orthogonal matrix satisfy* $|\lambda| = 1$*. Hint: keep in mind that complex eigenvalues of a real matrix come in complex-conjugate pairs. (d) Let* $B = (\boldsymbol{x}_1, \ldots, \boldsymbol{x}_n)$ *be the matrix whose columns* $\boldsymbol{x}_k$ *are normalized eigenvectors of a symmetric matrix* A*, that is,* $\boldsymbol{x}_k \cdot \boldsymbol{x}_k = 1$*,* $\boldsymbol{x}_k \cdot \boldsymbol{x}_m = 0$*,* $k \neq m$*. Show that the matrix* B^TAB *is diagonal, with eigenvalues of* A *along the diagonal. In other words,* A SYMMETRIC MATRIX IS DIAGONAL IN THE BASIS OF ITS EIGENVECTORS.

Let $f = f(z)$ be a function of a complex variable that is analytic for all z, so that $f(z) = \sum_{k\geq 0} a_k z^k$ and the power series converges for all $|z| < \infty$. One can then show that, for every square matrix A, the series $\sum_{k\geq 0} a_k A^k$ converges to some square matrix (convergence of this series means that each elements of the matrix $\sum_{k=0}^{N} a_k A^k$ converges to a limit as $N \to \infty$). Accordingly, for a square matrix A, we define $f(A) = \sum_{k\geq 0} a_k A^k$, where $A^0 = I$, the identity matrix. For example,

$$e^A = I + \sum_{k=1}^{\infty} \frac{A^k}{k!}. \tag{8.1.4}$$

EXERCISE 8.1.6. *(a)*[C] *Assume that* $A = B^{-1}DB$*, where* B *is an invertible matrix. Verify that* $f(A) = B^{-1}f(D)B$*. (b)*[A] *Verify that if* $AB = BA$*, then* $e^{A+B} = e^A e^B$*. Is the converse true? (c)*[B] *Verify that, for a symmetric matrix* A*,*

$$e^A = \lim_{k\to\infty} \left(I + \frac{A}{k}\right)^k,$$

where I *is the identity matrix of the same size as* A*. (d)*[A] *Does the result of part (c) hold for an arbitrary square matrix* A*?*

8.2 Ordinary Differential Equations

An initial value (Cauchy) problem for the system of n first-order ordinary differential equations $\boldsymbol{y}'(t) = \boldsymbol{f}(t, \boldsymbol{y}(t))$, $t_0 < t \le T$, $\boldsymbol{y}(t_0) = \boldsymbol{y}_0$, has a unique solution if each of the n component of $\boldsymbol{f}$ is a continuous function of $n+1$ variables, and there exists a number C so that, for all $t_0 \le t \le T$ and all $\boldsymbol{x}, \boldsymbol{y}$ in $\mathbb{R}^n$, we have $\|\boldsymbol{f}(t,\boldsymbol{x})\| \le C(1+\|\boldsymbol{x}\|)$ and $\|\boldsymbol{f}(t,\boldsymbol{x}) - \boldsymbol{f}(t,\boldsymbol{y})\| \le C\|\boldsymbol{x}-\boldsymbol{y}\|$. An equation $y^{(n)}(t) = F(t, y(t), y'(t), \ldots, y^{(n-1)}(t))$ of order n can be written as a system n first-order equation by setting $y_1(t) = y(t)$, $y_2(t) = y'(t)$, …, $y^{(n-1)}(t) = y_n(t)$, so that $y_k'(t) = y_{k+1}(t)$, $k = 1, \ldots, n-1$, and $y_n'(t) = F(t, y_1, \ldots, y_n)$. A system of equations is called `linear` if $\boldsymbol{f}(t,\boldsymbol{x}) = A(t)\boldsymbol{x}$ for some matrix $A = A(t)$. A linear ordinary differential equation of order n can be written as a linear system of n first-order ordinary differential equations.

EXERCISE 8.2.1. ^{C} A LINEAR SYSTEM OF ORDER n HAS A BASIS OF n LINEARLY INDEPENDENT SOLUTIONS.

Consider a linear system of n equations $\boldsymbol{y}'(t) = A(t)\boldsymbol{y}(t)$, $t > t_0$; $\boldsymbol{y}(t_0) = \boldsymbol{y}_0$. Let $\boldsymbol{y}_0 = \sum_{k=1}^n a_k \boldsymbol{u}_k$ where $\boldsymbol{u}_1, \ldots, \boldsymbol{u}_n$ is a basis in $\mathbb{R}^n$. Show that the solution of the system is $y(t) = \sum_{k=1}^n a_k \boldsymbol{y}_k(t)$, where $\boldsymbol{y}_k(t)$ solves $\boldsymbol{y}_k'(t) = A(t)\boldsymbol{y}_k(t)$, $t > t_0$; $\boldsymbol{y}_k(t_0) = \boldsymbol{u}_k$.

In particular, a linear second-order equation has two linearly independent solutions, and the general solution is an arbitrary linear combination of those two solutions. The general solution of the second-order linear equation with constant coefficients $y''(t) + by'(t) + cy(t) = 0$ is
(1) $y(t) = Ae^{r_1 t} + Be^{r_2 t}$, if r_1, r_2 are distinct real roots of $x^2 + bx + c = 0$;
(2) $y(t) = (A + Bt)e^{rt}$, if r is a double root: $x^2 + bx + c = (x-r)^2$;
(3) $y(t) = e^{rt}(A\cos\omega t + B\sin\omega t)$, if $r \pm i\omega$ are the complex conjugate roots of $x^2 + bx + c = 0$.
In particular, the solution of $y''(t) - b^2 y(t) = 0$ can be written either as $y(t) = Ae^{bx} + Be^{-bx}$ or as $y(t) = A_1 \sinh bx + B_1 \cosh bx$.

8.3 Tensors

Mathematical models of physics and engineering at an advanced level are impossible without tensors, and many problems discussed in this book have reached this level. We first encounter tensors in our study of classical mechanics (tensor of inertia, page 76). We rely heavily on tensors in our study of general relativity, starting with the metric tensor; see page 105; the main equation (2.4.22) of general relativity is written in terms of tensors. In our study of electromagnetism, we mentioned that both the dielectric constant and (magnetic) permeability are, in general, tensor fields. In what follows, we give the general definition of a tensor and show that it is a generalization of vectors and linear transformations. The main new idea is to define how a change of coordinates changes the look of a tensor.

Our presentation will be in $\mathbb{R}^n$ for an arbitrary n, and this $\mathbb{R}^n$ is not necessarily a Euclidean space in the sense that we do not assume that the distance between two points is measured according to the usual Euclidean metric. We will use `Einstein's summation convention`, which means summation over an index from 1 to n if the index appears twice in a product or single expression. For example, $B_i^i = \sum_{i=1}^n B_i^i$, $a_{ij}x^ix^j = \sum_{i,j=1}^n a_{ij}x^ix^j$. Often, one of the repeated indices appears as a subscript, and the other, as a superscript.

Consider two coordinate systems in $\mathbb{R}^n$, $\mathfrak{X} = (x^1, \ldots, x^n)$, $\widetilde{\mathfrak{X}} = (\widetilde{x}^1, \ldots, \widetilde{x}^n)$, and assume that there exists a one-to-one and onto transformation between the two coordinate systems. More precisely, we assume that there exist n smooth functions $x^i = x^i(\widetilde{x}^1, \ldots, \widetilde{x}^n)$, $i = 1, \ldots, n$, and n smooth functions $\widetilde{x}^j = \widetilde{x}^j(x^1, \ldots, x^n)$, $j = 1, \ldots, n$, with the following property: if a point $P \in \mathbb{R}^n$ has coordinates $(a^1, \ldots, a^n)$ in $\mathfrak{X}$ and $(b^1, \ldots, b^n)$ in $\widetilde{\mathfrak{X}}$, then $a^i = x^i(b^1, \ldots, b^n)$, $i = 1, \ldots, n$, and $b^j = \widetilde{x}^j(a^1, \ldots, a^n)$, $j = 1, \ldots, n$.

Let us emphasize that $\mathbb{R}^n$ in our presentation is not the usual Euclidean space: it is a collection of points and not vectors, with no postulates of Euclid to rely on. The coordinates $\mathfrak{X}$ and $\widetilde{\mathfrak{X}}$ are, in general, curvilinear. The only reason we use the notation $\mathbb{R}^n$ is not to complicate the presentation further with a more advanced mathematical notion of a manifold. Similar to our discussion of orthogonal curvilinear coordinates on page 141, we allow existence of some special points, where the functions x^i and $\widetilde{x}^j$ might be undefined or non-smooth. Unlike our discussion of curvilinear coordinates, we now use the same letters to denote both the coordinates and the corresponding functions of the transformation between the coordinate systems. As a result, the reader should not be confused by expressions such as $\partial\widetilde{x}^i/\partial x^k$, which is the partial derivative of the *function* $\widetilde{x}^i$ (expressing the i-th $\widetilde{\mathfrak{X}}$ coordinate of a point in terms of its $\mathfrak{X}$ coordinates) with respect to k-th variable.

EXERCISE 8.3.1.^{C} *Verify that*

$$\frac{\partial \widetilde{x}^i}{\partial x^k}\frac{\partial x^k}{\partial \widetilde{x}^j} = \frac{\partial x^i}{\partial \widetilde{x}^k}\frac{\partial \widetilde{x}^k}{\partial x^j} = \delta^i_j = \begin{cases} 1, & i=j; \\ 0, & i \neq j, \end{cases} \tag{8.3.1}$$

where δ^i_j *is the* `Kronecker symbol`*; see also page 107. Hint: by definition,*

$$x^i(\widetilde{x}^1(x^1,\ldots,x^n),\ldots,\widetilde{x}^n(x^1,\ldots,x^n)) = x^i,$$
$$x^j(x^1(x^1,\ldots,x^n),\ldots,x^n(x^1,\ldots,x^n)) = \widetilde{x}^j.$$

Definition 8.2 A `tensor of type` (p,q) at a point P in $\mathbb{R}^n$ is an object T, which can be defined without any reference to coordinate systems in such a way that, in every coordinate system, T is represented by n^{p+q} numbers $T^{i_1\cdots i_p}_{j_1\cdots j_q}$, where each upper index i_k and each lower index j_ℓ independently varies from 1 to n. Moreover, if $T^{i_1\cdots i_p}_{j_1\cdots j_q}$, $\widetilde{T}^{i_1\cdots i_p}_{j_1\cdots j_q}$ are representations of T in two coordinate systems $\mathfrak{X} = (x^1,\ldots,x^n)$, $\widetilde{\mathfrak{X}} = (\widetilde{x}^1,\ldots,\widetilde{x}^n)$, respectively, then, with the summation convention in force,

$$\widetilde{T}^{i_1\cdots i_p}_{j_1\cdots j_q} = T^{k_1\cdots k_p}_{\ell_1\cdots \ell_q}\frac{\partial \widetilde{x}^{i_1}}{\partial x^{k_1}}\cdots\frac{\partial \widetilde{x}^{i_p}}{\partial x^{k_p}}\frac{\partial x^{\ell_1}}{\partial \widetilde{x}^{j_1}}\cdots\frac{\partial x^{\ell_q}}{\partial \widetilde{x}^{j_q}}. \tag{8.3.2}$$

The number p is called the `contravariant valence` of the tensor, and q, its `covariant valence`. A `tensor field` of type (p,q) is a collection $(P, T(P))$, where P is a point in $\mathbb{R}^n$ and $T(P)$ is a tensor of type (p,q) defined at the point P. A sum of two tensors T, S of the same type (p,q) is, by definition, a tensor of type (p,q) with components $T^{i_1\cdots i_p}_{j_1\cdots j_q} + S^{i_1\cdots i_p}_{j_1\cdots j_q}$.

EXERCISE 8.3.2.^{C} *Let* $n=2$, $p+q=2$, $\widetilde{x}^1 = 2x^1+3x^2$, $\widetilde{x}^2 = x^1 - x^2$. *Write the corresponding four equalities in (8.3.2) when (i)* $p=2,\ q=0$; *(ii)* $p=q=1$; *(iii)* $p=0,\ q=2$. *Hint: in (i), you get* $\widetilde{T}^{11} = T^{11}\cdot 2\cdot 2 + T^{12}\cdot 2\cdot 3 + T^{21}\cdot 3\cdot 2 + T^{22}\cdot 3\cdot 3$, *etc. For (ii) and (iii), you first need to express* x^1, x^2 *in terms of* $\widetilde{x}^1$, $\widetilde{x}^2$.

The following exercise motivates the above definition of a tensor by showing that many familiar objects are particular cases of tensors.

EXERCISE 8.3.3.^{C} *(a) Verify that the gradient of a scalar field is a tensor of type* $(0,1)$ *and compare (8.3.2) with formula (3.1.42) on page 145. Hint:* $\widetilde{f}(\widetilde{x}^1,\ldots,\widetilde{x}^n) = f(x^1(\widetilde{x}^1,\ldots,\widetilde{x}^n),\ldots,x^n(\widetilde{x}^1,\ldots,\widetilde{x}^n))$; *by the chain rule,* $\partial\widetilde{f}/\partial\widetilde{x}^i = (\partial f/\partial x^k)(\partial x^k/\partial\widetilde{x}^i)$. *(b) Verify that a tangent vector to a curve is a tensor of type* $(1,0)$. *Hint: if* $x^k = x^k(t)$, $k = 1, \ldots, n$, *is a parametric equation of the curve in* $\mathfrak{X}$ *and* $\widetilde{x}^k = \widetilde{x}^k(t)$, $k = 1, \ldots, n$, *is a parametric equation of the curve in* $\widetilde{\mathfrak{X}}$, *then a tangent vector satisfies* $d\widetilde{x}^k(t)/dt = (dx^i(t)/dt)(\partial\widetilde{x}^k/\partial x^i)$. *(c) Verify that the Kronecker symbol* δ^i_j *is a tensor of type* $(1,1)$. *Hint: use (8.3.1).*

We will now establish a connection between tensors and linear transformations of $\mathbb{R}^n$. For that, we need to assume that $\mathbb{R}^n$ is a linear space. Let $\mathfrak{U} = (\boldsymbol{u}_1, \ldots, \boldsymbol{u}_n)$ be a basis in $\mathbb{R}^n$ of n linearly independent column vectors $\boldsymbol{u}_k$, $k = 1, \ldots, n$. In $\mathbb{R}^3$, the variables x, y, z correspond to the basis vectors $\hat{\imath}$, $\hat{\jmath}$, $\hat{\kappa}$, respectively. Accordingly, let us introduce a coordinate system $\mathfrak{X}$ corresponding to the basis $\mathfrak{U}$ so that x^i corresponds to $\boldsymbol{u}_i$: for every scalar field f on $\mathbb{R}^n$, define the function $F = F(x^1, \ldots, x^n)$ such that, if P is a point in $\mathbb{R}^n$ and $\overrightarrow{OP} = \sum_{k=1}^n x^k\,\boldsymbol{u}_k$, then $f(P) = F(x^1, \ldots, x^n)$.

EXERCISE 8.3.4.[C] *Verify that the function* F *is well defined. Hint: a point* P *is characterized by a unique combination of numbers* $x^1, \ldots, x^n$.

Recall (see Exercise 8.1.4 on page 453) that a linear transformation $\mathcal{A}$ of $\mathbb{R}^n$ is represented in the basis $\mathfrak{U}$ by an $n \times n$ matrix (a_{ij}, $i, j = 1, \ldots, n$). Note that when we work with a linear transformation, we require $\mathbb{R}^n$ to be a linear space and allow only a linear change of variables.

Theorem 8.3.1 *Let* $\mathfrak{U}$ *be a basis in the linear space* $\mathbb{R}^n$ *with origin* O, *and* $\mathfrak{X}$, *the corresponding coordinate system. Let* $\mathcal{A}$ *be a tensor of type* $(1,1)$ *at the point* O *whose components in* $\mathfrak{X}$ *are real numbers* $\mathcal{A}^i_j$. *Define the matrix* $A = (a_{ij},\ i,j = 1, \ldots, n)$, *so that* $a_{ij} = \mathcal{A}^i_j$. *Then the matrix* A *is a representation in* $\mathfrak{U}$ *of a linear transformation of* $\mathbb{R}^n$.

The proof is outlined in the following exercise; remember that we are using the summation convention.

EXERCISE 8.3.5.[C] *(a) Let* $\widetilde{\mathfrak{U}} = \{\widetilde{\boldsymbol{u}}_1, \ldots, \widetilde{\boldsymbol{u}}_n\}$ *be a different basis in* $\mathbb{R}^n$. *Verify that if* $\widetilde{\boldsymbol{u}}_j = b^i_j\,\boldsymbol{u}_i$ *for some real numbers* b^i_j, $i, j = 1, \ldots, n$, *and* $\widetilde{b}^i_j$ *are real numbers satisfying* $b^i_k\widetilde{b}^k_j = \widetilde{b}^i_k b^k_j = \delta^i_j$, *then* $\widetilde{x}^i = \widetilde{b}^i_j\,x^j$ *and* $x^j = b^j_k\,\widetilde{x}^k$, *so that* $\widetilde{b}^i_\ell = \partial\widetilde{x}^i/\partial x^\ell$ *and* $b^k_j = \partial x^k/\partial\widetilde{x}^j$. *(b) The definition of a tensor of type* $(1,1)$ *implies the relation* $\widetilde{\mathcal{A}}^i_j = \mathcal{A}^\ell_k\widetilde{b}^i_\ell b^k_j$. *To complete the proof, set* $a_{ij} = \mathcal{A}^i_j$ *and use the results of Exercise 8.1.4(b) on page 453.*

EXERCISE 8.3.6. [A+] *Let* $\mathfrak{U}$ *be a basis in* $\mathbb{R}^n$ *with origin* O, *and* $\mathfrak{X}$, *the corresponding coordinate system. Consider a tensor* T *of type* $(0,2)$ *at the point* O. *Denote by* T_{ij} *the components of* T *in* $\mathfrak{X}$. *What does the matrix*

$(T_{ij}, i,j = 1,\dots,n)$ *represent? Then answer the same question for a tensor of type* $(2,0)$. *Hint:* $(T_{ij}, i,j=1,\dots,n)$ *is a matrix of a quadratic form.*

Next, we discuss the metric tensor.

Definition 8.3 A metric tensor $\mathfrak{g}$ in $\mathbb{R}^n$ is a tensor field of type $(0,2)$ so that its components $\mathfrak{g}_{ij} = \mathfrak{g}_{ij}(P)$ at every point P in every coordinate system $\mathfrak{X}$ have the following properties: (i) $\mathfrak{g}_{ij} = \mathfrak{g}_{ji}$; (ii) the determinant of the matrix $(\mathfrak{g}_{ij},\ i,j=1,\dots,n)$ is non-zero; (iii) the `line element` $d\mathfrak{s}$ at P in $\mathfrak{X}$ coordinates is

$$(d\mathfrak{s})^2 = \mathfrak{g}_{ij}(P)dx^i dx^j. \tag{8.3.3}$$

EXERCISE 8.3.7. *(a)*[C] *Verify that (8.3.3) implies that* $\mathfrak{g}$ *is a tensor of type* $(0,2)$. *Hint:* $\mathfrak{g}_{ij}dx^i dx^j = (\mathfrak{g}_{ij}(\partial x^i/\partial \tilde{x}^k)(\partial x^j/\partial \tilde{x}^\ell))d\tilde{x}^k d\tilde{x}^\ell = \tilde{\mathfrak{g}}_{k\ell}d\tilde{x}^k d\tilde{x}^\ell$. *(b)*[B] *Verify that if* $\mathfrak{g}^{-1}$ *is a tensor whose components* $\mathfrak{g}^{ij}$ *in every* $\mathfrak{X}$ *satisfy* $\mathfrak{g}^{ik}\mathfrak{g}_{kj} = \delta^1_j$, *then* $\mathfrak{g}^{-1}$ *is a tensor of type* $(2,0)$. *(c)*[A] *Let* C *be a smooth curve defined in* $\mathfrak{X}$ *coordinates by* $x^i = x^i(t),\ i=1,\dots,n,\ a \le t \le b$. *Use (8.3.3) to define the length of the curve and verify that the length does not depend on the coordinate system. Hint: use (2.4.19) on page 104 as a starting point.*

If a metric tensor $\mathfrak{g}$ is defined on $\mathbb{R}^n$, then, for every tensor T of type $(1,1)$ in $\mathbb{R}^n$ we can define two other tensors,
- $T_{ij} = \mathfrak{g}_{ik}T^k_j$ (operation of lowering an index);
- $T^{ij} = \mathfrak{g}^{ik}T^j_k$ (operation of raising an index).

Note that both operations can be defined only if there is a metric in $\mathbb{R}^n$.

EXERCISE 8.3.8.[A] *(a) Define the operation of lowering an index for a general tensor of type* (p,q) *with* $p \ge 1$. *(b) Define the operation of rasing an index for a general tensor of type* (p,q) *with* $q \ge 1$. *(c) Verify that if* $\mathfrak{g}_{ij}$ *is the usual Euclidean metric, then the operations of raising and lowering an index do not change a tensor. Thus, there is no distinction between covariant (lower) and contravariant (upper) indices, and, by convention, all tensors on a Euclidean space are of type* $(0,q)$. *Hint: in cartesian coordinates,* $\mathfrak{g}_{ij} = \mathfrak{g}^{ij} = 1$ *if* $i=j$, *and all other components of* $\mathfrak{g}_{ij}$, $\mathfrak{g}^{ij}$ *are equal to zero.*

Next, we will discuss three applications of tensors: in special relativity, in electromagnetism, and in theory of elasticity. We begin with Lorenz transformation in special relativity, see page 100. Consider an inertial frame O and a frame O_1 moving relative to O in the $\hat{\imath}$-direction with speed v; see Figure 2.4.1 on page 99. Consider the relativistic space-time with the coordinate system (x^1, x^2, x^3, x^4) in frame O, where $x^1 = x$, $x^2 = y$, $x^3 = z$ represent the usual cartesian coordinates, and $x^4 = t$ represents time. Define the numbers Λ^i_j, $i, j = 1, \ldots, 4$, as follows:

$$\Lambda^1_1 = \Lambda^4_4 = \alpha,\ \Lambda^2_2 = \Lambda^3_3 = 1,\ \Lambda^4_1 = -v\alpha\, c^{-2},\ \Lambda^1_4 = -v\alpha, \tag{8.3.4}$$

and $\Lambda^i_j = 0$ otherwise, where $\alpha = (1 - (v/c)^2)^{-1/2}$.

Theorem 8.3.2 *(a) Let T be a type $(1,0)$ tensor with components T^i, $i = 1, \ldots, 4$ in frame O. Then the components $\widetilde{T}^i$ of the same tensor in frame O_1 are given by $\widetilde{T}^i = \Lambda^i_j T^j$. (b) Let T be a type $(2,0)$ tensor with components T^{ij}, $i, j = 1, \ldots, 4$, in frame O. Then the components $\widetilde{T}^{ij}$ of the same tensor in frame O_1 are given by $\widetilde{T}^{ij} = \Lambda^i_k \Lambda^j_\ell T^{k\ell}$.*

EXERCISE 8.3.9.[B] *Prove the above theorem. Hint: part (a) is a re-statement of relation (2.4.11). Part (b) follows from part (a).*

Let us now discuss some applications of tensors in the study of electromagnetic fields in vacuum. We will see how tensors lead to a more compact form of many equations and help to connect electromagnetism with special relativity. We continue to work in the relativistic space-time with the standard coordinate system (x^1, x^2, x^3, x^4). We combine the vector potential $\boldsymbol{A}$ and the scalar potential φ into a single tensor A^i, $i = 1, \ldots, 4$, of type $(1,0)$ as follows: $A = (\boldsymbol{A}, \varphi/c^2)$, that is, $\boldsymbol{A} = A^1\,\hat{\boldsymbol{\imath}} + A^2\,\hat{\boldsymbol{\kappa}} + A^3\,\hat{\boldsymbol{\kappa}}$ and $A^4 = \varphi/c^2$; see page 351 and remember that we are now using $\varepsilon = \varepsilon_0$ and $\mu = \mu_0$ so that $\varepsilon\mu = c^{-2}$, where c is the speed of light in vacuum. We also combine the current density $\boldsymbol{J} = \boldsymbol{J}_e$ and the charge density $\rho = \rho_f$ into a single tensor J^i, $i = 1, \ldots, 4$, in a similar way: $J = (\boldsymbol{J}, \rho)$. Next, we define the tensor operator $D_i = \partial/\partial x^i$, $i = 1, \ldots, 4$. This tensor is of type $(0,1)$ and combines the time and space derivatives.

EXERCISE 8.3.10.[C] *Verify that the Lorenz gauge relation (6.3.45) on page 352 becomes $D_i A^i = 0$.*

As in the theory of special relativity, consider the metric tensor $\mathfrak{g}_{ij}, i, j = 1, \ldots, 4$, with $\mathfrak{g}_{11} = \mathfrak{g}_{22} = \mathfrak{g}_{33} = 1$, $\mathfrak{g}_{44} = -c^2$, and $\mathfrak{g}_{ij} = 0$ otherwise (see page 104). Similarly, the inverse tensor $\mathfrak{g}^{ij}$ has components $\mathfrak{g}^{11} = \mathfrak{g}^{22} = \mathfrak{g}^{33} = 1$, $\mathfrak{g}^{44} = -c^{-2}$, and $\mathfrak{g}^{ij} = 0$ otherwise. We use this metric to raise and lower indices of tensors. In particular, we define $D^i = \mathfrak{g}^{ij} D_j$.

EXERCISE 8.3.11. [C] *(a) Verify that $D^i = \partial/\partial x^i$, $i = 1, 2, 3$, and $D^4 = -c^{-2}\partial/\partial t$. (b) Verify that the tensor relation $D_i D^i A^j = -\mu_0 J^j$, $j = 1, \ldots, 4$, combines the wave equations (6.3.46) and (6.3.47); see page*

352. (c) Verify that $D_i J^i = 0$ is the equation of continuity (3.3.12) on page 166.

Our next objective is to find a tensor form of Maxwell's equations. For that, define a tensor F_{ik}, $i, k = 1, \dots, 4$, of type $(0, 2)$ as follows:

$$F_{ik} = \frac{\partial A_k}{\partial x^i} - \frac{\partial A_i}{\partial x^k}, \tag{8.3.5}$$

where $A_k = \mathfrak{g}_{km} A^m$.

EXERCISE 8.3.12.^{C} *Write the components of F_{ik}. Hint: $F_{ik} = -F_{ki}$, in particular, $F_{11} = F_{22} = F_{33} = F_{44} = 0$. Also, F_{12} is the $\hat{\boldsymbol{k}}$-component of the magnetic field $\boldsymbol{B}$, F_{14} is the $\hat{\boldsymbol{\imath}}$-component of the electric field $\boldsymbol{E}$, etc.*

To proceed, we define the `permutation symbol`

$$e_{ijkm} = \begin{cases} 1, & \text{if } (ijkm) \text{ is an even permutation of } (1234); \\ -1 & \text{if } (ijkm) \text{ is an odd permutation of } (1234); \\ 0 & \text{otherwise.} \end{cases}$$

Recall that a permutation is even if it is a composition of an even number of transpositions (exchanges of two numbers). For example (2143) and (4321) are even permutation, each being a composition of two transposition, while (2134) and (3421) are odd. Using the permutation symbol and the inverse metric tensor, we define the tensor F_*^{ij} by

$$F_*^{ij} = \mathfrak{g}^{ii_1} \mathfrak{g}^{jj_1} \mathfrak{g}^{kk_1} \mathfrak{g}^{\ell\ell_1} e_{i_1 j_1 k_1 \ell_1} F_{k\ell}.$$

EXERCISE 8.3.13. *(a)^B Find the components of the tensor F_*. (b)^A Verify that Maxwell's equations (3.3.2)—(3.3.5) on page 164 can be written as*

$$D_i F^{ik} = -\mu_0 J^k, \quad D_i F_*^{ik} = 0, \quad k = 1, \dots, 4, \tag{8.3.6}$$

where $F^{ik} = \mathfrak{g}^{ij} \mathfrak{g}^{k\ell} F_{j\ell}$.

We conclude our discussion of tensors with the stress and strain tensors in the THEORY OF ELASTICITY. Consider an elastic solid material which occupies a region G in $\mathbb{R}^3$. Suppose G is subjected to external forces which deform it. Choose a cartesian coordinate system and let a small part of G be a rectangular box having a vertex at point $P = (x^1, x^2, x^3)$ and edges of lengths dx^1, dx^2, dx^3 in the coordinate directions. As a consequence of the elastic properties of G the neighboring parts of G will transmit stress to the box at point P. Let F_{ii} be the tensile stress acting in the x^i direction.

On the three faces that contain P there are shear, or twisting, stresses. On the face with sides dx^2, dx^3 there is a shear stress that is resolved into two components, F_{12} in the x^2 direction and F_{13} in the x^3 direction. On face dx^1, dx^3, there are shear stresses F_{21} in x^1 direction and F_{23} in the x^3 direction. On face dx^1, dx^2, there are shear stresses F_{31} in x^1 direction and F_{32} in the x^2 direction. The nine numbers F_{ij} represent the stress tensor F at P.

The strain, or deformation, tensor $D = (D_{ij})$ is defined as follows. Assume that point $P = (x^1, x^2, x^3)$ is displaced to point P', and $\overrightarrow{PP'} = y^1\,\hat{\imath} + y^2\,\hat{\jmath} + y^3\,\hat{\kappa}$. Then $D_{ii} = \partial y^i/\partial x^i$, $i = 1, 2, 3$, and $D_{ij} = (\partial y^i/\partial x^j) + (\partial y^j/\partial x^i)$, $i \neq j$. We call D_{ij} the shear strain in the (x^i, x^j) plane. We also define the tensor D'' with components

$$D''_{ij} = \begin{cases} D_{11} + D_{22} + D_{33}, & \text{if } i = j; \\ 0, & \text{if } i \neq j. \end{cases}$$

EXERCISE 8.3.14.^{A} *(a) Let $(ds)^2 = (dx^1)^2 + (dx^2)^2 + (dx^3)^2$ be the line element in the non-deformed region. Show that $(ds')^2 = D_{ij}dx^i dx^j$ is the line element in the deformed region (imagine a rubber ruler that is twisted and stretched.) (b) Verify that both F and D are symmetric tensor fields of type* $(0, 2)$.

In a *linear* material, the stress and strain tensors are connected by $F_{ij} = E_{ijkl}D_{kl}$, where E_{ijkl} is a tensor of type $(0, 4)$. This relation is known as `Hooke's Law`, after the English scientist ROBERT HOOKE (1635–1703). For a linear homogenous isotropic material, the stress and strain tensors are related by $F = 2\mu D + \lambda D''$, where μ, λ are positive constants, characterizing the elastic properties of the material. For more details, see the books *Theory of Elasticity* by L. D. Landau and E. M. Lifschitz, 1986, and *A Treatise on the Mathematical Theory of Elasticity* by A. E. H. Love, 1988. The book by Love, a classic first published at the end of the 19th century, also contains a detailed account of the history of the subject.

EXERCISE 8.3.15.^{B} *(a) Find the components of the tensor E_{ijkl} for a linear homogeneous isotropic material. (b) Verify that Hooke's Law is invariant under a coordinate transformation. (c) Assume that $\lambda = 0$ and the region G is deformed so that, for every point $P = (x^1, x^2, x^3)$ in the region, the displacement vector $\overrightarrow{PP'}$ is equal to $u\,\hat{\imath}$ for some function $u = u(x^1, x^2, x^3)$. (i) Show that the stress tensor F can be characterized by the vector $\boldsymbol{F} = F_{11}\,\hat{\imath} + F_{21}\,\hat{\jmath} + F_{31}\,\hat{\kappa}$ and $\boldsymbol{F} = 2\mu\,\mathrm{grad}\,u$. (ii) Let G be a spring of length L,* *placed along the* x^1 *axis, with one end fixed at point* $(0,0,0)$ *and the other moved from* $(L,0,0)$ *to* $(L+x,0,0)$. *Show that* $u(x^1,x^2,x^3) = (x/L)x^1$ *and derive the elementary Hooke's Law* $F = -Kx$, *where* $k > 0$ *is a constant and* F *is the force of the spring; note that* F *is acting in the direction opposite to the displacement.*

8.4 Lumped Electric Circuits

Maxwell's equations model electromagnetic phenomena in a region of a three-dimensional space; such a region is a continuum of points. An alternative mathematical model is obtained by discretising this continuum and replacing it by a finite collection of lumped components, connected by idealized zero-resistance wires. The resulting *lumped circuit* has passive elements (resistors, capacitors, and inductors) and active components (batteries, generators, etc.) In this section we present the main facts about the passive elements.

Conductors. In conductors, the free charges (usually electrons) move in an external electric field to create an electric current. Suppose an electric field $\boldsymbol{E} = -\boldsymbol{\nabla} U$ is applied to a conductor, where U is a potential. The force acting on a free charge e is $e\boldsymbol{E}$. If m is the mass of the charge, it is accelerated by the amount $e\boldsymbol{E}/m$ in a time interval τ before colliding with another particle. As a result, the charge reaches the drift velocity $\boldsymbol{v} = (\tau e/m)\,\boldsymbol{E}$. For copper, $\tau = 3 \cdot 10^{-14}$ seconds.

Now let the conductor be a small cylinder of length L and cross-section a. Let $V > 0$ be the voltage difference across L. Then $\|\boldsymbol{E}\| = \|\boldsymbol{\nabla} U\| = V/L$, and $\|\boldsymbol{v}\| = \tau eV/(Lm)$. Denote by d the number of electrons per unit volume. Then the flux of charge across a is $I = aed\|\boldsymbol{v}\| = (a/L)(V/\rho)$, where $\rho = m/(\tau e^2 d)$ is called the `resistivity` of the conductor. The quantity $R = \rho L/a$ is called the resistance of the cylinder. Therefore, we obtain `Ohm's Law`: $V = RI$. The SI units for V, R, I are volt (V), ohm (Ω), and ampere (A), respectively. The units for ρ are Ω·m; for copper under room temperature, $\rho = 1.7 \cdot 10^{-8}$ Ω·m.

In a lumped circuit, an ideal resistor R is assumed to be concentrated at a point. If a voltage $V(t)$ is changed by amount ΔV in an interval Δt at one end of a real resistor of length L, it takes time $\Delta t_1 = L/c$ to propagate to the other end, where c is the speed of light. Since Δt_1 is usually much smaller than Δt, the current $I(t) = V(t)/R$ in a real resistor can be assumed to change instantaneously, that is, without delay, from one end to the other.

Since the time τ is also much smaller than Δt, the drift velocity of the free charges is reached almost instantaneously, and it is reasonable to assume that ρ is constant throughout the resistor.

Capacitors. Consider two isolated conductors in a finite region of space, one having a charge q and the other a charge $-q$. In a typical example, the conductors can be two parallel metal plates separated by a thin layer of dielectric material. In the steady state, the charges distribute themselves over the surface of the conductors so that the potential due to the charges is constant throughout each conductor. By Coulomb's Law, the charges produce an electric field $\boldsymbol{E}$. Writing $\boldsymbol{E} = -\boldsymbol{\nabla} U$, we find $-\int_{\mathcal{C}(A,B)} \boldsymbol{E} \cdot d\boldsymbol{r} = U(B) - U(A) = V$, where A, B are two points on the two conductors, $\mathcal{C}(A, B)$ is a simple smooth curve connecting A and B, and V is the potential difference. By Coulomb's Law, $\|\boldsymbol{E}\|$ is proportional to q, and then V is also proportional to q, so that we write $V = q/C$, where C is a constant called the capacitance of the configuration of the two conductors. This configuration is called a capacitor. The SI units for V, q, and C are volt, coulomb, and farad, respectively.

If the field $\boldsymbol{E}$ changes in time, this change will cause charges to flow from the external region into one of the conductors and out of the other conductor. This flow will change V and produce a current $I(t) = dq(t)/dt = C\, dV(t)/dt$. If the capacitor is confined to a small region, the capacitor can be modelled as a lumped capacitor at a point of a circuit.

Inductors. By Faraday's Law, a changing current in a conductor induces a voltage difference V given by

$$V(t) = L\frac{dI(t)}{dt},$$

where L is a constant called the inductance. If the conductor is small and has a particular shape, for example, a coil of wire, then it can be modelled as a lumped inductor at a point in a circuit.

To analyze complex lumped circuits, two Kirchoff's Laws are used:

- The sum of voltage drops across the elements in a closed loop is zero (voltage law).
- The sum of the currents entering and leaving a point in a circuit is zero (current law).

8.5 Physical Units and Constants

BASIC SI UNITS: distance, meter [m]; electric current, ampere [A]; mass, kilogram [kg]; time, second [s]; temperature, kelvin [K]. Two other unit, not used in this book, are candela [cd] to measure luminosity, and mole [mol] to measure atomic weight.

DERIVED UNITS:

Quantity	Unit
frequency	Hertz [Hz]=[1/s]
force	Newton [N]=[kg·m/s^2]
energy	joule [J]=[N·m]=[kg·m^2/s^2]
power	watt [W]=[J/s]=[kg·m^2/s^3]
electric charge	Coulomb [C]=[A·s]
electric potential	volt [V]=[J/C]=[kg·m^2/(s^3·A)]
electric resistance	Ohm [Ω]=[V/A]=[kg·m^2/(s^3·A^2)]
electric capacitance	Farad [F]=[C/V]=[A^2·s^4/(kg·m^2)]
magnetic flux	weber [Wb]=[kg·m^2/(s^2·A)]
inductance	henry [H]=[Wb/A]=[kg·m^2/(s^2·A^2)]
electric field $\boldsymbol{E}$	[V/m]=[N/C]=[kg·m/(s^3·A)]
magnetic field $\boldsymbol{B}$	[Wb/m^2]=[kg/(s^2·A)]

EXERCISE 8.5.1.^{C} *Verify the following relations:*
[F·H]=[s^2], [V· A]= [W], [Ω·F]=[s].

PHYSICAL CONSTANTS:

Name	notation	approximate value	units
electron's charge	e	$-1.6 \cdot 10^{-19}$	[C]
electron's mass	m_e	$0.91 \cdot 10^{-30}$	[kg]
gravitational constant	G	$6.67 \cdot 10^{-11}$	[N·m^2/kg^2]
permeability of vacuum	μ_0	$1.26 \cdot 10^{-6}$	[N/A^2]
permittivity of vacuum	ε_0	$8.85 \cdot 10^{-12}$	[C^2/(N·m^2)
Planck's constant	h	$6.626 \cdot 10^{-34}$	[J·s]
	$\hbar = h/(2\pi)$	$1.05 \cdot 10^{-34}$	[J·s]
speed of light	c	$3 \cdot 10^8$	[m/s]

Bibliography

Artin, M. (1991). *Algebra.* Prentice Hall.

Beiser, A. (2002). *Concepts of Modern Physics.* 6th ed. McGraw-Hill.

Berman, G. P., Doolen, G. D., Mainieri, R. and Tsifrinovich, V. I. (1998). *Introduction to Quantum Computing.* World Scientific.

Birkhoff, G. and Rota, G.-C. (1989). *Ordinary Differential Equations*, 4th ed. Wiley.

Blum, E. K. (1972). *Numerical Analysis and Computation: Theory and Practice.* Addison-Wesley.

Bunge, M. (1967). *Foundations of Physics.* Springer.

Carleson, L. (1966). On convergence and growth of partial sums of Fourier series. *Acta Mathematica*, **116**, pp. 137–157.

Carrier, G. and Pearson, C. (1988). *Partial Differential Equations: Theory and Technique*, 2nd ed. Academic Press.

Coddington, E. A. and Levinson, N. (1955). *Theory of Ordinary Differential Equations.* McGraw-Hill.

Conway, J. H. and Smith, D. A. (2003). *On Quaternions and Octonions.* A.K. Peters, Natick, MA.

Colley, S. J. (1987). The Tumbling Box. *American Mathematical Monthly*, **94**(1), pp. 62–68.

Cooley, J. W. and Tukey, J. W. (1965). An algorithm for the machine calculation of complex Fourier series. *Mathematics of Computation*, **19**, pp. 297–301.

Devlin, K. (2002). *The Millenium Problems.* Basic Books, N.Y.

Dunford, N. and Schwartz, J. (1988). *Linear Operator, Part II: Spectral Theory.* Wiley.

Eddington, A. S. (2005). *Nature of the Physical World.* Kessinger Publishing, Whitefish, MT.

Evans, L. C. (1998). *Partial Differential Equations.* American Mathematical Society.

Feynman, R. P., Leighton, R. B. and Sands, M. (2005). *The Feynman Lectures on Physics*, Vol. 1, Addison - Wesley.

Forsythe, G. and Wasow, W. (2004). *Finite Difference Methods for Partial Differential Equations.* Dover.

Greenberg, M. J. (1993). *Euclidean and Non-Euclidean Geometries: Development and History,* 3rd ed. W.H. Freeman.

Griffiths, D. (1998). *Introduction to Electrodynamics*, 3rd ed. Prentice Hall.

Hardy, G. H. and Wright, E. M. (1979). *An Introduction to the Theory of Numbers*, 5th ed. Clarendon Press.

Hofstadter, D. R. (1999). *Gödel, Escher, Bach: An Eternal Golden Braid*, 20th ed. Basic Books, N.Y.

Holton, J. R. (2004). *An Introduction to Dynamic Meteorology*, 4th ed. Academic Press.

Hardy, Y. and Steeb, W. H. (2001). *Classical and Quantum Computing.* Birkhauser, 2002.

Johnson, C. V. (2003). *D-Branes.* Cambridge University Press.

Kellogg, O. D. (1969). *Foundations of Potential Theory.* Dover.

Kincaid, D. and Cheney, W. (1996). *Numerical Analysis: Mathematics of Scientific Computing*, 2nd ed. Brooks/Cole Publishing Co.

Knabner, P. and Angermann, L. (2003). *Numerical Methods for Elliptic and Parabolic Partial Differential Equations.* Springer.

Körner, T. W. (1989). *Fourier Analysis.* Cambridge University Press.

Kreyszig, E. (2006). *Advanced Engineering Mathematics*, 9th ed. Wiley.

Landau, L. D. and Lifschitz, E. M. (1986). *Theory of Elasticity*, Butterworth.

Levitan, B. M. and Sargsjan, I. S. (1991). *Sturm-Liouville and Dirac Operators.* Kluwer.

Liboff, R. L. (2002). *Introductory Quantum Mechanics.* Addison-Wesley.

Lorenz, E. N. (1967). *The Nature And Theory of The General Circulation of The Atmosphere.* World Meteorological Organization.

Love, A. E. H. (1988). *A Treatise on the Mathematical Theory of Elasticity.* Dover.

MacCluer C. R. (1994). *Boundary Value Problems and Orthogonal Expansions.* IEEE Press.

Margenau, H. and Lindsay, R. B. (1957). *Foundation of Physics.* Dover.

Mattuck, A. (1998). *Introduction to Analysis.* Prentice Hall.

Mirsky, L. (1990). *An Introduction to Linear Algebra.* Dover.

Mitchell, A. R. and Griffiths, D. F. (1980). *The Finite Difference Method in Partial Differential Equations.* Wiley.

Morse, P. M. and Feshbach, H. (1953). *Methods of Theoretical Physics, part I.* McGraw-Hill.

Moulton, F. R. (1984). *An Introduction to Celestial Mechanics.* Dover.

Nahin, P. J. (1998). *An Imaginary Tale: The Story of* $\sqrt{-1}$. Princeton University Press.

Nielsen, M. A. and Chuang, I. L. (2000). *Qunatum Computation and Quantum Information.* Cambridge University Press.

Pauli, W. (1981). *Theory of Relativity*, Dover.

Perko, L. (1991). *Differential Equations and Dynamical Systems.* Springer.

Pollack, G. and Stump, D. (2002). *Electromagnetism.* Addison Wesley.

Rudin, W. (1976). *Principles of Mathematical Analysis*, 3rd ed. McGraw-Hill.

Shu, S-S. (1987). *Boundary Value Problems of Linear Partial Differential Equations for Engineers and Scientists.* World Scientific.

Schwartz, L. (1966). *Mathematics for the Physical Sciences.* Addison-Wesley.

Sciama, D. (1969). *The Physical Foundations of General Relativity.* Doubleday, N.Y.

Strang, G. and Fix, G. (1973). *An Analysis of the Finite Element Method.* Prentice Hall.

Synge, J. L. and Griffith, B. (1949). *Principles of Mechanics*, 2nd ed. McGraw-Hill.

Szabo, R. J. (2004). *An Introduction to String Theory and D-Brane Dynamics.* World Scientific.

Szegö, G. (1975). *Orthogonal Polynomials.* American Mathematical Society.

Taylor, E. F. and Wheeler, J. A. (1992). *Spacetime Physics*, 2nd ed. W.H. Freeman.

Thaller, B. (1992). *The Dirac Equation.* Springer.

Tuckwell, H. (1988). *Theoretical Neurobiology, Volume 1: Cable Theory.* Cambridge University Press.

Weiss, E. (1998). *Algebraic Number Theory.* Dover.

Weinberger, H. (1995). *A First Course in Partial Differential Equations With Complex Variables and Transform Methods.* Dover.

Woodhouse, N. M. J. (2005). *Introduction to Analytical Dynamics.* Oxford University Press.

Zwillinger, D. (1997). *Handbook of Differential Equations*, 3rd ed. Academic Press.

List of Notations

Index